Stress Regimes in the Lithosphere

Stress Regimes in the Lithosphere

Terry Engelder

PRINCETON UNIVERSITY PRESS

PRINCETON, NEW JERSEY

Published by Princeton University Press, 41 William Street,
Princeton, New Jersey 08540
In the United Kingdom: Princeton University Press, Chichester, West Sussex

Library of Congress Cataloging-in-Publication Data

Engelder, T. (Terry)
Stress regimes in the lithosphere / Terry Engelder.
p. cm.
Includes bibliographical references and index.
ISBN 0-691-08555-2 (alk. paper)
1. Earth—Crust. 2. Strains and stresses—Measurement.
3. Rocks—Fracture. I. Title.
QE511.E54 1992
551 . 1′3—dc20 92-8604

This book has been composed in Linotron Times Roman and Helvetica
by Pine Tree Composition

Princeton University Press books are printed on acid-free paper,
and meet the guidelines for permanence and durability of the
Committee on Production Guidelines for Book Longevity of the
Council on Library Resources

Printed in the United States of America

10 9 8 7 6 5 4 3 2 1

Designed by Laury A. Egan

To

All my friends who have shared in the excitement of working
at the Lamont-Doherty Geological Observatory
of Columbia University, one of the world's finest
earth science institutes

Contents

Preface

The debate over the magnitude and orientation of stress in the lithosphere has continued for several decades. As a student I was impressed by the seemingly irreconcilable differences between the high differential stresses required by models for lithospheric flexure and the low differential stresses found during hot creep experiments on rock. More recently, in situ stress data from California suggest that the frictional strength of the San Andreas fault is lower than expected by predictions using the general rock friction law to model shear stress along fault zones. Other contentious issues concerned the choice of a reference state of stress, the extrapolation of near-surface measurements to depth, the origin of platewide stress fields, and the significance of residual stress. The purpose of this monograph is to acquaint the geoscientist with these and many other issues associated with the debate over stress in the lithosphere.

My goal is to provide a broad understanding of stress in the lithosphere while touching some of the specific details involved in the interpretation of stress data generated by the most commonly used measurement techniques. Although the discussion of stress measurement techniques lacks a cookbook form, I illustrate some of the subtle aspects of the measurements, often drawing upon my own experience in making stress measurements. Given the breadth of the subject and a limit on space, I had to strike a balance on several counts. I wrote for the senior undergraduate or first-year graduate student who has a modest background in either structural geology or mechanics including an introduction to stress as a tensor (e.g., Means, 1976; Suppe, 1985). Space limitations meant that an encyclopedic referencing of literature on the subject was impractical. Yet, the student will find enough material to easily continue a search for additional references. I attempted to balance the most recent references with some time-honored literature. The need for conciseness led to the introduction of many theories with the assumption that the student would return to the original reference for a complete development of the concepts. Nevertheless, some theories are treated in detail according to the following outline.

An understanding of stress in the lithosphere starts with an introduction to nomenclature based on three reference states of stress (chap. 1). Chapters 2 through 4 cover the role of rock strength as a governor for stress magnitude. Stress regimes in the lithosphere are identified according to the particular failure mechanism (crack propagation, shear rupture, ductile flow, or frictional slip) which controls the magnitude of stress at a particular time and place in the lithosphere. After introducing the various stress regimes, their extent in the upper crust is demarcated by direct measurements of four types: hydraulic-fracture; borehole-logging; strain-relaxation; and rigid-inclusion measurements (chaps. 5 through 8). The relationship between lithospheric stress and the properties of rocks is then presented in terms of microcrack-related phenomena (chap. 9) and residual stress (chap. 10). Chapter 11 deals with lithospheric stress as inferred from the analysis of earthquakes. Finally, lithospheric stress is placed in the context of large-scale stress fields and plate tectonics (chaps. 12 and 13).

Terry Engelder
Boalsburg, PA

Acknowledgments

This monograph is, in part, an account of my experience while attempting to measure lithospheric stress during my twelve years as a research scientist at the Lamont-Doherty Geological Observatory of Columbia University. My experience started during the summer of 1973 when Marc Sbar and I had a conversation which ended with a simple understanding that we were going to make a concerted effort to "measure" stress. Early encouragement from Chris Scholz and Lynn Sykes was important. Dick Plumb participated in several of the field experiments discussed within this monograph. One of Dick's important contributions was the development of the Lamont strain cell, a "doorstopper"-like module used in more than two hundred near-surface stress measurements. When funds were finally in hand for hydraulic fracture stress measurements, Keith Evans took the lead in the meticulous organization of several field experiments. Practical advice in many technical aspects of the field experiments was provided by Ted Koczynski. Others who helped with field work include Pat Barns (Lamont); John Barry (Penn State); Olaf Befeld (Bochem); Eric Bergman (Arizona); Steve Brown (Lamont); Tom Chen (Lamont); Peter Dahlgren (Arizona); Tom Engelder (Penn State); Zoe Engelder (Boalsburg); Chris Flaccus (Arizona); Peter Geiser (Connecticut); Larry Hooper (SUNY); Robert Kranz (Lamont); Tony Lomando (CUNY); Steve Marshak (Lamont); Irene Meglis (Penn State); Jim Mori (Lamont); Craig Nickolson (Lamont); Stu Nishenko (Lamont); Kevin Powell (Columbia); Randy Richardson (Arizona); Chuck Rine (Lamont); Fritz Rummel (Bochem); and Dave Stocker (Penn State).

Ideas rarely have a single source in the literature; usually they evolve through discussion among colleagues at national meetings and conferences. Ideas arising from such fora are often the most difficult to acknowledge or credit; my memory of the details of specific discussions are now fuzzy at best. Nevertheless, some of the richest discussions happened at the following meetings: the 1971 Penrose Conference on "Earthquake Source Mechanisms and Fracture Mechanics"; the 1974 Penrose Conference on "Earthquake Source Mechanisms and Fracture Mechanics"; the 1976 Chapman Conference on "Stress in the Lithosphere"; the 1979 U.S. Geological Survey Workshop on "Stress and Strain Measurements Related to Earthquake Prediction"; the 1979 U.S. Geological Survey Workshop on "Analysis of Actual Fault Zones in Bedrock"; the 1980 U.S. Geological Survey Workshop on "Magnitude of Deviatoric Stresses in the Earth's Crust and Upper Mantle"; the 1981 U.S. Geological Survey Workshop on "Hydraulic Fracture Stress Measurements"; the 1984

EPRI Workshops on "Tectonic Processes of Intraplate Stress Generation and Concentration"; and the 1988 NSF/GRI Workshop on "Hydraulic Fracturing Stress Measurements."

Although it is customary to acknowledge agencies supporting work on a particular project, the effort of individual contract monitors is often overlooked. Silent partners in my earth stress field experiments include: Jack Evernden for the U.S. Geological Survey; Jerry Harbor for the U.S. Nuclear Regulatory Commission; George Kolstad for the U.S. Department of Energy-Washington; Chuck Komar for the U.S. Department of Energy-Morgantown; Louis Lacy for EXXON Production Research; Dick Plumb for Schlumberger-Doll Research; Carl Stepp for the Electric Power Research Institute; Paul Westcott for the Gas Research Institute; and Leonard Johnson, Mike Mayhew, Dan Weill, and Thomas Wright for the National Science Foundation. Logistical support during the writing of this book came from EPRI contract R.P. 2556–24 and GRI contract 5088–260–1746.

Initial drafts of this book were written as a syllabus for a graduate-level course at The Penn State. One class exercise was the chapter-by-chapter review of a draft of this monograph by R. Cornelius, V. Lee, I. Meglis, P. Scott, and D. Srivastava. I am grateful to several colleagues for chapter reviews including Z. T. Bieniawaski; D. Dunn; K. Evans; M. Friedman; N. Gay; R. Goodman; M. Gross; B. Haimson; R. Hatcher; S. Hickman; J. Logan; S. Mackwell; S. Marshak; R. A. Plumb; D. Pollard; R. Richardson; M. Sbar; C. H. Scholz; H. Swolfs; C. Thornton; M. D. Zoback; and M. L. Zoback.

In addition to many colleagues who have allowed me to modify or reprint their figures herein, I would like to thank the following copyright holders for graciously permitting me to modify or reprint material in this book.

A.A. BALKEMA

From *Rock Joints* - Proceedings of a regional conference of the International Society for Rock Mechanics: Figures 1–5, 2–5 & 2–6, Engelder and Lacazette (1990) 35.

From *Rock mechanics design in mining and tunneling:* Figure 3–19, Bieniawski (1984) 272.

From *Rock Mechanics as a guide for efficient utilization of natural resources:* Proc. 30th U.S. Symposium: Figure 9–10, Teufel (1989) 269.

From *Rock at great depth - Rock mechanics & rock physics at great depth:* Figure 6–7, Plumb (1989) 761.

AMERICAN ASSOCIATION FOR THE ADVANCEMENT OF SCIENCES

From *Science:* Figure 8–2, Clark (1981) 211: 51.

AMERICAN JOURNAL OF SCIENCE

From *American Journal of Science:* Figure 4–13, Carter and Friedman (1965) 263: 747; Figure 10–6, Voight (1974) 274: 662.

AMERICAN GEOPHYSICAL UNION

From the *Geophysical Monographs:* Figure 7–3, Hoskins and Russell (1981) 24: 187.

From the *Geophysical Research Letters:* Figure 10–4, Savage (1978) 5: 633.

From the *Journal of Geophysical Research:* Figure 2–8, Brace and Bombolakis (1963) 68: 3709; Figure 3–11, Stock et al. (1985) 90: 8691; Figure 3–20, Hickman et al. (1985) 90: 5497; Figure 4–11, Kohlstedt and Weathers (1980) 85: 6269; Figure 4–15, Carter et al. (1982) 87: 9289; Figure 4–16, Handy (1990) 95: 8647; Figure–20, Mercier (1980) 85: 6293; Figures 5–3, 5–5, 5–9 & 5–10, Evans et al. (1989a) 94: 1729; Figure 5–11, Evans (1989) 94: 17619; Figure 5–13, Haimson and Rummel (1982) 87: 6631; Figure 5–15, Haimson and Doe (1983) 88: 7355; Figure 5–16, Hickman et al. (1988) 93: 15,183; Figure 6–3, Plumb and Cox (1987) 92: 4805; Figure 6–4, Plumb and Hickman (1985) 90: 5513; Figures 6–8, 6–9 & 6–10, Zoback (1985) 90: 5523; Figure 6–11, Zoback and Hickman (1982) 87: 6959; Figure 7–11, Sbar et al. (1984) 89: 9323; Figure 9–14, Carlson and Wang (1986) 91: 10,421; Figure 9–15, Plumb et al. (1984b) 89: 9333; Figures 9–16 & 9–17, Siegfried and Simmons (1978) 83: 1269; Figure 10–11, Engelder and Geiser (1984) 89: 9365; Figure 11–4, Fletcher and Sykes (1977) 82: 3767; Figure 11–4, Herrmann (1979) 84: 3543; Figure 12–4, Plumb et al. (1984a) 89: 9350; Figure 12–9, Zoback and Zoback (1980) 85: 6113; Figure 13–9, Watts et al. (1980) 85: 6369; Figures 13–10 & 13–11, Forsyth (1980) 85: 6364.

From *Tectonics:* Figure 13–6, Fleitout and Froidevaux (1983) 2: 315.

ANNUAL REVIEWS, INC.

From *Annual Review of Earth and Planetary Sciences:* Figure 4–17, Sibson (1986) 14: 149; Figure 6–5, Gough and Gough (1987) 15: 545; Figure 11–11, Simpson (1986) 14: 21.

BIRKHÄUSER VERLAG

From *Pure and Applied Geophysics:* Figures 2–11 & 2–12, Nakamura et al. (1977) 115: 87; Figure 3–10, Byerlee (1978) 116: 615; Figure 7–12, Engelder and Sbar (1977) 115: 41; Figure 10–8, Engelder (1977) 115: 27.

BLACKWELL SCIENTIFIC PUBLISHERS, LTD.

From *The Geophysical Journal of the Royal Astronomical Society:* Figure 5–8, Evans et al. (1988) 93: 251; Figures 13–1 & 13–4, Forsyth and Uyeda (1975) 43: 163.

CAMBRIDGE UNIVERSITY PRESS

From *Creep of crystals:* Figure 4–7, Poirier (1985) 260.
From *Earthquake mechanics:* Figure 11–1, Kasahara (1981) 248.
From *The mechanics of earthquakes and faulting:* Figure 11–8, Scholz (1990) 439.

ELSEVIER SCIENCE PUBLISHERS B.V.

From *Tectonophysics:* Figures 3–15 & 3–16, Angelier (1979) 56: T17; Figures 4–8 & 4–9, Carter and Tsenn (1987) 136: 27; Figure 10–5, Holzhausen and Johnson (1979) 58: 237; Figures 10–9 & 10–10, Engelder, (1979a) 55: 289; Figure 11–6, Cloetingh and Wortel (1986) 132: 49.

GEOLOGICAL SOCIETY OF AMERICA

From the *Geological Society of America Bulletin:* Figure 2–11, Johnson (1961) 72: 579; Figure 2–13, McHone (1978) 89:1645–55; Figures 4–2 & 4–10, Jamison and Spang (1976) 87: 868; Figures 4–3, 4–5 & 4–6, Carter and Raleigh (1969) 80: 1231; Figure 4–4, Groshong (1972) 82: 2025; Figure 9–3, Friedman et al. (1976) 87: 1049.
From the *Geological Society of America Memoir:* Figure 3–17, Zoback, (1983) 157: 3; Figure 12–8, Zoback and Zoback (1989) 172: 523.
From *Geology:* Figures 2–8 & 2–10, Olson and Pollard (1989) 17: 345; Figure 4–18, Wise et al. (1984) 12: 391; Figure 11–5, Gephart and Forsyth (1985) 13: 70.
From *Quantitative interpretation of joints and faults:* Geological Society of America Short Course Notes: Figure 2–4, Pollard (1989).

JOHN WILEY & SONS

From *Geodynamics—Applications of Continuum Physics to Geological Problems:* Figure 13–2, Turcotte and Schubert (1982) 450.

MACMILLAN MAGAZINES, LTD.

From *Nature:* Figure 12–9, Zoback et al. (1989) 341: 291.

NATIONAL ACADEMY PRESS

From *Hydraulic Fracturing Stress Measurements:* Figure 5–6, Hickman and Zoback (1983) 44; Figure 12–3, Haimson and Lee (1983) 107; Figure 12–1, Bell and Gough (1983) 201.

PERGAMON PRESS

From the *International Journal of Rock Mechanics and Mining Science:* Figure 5–12, Warpinski (1989) 26: 523; Figure 5–14, Baumgärtner and Zoback (1989) 26: 461; Figures 7–7, 9–11 & 9–12, Engelder (1984) 21: 63; Figures

9–20 & 9–21, Engelder and Plumb (1984) 21: 75; Figure 10–7, Hudson and Cooling (1988) 25: 363.

From the *Journal of Structural Geology:* Figure 2–9, Dyer (1988) 10) 685; Figure 2–14, Engelder (1985) 7: 459; Figure 3–18, Angelier (1981a) 3: 347.

SEISMOLOGICAL SOCIETY OF AMERICA

From the *Bulletin of the Seismological Society of America:* Figure 11–3, Sbar et al. (1972) 62: 1303; Figure 11–9, Kanamori and Anderson (1975) 65: 1073; Figure 11–10, Scholz et al. (1986) 76: 65.

SOCIETY OF PETROLEUM ENGINEERS

From *Journal of Petroleum Technology:* Figure 5–2, Hubbert and Willis (1957) 9: 153; Figure 5–4, Evans (1987) 109: 180; Figure 6–6, Hottman et al. (1979) 31: 1477.

SPRINGER-VERLAG, INC.

From *Rock Mechanics:* Figure 8–7, de la Cruz, (1977) 9: 27.

UNIVERSITY OF CHICAGO PRESS

From *Journal of Geology:* Figure 4–12, Friedman and Conger (1964) 72: 361.

List of Symbols

The following lists the symbols used in this book. Their order is alphabetical using the Latin and then Greek alphabets. The equation in which the symbol first appears or is first defined is given in brackets unless otherwise indicated.

A	area [2–26]; $\alpha_b \dfrac{1-2\upsilon}{1-\upsilon}$ [5–20]; real area of contact [11–3]
a	variable [6–10]; constant [11–3]
b	half length of the short axis of a crack [2–1]; variable [6–11]
$\mathbf{b}$	the Burgers vector [4–10]
C_{ijkl}	stiffness tensor[#] [1–11]
C_0	uniaxial compressive strength [3–12]
c	half length of the long axis of a crack [2–1]; variable [6–12]; half length of the slot [8–2]
c_0	half length of a flatjack [8–5]
D	average displacement along the faulted area [11–7]; flexural rigidity [13–27]
$\mathbf{D_o}$	normalized deviatoric tensor [3–35]
d	grain diameter [4–15]; variable [6–13]; lateral distance in lithosphere [13–1]
E	Young's modulus [1–10]; total work (energy) during an earthquake rupture [11–10]
$\hat{E}$	Young's modulus of a rock containing joints [2–31]
E_g	Young's modulus of the borehole inclusion device [8–1]
E_h	Young's modulus of the host rock [8–1]
E_s	energy dissipated as seismic waves [11–11]
$\mathbf{F}$	shear force [11–4]
F_i	components of force on lithospheric plates [13–9]
f	normalized stress intensity function [5–12]
G	energy release rate per unit length of crack tip [2–20]
G_c	critical energy release rate per unit length of crack tip [2–24]
g	gravitational acceleration [1–17]; normalized stress intensity function [5–12]
H^*	activation enthalpy [4–2]
h	topographic relief [13–19]
h_i	material constants [3–16]
h_o, h_a	normalized stress intensity functions [5–12]

I unit tensor [3–35]
i, j, k, l subscripts# [1–5]
i_h angle of emergence [11–1]
K_I stress intensity factor [2–16]
K_{Ic} fracture toughness [2–18]
k diffusivity constant (cm^2/sec) [7–22]
k_o ratio of S_h to S_v [1–15]
M mean of a data set of orientation data [12–1]; mass per unit area of a column of oceanic lithosphere [13–2]
M_b bending moment [13–24]
M_o seismic moment [11–8]
N normal force [11–3]
n amount of a substance [2–5]
n unit vector normal to the fault plane [3–41]
P_b breakdown pressure [5–9]
P_c confining pressure [1–13b]; radial pressure [7–19]
P_c^c crack closure pressure [9–3]
P_f fluid pressure inside a borehole [5–6]
P_f^H wellbore pressures at which horizontal fracture initiation is predicted [5–23]
P_f^V wellbore pressures at which vertical fracture initiation is predicted [5–24]
P_i crack driving pressure [2–15]
P_j flatjack pressure [8–7]
P_l pressure in the lidaosphere [13–12]
P_m magma pressure [1–13a]
P_p pore pressure [1–43]
P_{ro} fracture reopening pressure [5–13]
P_t fluid inclusion trapping pressure [2–27]
p_1 penetration hardness measured at unit time [11–3]
Q rock stress parallel to the flatjack [8–4]
R gas constant [2–5]; the stress ratio [3–36]; radius of a borehole [5–1]; radium of curvature [13–21]
R' the stress ratio [3–50]
R_e radius of the earth [13–17]
R_h least stress ratio, S_h/S_v [chap. 5]
r distance from the crack tip [2–16]; distance between dislocations [4–10]; distance from the center of a borehole [5–1]; characteristic length associated with the narrowest dimension of the fault [11–7]
r_c radius of curvature [2–1]
S stress tensor [3–32]
S_c actual stress under grain contacts with a crack or container wall under dry conditions [1–45]

S_c^a average rock stress normal to a crack or container wall under dry conditions{1–45]

S_H maximum horizontal stress [1–13]

$\bar{S}_H$ effective maximum horizontal stress [6–16]

S_h minimum horizontal stress [1–13]

S_h^T horizontally induced thermal stress [1–22]

S_i^* stress concentration at the end of a borehole [7–17]

S_N radial profile of normal stress across the plane of the fracture [5–18]

S_n rock stress normal to the flatjack [8–2]

S_R reduced stress tensor [3–44]

S_t circumferential stress tangent to the edge of an elliptical crack [2–1]

S_v vertical stress [1–23]

S_{vw} vertical stress at the borehole wall [5–23]

$\mathbf{s}$ slip lineation vector [3–45]

s_o standard deviation of a data set [12–3]

T temperature [1–22]; the travel time of the P-wave [11–1]

T_a amplitude of surface temperature variation [7–24]

T_e elastic thickness of the lithosphere [13–27]

T_m temperature of the mantle [13–6]

T_0 initial temperature [1–22]; uniaxial tensile strength [2–11] mean annual surface temperature [7–24]

T_s surface temperature [1–22]

T_z temperature variation at depth [7–22]

$\mathbf{T}$ torque generated by a lithosphere plate [13–8]

t period (seconds) [7–22]; time of maximum annual surface temperature [7–24]; time of contact [11–3]

$\mathbf{t}$ unit vector associated with τ [3–43]

U_E strain energy [2–6]

U_E^{el} strain energy due to an external load [2–12]

U_E^c strain energy due to crack wall displacement [2–13]

U_s surface energy [2–6]

U_T total energy [2–5]

u displacement [7–10]; velocity of spreading from midocean ridge [13–6]

u_y crack wall displacement [2–28]

V, V_i volume [2–5]

V_{out} volume of flowback {fig. 5–5]

W_c work is done by the fluid in the crack to move the crack wall [2–13]

W_j half displacement caused by raising jack pressure [8–5]

W_R work on rock surrounding the rock-crack system [2–6]

W_s work on expansion of a gas [2–5]

W_y half displacement across open flatjack slot [8–6]

W_0 half displacement during flatjack slot cutting [8–2]

W_1	half displacement due to finite slot width [8–3]
W_2	half displacement due to biaxial stress around a flatjack slot [8–4]
w	change in depth of isotatic compensation due to erosion [1–18]; the dislocation strain energy per unit length in the grain volume [4–18]
Y	crack modification factor [2–17]
x,y,z	subscripts representing orthogonal coordinates with z vertical for faults [fig. 1–3] or for vertical joints [fig. 2–2]
y	distance of measuring pins from the major axis of a flatjack slot [8–2]
y_0	half width of the slot [8–3]
z_1	thickness of the lithosphere [1–19]
z	depth within the earth [1–17]; thickness of eroded lithosphere [1–18]
α_b	Biot's poroelastic parameter [1–49]
α_m	volumetric thermal expansion coefficient of the mantle [13–5]
α_T	linear thermal expansion coefficient [$\mu\varepsilon$/°C) [1–22]
β	compressibility [1–14]; pore pressure coefficient [5–22]
β_b	the bulk compressibility of the solid with cracks and pores [1–49]
β_i	intrinsic compressibility [1–49]
β_{ij}	cosine between fault and stress tensor coordinate systems [3–37]
γ	free surface energy per unit surface area [2–10]; latitude relative to pole of rotation [13–7]
γ_{dis}	the dislocation strain energy per unit area in the grain boundary [4–8]
γ_τ	engineering shear strain [4–14]
Δ	symbol for differentiation [1–14]; the distance in degrees from the recording station to the epicenter [11–1]
ΔP_f	absolute magnitude of the difference between the formation pore pressure, P_p, and the borehole fluid pressure [6–3]
$\Delta\varepsilon$	volumetric strain [1–4]
$\Delta\sigma$	stress change in the host rock [8–1]
$\Delta\sigma_g$	stress change in the rigid-inclusion gauge [8–1]; stress change in a grain [10–6]
$\Delta\hat{\sigma}_h^*$	change in horizontal stress as a consequence of lithospheric thinning {1–21]
$\Delta\tau$	shear stress drop during stick slip [11–2]
δ	dip of a fault plane
δ_{ij}	Kronecker delta# [1–43]

$\delta\sigma_i$	deviatoric stress# [1–41]
$\varepsilon, \varepsilon_i, \varepsilon_{ij}$	strain# [1–1]
ε^i_{ij}	strain in the intrinsic rock (i.e., the uncracked solid) [9–4]
$\dot{\varepsilon}$	strain rate [4–2]
$\dot{\varepsilon}_s$	a steady-state creep rate [4–2]
ζ	Lame's constant which is also called the modulus of rigidity [1–1] and the shear modulus [4–10]
ζ_{ij}	cumulative crack strain [9–14]
ζ_v	total crack porosity [9–17]
η	seismic efficiency [11–11]
η_{ij}	strain due to the presence of cracks [9–4]
Θ	angular distance in polar coordinates [2–16]
θ	angle from plane of crack [2–39]; angle between σ_1 and normal to a plane [3–4]; angle measured clockwise from S_H at a borehole [5–1]
κ	bulk modulus [1–14]; the angles between σ_2 and the normal to the *e*-twin plane [4–6]; thermal diffusivity of the lithosphere [13–6]
λ	Lame's constant [1–1]; rake of the slip lineation on a fault; angle between σ_1 and the normal to the slip plane [4–1]
Λ	critical slip distance [11–6]
μ_o	initial coefficient of sliding friction [11–6]
μ_d	coefficient of dynamic friction [11–2]
μ_s	coefficient of static friction [3–24]
$\overline{\mu}_s$	static friction coefficient at unit time of contact [11–5]
μ^*	coefficient of internal friction [3–2]
$\mu\varepsilon$	microstrain [7–1]
ν	Poisson's ratio [1–12]
$\xi_{ij}(P_c)$	total crack spectra [9–8]
ρ	density of the overburden [1–17]
ρ_c	spatial density of cracks [2–35]
ρ_{dis}	the steady state dislocation density in the grain volume [4–18]
ρ_l	density of lithosphere [1–18]
ρ_m	density of mantle [1–18]
ρ_w	density of water [13–2]
Σ	Surface traction [3–41]
$\lvert\overline{\sigma}^c_3\rvert$	critical effective stress for crack propagation [2–11]
σ_i, σ_{ij}	applied stress # [1–1]; total stress # {1–44]
$\overline{\sigma}, \overline{\sigma}_{ij}$	effective stress # [1–43]
σ_d	differential stress [1–40]
σ^H_d	differential stress in the horizontal plane [5–27]
σ^s_d	steady-state flow stress [4–2]
σ_e	excess stress [1–32]
σ^i_c	cement stress just after cementation [10–5]
σ^i_g	intragranular stress after cementation [10–5]

$\hat{\sigma}_h$ average horizontal stress in the lithosphere [1–19]
$\hat{\sigma}_h^*$ average horizontal stress in a thinned lithosphere [1–20]
$'\sigma_{ij}$ stress in the vicinity of a crack tip [2–16]
σ_{int} internal stress within the host grain [4–9]
σ_m lithostatic stress in the mantle [1–19]; mean stress [1–42]
σ_n normal stress [3–2]
σ_r radial stress near a borehole [5–1]
σ_θ circumferential stress tangent to a borehole [5–1]
σ_t tectonic stress added to the uniaxial-strain reference state [1–30]
σ_t^* tectonic stress added to the lithostatic reference state [1–29]
τ shear stress [3–2]; shear stress from viscous drag [13–1]
$\boldsymbol{\tau}$ shear traction on a fault plane [3–42]
τ_a average stress along the fault [11–10]
τ_c critical resolved shear stress [4–1]
τ_f shear stress on a fault plane after an earthquake rupture [11–10]
τ_i shear stress on the fault plane before an earthquake [11–10]
τ_m maximum shear stress [3–13]
τ_o cohesion of the material [3–2]; cohesive strength of a fault zone [3–25]
τ_{oct} octahedral shear stress [6–1]
τ_r resolved shear stress [4–1]
$\tau_{r\theta}$ shear stress near a borehole [5–3]
ϕ rock porosity [1–46]; angle of internal friction [3–15]; angle between σ_1 and the slip direction [4–1]
ϕ_p the angle measured in the counterclockwise direction from the $\theta = 0°$ axis of the strain rosette [7–6]
φ relative magnitude of stress [3–42]; crack aspect ratio [9–3]
χ the angle between σ_2 and the *e*-twin glide line [4–6]; material property [10–13]
Ω_o $\cos\lambda\cos\phi$ [4–1]

#—Subscripts have ranges from one to three: i, j, k, l = 1, 2, 3.

Kronecker delta δ_{ij} is defined as $\delta_{ij} = \begin{bmatrix} 1, i = j \\ 0, i \neq j \end{bmatrix}$.

Einstein summation convention is followed. Repeated indices in any single term mean that the term is to be summed over the full range of the term. For example, $\sigma_{ij} = \sigma_{i1}x_1 + \sigma_{i2}x_2 + \sigma_{i3}x_3$.

Stress Regimes in the Lithosphere

— 1 —

Basic Concepts

> One of the most interesting and unexpected results yielded by rock (stress) measurements . . . is the demonstration of large horizontal stresses in the earth's crust. . . . In the virgin rock and at depths of about 400 m measurements have revealed horizontal [stresses] as great as 1.5–3.5 times the dead weight of the overlying strata.
>
> —Nils Hast (1958) during his survey of earth stress within the Scandinavian Peninsula

The dynamic nature of our earth is apparent in the deformation of its lithosphere as manifest by mountain belts, continental rifts, rapidly subsiding basins, high plateaus, and deep oceanic trenches. Evidence from magnetic anomalies, earthquake distribution, and continental shape suggests that large-scale deformation is a consequence of plate tectonics. On a more esoteric level, a geoscientist might explain that large-scale deformation occurs in response to stress within the lithosphere. But, what is meant by the expression, stress within the lithosphere? Although the word, stress, commonly appears in the geoscience literature, doubt about its meaning arises because its definition, simply stated as force per unit area, is too vague for the context of the discussion. In the use of the word, stress, there are often subtle nuances which are either missed by the reader or never clarified by the author. The purpose of this book is to clarify some of the subtle nuances associated with the use of the term, stress, through a discussion of in situ stress measurements and their interpretation.

Starting in the early part of the twentieth century, geoscientists began making some insightful inferences about the state of stress[1] in the lithosphere. For example, Anderson (1905) deduced that $S_H > S_h > S_v$[2] is necessary for thrust faulting during mountain building. Later, more certain knowledge about stress within the lithosphere was gained through in situ measurements. Fifty years after Anderson's prediction, systematic in situ stress measurements began when Hast (1958) found horizontal stresses (maximum, S_H, and minimum, S_h) exceeding the stress developed under the weight of the overburden (S_v) in the shallow continental crust of Scandinavia. With his seminal work Hast (1958) was the first to confirm, using in situ measurements over a broad region, that the relative magnitudes of the principal stresses in the upper crust met conditions specified by Anderson (1905) for thrust faulting. Because Hast's (1958) in situ measurements preceded the theory of plate tectonics, interpretation of his stress data in the context of large-scale deformation was difficult.

Prior to the formulation of plate tectonic theory, interpretation of stress data was based on a few rather simple models for the response of the earth (i.e., the lithosphere) to stress. For example, Seager (1964) was skeptical that Hast's in situ stress data had anything to do with tectonics and chose, instead, to use a simple elastic-plastic model to interpret the data. In Seager's model, rocks are subject to *Poisson's effect* which is a tendency for an elastic rock to expand laterally under its own weight. According to Seager's assumption, horizontal stresses build because adjacent rock serves as a rigid boundary and prevents lateral strain (i.e., $\varepsilon_H = \varepsilon_h = 0$). Because rocks generally expand less than they shorten from overburden weight, horizontal stress generated through Poisson's effect remains a fraction of overburden stress. In Seager's (1964) model, differential stress, the difference between S_v and S_h, increases as more overburden is added until the rock deforms plastically. Even though tectonic stress is left out of Seager's model, such models invariably lead to the conclusion that rock strength places definite limits on differential stress in the lithosphere, a concept developed in detail in chapters 2, 3, and 4. Although the theory that horizontal stresses arise solely from Poisson's effect does not explain upper crustal stresses, a state of stress represented by this model is convenient as one of three reference states.

In introducing some basic concepts concerning stress in the lithosphere, the first step is to discuss three convenient reference states of stress. These reference states are based on boundary conditions expected within planetary bodies with globally continuous lithospheres (i.e., one-plate planetary bodies such as Mercury and the Moon). From the equations of linear elasticity we will derive a reference state of stress based on a state of uniaxial strain (Price, 1959). Another reference state of stress arises from a constant stress boundary condition and is independent of elasticity (Artyushkov, 1973; McGarr, 1988). A third reference state of stress assumes all principal stresses are equal. A second step in the introduction of basic concepts is to add tectonic stress to the lithospheric reference states of stress. The third step is to define several terms which are used in the literature as expressions for state of stress. Commonly, the word, "stress," appears in the literature when the author could have been more precise. While there are precise definitions for total stress, effective stress, deviatoric stress, and differential stress, there is no consensus for a definition of tectonic stress and it is often used in a context in which the terms, deviatoric or differential stress, are more appropriate. Because there is more than one lithospheric reference state of stress there is more than one definition for tectonic stress.

Elasticity and Lithospheric Stress

Discussion of three reference states is preceded by an introduction to elementary elasticity. Such an introduction is appropriate because many in situ stress measurement techniques capitalize on the elastic behavior of rocks. Furthermore, many notions concerning state of stress in the lithosphere arise from the assumption that the upper crust behaves as a linear elastic body.

On scales ranging from granitoid plutons and salt domes (≈ 1–5 km) to the thickness of lithospheric plates (≈ 100 km), the earth is approximately isotropic with elastic properties independent of direction. If we assume that the lithosphere is subject to small strains, as is the case for elastic behavior, principal stress axes must coincide with the principal strain axes. Then, elastic behavior of the lithosphere is represented by the equations of linear elasticity which define the *principal stresses*[3] of the three-dimensional stress tensor as linear functions of the principal strains (Jaeger and Cook, 1969, section 5.2):

$$\sigma_1 = (\lambda + 2\zeta)\varepsilon_1 + \lambda\varepsilon_2 + \lambda\varepsilon_3 \qquad (1\text{–}1)$$

$$\sigma_2 = \lambda\varepsilon_1 + (\lambda + 2\zeta)\varepsilon_2 + \lambda\varepsilon_3 \qquad (1\text{–}2)$$

$$\sigma_3 = \lambda\varepsilon_1 + \lambda\varepsilon_2 + (\lambda + 2\zeta)\varepsilon_3 \qquad (1\text{–}3)$$

where λ and ζ are elastic properties of the rock, known as the Lame's constants. ζ is commonly known as the modulus of rigidity, the ratio of shear stress to shear strain. $\lambda + 2\zeta$ relates stress and strain in the same direction, and λ relates stress with strain in two perpendicular directions.

If *volumetric strain* is defined as

$$\Delta\varepsilon = \varepsilon_1 + \varepsilon_2 + \varepsilon_3, \qquad (1\text{–}4)$$

then we may combine equations 1–1 to 1–3 as

$$\sigma_i = \lambda\Delta\varepsilon + 2\zeta\varepsilon_i. \qquad (1\text{–}5)$$

The uniaxial stress state is of immediate interest in developing an understanding of the elastic constants, Young's modulus (E) and Poisson's ratio (υ), both of which will appear several times during discussions of lithospheric stress in this book. *Uniaxial stress*, a stress state for which only one component of principal stress is not zero, is rare in the earth's crust except in the pillars of underground mines. For uniaxial stress, $\sigma_1 \neq 0$, $\sigma_2 = \sigma_3 = 0$, and the three equations of elasticity are written as

$$\sigma_1 = (\lambda + 2\zeta)\varepsilon_1 + \lambda\varepsilon_2 + \lambda\varepsilon_3, \qquad (1\text{–}6)$$

$$0 = \lambda\varepsilon_1 + (\lambda + 2\zeta)\varepsilon_2 + \lambda\varepsilon_3, \qquad (1\text{–}7)$$

$$0 = \lambda\varepsilon_1 + \lambda\varepsilon_2 + (\lambda + 2\zeta)\varepsilon_3, \qquad (1\text{–}8)$$

Using equations 1–7 and 1–8, strain parallel to the applied stress, ε_1, is related to strain in the directions of zero stress, ε_2 and ε_3,

$$\varepsilon_2 = \varepsilon_3 = -\left(\frac{\lambda}{2(\lambda + \zeta)}\right)\varepsilon_1 . \tag{1–9}$$

To derive the ratio between stress and strain for uniaxial stress, values for ε_2 and ε_3 are substituted back into equation 1–6. This ratio is Young's modulus,

$$E = \frac{\sigma_1}{\varepsilon_1} = \frac{\zeta(3\lambda + 2\zeta)}{\lambda + \zeta} . \tag{1–10}$$

Equation 1–10 is a simplified form of Hooke's Law,

$$\sigma_{ij} = C_{ijkl}\varepsilon_{kl} \tag{1–11}$$

where C_{ijkl} is the stiffness tensor. Under uniaxial stress, an elastic rock will shorten under a compressive stress in one direction while expanding in orthogonal directions. The ratio of the lateral expansion to the longitudinal shortening is Poisson's ratio,

$$\nu = -\frac{\varepsilon_2}{\varepsilon_1} = \left(\frac{\lambda}{2(\lambda + \zeta)}\right). \tag{1–12}$$

ν for rocks is typically in the range of .15 to .30.

Three Reference States of Stress

While developing an understanding of lithospheric stress, it is convenient to start with reference states which may occur in a planet with just one lithospheric plate and thus devoid of plate tectonics. The next step is to describe the difference between these reference states and the actual state of stress generated by plate tectonic processes. In the earth, geologically appropriate reference states are those expected in "young" rocks shortly after lithification. Two general types of "young" rocks are intrusive igneous rocks, shortly after solidification within large plutons deep in the crust, and sedimentary rocks, shortly after the onset of burial and diagenesis within large basins.

Lithostatic Reference State

The simplest reference state is that of *lithostatic stress* found in a magma which has no shear strength and, therefore, behaves like a fluid with

$$\sigma_1 = \sigma_2 = \sigma_3 \Rightarrow S_H = S_h = S_v \Rightarrow P_m \tag{1–13a}$$

where P_m is the pressure within the magma. At the time enough crystals have solidified from the magma to form a rigid skeleton and support earth stress, the "young" rock, an igneous intrusion, is subject to a lithostatic state of stress

$$\sigma_1 = \sigma_2 = \sigma_3 \Rightarrow S_H = S_h = S_v \Rightarrow P_c \qquad (1\text{–}13b)$$

where P_c is confining pressure. This, presumably, is the state of stress at the start of polyaxial strength experiments in the laboratory. Strictly speaking, there are no principal stresses in this case, because the lithostatic state of stress is isotropic. Equation 1–4 applies to calculate volumetric elastic strain accompanying the complete erosion of an igneous intrusion assuming no temperature change. Adding equations 1–1 to 1–3 gives an expression for the response of rocks, including laboratory test specimens, to changes in confining pressure,

$$\Delta P_c = \left(\frac{3\lambda + 2\zeta}{3}\right)\Delta\varepsilon = \kappa\Delta\varepsilon = \frac{1}{\beta}\Delta\varepsilon \qquad (1\text{–}14)$$

where κ is the bulk modulus and its reciprocal, β, is the compressibility.

Lithostatic stress will develop if a rock has no long-term shear strength. Although some rocks such as weak shales and halite have very little shear strength, experiments suggest that all rocks support at least a small differential stress for very long periods (Kirby, 1983). Indeed, heat will cause rocks to relax, but they never reach a lithostatic stress state. Although metamorphic rocks retain a shear strength during deformation, the metamorphic petrologist commonly uses the term pressure when actually referring to a state of stress which may approximate lithostatic stress (e.g., Philpotts, 1990). A structural geologist reserves the term *pressure* for describing confined pore fluid or other material with no shear strength. The terms signifying states of stress, of which lithostatic stress is one, apply to rock and other materials that can support a shear stress. Although the lithostatic state of stress is rare in the lithosphere, it is a convenient reference state.

Uniaxial-Strain Reference State

A second reference state is based on the postulated boundary condition that strain is constrained at zero across all fixed vertical planes (Terzaghi and Richart, 1952; Price, 1966, 1974; Savage et al., 1985). Such a boundary condition leads to a stress state which approximates newly deposited sediments in a sedimentary basin: the state of stress arising from *uniaxial strain* (fig. 1–1). Prior to complete diagenesis and lithificaiton "young" sediments compact by a process called uniaxial consolidation (Karig and Hou, 1992). During uniaxial consolidation horizontal stress, $S_H = S_h$, increases as function of depth of burial but at a rate less than the vertical stress, S_v. Diagenesis under uniaxial strain conditions leads to a stress ratio,

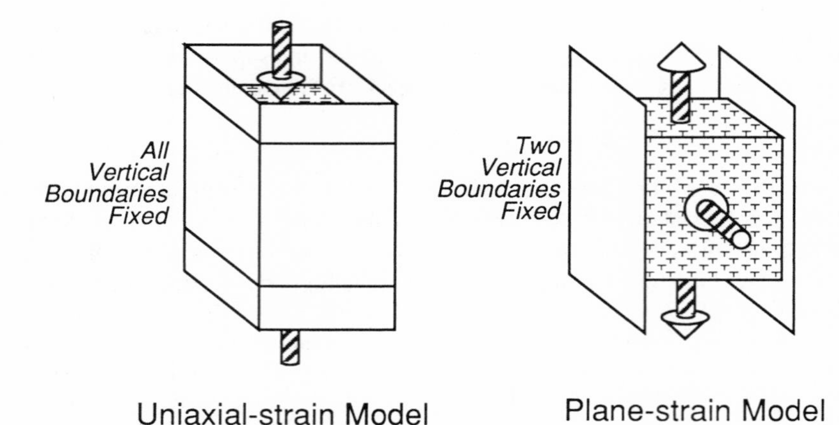

Fig. 1–1. Models for plane strain and uniaxial strain in the crust of the earth.

$$k_o = \frac{S_h}{S_v} < 1, \qquad (1\text{–}15)$$

which develops independently of the elastic properties of the rock.

Uniaxial strain is also used to constrain changes in horizontal stress as a function of changing overburden load, assuming that rocks develop fixed elastic properties at some point after deposition. Under elastic conditions, σ_1 is still the vertical stress, S_v, arising from the weight of overburden. If rocks were unconfined in the horizontal direction the response to an addition of overburden weight would be a horizontal expansion. Because rocks are confined at depth in the crust, Price (1966) suggests that horizontal expansion is restricted by adjacent rock so that in the ideal case, $\varepsilon_2 = \varepsilon_3 = 0$. For the case of uniaxial strain, $\varepsilon_1 \neq 0$, $\varepsilon_2 = \varepsilon = 0$, the equations of elasticity (1–1 to 1–3) are written (Jaeger and Cook, 1969, sec. 5.3):

$$\sigma_1 = (\lambda + 2\zeta)\varepsilon_1 \qquad (1\text{–}16a)$$

$$\sigma_2 = \sigma_3 = \lambda\varepsilon_1 \qquad (1\text{–}16b)$$

From equations 1–16a and b, the relationship between vertical ($S_v = \rho g z = \sigma_1$) and horizontal stresses ($S_H = S_h = \sigma_2 = \sigma_3$) are given in terms of the Poisson's ratio

$$S_H = S_h = \left[\frac{\nu}{(1-\nu)}\right] S_v = \left[\frac{\nu}{(1-\nu)}\right] \rho g z \qquad (1\text{–}17)$$

where ρ is the integrated density of the overburden, g is the gravitational acceleration, and z is the depth within the earth. Assuming that the rock in question is fully lithified, one application of equation 1–17 is the calculation of changes in horizontal stress during sedimentation and erosion. During erosion and removal of overburden weight, lack of contraction in the horizontal direction by uniaxial-strain behavior also leads to large changes in horizontal stresses. Major deviations from the functional relationship between S_h and S_v predicted by equation 1–17 signal that elastic behavior using the uniaxial-strain model is not a complete representation for state of stress in the lithosphere.

In practice, it is understood that horizontal stress arising from a uniaxial strain state in a sedimentary basin is modified by changes in elastic properties during diagenesis and creep relaxation. Each of these processes bring the state of stress in sedimentary basins closer to lithostatic. Mandl (1988) refers to a horizontal stress arising from such inelastic deformation mechanisms as a *prestress* or a stress which is not explained by simple incremental elastic deformation. As discussed in chapter 10, remnant stress is a version of prestress. Of course, upon initiation of tectonic processes such as slumping along listric normal faults, horizontal stresses can vary widely from those calculated using the uniaxial-strain reference state. The analysis of tectonic stress presented in the next section deals with the case where horizontal stresses under uniaxial strain at depth arise from elastic behavior rather than uniaxial consolidation.

Constant-Horizontal-Stress Reference State

If the uniaxial-strain model applies to stress conditions in the lithosphere, then extensional stress regimes (i.e., $S_v > S_H > S_h$) should be very common (McGarr, 1988). Contrary to this prediction, the compressional stress regime is unusually common (i.e., $S_H > S_h > S_v$) in continental interiors (McGarr and Gay, 1978). Furthermore, the uniaxial-strain reference state might not apply to the lithosphere because that model predicts extreme changes in horizontal stress during thermal cooling or removal of overburden. As an alternative McGarr (1988) proposed that the constant-horizontal-stress reference state is ideal for the lithosphere. The essence of this model is that, regardless of the local thickness of the lithosphere, the average stress in the lithosphere is everywhere the same to the depth of isostatic compensation under the thickest lithosphere (fig. 1–2).

McGarr's model for a constant-horizontal-stress boundary condition is based on Artyushkov's (1973) approach using the constraints of isostasy and static horizontal force equilibrium. The effect of erosion is seen by assuming that the lithosphere is isostatically compensated at a depth, z_l, below which the earth is assumed to act like a fluid substrate with no long-term shear strength. The state of stress in the substrate is lithostatic with $S_h = S_v = \sigma_m$ arising from the weight of the overburden. The average density of the lithosphere is ρ_l and the density of the substrate is ρ_m. Assume that a thickness, z, is eroded from a portion of the lithosphere. Isostatic compensation occurs when the base of the eroded lithosphere is elevated by the amount

EROSION OF THE LITHOSPHERE

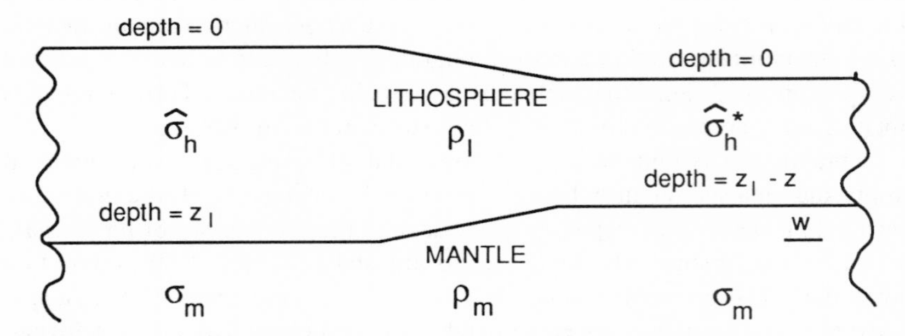

Fig. 1–2. An illustration of the effect of erosion of the lithosphere on thickness of the lithosphere. An amount z was eroded from the right-hand side of this cross section so that the level of isostatic compensation moved up by the amount w. Such isostatic compensation leads to the constant-horizontal-stress reference state.

$$w = \frac{\rho_l}{\rho_m} z \tag{1–18}$$

relative to the lithosphere which was not eroded. In the thicker portion of the lithosphere the average horizontal stress is denoted as $\hat{\sigma}$. Under the assumption that horizontal forces must balance after erosion, the average horizontal stress, $\hat{\sigma}^*_h$, in the thinner portion of the lithosphere is higher. In the constant-horizontal-stress model the state of stress in the original lithosphere (left, in fig. 1–2) is unaffected by erosion of a portion of the lithosphere (right). So, by force balance

$$\hat{\sigma}_h \, z_l = \hat{\sigma}^*_h(z_l - z) + \sigma_m w. \tag{1–19}$$

Equations 1–18 and 1–19 are combined to solve for $\hat{\sigma}^*_h$, where

$$\hat{\sigma}^*_h = \hat{\sigma}_h \left(\frac{z_l}{z_l - z} \right) - \rho_l g z \left(\frac{\rho_l}{\rho_m} \right) \frac{\left(z_l - \frac{z}{2} \right)}{(z_l - z)}. \tag{1–20}$$

The first term on the right side of equation 1–20 corresponds to an increase horizontal lithospheric stress from crustal thinning and the second term is a decrease arising from isostasy. The more general solution of equation 1–20 for sedimentation or erosion is found by setting Δz equal to thickness changes (negative values for erosion). Then

$$\Delta\hat{\sigma}_h = \hat{\sigma}^*_h - \hat{\sigma}_h = \left(\frac{\Delta z}{z_l - \Delta z_l} \right) \left[-\hat{\sigma}_h + \rho_l g \left(z_l + \frac{\Delta z}{2} \right) \left(\frac{\rho_l}{\rho_m} \right) \right]. \tag{1–21}$$

In calculating the relative changes in horizontal and vertical stress during erosion, McGarr (1988) assumes a lithostatic state for lithosphere of the initial thickness z_l. Using the constant horizontal stress boundary condition, McGarr (1988) shows that, upon removal of overburden, horizontal stress changes are significantly less than changes calculated using the uniaxial-strain boundary condition.

THE EFFECT OF TEMPERATURE CHANGES ON THE REFERENCE STATE

So far we have operated as if the reference state of stress is related to depth or lithosphere thickness but independent of temperature. However, temperature changes have a large effect on horizontal stress, particularly if we assume a uniaxial-strain reference state. Uniaxial-strain models for cooling during uplift and erosion under elastic conditions predict large tensile stresses, sometimes at depths in excess of 1 km (Voight and St. Pierre, 1974; Haxby and Turcotte, 1976). The magnitude for thermally generated horizontal stresses are estimated using the following equation for uniaxial strain behavior

$$\Delta S_h^T = \frac{E\alpha_T(T_s - T_0)}{1 - \nu}. \tag{1–22}$$

Thermal stresses develop in the horizontal direction with the vertical component being relieved due to lack of a fixed boundary at the earth's surface. Thermal stresses are axially symmetric. Large tensile stress are generated if the temperature change goes from T_0, the temperature of the rock at which creep becomes significant, to T_s, the surface temperature. If $E = 100$ GPa, $\nu = .25$, and $\alpha_T = 7 \times 10^{-6}\ °C^{-1}$ for a granite, then a less drastic cooling of, say, $\Delta T = 100°C$ will reduce S_h by 93 MPa. If this stress decrease is enough to overcome the ambient compressive stress, then conditions will favor the propagation of tensile cracks.

For uniaxial strain boundary conditions horizontal stresses induced by the removal of overburden are proportional to the Poisson's ratio through equation 1–17. If a rock has a $\nu = .25$, removal of 3 km of overburden, about the thickness of rock for a 100°C temperature change from top to bottom in a normal geothermal gradient, will induce a tensile component of about 25 MPa to S_h. Although thermally induced changes in S_h are larger than accompanying changes induced by removal of overburden both effects may act in concert to generate large components of tensile stress. Assuming the constant-horizontal-stress boundary condition, stress changes are much less for the same temperature changes (McGarr, 1988).

Because temperature changes affect horizontal stress, there is no unique reference state at any depth in the lithosphere. But again, relative reference states are a convenient point of departure for defining tectonic stress.

Tectonic Stress

Stresses can vary from reference states as a consequence of either natural or man-made processes. Those components of the in situ stress field which are a deviation from a reference state as a consequence of natural processes include tectonic stresses, residual stresses (chap. 10), and near-surface thermal stresses induced by diurnal and annual heating. Large deviations from a reference state of stress arising from mining, drilling, excavation, and other societal activities are man-made and, thus, unrelated to natural processes in the lithosphere. Tectonic stresses are often assumed to arise from the largest-scale natural boundary conditions such as plate-boundary tractions (e.g., Zoback et al., 1989; Hickman, 1991). Local stresses arising from topographic loading, thermoelastic loading, and unloading due to erosion are considered nontectonic. The term *contemporary tectonic stress* grew from Sbar and Sykes's (1973) paper to mean that component of the present stress field which is reasonably homogeneous on a regional or platewide scale. However, the distinction between platewide and local stresses breaks down in certain geological contexts. For example, the orientation of some late-formed joint sets is controlled by platewide stress fields, yet they propagate in response to a component of the stress field developed during local unloading due to erosion (e.g., Hancock and Engelder, 1989). Thus, the definition of tectonic stress is simplified if most constraints concerning scale and source are removed from the definition. *Tectonic stresses* are usually horizontal components of the in situ stress field which are a deviation from a reference state as a consequence of natural processes on all scales from platewide to local. By this definition local stresses developed under topographic loading are considered tectonic. Unfortunately, this definition is still so loose that geoscientists will continue to quibble about what qualifies as a tectonic stress. In the extreme some will argue that residual and remnant stresses are tectonic, whereas most will agree that near-surface thermal stresses are not tectonic. This book will continue to use the term "contemporary tectonic stress" in its narrower sense to indicate that component of the earth's stress field arising from plate-scale boundary tractions.

Tectonic stress is that quantity of stress added to or subtracted from a horizontal component of stress to cause the state of stress to deviate from a reference state. The most famous classification of tectonic stress states is Anderson's (1951) where S_h and S_H vary in magnitude relative to S_v. In Anderson's scheme the S_v arises from the overburden load

$$S_z = S_v = \rho g z. \qquad (1\text{–}23)$$

The vertical component[4] is generally invariant, unless z changes by deposition, thrust imbrication, or erosion, because the surface of the crust is a free boundary which will not support an additional component of stress in the vertical direction. S_v is generally a principal stress unless the principal stresses rotate to an off vertical direction as a consequence of unusual boundary conditions. Under unusual circumstances buckling will lift part of the overburden load to create a condition where a stress is subtracted from the vertical component of stress (e.g., Evans et al., 1988). Three arrangements of the principal stresses are possible in Anderson's scheme with each arrangement expressing the relative stresses found during one of three major types of faulting in the lithosphere (fig. 1–3). One classification of stress regimes in the lithosphere involves these three arrangements of principal stresses where the stress regimes are the normal-fault regime ($S_v > S_H > S_h$), the strike-slip fault regime ($S_H > S_v > S_h$), and the thrust-fault regime ($S_H > S_h > S_v$). When referring to these regimes, an author indicates knowledge of the relative magnitudes of the three principal stresses but expresses no information on the absolute magnitudes of the stresses.

For two definitions of tectonic stress during deformation associated with the three fault regimes in the lithosphere, we may use a plane-strain model where $\varepsilon_1 > 0$, $\varepsilon_2 = 0$, $\varepsilon_3 < 0$ under volume-constant strain (fig. 1–1). For these definitions of tectonic stress, we use the uniaxial and lithostatic reference states. The distinction between uniaxial strain and plane strain is that, in the latter case, horizontal strain occurs in one direction but not in the orthogonal horizontal direction. Plane strain applies to orogenic events where either shortening or

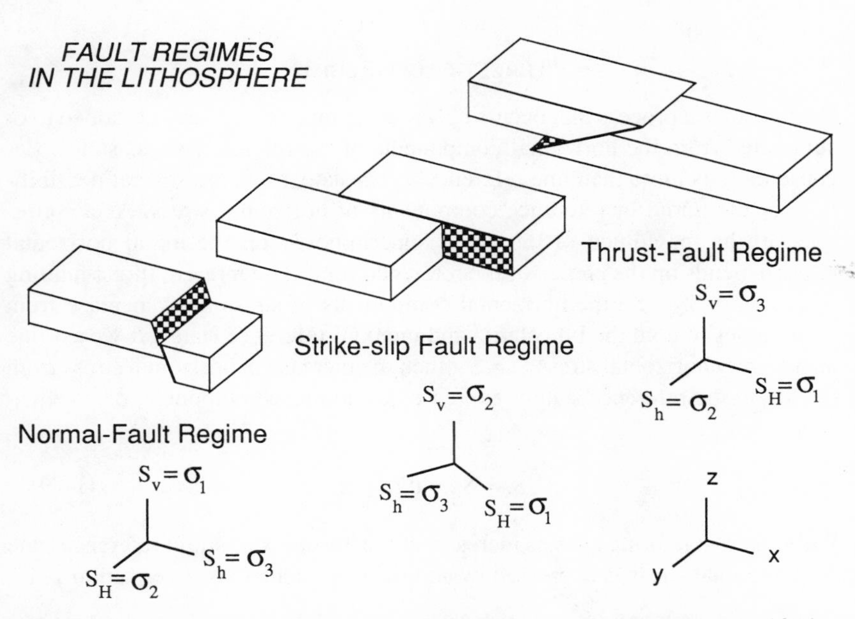

Fig. 1–3. The three states of stress associated with thrust, strike-slip, and normal faulting. These three stress states, known as the Andersonian stress states, are referred to as the thrust-fault, strike-slip-fault, and normal-fault regimes, respectively.

extension take place parallel to the surface of the earth. Rewriting the equations of elasticity, 1–1 to 1–3, for plane strain

$$\sigma_1 = (\lambda + 2\zeta)\varepsilon_1 + \lambda\varepsilon_3 \quad (1\text{–}24)$$

$$\sigma_2 = \lambda(\varepsilon_1 + \varepsilon_3) \quad (1\text{–}25)$$

$$\sigma_3 = \lambda\varepsilon_1 + (\lambda + 2\zeta)\varepsilon_3. \quad (1\text{–}26)$$

In this model, using these equations of elasticity, we derive stress in the horizontal direction of no strain as

$$S_h = \left[\frac{\lambda}{2(\lambda + \zeta)}\right](S_v + S_H) = \nu\,(S_v + S_H), \quad (1\text{–}27)$$

if ε_1 is horizontal as is the case for the normal-fault regime and

$$S_H = \left[\frac{\lambda}{2(\lambda + \zeta)}\right](S_v + S_h) = \nu\,(S_v + S_h), \quad (1\text{–}28)$$

if ε_3 is horizontal as is the case for the thrust-fault regime. Unlike the case for uniaxial strain, stress against the fixed vertical boundary is a function of both vertical and horizontal stresses.

Thrust-Fault Regime

Tectonism is a process that occurs because a component of stress is added to or subtracted from the horizontal components of the reference stress states. Because there is more than one reference stress state, there are several possibilities for the initial or reference components of horizontal stress. As a consequence, the magnitude of the stress superimposed on the initial horizontal stress depends on the choice of the reference state. To represent thrust faulting where $S_H > S_h > S_v$, the horizontal components of stress must increase from their values in both the lithostatic- and uniaxial-reference states. If we assume an arbitrary horizontal stress $S_x = S_H$, then the increase in horizontal stress from the lithostatic reference state is expressed by adding a component, σ_t^*, to equation 1–13

$$S_x = S_H = \rho g z + \sigma_t^* . \quad (1\text{–}29)$$

During thrust faulting a stress increase above the uniaxial-strain reference state for horizontal stress is expressed by adding a component, σ_t, to equation 1–17

$$S_x = S_H = \left(\frac{\nu}{1 - \nu}\right)\rho g z + \sigma_t \quad (1\text{–}30)$$

where $\sigma_t > \sigma_t^*$. σ_t and σ_t^* are both expressions of *tectonic stress*, and yet they are very different quantities because they represent deviations of horizontal stress from different reference states, the uniaxial and lithostatic, respectively. If the reference state for horizontal stress is the lithostatic stress, then any deviation from the reference state is appropriately named the tectonic stress, σ_t^*. When the reference state for horizontal stress is taken as that generated by overburden weight under uniaxial strain conditions, then the tectonic stress, σ_t, is that component of horizontal stress which deviates from the reference state. The tectonic stress, σ_t^*, is equivalent to Turcotte's and Schubert's (1982) "deviatoric normal stress," whereas σ_t corresponds, in part, to Hafner's (1951) "supplementary" stress. The tectonic stress, σ_t, is defined as the difference between the actual value of S_H which includes a component of stress arising from plate tectonic processes and that component of S_H derived from an overburden load using the uniaxial-reference state. At some places in the lithosphere where rocks have a υ which approaches .5, then equation 1–30 reduces to equation 1–29 and $\sigma_t = \sigma_t^*$.

A simple model for the onset of mountain building by thrust faulting assumes plane strain which permits no strain in the y direction but contraction or expansion in the x direction (Turcotte and Schubert, 1982). Starting with the lithostatic-reference state, the additional stress in the y direction is $\nu\sigma_t^*$. Stress in the y direction is now written in terms of both reference states as

$$S_y = S_h = \nu(\rho gz + \sigma_t) = \rho gz + \upsilon\sigma_t^* . \quad (1\text{–}31)$$

In the y direction *excess stress*, σ_{ex} , above the reference for overburden loading in uniaxial-reference state is

$$\sigma_{ex} = \frac{\nu^2}{1-\nu}\rho gz = \nu\sigma_t. \quad (1\text{–}32)$$

Thus the horizontal stress in the y direction exceeds the overburden stress but is less than the horizontal stress in the x direction. This is one possible example of Anderson's model for the stress condition

$$S_x > S_y > S_z \text{ or } S_H > S_h > S_v \quad (1\text{–}33)$$

which represents the thrust-fault regime.

Normal-Fault Regime

For the case of normal faulting the overburden stress remains the same as above but the horizontal stresses are less than the vertical stress. Whether or not the tectonic stress in the horizontal direction is tensional depends on the initial reference state. If the reference state is lithostatic, then

$$S_h = S_x = \rho gz - \sigma_t^*. \tag{1–34}$$

and σ_t^* is always tensional. Under some situations in the normal-fault regime the tectonic stress is still compressive (i.e., positive) relative to the uniaxial-reference state. σ_t is most likely to become tensional where υ » 0.3 in areas of active normal faulting. Even though σ_t is tensional, S_h will remain compressive at depth. A situation where $| -\sigma_t | > \rho gz$ is most unlikely in the lithosphere largely because rock strength will not permit such a state to develop. In a normal-fault regime, regardless of the sign of σ_t, it is low enough in magnitude so all horizontal stress components are less than the vertical component.

$$S_x = S_h = \left(\frac{\nu}{1-\nu}\right) \rho gz \pm \sigma_t. \tag{1–35}$$

Again, horizontal stress in the y direction depends on the reference state and is

$$S_y = S_H = \nu(\rho gz \pm \sigma_t) = \rho gz - \upsilon\sigma_t^* . \tag{1–36}$$

Using the plane strain assumption to calculate the S_y, the relationship among the three principal stresses is

$$S_z > S_y > S_x \text{ or } S_v > S_H > S_h \tag{1–37}$$

which represents the normal-fault regime. By the form of equation 1–35 the student is reminded that in parts of the normal-fault stress regime, σ_t is compressive if the reference state is uniaxial strain. A stress state representative of the thrust-fault regime is found only if σ_t is relatively large.

In the extreme case for both the thrust-fault and normal-fault regimes tectonic stress is uniaxial. The orthogonal horizontal stress, an excess stress, arises as a consequence of the elastic properties of the rock. If stress in the orthogonal direction contains a component which is not a consequence of the elastic properties of the rock, then we have a situation where the tectonic stress is biaxial. The secondary component of tectonic stress is relatively small compared to the primary component. If the secondary component becomes large but has the same sign as the primary component, then the stress regime remains the same. However, if the secondary component has the opposite sign from the primary component of tectonic stress, then we enter the strike-slip fault regime.

Strike-Slip Fault Regime

The strike-slip fault regime is considered a hybrid, a combination of the uniaxial thrust-fault regime at right angles to the uniaxial normal-fault regime and, hence, tectonic stress is biaxial. For strike-slip faulting the vertical overburden stress remains the same as above but the horizontal tectonic stresses are positive in one direction and negative in the other direction relative to the lithostatic reference state:

$$S_x > S_v = S_z \text{ so } (\sigma_t^*)_x > 0 \qquad (1\text{–}38)$$

and

$$S_y < S_v = S_z \text{ so } (\sigma_t^*)_y < 0. \qquad (1\text{–}39)$$

The same relationship holds if x and y are reversed in equations 1–38 and 1–39. Relative to the uniaxial-strain reference state, σ_t is likely to remain positive in orthogonal horizontal directions, and only become negative in one direction under extreme circumstances.

In summary, a definition of tectonic stress is dependent on choice of reference state. σ_t is positive (compressive) under all but the most extreme cases of extension in the normal faulting regime. σ_t^* is positive or negative for the thrust-fault and normal-fault regimes, respectively. In thrust faulting regimes the absolute value of σ_t^* is always smaller than absolute value of σ_t, however, the opposite is not necessarily true for normal faulting regimes.

Differential Stress

The uniaxial-strain reference state for stress (equation 1–17) postulates the initial condition that $S_H = S_h < S_v$. In this reference state the rock system is subject to a *differential stress*, σ_d, where

$$\sigma_d = \sigma_1 - \sigma_3 = S_v - S_h = S_v\left(\frac{1-2v}{1-v}\right). \qquad (1\text{–}40)$$

For small ν, the σ_d approaches the weight of the overburden.

In the lithosphere S_v is generally constrained by the weight of the overburden, whereas S_H and S_h are constrained by rock strength to remain within definite limits of S_v. A rock, subject to a large σ_d, will fail to sustain such an awkward stress state and will respond by deforming in either a brittle or ductile manner. Rock strength is then a governor keeping σ_d within certain bounds. Rock strength varies depending on conditions such as confining pressure, temperature, strain rate, and pore fluid pressure (Paterson, 1978; Poirier, 1985). For intact rock, strength is limited by the activation of either a brittle or ductile deformation mechanism. At higher strain rates (> 10^{-10} sec^{-1}) brittle shear rupture is a likely mechanism acting to relieve σ_d in most intact rocks. Where water chemistry is active, temperature is higher (> 100°C), and strain rate is lower (< 10^{-10} sec^{-1}), ductile deformation mechanisms such as creep by diffusion-mass transfer (i.e., stress solution) become the dominant deformation mechanism and act to moderate σ_d at a level much lower than that necessary for brittle shear rupture. Rock strength under conditions favoring shear rupture is much higher than under conditions favoring diffusion-mass transfer. However, because the upper crust is cut by many joints and fractures, the strength of intact rock is not the sole governor of σ_d. Frictional slip on preexisting joints, fractures, and faults also acts to weaken rock and, thus, limit σ_d to values less than necessary for shear rupture. It is important to appreciate that rock deformation occurs as a function of σ_d, not σ_t. If the uniaxial-strain reference state applies to the lithosphere, brittle fracture or creep are both possible under the high σ_d generated according to equation 1–40 without the superposition of a component of σ_t. Yet, if the constant horizontal stress reference state is a more appropriate representative of conditions in the lithosphere, a high σ_d may occur only with the superposition of a component of σ_t. Regardless of which reference state is favored, it is clear that σ_t is a necessary component of the state of stress for many tectonic processes in the lithosphere.

Deviatoric stress is commonly used in the literature when the authors wish to discuss stress states leading to crustal deformation (e.g., Hanks and Raleigh, 1980). Furthermore, it is a common misconception that any deviation (i.e., σ_t^*) from lithostatic stress, ρgz, is a deviatoric stress. *Deviatoric stress*, $\delta\sigma_i$, has a precise definition where a component of the deviatoric stress tensor is the difference between a normal stress component of the stress tensor and the mean stress, σ_m

$$\delta\sigma_i = \sigma_i - \sigma_m \tag{1–41}$$

where *mean stress* is

$$\sigma_m = \frac{S_x + S_y + S_z}{3} = \frac{\sigma_1 + \sigma_2 + \sigma_3}{3}. \tag{1–42}$$

Any deviation from the lithostatic stress state causes a change in the mean stress and so, by definition, the mean stress and lithostatic stress are not the same. The mean stress is the isotropic component of the total stress tensor at a point, whereas the deviatoric stress tensor is the anisotropic component of the total stress tensor at the same point. It is the anisotropic component that is largely responsible for rock deformation at a point.

Despite its common use in the literature, deviatoric stress is exceedingly difficult to visualize and is not very useful in general discussions of lithospheric stress. To say that a rock is subject to a high deviatoric stress, the author means that there is a large difference between a particular component of the stress tensor and the mean stress. Quantitatively, reference to a value for a component of the deviatoric stress tensor is possible only if the absolute magnitude of all three principal stresses are known. Because the magnitude of σ_2 is a second-order effect, a more convenient and easily visualized way to express an anisotropic stress state is to use differential stress, σ_d.

The expression, deviatoric stress, does have a very practical use during general discussions. The extreme components of the deviatoric stress tensor are positive ($\sigma_1 - \sigma_m$) and negative ($\sigma_3 - \sigma_m$). The positive component, $\delta\sigma_1$, is the direction of deviatoric compression and the negative component, $\delta\sigma_3$, is the direction of deviatoric tension. In a normal-fault regime, it is appropriate to refer to the direction of extension as the direction of deviatoric tension and, thus, the vertical direction is the direction of deviatoric compression. Note that because the isotropic component of the total stress tensor is generally positive in the lithosphere, the value of the principal component of the total stress tensor which is in the direction of deviatoric tension is generally positive (i.e., compressive).

The Effect of Pore Pressure on Stress

Effective Stress

So far our discussion has not treated water-saturated rock. The introduction of pore fluid under pressure has a profound affect on the physical properties of porous solids (Terzaghi, 1943; Hubbert and Rubey, 1959). Most physical properties of porous rock obey a law of effective stress which depends on the difference between pore pressure, p_p, and applied stress, σ_{ij},

$$\overline{\sigma}_{ij} = \sigma_{ij} - P_p\delta_{ij}. \tag{1–43}$$

The application of the law of effective stress to predict rock strength in shear rupture is discussed in chapter 3. In brief, an increase in P_p will weaken the rock so that it will sustain a relatively smaller σ_d.

Like σ_d, tensile effective stress ($\overline{\sigma}_3 = \sigma_3 - P_p < 0$) in the earth is also maintained within certain bounds by the rock strength. In this case the limiting strength parameter is the fracture toughness (see chap. 2) rather than the shear strength of intact rocks or frictional strength along existing fractures. Failure under $\overline{\sigma}_3 < 0$ is by tensile crack propagation rather than shear crack propagation or frictional slip. Two general situations lead to the development of tensile conditions. First, $\overline{\sigma}_3 < 0$ may develop if P_p approaches lithostatic stress by one of several mechanisms (Hubbert and Rubey, 1959). Second, $\overline{\sigma}_3 < 0$ may also develop if P_p remains at hydrostatic levels but $\overline{\sigma}_3 = S_h$ is reduced by thermal cooling and lateral contraction accompanying uplift and erosion (Haxby and Turcotte, 1976; Narr and Currie, 1982). $\overline{\sigma}_3 < 0$ is most commonly achieved in the upper crust in the presence of a significant P_p while σ_3 is compressive. σ_d does not factor directly into limiting $\overline{\sigma}_3$, nor does tensile crack propagation act to limit or reduce σ_d. However, σ_d must be relatively low as P_p increases or the rock will fail in shear before tensile conditions develop (see chap. 3).

Poroelastic Behavior

The influence of pore pressure on total stress is best described in the context of poroelastic behavior. *Total stress* is defined by rearranging equation 1–43:

$$\sigma_{ij} = \overline{\sigma}_{ij} + P_p\delta_{ij}. \qquad (1\text{–}44)$$

Such a rearrangement indicates that the term total stress and applied stress have the same meaning. Applied stress is used in the laboratory to indicate the stress a piston transfers to a test cylinder of rock. That piston not only has to push on the sample but it also has to resist the outward force exerted by any pore fluid. A change in P_p within the rock cylinder is counterbalanced by a change in load on the piston to maintain the original shape of the sample. Total stress is often used if one wants to make it clear that stress across some boundary is a combination of rock matrix stress plus pore pressure.

Equation 1–44 suggests that pore pressure is totally transferred to an arbitrary boundary from the internal portion of an aggregate. Such is indeed the case for sand but not so for rock consisting of grains with elastic grain-grain contacts and pore space between the grains. Imagine a rock of grains, elastic grain-grain contacts, and pore space connected so that the rock has a relatively high permeability. The rock is placed in a container with rigid walls but open at the top (i.e., the uniaxial-strain model applies) (fig. 1–4). Assume that contacts between grains and the rigid wall are totally impermeable. Initially the rock is dry so that the grains press against the rigid walls of the container with an average stress, S_c^a. The actual stress under the grain contacts, S_c, is larger

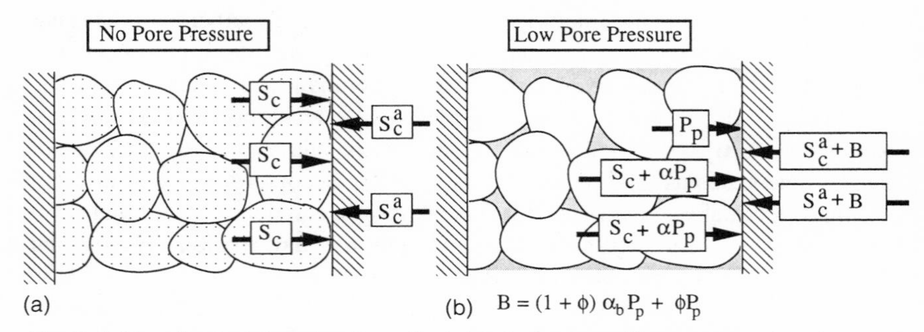

Fig. 1–4. The effect of changing P_p on the total stress exerted by a rock against the rigid walls of a container. (a) Dry rock; (b) Rock with pore fluid at pressure = P_p.

$$S_c = \left(\frac{1}{1-\phi}\right) S_c^a \qquad (1\text{–}45)$$

where ϕ is the porosity of the rock which represents the percentage of open pore space against the wall of the container. The container is then filled with a pore fluid at a relatively low pressure, P_p. Assume that pore fluid cannot penetrate the grain-container wall contact. The addition of pore fluid will not cause a uniform increase in stress along the wall of the container. Where pore fluid is in direct contact with the wall, the pressure on the wall will increase by P_p. Where grains are in contact with the wall the normal stress on the wall will increase by a fractional portion of P_p. The average total stress, S_h, on the wall of the container will increase by less than P_p according to

$$\Delta S_h = \phi P_p + (1-\phi)\,\alpha_b P_p. \qquad (1\text{–}46)$$

This fractional increase in normal stress arises because the grains are connected by cement so that the elastic contacts take up part of the force exerted by pore fluid inside pores. This partial transfer of pore pressure to the container wall is known as the *poroelastic effect* (Biot, 1941). This analysis is further developed in chapter 2 during a discussion of crack propagation. [5]

In saturated rocks total horizontal stress is divided into two components: the stress carried across grain contacts under dry conditions and stress generated by fluid pressure within pore space of the rock. By the poroelastic effect, an increase in P_p will cause an increase in total horizontal stress by some fraction of the change in P_p provided that the rock is constrained by rigid boundaries. Such a uniaxial strain model commonly represents the behavior of stress in sedimentary basins (Geertsma, 1957). The effect of a change in pore pressure in sedimentary basins is seen by solving Biot's (1941) elasticity equations for uniaxial strain. Biot's elasticity equations are given by Rice and Cleary (1976) as

$$2\zeta\varepsilon_{ij} = \sigma_{ij} - \left(\frac{\nu}{1-\nu}\right)\sigma_{kk}\delta_{ij} \qquad (1\text{–}47)$$

where

$$\sigma_{ij} = \bar{\sigma}_{ij} + \alpha_b P_p \qquad (1\text{–}48)$$

and

$$\alpha_b = \left(1 - \frac{\beta_i}{\beta_b}\right). \qquad (1\text{–}49)$$

In Biot's equations α_b is Biot's poroelastic term which is inversely proportional to the ratio of the intrinsic compressibility of the uncracked solid, β_i, and the bulk compressibility of the solid with cracks and pores, β_b (Nur and Byerlee,

1971). ζ and υ are the shear modulus and Poisson's ratio of the rock when it is deformed under "drained" conditions.

In basins subject to high P_p, the total vertical stress is still calculated using equation 1–23. The poroelastic effect does not apply to the vertical stress largely because the surface of the earth does not act as a fixed boundary and strain is allowed to absorb any increase in pore pressure. To look at the effect of P_p on horizontal stress, one defines a tectonically relaxed basin as one in which S_h is proportional to S_v through the uniaxial elastic strain model in equation 1–17. Solving Biot's elasticity equations for uniaxial strain we derive

$$S_h = \frac{\nu}{1-\nu} S_v + \frac{(1-2\nu)}{(1-\nu)} \alpha_b P_p . \tag{1–50}$$

We can rearrange the terms in this equation to write it in the same form as Anderson's et al. (1973) equation for fracture pressure at a borehole

$$S_h = \frac{\nu}{1-\nu} (S_v - \alpha_b P_p) + \alpha_b P_p . \tag{1–51}$$

Equation 1–50 demonstrates that total horizontal stress will increase with an increase in pore pressure. Earth stress measurements generally detect total horizontal stress rather than effective horizontal stress. There are a number of geological situations which suggest that poroelastic behavior occurs within the lithosphere. For example, the correlation between an increased seismic activity in the vicinity of a reservoir and a pore-pressure decline during fluid extraction suggests that σ_d has increased due to a poroelastic contraction of the reservoir (Segall, 1989).

Equation 1–40 suggests that σ_d might become very large in a subsiding basin. In practice, this does not happen because poorly lithified sediments are likely to creep and undergo changes in elastic properties through diagenesis during burial. Pore fluid also plays a major role in moderating σ_d. Substituting equation 1–51 into equation 1–40 we derive the equation governing the poroelastic behavior of σ_d in a subsiding basin, assuming the horizontal stresses have not relaxed due to creep and diagenesis

$$\sigma_d = \left(\frac{1-2\nu}{1-\nu}\right)(S_v - \alpha P_p). \tag{1–52}$$

The variation of S_h due to an increase in P_p in a tectonically relaxed basin is shown in figure 1–5. Calculations for figure 1–5 assume 3 km of overburden which is the average depth for high P_p in the Gulf of Mexico and many other basins. Assume overburden has a density of 2.7 g/cc so that $S_v = 79.5$ MPa and $\alpha_b = 0.7$. S_h can vary as much as 50% of the overburden weight depending on ν and P_p. A larger S_h is generated in rock with a higher ν. In rocks with a very low

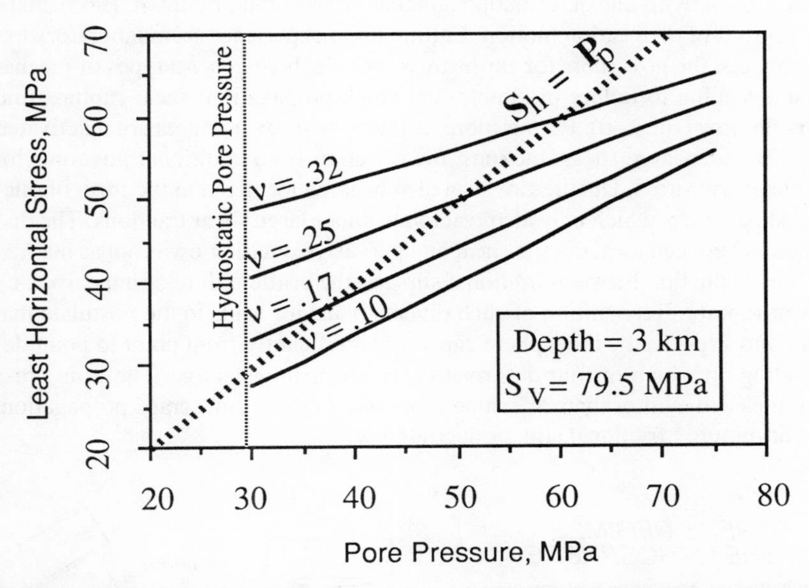

Fig. 1–5. The relationship between S_h and P_p in a tectonically relaxed basin assuming poroelastic behavior. In a tectonically relaxed basin horizontal stresses are due solely to the overburden load. The poroelastic effect is strongly dependent on Poisson's ratio, ν, as indicated by the four curves for various values of ν. This calculation assumes conditions at a depth of burial of 3 km (from Engelder and Lacazette, 1990).

ν, conditions favoring crack propagation (i.e., $\overline{\sigma}_3 < 0$ where $P_p > S_h$) are found even at hydrostatic pore pressure.

Stress Regimes in the Lithosphere

The active role played by rock strength as a governor for both σ_d and $\overline{\sigma}_3$ in the lithosphere is indicated by the abundance of both brittle and ductile structures evident in outcrops. Major classes of brittle structures seen in outcrops include *joints* (i.e., mesoscopic cracks in rock along which there has been no appreciable shear displacement), *shear fractures* (i.e., primary ruptures with appreciable displacement parallel to the plane of the rupture), and *faults* (i.e., joints and shear fractures reactivated in shear displacement by frictional slip). Other brittle structures include sheet fractures (Holzhausen and Johnson, 1979), host fractures with pinnate joints (Engelder, 1989), and en échelon vein arrays (Beach, 1975). Also evident in outcrops and thin sections of rocks taken from outcrops are common low-temperature ductile structures including disjunctive cleavage (Engelder and Marshak, 1985), a manifestation of stress solution, and mechanical twins and deformation lamellae (Carter and Friedman, 1965), manifestations of dislocation motion. Calling upon experience from laboratory experiments, the governors for earth stress include three general types of mechanisms leading to failure of intact rock: crack propagation; shear rupture; and ductile flow (fig. 1–6). Furthermore, if joints or shear fractures are reactivated to slip under large shear tractions, then friction is an additional governor for lithospheric stress. Ductile flow can also become localized in the form of ductile shear zones which may also reactivate under large shear tractions. The distinction between local ductile shear and pervasive ductile flow is not as marked as the distinction between frictional slip and the brittle failure of intact rock by shear rupture. Recognition of such diverse behavior leads to the postulate that tectonic stress in the lithosphere can vary significantly from point to point depending on which of four deformation mechanisms is active. The four stress regimes in the lithosphere are named for these mechanisms: crack propagation, shear rupture,[6] frictional slip, or ductile flow.

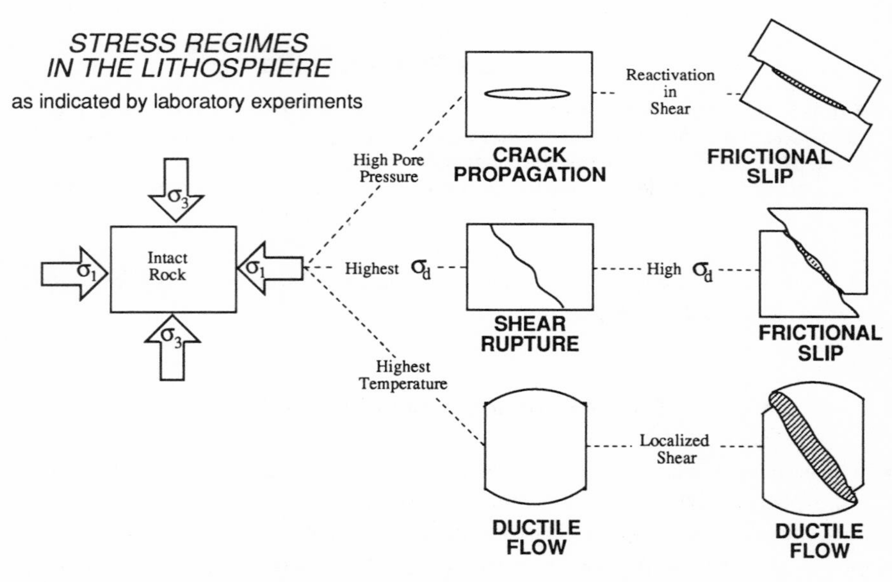

Fig. 1–6. The definition of stress regimes in the lithosphere based on the general types of failure encountered during laboratory rock mechanics experiments. Each rectangular box represents the cross section of a cylindrical sample during a polyaxial rock deformation experiment. In the laboratory, intact rock fails by three general mechanisms: crack propagation; shear rupture; and ductile flow. When subject to large shear stresses, joints and shear fractures are reactivated by frictional slip. Shear zones develop if ductile flow is localized. The four stress regimes of the lithosphere are identified by bold letters.

Although somewhat arbitrary, the rationale for the order of treatment and grouping of these four stress regimes is as follows. Rare is the outcrop in which joints or shear fractures are spaced more than a meter or two apart. This is so regardless of whether rocks within the outcrop reflect any major ductile deformation. Brittle deformation figures more prominently in rock behavior in the near surface, the region of the lithosphere which is most accessible to man through in situ earth stress measurements. For these reasons, a discussion of stress regimes associated with brittle behavior comes first. Brittle failure mechanisms reflect a range of σ_d from tensile crack propagation at low σ_d to shear rupture at relatively high σ_d . Joints, veins, and dikes are a manifestation of the former type of brittle failure, whereas shear fractures are a manifestation of the latter. Upon examining outcrops in terrain including the craton, the foreland, and the core of mountain ranges, I am impressed with abundance of joints and veins relative to shear fractures. Furthermore, it is apparent that many discontinuities covered with slickensides or structures characteristic of shear offset initially propagated as joints. Because of the distinction between opening-mode crack propagation at low σ_d under the influence $\overline{\sigma}_3 < 0$ and shear rupture under high σ_d, the stress regime favoring joint, vein, and dike development, the *crack-propagation regime* (chap. 2), is treated separately from the *shear-rupture regime* (chap. 3). Reactivation by frictional slip of both joints and veins requires relatively high σ_d and other conditions similar to those required for shear rupture. Therefore, the *frictional-slip regime* is treated along with the shear-rupture regime in chapter 3. A major reason for addressing questions associated with the crack propagation regime before the frictional-slip regime is that brittle fracture must occur before frictional sliding is possible and outcrop evidence suggests that a significant number, if not a majority, of brittle fractures in the lithosphere initially formed by tensile crack propagation under lower σ_d. Discussion of stress in the *ductile-flow regime* within the lithosphere follows in chapter 4.

In a book on the mechanics of earthquakes, Scholz (1990) divides the lithosphere in two parts with distinctly different rheological properties: the *schizosphere* or brittle region and the *plastosphere* or ductile region. The extent of the schizosphere is demarcated by earthquakes which are regarded as a manifestation of brittle behavior. However, because some ductile deformation mechanisms such as stress solution are found throughout the schizosphere, the boundary between schizosphere and plastosphere is somewhat blurred. In addition to being distinct because of its unique association with earthquakes, the schizosphere is distinct as the host for a higher σ_d than is found in the plastosphere. For this latter reason, Scholz's (1990) division of the lithosphere into a schizosphere and plastosphere lends itself well to a treatment of state of stress throughout the lithosphere and, so, the use of the terms will continue in this book.

2

Stress in the Crack-Propagation Regime

If the [stresses] which affect the crust are variable in direction, the same must apply to the fractures which they produce. The faults and dikes which were formed in this country [Great Britain] during any one geological epoch, are, however, more or less parallel. . . . [The faults and dikes suggest] that there has been a certain uniformity in the directions of stress.
—Ernest M. Anderson's (1951) assessment of regional stress in the crust of the earth

Anderson (1951) recognized that dikes, a manifestation of crack propagation in the schizosphere, have a consistent orientation relative to the earth's stress field (intrusion is perpendicular to S_h) and that they intrude instantaneously relative to the geological time scale. Through careful observation Anderson reached the profound conclusion that the orientation of the earth's stress field is quite uniform on a regional scale. At the same time Anderson showed that tracing the path of crack (dike) propagation is a powerful tool for mapping the orientation of the earth's stress fields, both present and past. Anderson's work spawned a number of developments in our understanding of crack propagation, the most sophisticated of which involves the theory of crack propagation in an elastic medium to estimate stress magnitudes (e.g., Pollard and Segall, 1987). This chapter develops the theory needed for the study of orientation and magnitude of stress in the crack-propagation regime of the schizosphere.

A crack may propagate by any one of three modes of surface displacement (Lawn and Wilshaw, 1975). Opening mode (mode I) cracks form by separation of the crack walls without shear, under the action of effective tensile stresses (fig. 2–1). Sliding mode (mode II) cracks propagate upon mutual shearing of the crack walls with the shear couple oriented in the direction normal to the crack front. Tearing mode (mode III) cracks advance when the crack walls are subject to a shear couple aligned parallel to the crack front. Many cracks in rocks, visible on the mesoscopic to macroscopic scale, show opening displacements with no appreciable shear (Engelder, 1982b; Segall and Pollard, 1983). If they are unfilled, these cracks are variously referred to as joints (Badgley, 1965), extension fractures (Griggs and Handin, 1960), or tension gashes (Blés and Feuge, 1986). The term, joint, is preferred, whereas the term *extension*

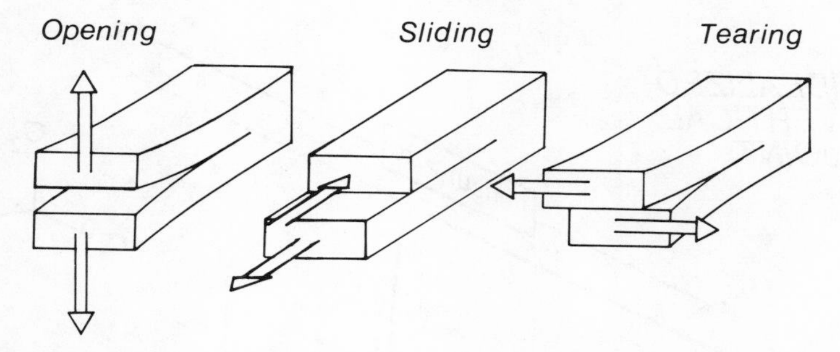

Fig. 2–1. Three modes of crack displacement.

fracture is restricted to those features produced during laboratory compression experiments at low P_c. Examples of filled cracks include veins (Ramsay, 1980), en échelon veins (Beach, 1975, 1977), and dikes (Anderson, 1951). Crack filling material ranges from solution-precipitated zeolite, calcite, or quartz (i.e., veins), to solidified magmas (i.e., dikes).[1]

Joints suggested to geologists almost a century ago that "tension" was common in the crust. Examples include joints along the crests of anticlines (e.g., Hiwassee section of the Ocoee series in the southern Appalachians—Van Hise, 1896), in cooling igneous bodies (e.g., Crosby, 1882), in dried sediment (e.g., Le Conte, 1882), and as tension gashes in fault zones (e.g., Leith, 1913). Other evidence suggested that joints were subject to net tension under opening-mode loading and, hence, propagated normal to a principal stress. Such evidence includes: (1) joints cutting objects such as individual grains, fossils, or veins that show no shear offset (Engelder, 1982b, Segall and Pollard, 1983); (2) microcracks and joints propagating in the plane containing σ_1 as indicated by dynamic analysis of calcite, the orientation of sutures on stylolitic cleavage planes, and strain relaxation tests (Friedman, 1964; Engelder, 1982b); and (3) joints orthogonal to the contemporary tectonic stress field as mapped by earthquake focal mechanisms and deep in situ stress measurements (Engelder 1982a; Hancock and Engelder, 1989).

The relationship between state of stress in the lithosphere and the propagation of mode I cracks is the focus of this chapter. Mode II and III crack propagation are addressed in chapter 3. Macroscopic mode I cracks in rocks, including dikes, joints, and veins, all tend to have a general form characterized by a thickness (2b) much smaller than a width (2c) (fig. 2–2). Veins and joints in sedimentary rocks often have a length which far exceeds their width. As a crack propagates, primary displacement of its walls is in the u_y direction with some displacement in the u_z direction (Pollard and Segall, 1987). For the analyses in this chapter the crack displacement vector u_x acting parallel to the length of the joint is zero everywhere. With this condition the crack tip is in a state of plane strain and crack propagation is treated as a two-dimensional

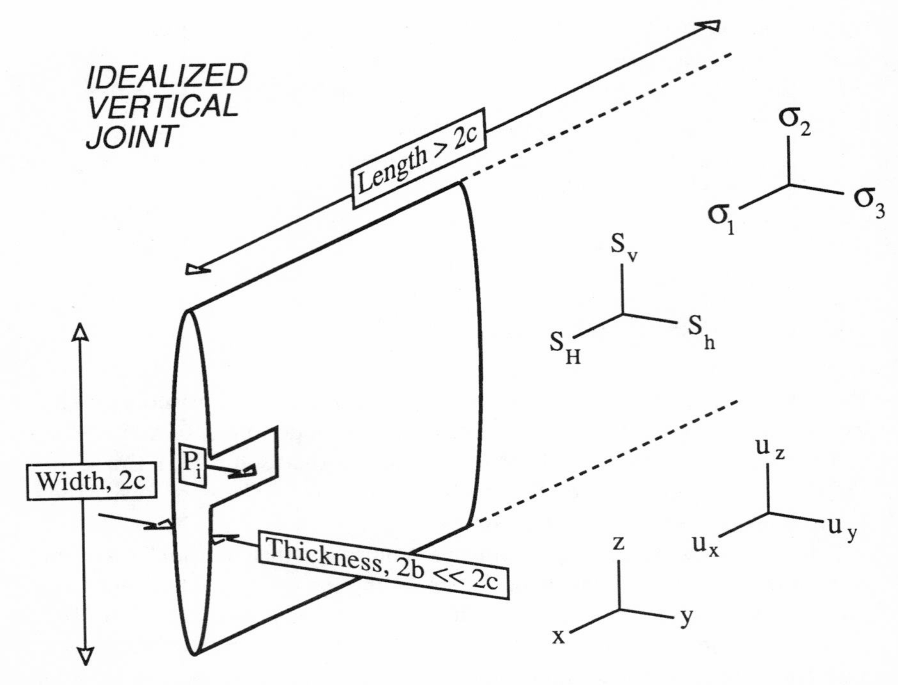

Fig. 2–2. Sketch of an idealized joint of length much greater than width and thickness much less than width. The subscript notation follows the convention that x is in the direction of σ_1. The remote stress normal to the joint plane is compressive, $\sigma_3 = S_h$, whereas the stress on the internal wall of the crack is the fluid pressure P_i. The driving stress for such a joint is $P_i - \sigma_3 > 0$.

problem (Broek, 1987). A crack in rock is subject to plane strain because the rock is thick in the *x* direction and its strength prevents displacement in that direction.[2]

Stress Concentrations

A rock is a composite material with many holes (i.e., pores). By the late 1800s engineers recognized that holes affect the strength of various building materials and structures. Holes modify the distribution of stress within a solid so that stress is concentrated in the vicinity of the hole where there is a much larger stress than applied at a boundary some distance away. Stress concentrations around pore spaces and other inclusions in intact rock serve as the initiation point for the propagation of joints. Kirsch (1898) provided a solution for the stress distribution about a circular hole in a uniformly stressed elastic plate to show that, at two points on the edge of the hole, circumferential stress, $-S_t$, is three times larger than the remote stress, $-S_h$ (fig. 2–3). Other details of Kirsch's (1898) solution are discussed in chapters 5 and 6. Later, Inglis (1913) recognized that an elliptical hole in an elastic plate further concentrates stress in proportion to the ratio, c/b, of the semi-axes of the elliptical hole

$$-S_t = -S_h\left(1 + \frac{2c}{b}\right). \tag{2–1}$$

For a circular hole where $c/b = 1$ and $S_h > 0$, $S_t = 3S_h$.

Several details concerning stress concentrations are important. First, stress at the edge of an elliptical hole can become much larger than the average stress to which a material is subject. Second, the stress concentration depends on the shape of the hole, not its size. Stress magnification at the edge of a circular hole will never exceed three times the remote stress. Third, the major stress perturbation is found within a distance *c* from either a circular or elliptical hole (fig. 2–3). However, the depth of the stress magnification into the host material increases with hole size. Fourth, for an elliptical hole the stress magnification increases in inverse proportion to the radius of curvature, r_c, at the narrow end of the ellipse where

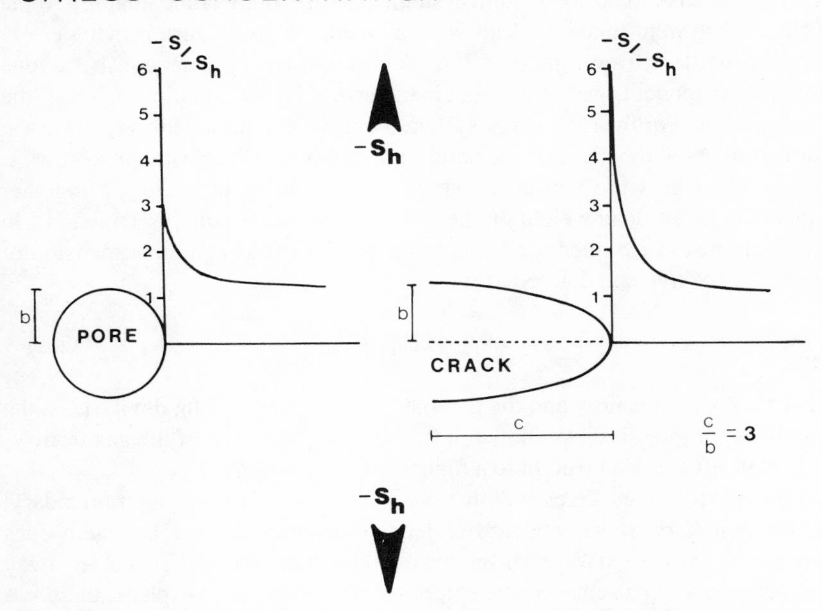

Fig. 2–3. Stress concentrations around a pore (i.e., a circular hole) and a crack (i.e., an elliptical hole). In this model the far-field stress ($-S_h$) is tensile.

$$r_c = \frac{b^2}{c} \tag{2–2}$$

Stress concentration at the tip of the elliptical hole is

$$S_t = S_h\left(1 + 2\sqrt{\frac{c}{r_c}}\right) \tag{2–3}$$

If b << c, then equation 2–3 reduces to

$$\frac{S_t}{S_h} \approx \frac{2c}{b} = 2\sqrt{\frac{c}{r_c}}. \tag{2–4}$$

This latter case represents the situation around the tips of cracks on all scales where the stress concentration increases with a decrease in r_c. Again, the greatest stress concentrations are found within a distance of r_c from the tip of the crack.

Griffith's Analysis of Crack Propagation

To understand the interaction between earth stress and mode I crack propagation in rock, the next step is to examine Griffith's (1920; 1924) proposal that a cracked rock is a thermodynamic system. The thermodynamic analogy to the rock-crack system under a "dead-weight" load is a steam piston-cylinder expanding to lift such a weight (fig. 2–4). A steam piston-cylinder consists of one internal component, gas, which is characterized by an equation of state, the ideal gas law. For the rock-crack system the internal components are the crack defined by its width, 2c, and the solid rock defined by its elastic properties, E and υ. Fluid pressure within the crack and confining pressure are together equivalent to the dead weight on the piston of the steam cylinder (fig. 2–4). In the steam piston-cylinder adiabatic work, W_S, by the system on expansion of the gas is positive and defined as

$$U_T = W_S = nRT \int_{V_a}^{V_b} \frac{dV}{V} \tag{2–5}$$

where U_T is total energy and the external force on the loading device (i.e., the piston) is compressive. As shown in figure 2–4, expansion of the gas from V_a to V_b will lift the dead weight to a higher potential energy.

One way to drive a crack within a dry rock is to pull on the outer boundary. As the boundary moves outward under tension, the rock-crack system does negative work (i.e., $dW_R < 0$) on surrounding rock across boundaries away from the crack. In a sense, as the exterior walls of the rock displace, there is a decrease in potential energy of the loading device, which in a geological setting, is any combination of boundary tractions that cause the generation of ef-

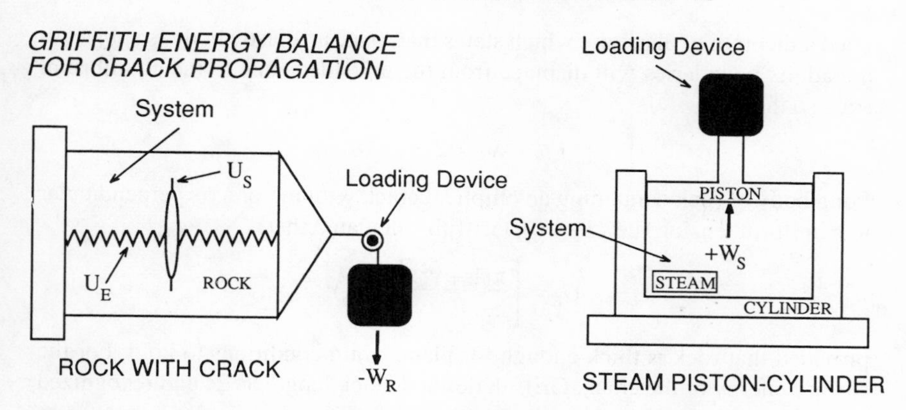

Fig. 2–4. The components of total energy in a steam piston-cylinder and a rock with a crack (adapted from Pollard, 1989). The steam piston-cylinder has one component, W_S, work done by the expanding steam. The components for the Griffith energy balance concept are listed as the work term, W_R, a strain energy term, U_E, and a surface energy term, U_S.

fective tensile stresses. For example, in the outer arc of a fold, a loading device is the boundary condition causing the stretching of a rock layer. The work to propagate the crack is positive and defined as the increase in surface energy, dU_S. As the crack propagates, the rock will undergo a change in strain energy, dU_E. The total change in energy for crack propagation is

$$dU_T = dU_S - dW_R + dU_E. \tag{2–6}$$

Unlike the steam piston-cylinder, crack propagation may take place without changing the total energy of the rock-crack system. This is known as the Griffith energy-balance concept where the standard equilibrium requirement is that for an increment of crack extension dc,

$$\frac{dU_T}{dc} = 0. \tag{2–7}$$

The mechanical ($-dW_R + dU_E$) and surface energy (dU_S) terms within the rock-crack system must balance over a crack extension, dc. During crack propagation the crack walls move outward to some new lower energy configuration upon removal of the restraining tractions across an increment of crack. In effect, the motion of the crack walls represents a decrease in mechanical energy while work is expended to remove the restraints across the crack increment. The work to remove the restraints is the surface energy for incremental crack propagation.

To evaluate the three energy terms relating to crack propagation, Griffith cited a theorem of elasticity which states that, once an elastic body is subject to cited a theorem of elasticity which states that, once an elastic body is subject to a load, its boundaries will displace from the unloaded state to the equilibrium state so that

$$W_R = 2U_E. \tag{2–8}$$

For a body of rock containing an elliptical crack with major axis perpendicular to a uniform tension (i.e., $S_h < 0$), Griffith calculated that

$$U_E = \left[\frac{\pi(1-\nu^2)}{E}\right] c^2(S_h)^2 \tag{2–9}$$

provided that rock is thick enough for plane strain conditions to hold. For the surface energy of the crack, Griffith defined crack length as 2c and recognized the crack propagation produces two crack faces. Therefore,

$$U_S = 4c\gamma \tag{2–10}$$

where γ is the free surface energy per unit area. Substituting equations 2–8, 2–9, and 2–10 into 2–6 and then applying equation 2–7, Griffith solved for the critical stress for crack propagation

$$|\overline{\sigma}^c_3| = T_0 = \left[\frac{2E\gamma}{\pi(1-\nu^2)c}\right]^{\frac{1}{2}} \tag{2–11}$$

where T_0 is the uniaxial tensile strength of the rock. Here $|\overline{\sigma}^c_3| = -S_h$ for a dry rock in the near surface.

Equation 2–11 states that crack propagation will take place only in the presence of a net tensile stress. Yet, the tremendous weight of overburden favors $S_h > 0$ within the schizosphere and the condition that $S_h < 0$ is rare. Nevertheless, laboratory experiments have shown that in rock subject to a compressive stress field local tensile stresses develop and can induce microcracks (e.g., Hoek and Bieniawski, 1965; Scholz, 1968). In particular, the mismatch of elastic properties across grain boundaries or sliding of walls of microcracks cause tensile stresses which lead to further microcrack propagation in the direction of σ_1. These microcracks will grow to mode I cracks much larger than initial pores and grain boundaries only if $\sigma_3 \approx 0$ in which case the feature is called an extension fracture (Griggs and Handin, 1960) or axial splitting crack (Nemat-Nasar and Horii, 1982). Under high compressive stress these cracks tend to stop, and therefore do not lead to the propagation of large joints (e.g., Brace and Bombolakis, 1963).

Vertical Joint Propagation at Depth

Debate over the paradox of the propagation of joints in a compressive stress field dates from the latter part of the nineteenth century. During this period, tensile stresses, and others propagated under compressive stresses (Van Hise, 1896). However, stress conditions leading to the formation of joints at great depth were not obvious. Crosby (1882) argued that tensile stresses did not exist in deeply buried rocks because heat and the enormous pressure of the overlying strata caused lateral expansion and prevented the necessary contraction. The paradox was resolved with Secor's (1965) proposal that joints develop at great depths in the crust of the earth if the ratio of fluid to overburden pressure is near one and σ_d is low. Concurrently, Price (1966) presented an analysis for jointing upon uplift associated with removal of overburden where the effective tensile stress for jointing arises from the thermal-elastic contraction of rock upon uplift. We can apply Secor's solution to Griffith's analysis.

One of two end member cases for vertical joint propagation in rocks is a crack driven by "dead-weight" loading on an external boundary where σ_3 is tensile. The other end member is a crack driven by an internal fluid pressure. In this latter case, the rock is often in a state of compression from boundary tractions and would, therefore, contain a strain energy $U^{el}{}_E$ arising from the external load, S_h

$$U^{el}{}_E = \frac{(S_h)^2}{2E}. \qquad (2\text{–}12)$$

For a vertical crack driven by internal fluid pressure, the vertical, outer boundaries of the local rock-crack system do not necessarily displace during crack propagation because the boundaries are restrained by the adjacent rock mass (the uniaxial strain model applies). If the outer boundary does not displace then the rock-crack system does no work against remote S_h during crack propagation, so $dW_R = 0$. Therefore, the strain energy term from remote boundary tractions (eq. 2–12) does not enter into Griffith's thermodynamic calculation. This situation may seem like the "fixed-grips" case in fracture mechanics where the loading device does not move the outer boundary of the dry-cracked solid (e.g., Lawn and Wilshaw, 1975). It is not a "fixed-grips" case, because fluid pressure moves the crack walls.

If we assume that the crack in the rock is internally pressurized but that the pore space behind the crack remains dry by virtue of an impermeable membrane at the wall of the crack, then we can solve for the crack driving pressure, P_i. When the crack propagates, the crack walls displace, as was the case for crack propagation under "dead-weight loading." However, in the latter case there was no internal pressure on the crack wall so that no work was done during the displacement of the crack wall. For the case with internal fluid pressure, work is done by the fluid in the crack to move the crack wall

$$W_c = 2U^c_E. \qquad (2\text{–}13)$$

Note that because work was done on the rock-crack system by the crack fluid, the sign of the work term is again negative. Here the internal fluid is part of the loading device, whereas in the example of the steam piston-cylinder, the fluid (gas) was the system under consideration and not the loading device. Examination of the forces acting on the crack suggest that the crack walls will not part under the influence of P_i until there is a net outward force which happens only when $P_i > |S_h|$. The strain energy for movement of the crack wall depends on the net outward force or effective stress within the crack which is the difference between fluid pressure inside the crack, P_i, and the total stress on the outer boundary of the crack system, S_h,

$$U^c{}_E = \frac{(P_i - S_h)^2}{2E}. \tag{2–14}$$

With fluid driving the crack wall, there is a decrease in potential energy of the loading system, the fluid, and an increase in strain energy in the rock. Such change in energy is associated with "dead-weight" loading where the "dead-weight" is the pressure of the fluid against the wall of the crack. We now solve equation 2–6 for the internal pressure necessary to drive the crack when the outer boundary of the rock-crack system is held in a fixed position

$$P_i = \left[\frac{2E\gamma}{\pi(1 - \nu^2)c}\right]^{\frac{1}{2}} + S_h \tag{2–15}$$

where this equation is virtually identical to equation 2–11 except that a compressive S_h is counter balanced by P_i. This is Secor's (1969) solution to crack propagation under the influence of very high fluid pressures. However, this equation applies only if there is an impermeable membrane between the crack wall and pore space behind the crack, so that fluid from the crack is not allowed to drain into the pore space. Such a situation is geologically unrealistic as is shown below.

Linear Elastic Fracture Mechanics

Conditions defining tensile strength, shear fracturing, and frictional slip on faults are illustrated using the Coulomb-Mohr failure envelope (e.g., Engelder and Marshak, 1988). Although the Coulomb-Mohr failure envelope serves as a good empirical gauge for the stresses at shear failure, it provides information on neither the rupture path nor postfailure behavior. Likewise, Griffith's analysis using Inglis's stress concentration factor served well to predict the initiation of crack propagation but proved unsatisfactory as a propagation criterion. This is largely because both the stress field in the vicinity of the crack tip and the radius of the crack tip are poorly defined.

IRWIN'S CONTRIBUTION

To satisfy the need for a propagation criterion, Irwin (1957; 1958) noted that the stress field, $'\sigma_{ij}$, in the vicinity of a sharp crack tip in an elastic body was approximately proportional to K_I, the stress intensity factor[3]

$$'\sigma_{ij} \approx \frac{K_i f_{ij}(\Theta)}{\sqrt{2\pi r}} \tag{2–16}$$

where r and Θ are polar coordinates centered at the crack tip. The trigonometric functions, $f_{ij}(\Theta)$, vary slightly from unity near the crack tip. These lengthy functions are given later in this chapter when stress away from the crack tip is discussed (eq. 2–39). Equation 2–16 for the stress field in the vicinity of the crack tip is remarkable in that the effect of the applied load and geometry of the crack are both incorporated in the stress intensity factor, K_I. For dry conditions in rock, K_I at the crack tip depends on the loading stresses, $-S_h$, the length of the crack, 2c, and the geometry of the the crack which is specified by the dimensionless crack-shape factor, Y. These three parameters are multiplied so that

$$K_I = | -S_h | \, Y \sqrt{c}\,. \tag{2–17}$$

The absolute value of $-S_h$ is used because linear elastic fracture mechanics assumes a positive sign for remote tensile stress. For a penny-shaped crack, $Y = \frac{2}{\sqrt{\pi}}$, whereas for a tunnel (i.e., blade-shaped as in fig. 2–2) crack $Y = \sqrt{\pi}$. Other shape factors are listed in Sih (1973). Note that for $r << c$, then $'\sigma_{ij} >> |-S_h|$.

Griffith defined T_0 of a rock in terms of a balance among the work by the loading system (i.e., the boundary conditions), the strain energy within the rock, and the crack surface energy. T_0 is also specified in terms of the value for the stress intensity factor at the time of crack propagation

$$K_{Ic} = K_I \tag{2–18}$$

where K_{Ic} is called the *fracture toughness* of the rock. K_{Ic} is a more precise measure of tensile strength than T_0, because K_{Ic} takes into account shape and size of initial cracks in porous rocks. If K_{Ic} is known, then the internal crack fluid pressure necessary to initiate crack propagation is defined by rewriting equation 2–17

$$P_i = \frac{K_{Ic}}{Y\sqrt{c}} + S_h \tag{2–19}$$

where P_i, which produces a net tension on the initial crack, must counterbalance the remote compressive stress, S_h, as well as fracture toughness. Equation 2–19 assumes that the wall of the crack is impermeable. Experiments on crack propagation have shown that cracks will propagate when $K < K_{Ic}$ at the crack tip (Anderson and Grew, 1977; Atkinson, 1984). This phenomenon is a process called *subcritical crack growth*. Subcritical crack growth is permitted by a chemical reaction at the crack tip, known as stress corrosion, which acts to weaken atomic bonds in the vicinity of the tip. Under the influence of stress corrosion, crack propagation may take place at velocities less than 1 mm/sec. When cracks propagate under tip stresses equal to K_{Ic}, the cracks travel unstably at speeds which may approach the shear wave velocity of the host rock.

Equation 2–19 for crack propagation using K_{Ic} should reconcile with Griffith's energy balance (eq. 2–11) (Irwin, 1957). If no remote displacements are assumed, the "fixed-grips" case for crack propagation, then Irwin (1957) showed that the reduction of strain energy in the rock with respect to an increase in crack length is a measure of the energy available for crack propagation

$$G = -\frac{\partial U_E}{\partial c} \tag{2–20}$$

where G is the energy release rate per unit length of crack tip. For a crack subject to plane strain and propagating in its own plane, G is related to K_{Ic} at the crack tip (Irwin, 1957)

$$G = K_{Ic}^2 \left[\frac{(1 - \nu^2)}{E}\right]. \tag{2–21}$$

The propagation of a crack is resisted by a surface tension force, 2γ, where the cutting of such a crack requires the supply of an amount of energy equivalent to

$$dU_s = 2\gamma dc. \tag{2–22}$$

For the "fixed-grips" case under dry conditions where $W_R = 0$, equations 2–20 and 2–22 may be substituted into the Griffith energy-balance equation 2–6

$$dU_T = -Gdc = 2\gamma dc. \tag{2–23}$$

At Griffith equilibrium $dU_T = 0$, crack propagation begins when $G = G_c$, so the critical energy release rate is related to the surface energy of the rock

$$G_c = 2\gamma. \tag{2–24}$$

In cracking rocks γ is often larger than the surface energy associated with a single crack surface (Brace and Walsh, 1962). This is so because crack propagation in rocks and other polycrystalline aggregates consumes additional energy by processes at the crack tip such as acoustic emission, out-of-plane cracking, and microplasticity (Friedman et al., 1972).

The Stress-Pore Pressure Relationship at Joint Initiation

Now we examine the situation associated with the initial propagation of a joint within an intact body of rock. Here we want to understand the stress conditions that favor joint propagation within joint-free "young" rocks. Except for joints propagating above the water table where total stress may be tensile (e.g., Schmitt, 1979), joint propagation takes place when fluid pressure (P_i) in the initial flaw or preexisting crack is larger than the total stress (σ_3) acting to close that flaw or initial crack. The amount by which P_i exceeds σ_3 depends on fracture toughness, crack size, and shape of the crack (eq. 2–19). While the rules for joint initiation apply even if P_p is hydrostatic, the following discussion focuses on circumstances surrounding the development of abnormally high P_p.

There are a number of mechanisms for the development of abnormally high fluid pressure in the schizosphere. Two basic classes of mechanisms are dynamic mechanisms which lead to steady-state high pressures through continual recharge and transient mechanisms which cause a momentary production of fluids or pressure followed by a pressure drop accompanying leakage. Note that all rocks have an intrinsic permeability which will eventually allow abnormal pressures to bleed to hydrostatic unless they are renewed. Dynamic mechanisms include artesian flow from topographic highs (Tôth, 1980). Quasi-dynamic mechanisms include the release of gas during the maturation of hydrocarbons (Tissot and Welte, 1978; Hunt, 1989). Transient mechanisms include pore collapse during compaction under overburden load (Magara, 1978, Bethke et al., 1988); tectonic compaction (Berry, 1973), clay dehydration (Powers, 1967); and aquathermal pressuring (Barker, 1972). Rapid uplift and decompaction can lead to abnormally low pressures (Russell, 1972).

Joints and veins are sometimes referred to as natural hydraulic fractures (e.g., Beach, 1977, Engelder, 1985). Like commercial hydraulic fractures, some joints propagate under high fluid pressures, but joint initiation under high fluid pressures is different from commercial hydraulic fracturing (see chap. 5) for several important reasons. To a rough approximation, hydraulic fracturing boosts the internal pressure of a borehole (i.e., the initial flaw) but it does not affect a change in P_p in the remote portions of the rock so that P_b (breakdown pressure) > P_p at the start of crack propagation.[4] Although there is some infiltration through the mudcake on the wall of the borehole during hydraulic fracturing, infiltration is confined to the immediate vicinity of the borehole. A hydraulic fracture is driven by high pressure fluids pumped into the borehole from a large reservoir. In contrast, a joint is driven by a slow increase in internal pressure within the initial flaw. Because the source of fluid is pore space behind the flaw, pressure in the flaw increases at the same rate as P_p in the rock behind the wall of the flaw. As a consequence, $P_i = P_p$ at the start of crack propagation. For a hydraulic fracture, fluid infiltration (i.e., a P_p increase) will decrease P_b necessary for crack initiation whereas the opposite is true for joint initiation (i.e., P_i necessary for joint initiation goes up as P_p increases within the rock surrounding the flaw). The crack-driving pressure for a hydraulic fracture does not drop abruptly after crack initiation because the borehole is continually being charged from surface pumps. In contrast, crack driving pressure for joints drops either immediately or soon after crack initiation (Secor, 1969). Surface morphology on joint faces indicates that some natural joints propagate in increments as a consequence of a drop in crack-driving pressure (Lacazette and Engelder, 1992). These differences between commercial hydraulic fracturing and joint initiation under high P_p impact on our understanding of stresses associated with joint initiation and propagation.

There are a number of reasons why P_i must equal P_p at the time of the initial propagation of a joint from a pore (i.e., flaw) contained within an intact rock. In sedimentary rocks, even with a very low permeability, fluid pressure in one pore will tend to equilibrate with pressure in surrounding pores provided that the pores are interconnected. There is no known mechanism for suddenly increasing pressure in one pore relative to its neighbors as is the case for a hydraulic fracturing with $P_b > P_p$. As abnormally high pressures develop in sedimentary basins, pore pressure increases uniformly in the immediate vicinity of the incipient joint and so the mechanism for the initiation of joints must operate even though P_i and P_p increase at the same rate.

It is not intuitively obvious that a net tensile stress (i.e., a high crack-driving pressure) can be generated within an initial flaw for several reasons: (1) if uniaxial-strain behavior is assumed, a normal stress acting to close the flaw increases as a function of P_p; (2) excess fluid pressure can readily drain from the initial flaw into other pore space; and so (3) pores of the rock behind the flaw are also subject to the same pressure. One explanation for joint initiation is found in the poroelastic behavior of rock (see chap. 1) which is responsible for the generation of a net tensile stress against the face of a flaw. Poroelastic behavior is illustrated using a model of the force-balance along the flaw-rock interface where the initial condition is that the average normal stress across the rock-flaw interface exceeds the fluid pressure within the flaw and, therefore, the flaw is closed (fig. 2–5). Consider the effect of increasing pore pressure by an amount, ΔP_p, so that fluid pressure in the flaw, $P_f = P_p + \Delta P_p$. P_f is balanced by $P_p + \Delta P_p$ where pore space is against the interface. But by the poroelastic effect (eq. 1–46), the pore pressure increase does not cause the average normal stress on the rock side of the interface to increase at the same rate as P_f. Hence, as P_p continues to increase, P_f will eventually exceed the average normal stress exerted by grains on the interface. Once this condition is reached a net tensile force will act to compress the rock and push the joint-rock interfaces apart. At this point $P_f = P_i = P_p$. Joint initiation occurs only after the crack (i.e., initial flaw) walls are pulled apart or subject to a net tensile stress as a consequence of the poroelastic effect. For the case of initial propagation of a vertical joint,

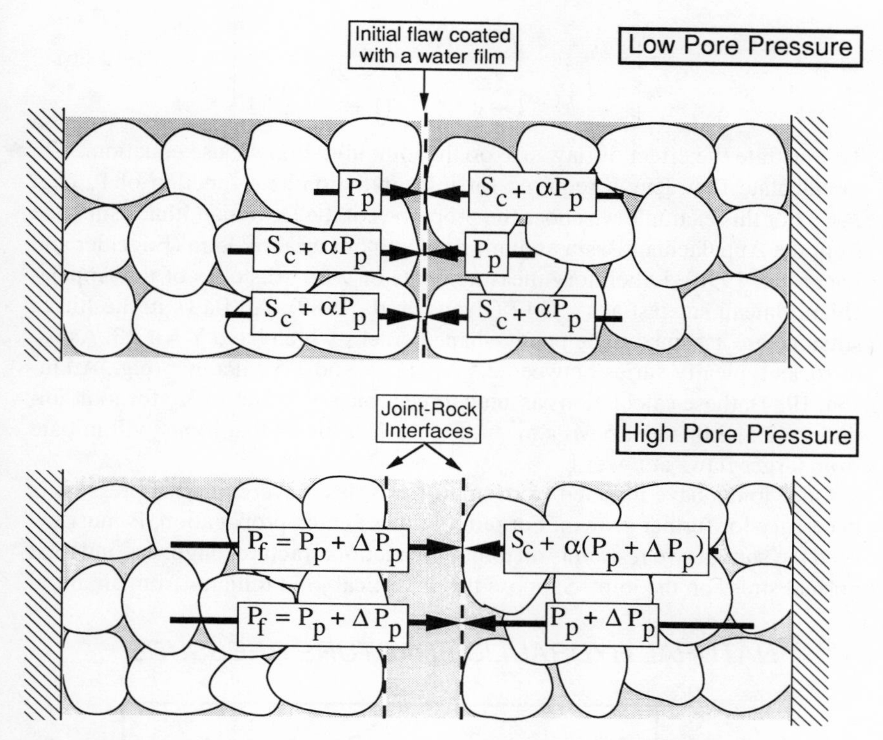

Fig. 2–5. Poroelastic model for a rock with an initial flaw and constrained by rigid boundaries on all sides. The model consists of grains, elastic grain-grain contacts, and interconnected pore space. Vectors are shown to represent the balance of forces along the initial flaw interface and the joint-rock interface (from Engelder and Lacazette, 1990).

equations 2–19 and 1–50 are combined to account for fracture toughness, stresses arising from fixed vertical boundaries, and poroelastic behavior

$$P_i = \left\{ \frac{K_{Ic}}{Y\sqrt{c}} \right\} + \frac{\nu}{1-\nu} S_v + \frac{(1-2\nu)}{(1-\nu)} \alpha_b P_p. \qquad (2\text{–}25)$$

Equation 2–25 indicates that P_i can vary significantly depending on the size of the preexisting crack. Again, for the initial propagation of a joint $P_i = P_p$.

With measured rock properties we now constrain the stress and pore pressure conditions for the initiation of joints within intact rock. Under the assumption that joints start from small cracks or flaws when $P_i = P_p$, equation 2–25 is rewritten to give an indication of the flaw length leading to initiation of joints

$$c = \left[\frac{K_{IC}}{Y \left\{ P_i - \frac{\nu}{1-\nu} S_v - \frac{(1-2\nu)}{(1-\nu)} \alpha_b P_p \right\}} \right]^2 . \qquad (2–26)$$

To illustrate the effect of flaw size on the joint initiation we use equation 2–26 to calculate flaw length required for joint initiation as a function of P_p (fig. 2–6). For this example we chose the properties of the Devonian Ithaca siltstone from the Appalachian Basin assuming a depth of burial of 3 km (Engelder and Lacazette, 1990). Laboratory measurements of ν for siltstones of the Appalachian Plateau suggest a $\nu = 0.17$ (Evans et al., 1989). All flaws in the Ithaca siltstone are assumed to be penny-shaped cracks which have $Y = 1.13$. As K_{Ic} of rocks typically varies between 2.5 $MPa.m^{1/2}$ and 1.5 $MPa.m^{1/2}$ (e.g., Atkinson, 1984), these calculations assume three arbitrary values of K_{Ic} for joint initiation (2.5, 2.0, and 1.5 $MPa.m^{1/2}$)[5]. Figure 2–6 shows that joints will initiate from larger flaws at lower P_p.

Once joints have initiated from small flaws, less severe internal pressure is necessary for further growth. For reinitiation of crack propagation, P_i must exceed the sum of two restraints on joint propagation, fracture toughness and total normal stress on the joint. Suppose that a vertical joint initiates from an initial

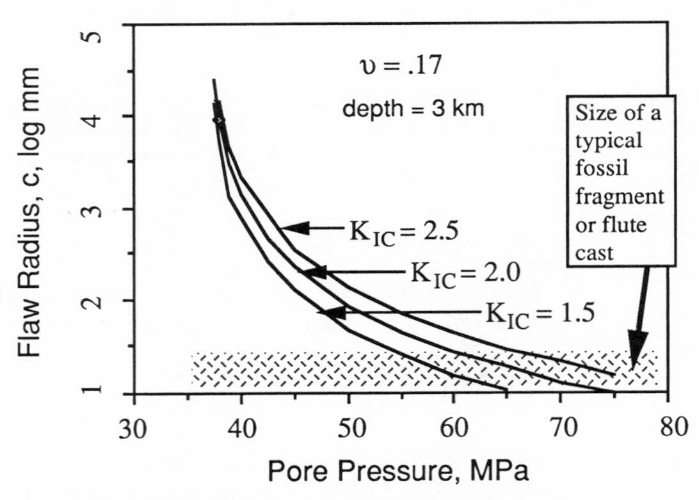

Fig. 2–6. The relationship between flaw or crack radius, c, and pore pressure required to initiate crack propagation within a siltstone of various fracture toughnesses, K_{IC}. The flaw length at initiation of vertical joints is strongly dependent on the Poisson's ratio of the rock within which crack propagation takes place (fig. 1–5). This calculation assumes a penny-shaped flaw, a tectonically relaxed basin with a poroelastic response to changes in pore pressure, $\nu = 0.17$, and a burial depth of 3 km.

flaw with a radius of about 1 cm when $P_p = P_i = 68$ MPa (fig. 2–6). By the time the joint has run to a length of 30 cm, P_i for reinitiation of joint propagation is 20 MPa less than P_p within surrounding pore space. Although fracture toughness does not change with ever increasing joint length, the fracture toughness term, $\frac{K_{Ic}}{Y\sqrt{c}}$, becomes even smaller and with it P_i for joint reinitiation. This drop in P_i during joint growth leads to a fluid pressure gradient which allows for the spontaneous recharging of the joint from surrounding pore space.

Once a joint propagates to a length of more than one meter, reinitiation pressure, P_i, decreases to approach total stress normal to the crack wall. The correlation between P_i and crack normal stress allows us to devise techniques for measuring σ_3 during crack propagation. Two examples of the use of P_i to infer total crack normal stress include the trapping pressure of fluid inclusions and abnormally high pore pressure in sedimentary basins.

Fluid Pressure during Crack Propagation

There is some question about the depth in the schizosphere to which the crack propagation might extend. If depths exceed 2–3 km where σ_3 is highly compressive, then linear elastic fracture mechanics predicts that abnormally high fluid pressures are necessary for crack propagation. Confirmation of abnormally high fluid pressures deep within the schizosphere comes from the data on trapping pressures of fluid inclusions in veins (Mullis, 1979, 1988; Vrolijk, 1987; Vrolijk et al., 1988; Lacazette, 1991). Theoretically, the trapping pressure of some fluid inclusions is a measure of absolute stress normal to the vein walls at the time of vein propagation, and hence fluid inclusions are prospective stressmeters (Lacazette and Engelder, 1987).[6]

One technique for deducing trapping pressures, P_t, requires the synchronous trapping of immiscible fluids in the same vein; methane-saturated brines and water-poor methane are immiscible fluids commonly found as fluid inclusions. Three lines of evidence strongly suggest the synchronous trapping of the two immiscible fluids. First, both types of fluid inclusions must occur within the same growth zones. Second, within one growth zone, all the water-rich inclusions must show consistent liquid/vapor ratios at room temperatures and homogenized within narrow ranges of temperatures (± 2 °C). Third, all the methane-rich inclusions of one growth zone must homogenize within a range of ± 5 °C.

If there was simultaneous trapping of the two immiscible fluids, then the homogenization temperature of water-rich inclusions is equal to the trapping temperature of the immiscible system (Roedder, 1984). The homogenization temperature of each methane-rich inclusion is used to calculate the density of that inclusion. By plugging density into an equation of state from Angus et al. (1978), a line of equal density in P-T space (i.e., an isochore) for each methane-rich inclusion is calculated. This isochore, in P-T coordinates, is a line containing the point representing P_t and trapping temperature of the immiscible system. The actual P_t point on the methane-isochore line is located using the mean trapping temperature for the brine inclusions. Methane-rich inclusions within a single vein have a variety of densities, each of which gives a different P_t by using the mean trapping temperature. Lacazette's (1991) technique for fluid inclusion stressmeters assumes that the densest of the methane-rich inclusions gives a P_t equal to P_i for the host vein.

Calibration of the Fluid-Inclusion Stressmeter Technique

P_t is a measure of crack-normal stress according to equation 2–25, if we assume that $P_t = P_i$. To assure that P_t is a measure of crack-normal stress at the time of crack propagation, a calibration is necessary. One technique is to calibrate P_t against overburden load by using data from horizontal veins which are bedding-parallel in foreland fold-thrust belts (Lacazette and Engelder, 1987). In horizontal veins, crack propagation takes place when fluid pressures (P_i) are large enough to overcome fracture toughness of the rock (K_{Ic}) and lift overburden (ρgz) which is the total stress in the vertical direction

$$P_i = P_t = \frac{K_{Ic}}{Y\sqrt{c}} + \rho gz. \qquad (2\text{–}27)$$

With the use of inclusions from veins > 1 m in radius, the contribution from the fracture toughness term is 3% or less depending on the width of the vein. If overburden is estimated using veins less than 10 cm in radius, the fracture toughness of the rock should be factored into the calculation; otherwise the estimate of overburden will be in error by more than 10% largely because there will always be great uncertainty concerning the actual fracture toughness.

By applying equation 2–27 for P_t in bedding-parallel veins in the Ordovician Bald Eagle Formation, Pennsylvania, Lacazette and Engelder (1987) confirmed that $P_t \approx \sigma_3 = \rho gz$ (i.e., fluid pressures were lithostatic). The accuracy of this exercise was uncertain because the exact thickness of the stratigraphic column at the time of crack propagation is unknown due to erosion. To improve confidence in the fluid-inclusion stressmeter, Srivastava et al. (1992) proposed a two-formation technique by measuring trapping pressures in two formations whose stratigraphic separation is well known. The section between the middle of the Ordovician Coburn and the middle of the Ordovician Bald Eagle is somewhere between 500 and 700 m. Work on the Bald Eagle has estimated an overburden of 5.85 km (Lacazette and Engelder, 1987) whereas an estimate of overburden above the Coburn is 6.5 km (Srivastava et al., 1992). The difference of overburden between the Bald Eagle and Coburn is 650 m based on bedding-parallel vein fluid inclusion data. From this exercise Srivastava et al. (1992) conclude that P_t of the densest fluid inclusions give a reasonable estimate of σ_3 provided large veins are sampled.

Limits to Abnormal Pore Pressure

The previous section suggests that fluid inclusion P_t is proportional to crack-normal stress during crack propagation. From equation 2–27 we see that, when a crack grows to widths greater than a meter, the driving pressure is within 3% of σ_3. Theoretically, P_p is reduced to near σ_3 by crack propagation in a sedimentary basin. This simple hypothesis is testable in regions where $S_h < S_v$, for in such regions crack propagation should modulate the buildup of P_p at a level near S_h but certainly less than S_v.

The Gulf of Mexico is famous for abnormally high pore pressures. Regional compilations show that P_p is generally hydrostatic down to a depth of about 2 km, where P_p begins to climb rapidly to a level of about 85% of lithostatic by a depth of 3 km (Dickinson, 1953). While figure 2–7 is a schematic summary of the P_p-z trend found in the Gulf of Mexico, the student should understand that the top of overpressure may vary for depths of 1 km south of Port Arthur, Texas, to as much as 5 km under the sandier portions of the Mississippi Delta. Contrary to the situation in Appalachian fold-thrust belt where the condition, $S_h > S_v$, favored the propagation of bedding-parallel veins at $P_p = \rho gz$, the highest P_p do not reach lithostatic pressures in many portions of the Gulf of Mexico. The modulation of $P_p < \rho gz$ is tied to the fact that the clastic wedge of the Gulf of Mexico is currently slumping toward the deep basin causing $S_h < S_v$. On average maximum P_p is fixed at 85% of S_v because the current S_h will not support larger P_p in fractured rock and the tendency for P_p to increase above this level is relieved by further crack propagation. In a sense, a measure of maximum P_p in the Gulf of Mexico and other sedimentary basins is a direct measure of σ_3[7]. Commonly, this maximum P_p is also equivalent to the fracture pressure gradient, a pressure at which hydraulic fractures are produced during drilling in the Gulf of Mexico (Anderson et al., 1973).

Crack-Driving Stress Based on Displacement of Dike and Vein Walls

Another interesting result of the analysis of crack-normal stress during crack propagation is that displacement of the crack wall is a function of the elastic properties of the host rock, the length of the crack, and the net tensile stress on the crack wall. Pollard and Segall (1987) derive the relationship for crack wall displacement, $u_y(z)$, as

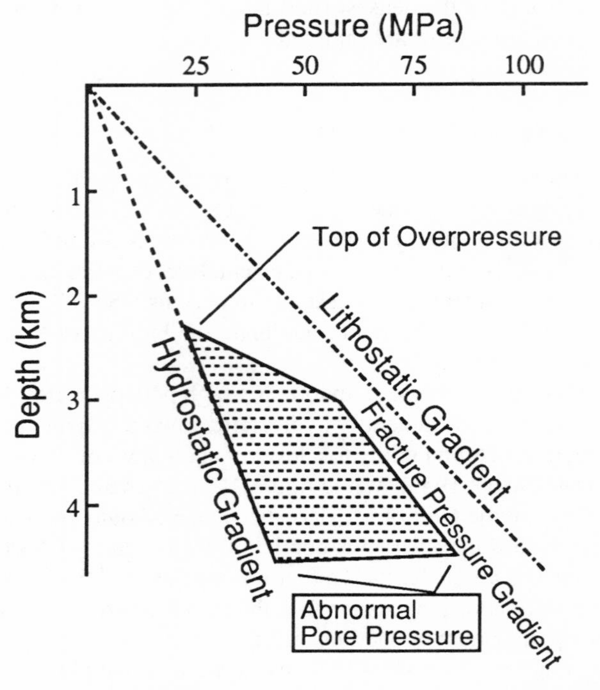

Fig. 2–7. A schematic diagram showing the variation of pore pressure with depth in a sedimentary basin based on Dickinson's (1953) compilation for the Gulf of Mexico. P_p is hydrostatic to a depth of about 2 km. Below a depth of 3 km, P_p is found to vary between hydrostatic pressure and 85% of lithostatic stress (dashed lines).

$$u_y(z) = \frac{4(1 - \nu^2)}{E} (P_i - \sigma_3) \sqrt{c^2 - z^2} \tag{2–28}$$

where z is the distance from the center of the crack. Crack wall displacement is well preserved in the cross-sectional shape of dikes (Delaney and Pollard, 1981). Note that in Delaney's and Pollard's (1981) analysis, their dike is oriented so that u_y is a function of x and not a function of z as is shown in figure 2–2. For cracking at depth in the crust, the cross-sectional shape of the final crack reflects the amount that the crack fluid pressure exceeds the total stress on the crack wall and this amount is called the *crack-driving stress* by Delaney and Pollard (1981). The shape of dikes intruded into shale in northwest New Mexico reflects the effect of a crack-driving stress (Delaney and Pollard, 1981; Delaney et al., 1986). These dikes are elliptical in cross section, leading Delaney and Pollard (1981) to conclude that the dike profile does reflect the effect of magma pressure dilating the dike wall against the elastic host rock. Picking appropriate elastic constants and dike length, Pollard and Segall (1987) estimate that the driving stress on the New Mexico dikes was between 11 MPa and 1.1 MPa. Many veins such as those cutting Lower Paleozoic beds of the Appalachian Valley and Ridge have a thickness-to-width ratio on the order of 1%. Using appropriate numbers for E and ν, the driving stress on these veins is on the order of 1 MPa.

Net Tensile Stress Necessary for the Initiation of Joint Propagation

Another means of estimating crack driving stress (i.e., the difference between the internal fluid pressure, P_i, and S_h at the initiation of jointing) is based on the amount of strain associated with the formation of a joint set. This is illustrated using multiple subparallel en échelon segments of composite joints in outcrops of the Mount Givens Granodiorite, central Sierra Nevada, California (Segall and Pollard, 1983). Each en échelon segment ranges in length from a few cm to 1 m with the composite joints 1 to 70 m in length. Strain associated with opening of these joints is estimated to be on the order of 10^{-4} or 0.01%. Following Segall and Pollard (1983) the total or final strain (ε_f) after cracking is the sum of crack strain (ε^c_f) and elastic strain (ε^e_f)

$$\varepsilon_f = \varepsilon^e_f + \varepsilon^c_f . \qquad (2.29)$$

Initial strain, ε_i, at the initiation of joint propagation is

$$\varepsilon_i = \varepsilon^e_i . \qquad (2\text{–}30)$$

When multiplied by Young's Modulus E, an estimate of ε^e_i will give the driving stress for the initiation of joint growth. Segall and Pollard (1983) require two postulates in order to estimate ε^e_i from the available data: $\varepsilon_i < \varepsilon_f$ and $\varepsilon^e_f < \varepsilon^e_i$. The first postulate states that total strain after jointing exceeds elastic strain at the initiation of jointing. The second postulate states that the elastic strain after jointing is less than the elastic strain at the onset of jointing. As joints propagate, the effective Young's modulus ($\hat{E}$) of the rock containing the joints decreases and is always less than E of the intact rock so that

$$\varepsilon^e_f = \left(\frac{\hat{E}}{E}\right)\varepsilon_f . \qquad (2\text{–}31)$$

$\hat{E}$ of the rock in its final state is estimated from the density of the joints in the following manner. The combination of equations 2–29 and 2–31 is

$$\varepsilon_f = \varepsilon^c_f \left[1 - \frac{\hat{E}}{E}\right]^{-1} . \qquad (2\text{–}32)$$

The two postulates then become

$$\varepsilon^{e}_{i} < \varepsilon^{c}_{f}\left[1 - \frac{\hat{E}}{E}\right]^{-1} \tag{2–33}$$

$$\varepsilon^{e}_{i} > \varepsilon^{c}_{f}\left[\frac{E}{\hat{E}} - 1\right]^{-1} \tag{2–34}$$

These two equations relate initial elastic strain to crack strain and effective modulus in the final state.

Walsh (1965) derived the effective Young's modulus for a two-dimensional body containing an array of cracks as

$$\frac{\hat{E}}{E} = \left[1 + 2\pi(1 - \nu^2)\rho_c\right]^{-1} \tag{2–35}$$

where ρ_c is spatial density of cracks defined as

$$\rho_c = \left(\frac{1}{A}\right)\sum_{n=1}^{N} c_2 \tag{2–36}$$

where N is the total number of cracks having half-length c in an area A. Substituting into the two postulates (eqs. 2–33 and 2–34) the driving stress for crack propagation may be calculated

$$P_i - S_h \leq E\varepsilon^{c}_{f}\left[\frac{1 + 2\pi(1 - \nu^2)\rho_f}{2\pi(1 - \nu^2)\rho_f}\right] \tag{2–37}$$

$$P_i - S_h \geq E\varepsilon^{c}_{f}\left[2\pi(1 - \nu^2)\rho_f\right]^{-1} \tag{2–38}$$

giving a maximum and minimum estimate of initial crack-driving stress. ρ_f is the crack density in the final state. In calculating the driving stress, the values of E (38 GPa) and ν (0.08) are taken from Hawkes et al. (1973) for fluid saturated Barre granite in uniaxial tension. For two outcrops of the Mount Givens Granodiorite, Segall and Pollard (1983) estimate maximum driving stress as 26 and 10 MPa, respectively. The minimum estimate for driving stress is 1.7 and 1.2, respectively.[8]

Crack Paths

The crack path is the direction that a crack grows relative to its plane. The direction of crack propagation is controlled by stress in the vicinity of the tip of the crack. Early experiments on the propagation path of cracks in rocks concerned the mechanism leading to shear fracture when rock was compressed (e.g., Brace and Bombolakis, 1963). These experiments were designed to test the hypothesis that cracks would propagate in their own plane. However, plexiglass models showed that despite a large component of sliding-mode loading, crack propagation was not coplanar (fig. 2–8a). The tendency was for cracks to follow curved paths so that the crack would become reoriented to maximize the decrease in total system energy (Lawn and Wilshaw, 1975). On reorientation, the growing crack favors opening mode loading and, so, becomes normal to σ_3 and, hence, parallel to σ_1.

Crack propagation is best understood by analyzing the stresses at or in the vicinity of the crack tip. Two approaches to this problem, both two-dimensional, differ by the assumption made about the shape of the tip of the crack. The first assumption is that linear elasticity holds everywhere in the vicinity of a crack tip which has an elliptical shape (fig. 2–3). Crack extension will initiate at a point near the tip of an existing elliptical crack where net tensile stresses are largest. If an elliptical crack is tilted relative to σ_1, sliding par-

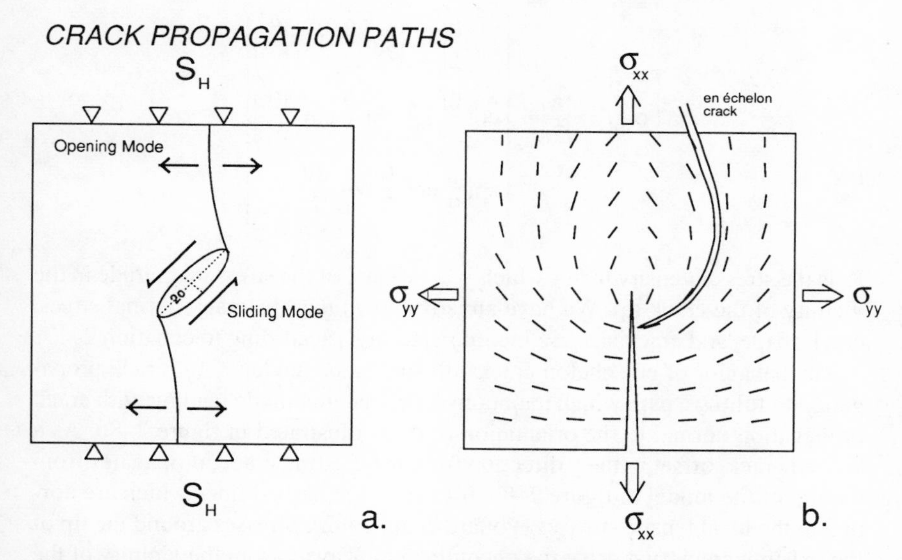

Fig. 2–8. (a) Plan view of crack development in a strike-slip faulting regime with the crack path (solid line) starting at a preexisting elliptical crack (adapted from Brace and Bombolakis, 1963). Note that along the existing crack, sliding mode loading is relatively large compared with opening mode loading. However, once the crack extension starts, opening mode loading dominates in the newly formed portions of the crack. (b) The orientation of tensile stresses in the vicinity of a crack tip as indicated by trend lines which are drawn normal to the local tensile stresses (adapted from Olson and Pollard, 1989). If an en échelon crack were to propagate in the vicinity of the existing crack, the en échelon crack would follow the trend lines. Coordinate system as in figure 2–2.

allel to the crack will cause tensile stresses to develop in the vicinity of the crack tip but not at the point of smallest radius (fig. 2–8a). Linear elastic fracture mechanics shows that initial crack extension will occur at an angle of 71° to the initial planar crack subject to mode II loading (Ingraffea, 1987). Tensile stresses will develop in such a manner even when all far-field stresses are compressive. This approach predicts the point from where incremental crack extension takes place but contributes nothing to predicting the crack propagation path once the new crack leaves the tip of the existing crack.

To predict the propagation direction of the new crack, information on the stress field within the rock near the crack tip is necessary. Analytical solutions to the problem are bulky and difficult to apply. To simplify matters, higher order terms in the solution for stress expanded about the crack tip are dropped. With this assumption solutions for the stress field about the crack tip take a simple analytic form for the three modes of crack loading shown in figure 2–1. The solutions for the stress field in the vicinity of a crack tip subject to opening mode loading follow equation 2–16

$$\begin{bmatrix} \sigma_{zz} \\ \sigma_{yy} \\ \sigma_{yz} \end{bmatrix} = \frac{K_I}{\sqrt{2\pi r}} \begin{bmatrix} \cos\frac{\theta}{2}\left(1 - \sin\frac{\theta}{2}\sin\frac{3\theta}{2}\right) \\ \cos\frac{\theta}{2}\left(1 + \sin\frac{\theta}{2}\sin\frac{3\theta}{2}\right) \\ \sin\frac{\theta}{2}\cos\frac{\theta}{2}\cos\frac{3\theta}{2} \end{bmatrix}. \qquad (2\text{–}39)$$

K_I is the stress intensity factor which is a measure of the stress magnitude in the vicinity of the crack tip. We have already shown that the crack normal stress, crack shape, and crack size are incorporated in K_I according to equation 2–17.

The behavior of en échelon cracks illustrates the tendency for crack propagation to follow a path which maintains local opening mode loading with crack propagation normal to the orientation of σ_3 as illustrated in figure 2–8b. As a second crack, offset in the y direction from the existing crack, propagates from the top of the model in figure 2–8b, it follows the dashed lines which are normal to the local tensile stresses (Pollard et al., 1982). Stresses around the tip of the existing crack first drive the oncoming crack away from the vicinity of the crack tip, but once the oncoming crack passes the existing crack tip, local stress causes the en échelon crack to rotate sharply into the existing crack. With continued propagation the en échelon crack abuts the existing crack (fig. 2–8b). As the en échelon crack propagates, it will follow a path that reduces the sliding-mode loading. Lawn and Wilshaw (1975) stated, "The action of the imposed shear is to deflect the crack away from plane geometry (i.e., a directional instability develops). Moreover, this deflection always tends towards the orientation of minimum shear loading; in this sense the shear stresses may be seen in a corrective role, restoring deviant cracks to a stable path of orthogonality to the greatest principal tensile stresses in the applied field." Because cracks exhibit this directional instability when subject to sliding-mode loading, they are extraordinarily reliable markers for a stress trajectory normal to σ_3 at the time of propagation.

Although the crack propagation path for a mesoscopic joint appears planar, on a microscopic scale the crack tip is continually deflected by small pores and grain boundaries. This microscopic deflection leads to a distinctive surface morphology called a *plumose pattern* (e.g., Woodworth, 1896; Lutton, 1971) which is used for an interpretation of crack initiation, propagation, and arrest (Kulander et al., 1979; 1990). Crack initiation is usually at the point of a stress riser (Pollard and Aydin, 1988) such as an existing crack, an inclusion, or at a bedding plane (fig. 2–8a). Propagation is characterized by a mirror zone, a mist zone, and a hackle zone. The hackle zone, and accompanying plumose pattern, develops largely because of local twists and tilts (i.e., directional instabilities) during propagation that otherwise would be planar. Such directional instabilities occur around local microcracks and grain-boundary discontinuities on a small scale and do not interfere with the overall planar growth of a joint. The en échelon crack shown in figure 2–8b is an example of how the tip of a joint might be tilted on a microscopic scale as it attempted to pass an existing microcrack. Arrest is usually marked by a sudden tilting or branching.

Local State of Stress Inferred from Crack Paths

The presence of a crack interrupts a homogeneous stress field in an elastic medium at two distinct locations around the crack: the crack tip and the crack wall. Stresses in the vicinity of the crack tip are highly inhomogeneous according to equation 2–39. In contrast, the crack wall acts like a free surface if open or it carries tractions if closed. At a free surface principal stresses are parallel and perpendicular to the surface. The local stress state both in the vicinity of the crack tip and near a crack wall will have a profound influence on the direction of crack propagation. For stress analyses, the path of a younger crack approaching an older crack tip or wall is a record of the state of stress at the time of crack propagation.

Dating the relative sequence of joint development is commonly accomplished using abutting relationships. The wall of a preexisting joint may act as a free surface which prevents the transmission of the elastic crack-tip stress field of an approaching crack. As a consequence many younger joints in outcrops terminate at preexisting joints. This rule does not always apply; some joints and many veins do cross cut. In these latter cases tractions on the crack wall permitted the transmission of crack-tip stresses across the preexisting crack.

STRESS NEAR A CRACK WALL

In a study of the joints in the Entrada Sandstone in Arches National Park, Utah, Dyer (1988) observed that, on approaching older joints, younger joints follow one of two distinct propagation paths. Younger joints rotate to become either parallel or perpendicular to the older joint (fig. 2–9). Dyer's interpretation is that the remote stress field changed orientation after propagation of a primary joint set. Upon realignment of the remote stress field, a second set of joints propagated from the mid region between the primary joints. In the mid region the younger joints propagated normal to the remote S_h in its new orientation. The change in orientation of the younger joints near the primary joints indicates a local rotation of the principal stresses in the vicinity of the wall of the preexisting joints. To solve for the stress field about the primary joints, Dyer (1988) resolves the remote stresses into stresses parallel and perpendicular to the joint wall. He then assumes that the primary joints are either open (no tractions) or closed (carries tractions). For the closed joint two cases are possible: a frictionless joint and a joint carrying shear tractions. In comparing analytic stress field solutions with field examples, Dyer (1988) concluded that curving-parallel younger joints are indicative of a stress field in which $-3 < \frac{S_H}{S_h} < -\frac{1}{3}$. In this case a large negative number means that the S_H is large and compressive, whereas S_h is an effective tensile stress. A curving-perpendicular geometry re-

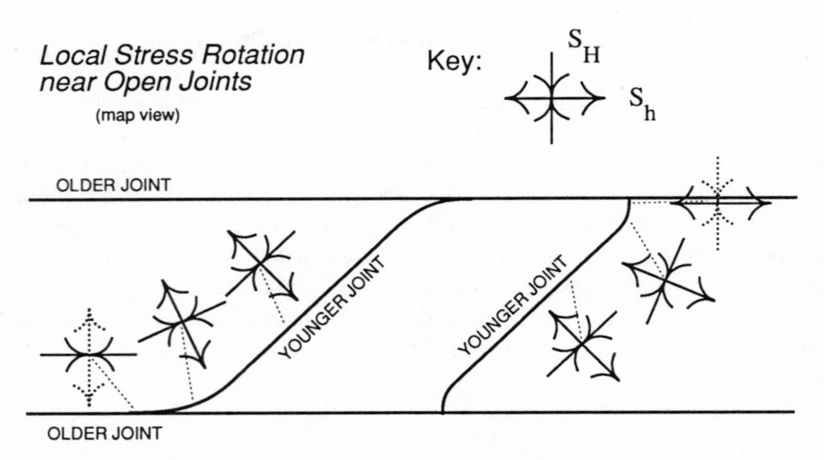

Fig. 2–9. A map view of the two ways in which the state of stress may rotate to accommodate a free surface at a through-going joint. Orientation of magnitude of principal stresses in midregion between older joints approximates the remote stresses at the time of propagation of the younger joints. Younger joints propagate outward from the midregion between older, through-going joints, with each growth increment perpendicular to the local σ_3. Sigmoidal geometry develops due to rotation of principal stresses near the free surfaces (adapted from Dyer, 1988). Dashed arrow equals zero stress.

flects far-field stresses where $-\frac{1}{3} < \frac{S_H}{S_h} < 1$. In this case, the two horizontal stresses are close to the same value and both may be effective tensile stresses. In summary, curving-parallel younger joints indicate a compressive stress parallel to the walls of the primary joints, whereas the curving-perpendicular geometry indicates an effective tension parallel to the walls of the primary joints.

Stress in the Vicinity of a Crack Tip

The stresses in the vicinity of a crack tip are reflected by the very complicated path which a propagating en échelon crack may take as it approaches the tip of a stationary crack. Typically, the younger crack exhibits a hooked shape geometry where it diverges slightly and then curves sharply to abut the stationary crack behind the latter's tip (Melin, 1982; Pollard et al., 1982). Such an en échelon crack geometry is described in detail by Beach (1975), Kranz (1979), and Nicholson and Pollard (1985).

By analyzing the superposition of remote stresses on the crack-tip stress field, $\Delta\sigma = \sigma_x - \sigma_y$ may be estimated from the shape of overlapping joint traces. Olson's and Pollard's (1989) analysis is based on the premise that propagation paths for opening mode cracks (joints, veins, and dikes) in certain geological settings are strongly influenced by the state of stress and are only weakly dependent on the material properties of rock. In their analysis, cracks are viewed in two dimensions where there is no information on stress in the third dimension. Figure 2–10 shows various shapes for nonstationary en échelon cracks

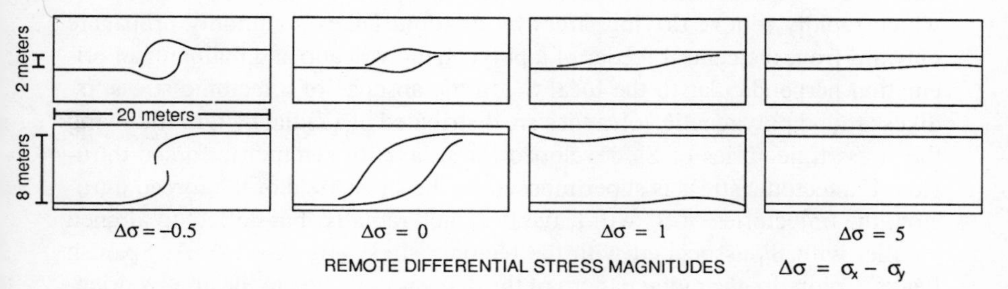

Fig. 2–10. A map view for theoretical crack paths for en échelon vertical cracks propagating from the left and right, respectively (adapted from Olson and Pollard, 1989). Each row represents cracks passing at different distances apart. Each column shows the effect of remote differential stress indicated as $\Delta\sigma$ at the bottom of the diagram. For the situation where $\Delta\sigma = -0.5$, the passing cracks are constrained to grow in the x direction until they are 10 meters apart and then this constraint is removed even though $\sigma_y > \sigma_x$. Coordinate system as in figure 2–2.

passing under various remote stress conditions where σ_x is the remote stress parallel to the crack and σ_y is the remote stress normal to the crack. For the remote stress with a small differential tension, $\Delta\sigma = \sigma_x - \sigma_y = -0.5$ MPa, the path curvature is exaggerated with an initial divergence and then a strong convergence. This case is equivalent to the curving-perpendicular joints at Arches National Park. Here the crack tips approach the neighboring crack plane at the greatest angle of the four examples. For the isotropic case, $\Delta\sigma = 0$ MPa, the closely spaced cracks exhibit a slight initial divergence of paths followed by convergence. The cracks do not approach tip to plane but turn to approach each other asymptotically. A differential stress with some crack-parallel compression ($\Delta\sigma = 1.0$ MPa) eliminates any divergence and reduces convergence. A crack parallel compression with $\Delta\sigma = 5$ MPa produces nearly planar paths as is the case for curving-parallel joints at Arches National Park. This analysis reflects the relatively low $\Delta\sigma$ present during crack propagation.

Stress Trajectories

Vertical cracks are useful for mapping the trajectories of S_H at the time of crack propagation. Depending on the age of the cracks we may map either paleostress fields or the contemporary tectonic stress field. One of the most compelling field checks of the hypothesis that vertical cracks follow principal *stress trajectories*[9] is based on an analysis of the pattern of dikes at Spanish Peaks, Colorado (Odé, 1957; Muller and Pollard, 1977). Igneous dikes are believed to fill cracks propagating under opening-mode loading during the forced intrusion of magma. Because they are continuously driven by very large fluid reservoirs, dikes behave more like industrial hydraulic fractures (chap. 5) than joints, which rapidly relieve driving stress by dilation. Dikes commonly propagate outward from the central feeder of a polygenetic volcano and maintain an orientation perpendicular to the local S_h. In the absence of a tectonic stress, σ_t, dikes around polygenetic volcanoes are distributed in a radial pattern following the stress trajectories of S_H developed for an axially symmetric forced intrusion. If a tectonic stress is superimposed on the stress field of the forced intrusion, the trajectories of S_H will leave the stock radially, but deflect to aligned parallel with S_H associated with the tectonic stress (fig. 2–11a). At Spanish Peaks, Colorado, the radial pattern of the dikes is deflected to the an E-W orientation to follow the trajectories of S_H associated with west to east thrusting in the Sangre de Cristo Mountains (fig. 2–11b). The shape of the stress trajectories around the Spanish Peaks intrusion was modeled by solving an elastic boundary value problem (Odé, 1957; Muller and Pollard, 1977). The similarity in the shape of the dike pattern around the Spanish Peaks and the stress trajectories calculated for the elastic boundary value problem devised by Odé (1957)

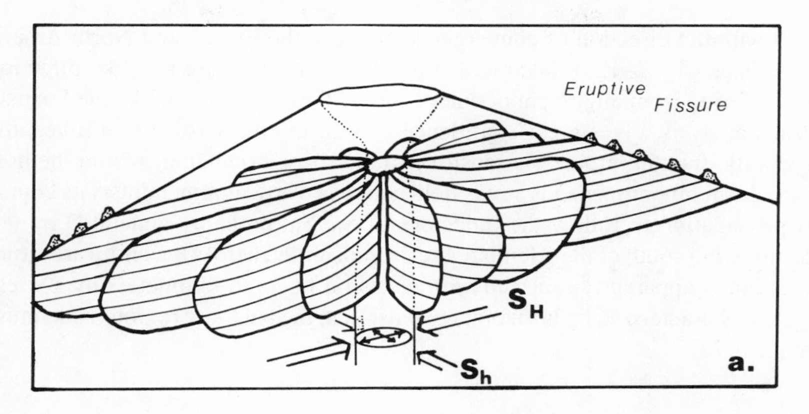

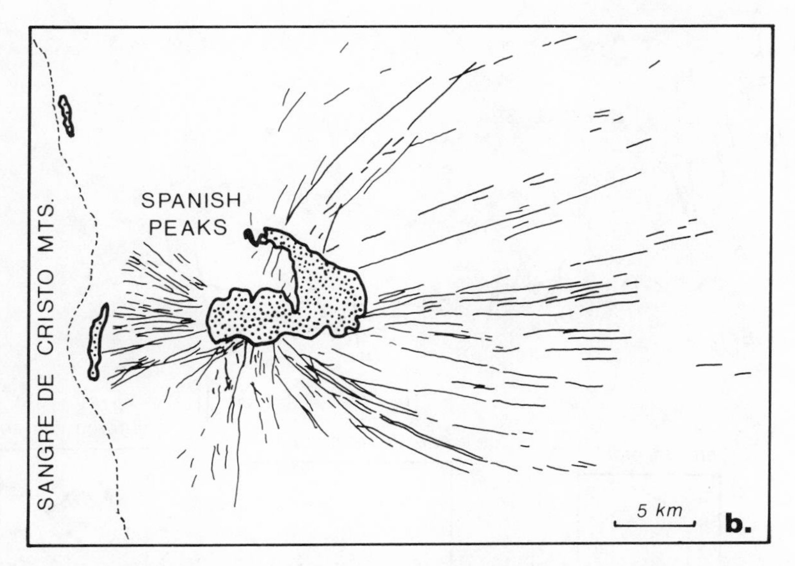

Fig. 2–11. (a) The architecture of a volcano showing the orientation of dikes and eruptive fissures in a stress field where $S_H > S_h$ (adapted from Nakamura et al., 1977). Note the tendency of the dikes to rotate into the direction of S_H after propagating from the main feeder of the volcano. (b) The radial dike pattern around the Spanish Peaks (adapted from Johnson, 1961). This pattern is a cross section in plan view of the volcano shown in figure 2–11a.

support the notion that joints, veins, and dikes follow stress trajectories in an elastic upper crust.

The orientation of dikes are useful for mapping the orientation of S_H of the contemporary tectonic stress fields over large regions such as the Aleutian Arc. The direction of S_H obtained at twenty volcanoes in the Aleutian Arc coincides well with the direction of convergence between the Pacific and North American Plates (fig. 2–12) (Nakamura et al., 1977). In this case the dike direction was inferred from alignment of cinder cones. This correlation, like the Spanish Peaks analysis, gives great confidence that during the intrusion of dikes, the hydraulic fracture process is sensitive to the stress orientation within the host rock. In an inhomogeneous stress field a dike or vertical joint adjusts its course of propagation to follow the trajectory of S_H. S_H, probably generated by underthrusting south of the Aleutian Arc, is transmitted across the entire arc structure, but is apparently replaced within several hundred kilometers by a stress system characterized by horizontal extension in the back-arc region (Nakamura et al., 1977).

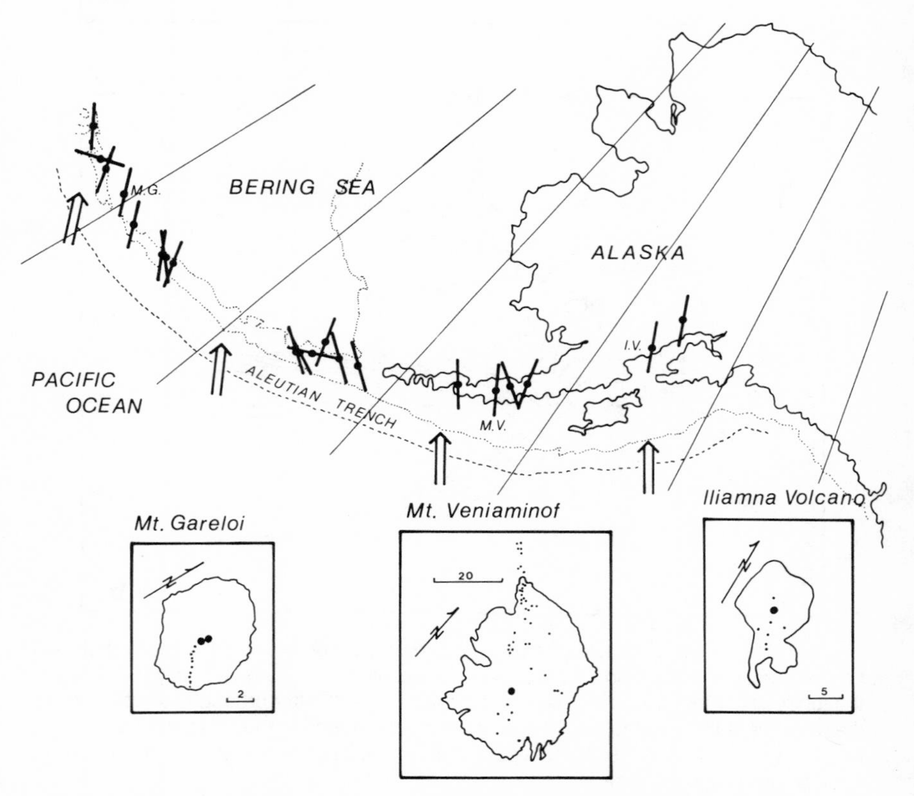

Fig. 2–12. Map of Alaska showing the orientation of eruptive fissures (black lines) on volcanoes of the Aleutian Arc (adapted from Nakamura et al., 1977). The direction of convergence of the Pacific Plate relative to the North American Plate is shown with open arrows. Other features include the Aleutian Trench (dashed line), the 500 m below sea level depth contour (dotted line), and lines of longitude (straight lines). Detailed maps of three volcanoes show the orientation of cinder cones (dots) associated with eruptive fissures.

Using the same rationale developed for the Aleutian Arc, we can track the orientation of stress through time. For example, post-metamorphic dikes in New England and adjacent areas fall into three groups depending on their age (McHone, 1978). Each group of dikes has a distinct orientation from which we infer that the Mesozoic stress field in New England rotated clockwise so that S_H moves from NE-SW to E-W to WNW-ESE between 220 m.y. and 95 m.y. (fig. 2–13). This sequence accompanies the rift to shift phase of Mesozoic tectonics for the New England Appalachians (de Boer et al., 1988). During the transition from the shift to drift phases, S_H rotated counterclockwise from WNW through E-W to ENE-WSW which is the present orientation of the tectonic stress field for much of eastern North America. The point made here is that we can map the orientation of ancient stress fields over very large regions provided that we can correlate and date contemporaneous cracks of various kinds.

Joint patterns are used in the same manner as dikes for mapping stress trajectories. However, the isotopic techniques used to date and correlate dikes is not available for joints. As a consequence tracking the orientation of S_H through time is complicated by the lack of confidence in estimating the relative age of joint sets. Correlating regional joint sets depends on assumptions about the way joint patterns repeat from outcrop to outcrop. Hence, mapping the stress trajectories of more than one ancient tectonic event remains subjective as illustrated by the differences between interpretations of Nickelsen and Hough (1967) and Engelder and Geiser (1980) for Alleghanian cross-fold joints[10] of the Appalachian Plateau. Nickelsen and Hough's (1967) outcrop-to-outcrop correlation of

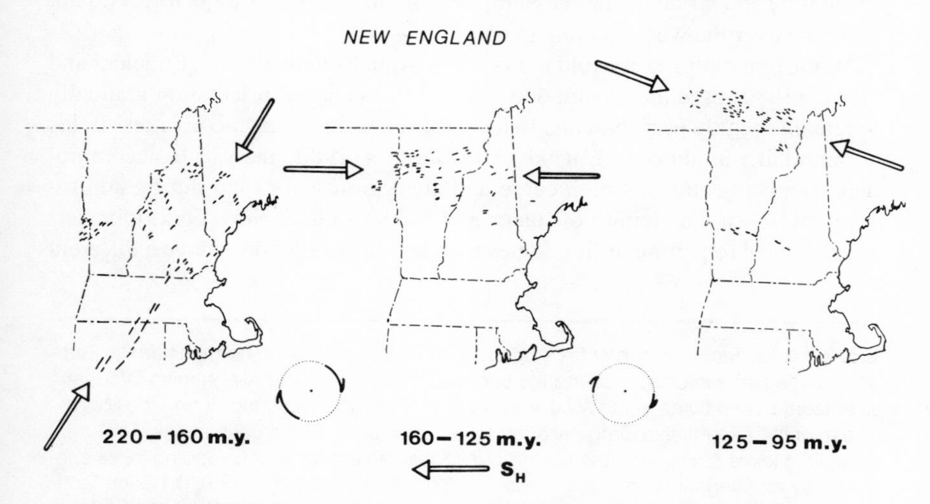

Fig. 2–13. Proposed directions of S_H (open arrows) as inferred from Mesozoic dike trends (solid black lines) in central New England, U.S.A., for three periods of time in the Mesozoic (adapted from McHone, 1978). S_H rotates clockwise during the Mesozoic.

joints depends largely on the assumption that the earth's stress field is nearly homogeneous and, therefore, joints of one set have similar orientations over large regions. If a set of joints at an outcrop is misoriented by, say, 15° from a joint set established at other outcrops, then this local suite of joints belongs to a second joint set, regardless of its orientation with respect to local structures. A consequence of Nickelsen and Hough's assumption is that members of one joint set do not change orientation even as fold axes swing through a mountain range such as the central Appalachian Plateau (fig. 2–14). On a regional basis, the change in strike of fold axes is accommodated by the overlap of joint sets of different orientations according to Nickelsen and Hough (1967). The notion for overlapping joint sets is supported by outcrops containing more than one cross-fold joint set.

Nickelsen and Hough's (1967) correlation strategy was developed in response to the observation that first-order folds in the Pennsylvania Valley and Ridge are kink folds (e.g., Faill, 1973) with straight axes (Nickelsen, 1987, personal communication). Nickelsen's idea is that the Valley and Ridge developed as overlapping thrust sheets cored with duplexes moving toward the craton with straight-axis kink folds delimiting the thrust duplexes. The curvature of the central Appalachian Valley and Ridge is accommodated by abrupt changes in the orientation of first-order folds. Strictly parallel joint sets reflect the kinematics of thrust sheets associated with straight-axis kink folds. Presumably the motion of various sheets is independent, thereby setting up homogeneous stress fields responsible for parallel joint sets which are unrelated to others in time and space. If these assumptions hold, then stress trajectories do not correlate over the whole mountain belt.

While remapping cross-fold joints on the Appalachian Plateau, Engelder and Geiser (1980) used the assumption that joint sets change orientation gradually to remain roughly perpendicular to local fold axes. Inhomogeneous stress fields are implicitly assumed in Engelder and Geiser's (1980) analysis to accommodate stress trajectories which curve and are regional in extent during an orogenic pulse. In the vicinity of Ithaca and Watkins Glen, New York, joint sets are restricted to particular lithologies where joints in siltstones have a different

---

Fig. 2–14. (a) Tectonic map of the Appalachian Plateau within the northeastern United States. Cleavage imprinted during the Lackawanna Phase of the Alleghanian Orogeny affected the area south of the solid line, whereas cleavage imprinted during the Main Phase of the Alleghanian Orogeny affected the area south of the dashed line. The cross-fold joints are either tectonic and of Alleghanian age or later unloading joints controlled by an Alleghanian residual stress (adapted from Engelder, 1985). (b) Orkan and Voight's (1985) map of the trend lines for regional joint sets within the Central Appalachian fold-thrust belt. Sets A through E are those of Nickelsen and Hough (1967). Regional joint sets are based on the data of Nickelsen and Hough (1967), and Engelder and Geiser (1980).

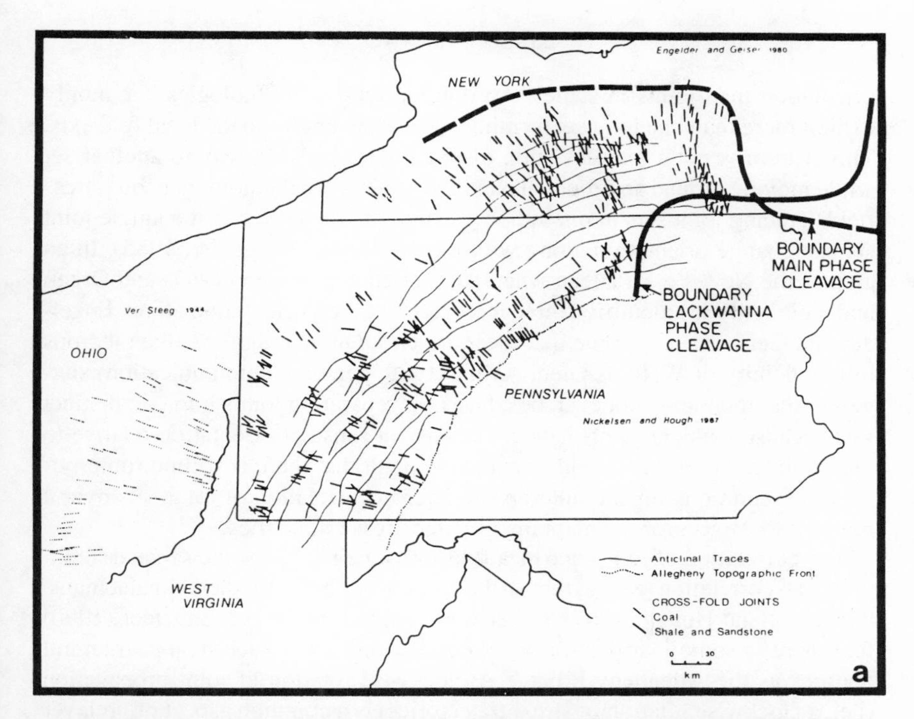
Engelder and Geiser 1980
NEW YORK
BOUNDARY
MAIN PHASE
CLEAVAGE
BOUNDARY
LACKWANNA
PHASE
CLEAVAGE
Ver Steeg 1944
OHIO
PENNSYLVANIA
Nickelsen and Hough 1967
Anticlinal Traces
Allegheny Topographic Front
CROSS-FOLD JOINTS
Coal
Shale and Sandstone
WEST
VIRGINIA
0 30
km
a

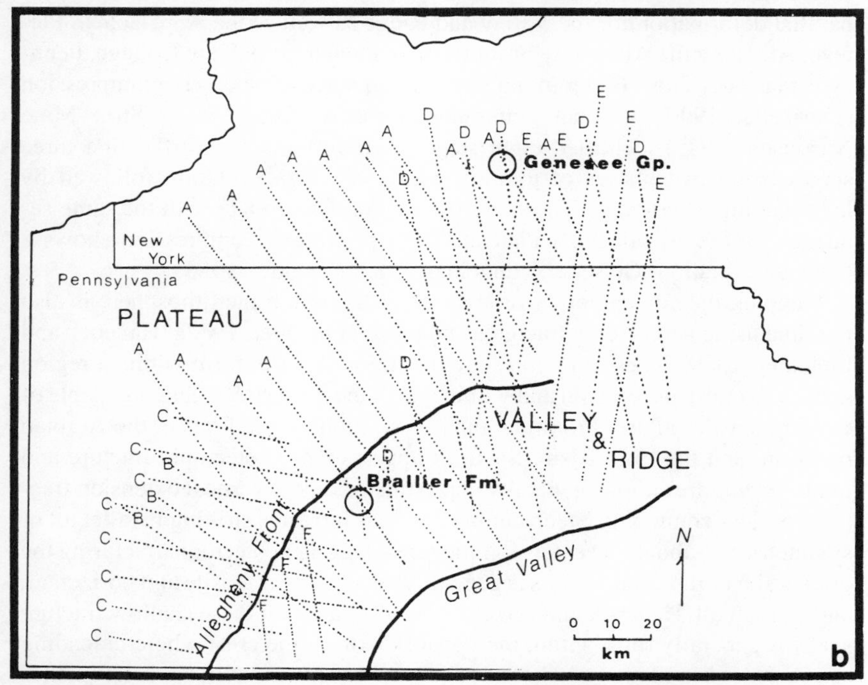
Genesee Gp.
New
York
Pennsylvania
PLATEAU
VALLEY
&
RIDGE
Brallier Fm.
Allegheny Front
Great Valley
N
0 10 20
km
b

orientation than joints in shales. An outcrop with two lithologies commonly exhibit more than one joint set forming at different angles to the local fold axis. This pattern is not a manifestation of one joint set giving way to another set while moving around an oroclinal bend but rather a local inhomogeneous stress field. Tracing joint sets in one lithology supports the notion that a single joint set can change orientation along with local fold axes (Engelder, 1985). In an area of the New York Plateau where Nickelsen and Hough (1967) and Orkan and Voight (1985) identified three joint sets based on orientation alone, Engelder and Geiser (1980) argue that there are two. On visiting the same outcrops in the vicinity of Watkins Glen, Aydin (1989, personal communication) suggested that there may not even be a basis for separating joints into two distinct sets because younger joints have no consistent sense of orientation relative to older joints. All of this is said to make the point that there is still no foolproof set of assumptions for the outcrop-by-outcrop correlation of joint sets over a region and, hence, for the mapping of paleostress trajectories.

The correlation of joints across a fold-thrust belt requires the same assumptions as correlation along the strike of a fold belt. In the Appalachians, Nickelsen and Hough (1967), Orkan and Voight (1985), and Engelder (1990) feel there is some justification for a correlation across such major structural features as the Allegheny Front.[11] A clockwise rotation of joint propagation (i.e., a clockwise rotation of stress trajectories) is a common aspect of prelayer parallel deformation for the Valley and Ridge as well as the Appalachian Plateau. All along the Allegheny Front from Williamsport to State College, Pennsylvania, early cross-fold jointing shows a sequence of clockwise compression (Lacazette, 1990, personal communication). At Bear Valley Strip Mine Nickelsen (1979) identified eight stages of deformation with the first three stages being two phases of jointing with a clockwise rotation followed by layer-parallel shortening. These prefolding events correlate with the same sequence on the Appalachian Plateau where prefolding compression shows a clockwise rotation (Geiser and Engelder, 1983; Engelder, 1985).

Mapping the contemporary orientation of S_H within the lithosphere is also possible using neotectonic fractures and joints (Engelder, 1982a; Hancock and Engelder, 1989). These structures are the most recent to form within a region subject to uplift and erosion and within which the local stress field is capable of causing tensile failure. From mapping in SE England-NE France, the Arabian platform, and the Ebro basin, Spain, it appears that neotectonic fracture and joint systems are simple, generally consisting of sets of vertical extension fractures or, less commonly, steep conjugate shear fractures striking parallel to, or symmetrically about, the extension fractures. During neotectonic fracturing the orientation of the σ_1 stress axis is generally vertical, σ_3 is tensile and horizontal, and σ_d is small. Hancock and Engelder (1989) infer that these shallow fracture systems generally form within the upper 0.5 km of the crust where unloading as a result of denudation and lateral relief consequent on uplift are prerequisites for their propagation. In structurally complex terrains, neotectonic joints post-date and are generally nonorthogonal to preexisting structures. Some regional joints, interpreted as neotectonic on the basis of field relationships, strike parallel to or approximately parallel to directions of contemporary horizontal maximum stress (S_H) known from *in situ stress* measurements or fault plane solutions of earthquakes. For this reason, they are structures of potential value for tracking the contemporary tectonic stress trajectories in regions where *in situ stress* measurements are not available. This technique was applied in the Valley and Ridge Province of central Pennsylvania where it was found that the most recent joints and, hence, the contemporary tectonic stress, S_H, trends N75°–90°E (Hancock and Engelder, 1989). Figure 2–15 is a compilation of those joint sets in the northeastern United States which are thought to be late-

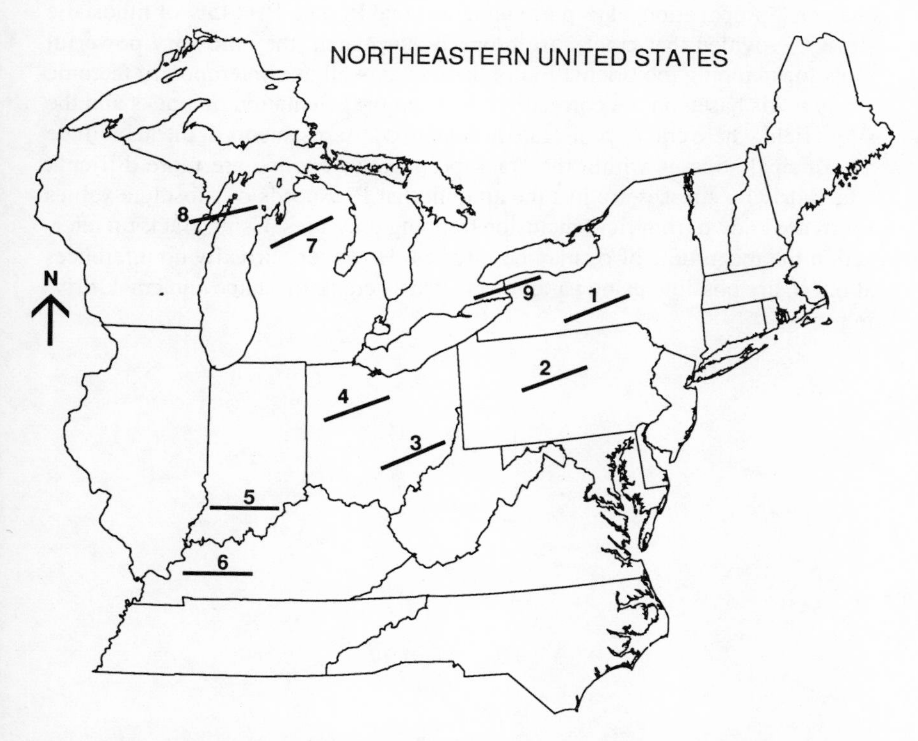

Fig. 2–15. The orientation of late-formed or neotectonic joints in the northeastern United States. The strike of these points is parallel to S_H of the platewide stress field found in eastern North America. Data on late-formed joints come from: 1 - Engelder, 1982a; 2 - Hancock and Engelder, 1989; and 9 - Gross and Engelder, 1990. Other joint sets are possible candidates for late-formed joints but lack documentation that such is the case. These include: 3 & 4 - Ver Steeg, 1942; 5 - Powell, 1976; 6 - Swenson, 1989; 7 - Holst and Foote, 1982; and 8 - Bretz (1942).

formed joints. The correlation between the strike of these joints and S_H of the contemporary tectonic stress field as first mapped by Sbar and Sykes (1973) is unmistakable.

Concluding Remarks

Throughout the upper portion of the lithosphere veins, dikes, and joints are so common that it is appropriate to recognize a separate stress regime, the crack-propagation regime, where the state of stress favors the propagation of opening mode cracks. In general, the crack-propagation regime of the lithosphere is characterized by a low σ_d and P_p at or above hydrostatic pressure. At places in the uppermost portion of the lithosphere the requirement of high P_p is relaxed and crack propagation takes place under normal P_p (i.e., $P_p \approx 45\%$ of lithostatic stress). Provided that cracks are properly correlated, they are very powerful tools for mapping the orientation of ancient as well as contemporary tectonic stress fields based on the correlation between the orientation of cracks and the stress field where cracks propagate normal to σ_3. Assessments of the magnitude of principal stresses within the crack-propagation regime are more difficult. Crack-driving stresses are just the amount that P_i exceeds σ_3. Absolute values for σ_3 are known from fluid inclusion trapping pressures. Using crack-tip interaction the magnitude of σ_d may be inferred. However, virtually no inferences about σ_1 are possible using parameters such as crack-tip shape and crack-driving stress.

3

Stress in the Shear-Rupture and Frictional-Slip Regimes

Rocks, which were bent or sheared at depths in the crust where they were subjected to great confining pressures, were materially stronger than they are when exposed at the surface. The unbalanced [stress] required to move [the rocks] was, therefore, much greater and was proportionately greater as [the rocks] lay deeper.

—Bailey Willis (1923) in discussing the relationship between geologic structures and the mechanics of rock deformation

A shear fracture that started as a primary shear rupture is difficult to recognize in outcrop. One of the best field criterion for the primary shear rupture of rock is a band of cataclastic material called a braided shear fracture (Engelder, 1974) or a deformation band (Aydin and Johnson, 1978). Confusion over identification of primary shear fractures arises because joints are often reactivated in shear[1] and, thus, give the impression of starting as a primary shear rupture. For example, Martel et al. (1988) show that some strike-slip fault zones in granite develop through three stages, the first of which involves the propagation of joints in en échelon patterns. Adjacent joints then slip to form boundary faults separated by tabular volumes of fractured rock. Nevertheless, if present, primary shear fractures are reminders that the host rock was at one time subjected to a very large σ_d.

Early in the twentieth century, the interpretation of fractures in the schizosphere was influenced by the polyaxial compression experiments of Daubree (1879), Becker (1893), Adams (1910), and von Karmon (1911). Rock failure was by shear rupture in these experiments performed under conditions found deep in the schizosphere. Such experiments gave the false impression that joints were near-surface phenomena and that fracturing deep in the crust was by shear rupture under high σ_d (e.g., Badgley, 1965). A concomitant misconception arose about the significance of two or more sets of fractures in an outcrop. Many would mistakenly interpret two sets of joints as conjugate shear fractures (e.g., Bucher, 1920).

Unlike joint sets, shear fractures are not common as regional sets but are localized near fault zones and folds (e.g., Stearns, 1972). The sparse occurrence of shear fractures over broad regions of foreland basins suggests that significant portions of the upper crust, particularly the cratonic cover, were rarely, if ever, subjected to a σ_d large enough to induce a primary shear rupture. Even in tectonically active terrain, shear fractures are sometimes less common than might be expected. This is graphically illustrated in the Miocene Punchbowl Formation, a sandstone adjacent to the San Andreas fault between Cajon Pass and Palmdale, California. One imagines that this sandstone was subject to large σ_d during earthquake cycles on the San Andreas fault. Yet, the sandstone contains few shear fractures, spaced dozens of meters apart. The other extreme from the Punchbowl Formation is found in foreland fold-thrust belts where well-developed shear fracture sets give the impression that large σ_d was relatively common. In select areas on fold limbs, the Ordovician Bald Eagle Formation of the Appalachian Valley and Ridge of Pennsylvania is riddled with primary shear fractures (Lacazette, 1991).

The association between primary shear fractures and foreland fold limbs suggests that stress is concentrated at the tips of laterally propagating thrust faults (Pavlis and Bruhn, 1989). Complex fault geometries such as thrust ramps crossing bedding and bends in strike-slip faults can lead to rotation of stress fields and concentration of stress (Kilsdonk and Fletcher, 1989; Bilham and King, 1989). The correlation between seismicity and active folding at Coalinga, California (Stein and King, 1984) and El Asnam, Algeria (King and Vita-Finzi, 1980) also suggests that cores of active folds are a likely loci for primary shear rupture. Such folds are cored with earthquake faults associated with a mainshock and a widely distributed pattern of aftershocks. These aftershocks are diffuse rather than aligned with a major fault plane, implying that stress concentration throughout the core of the fold is large enough to cause the primary shear rupture of intact rock. Furthermore, detailed aftershock studies suggest that the orientation of the ambient stress field can change due to slip during major earthquake faulting (Michael, 1987; Hauksson and Jones, 1988). Another common example of primary shear rupture is associated with the stress concentrations developed during deep mining (McGarr et al., 1979). The restricted distribution of primary shear fractures graphically illustrates the point that regional σ_d is not large enough to favor shear rupture. Primary shear fractures occur only when regional tectonic stress is concentrated by local structures, primarily faults and folds.

Specific details concerning the magnitude of σ_d during primary shear rupture of rock (i.e., brittle rock strength) come from laboratory rock mechanics. From laboratory data we know that rock strength varies as a function of confining pressure, pore pressure, strain rate, and temperature. These parameters define the limits to σ_d in the shear-rupture regime.

Stress within the Shear-Rupture Regime

MACROSCOPIC FAILURE CONDITIONS

The behavior of rock in the shear-rupture regime is studied in laboratory experiments using polyaxial rock deformation machines and cylindrical samples (fig. 3–1). One purpose of rock deformation experiments is to map rock strength as a function of the parameters mentioned above. A detailed review of experimental rock deformation techniques is given by Tullis and Tullis (1986). Briefly, a rock-strength experiment starts with the application of confining pressure which, under ideal circumstances, subjects the cylindrical sample to a state of lithostatic stress. The sample is then subjected to a σ_d with the initiation of axial compression. On initial loading, rock will compress in response to microcrack closure and then behave as a linear elastic material. After a certain amount of elastic strain, rock will deform in some other manner (fig. 3–2). At about half the fracture strength, the rock will deviate slightly from elastic behavior and dilate with the onset of new microcracking (Brace et al., 1966). The point of significant nonrecoverable deformation by a localization of microcracking (e.g., Scholz, 1968) is called the yield point with two common modes of behavior after yield being a partial to complete loss in strength (i.e., brittle fracture) and a maintenance of strength during large strains (i.e., ductile flow). Brittle fracture is common when a sample is subjected to temperature and confining pressures found in the schizosphere. Shear rupture is the dominant mode of macroscopic brittle fracture in polyaxial compression tests at all but the lowest confining pressures where some samples will fail by axial splitting (Griggs, 1936; Paterson, 1958).

Rock strength increases directly with confining pressure (Handin and Hager, 1957) which means that the range of σ_d is likely to increase with depth in the schizosphere. Between depths of 4 and 8 km (P_p = hydrostatic), the strength of crystalline rock increases more than 200 MPa (Ohnaka, 1973). These laboratory data set an upper limit to the σ_d achieved under any circumstances in the crust of the earth. Assuming hydrostatic P_p, this limit could be 900 to 1000 MPa at a depth of 15 km. Changes in temperature and strain rate have much smaller effects on the brittle strength of rock in the upper 10 km of crust (Paterson, 1978). However, strength of rocks is strongly dependent on lithology as indicated by the comparison of data on crystalline rocks and carbonates (fig. 3–3). As temperature increases with depth, thermally activated processes start to influence rock strength to the point that ductile deformation moderates σ_d at depths starting in the range of 10–15 km in the crust (see chap. 4).

To map the limit to σ_d in the lithosphere we need failure criteria which are equations for rock strength at brittle failure expressed as the relation between the magnitude of shear stress (τ) and normal stress (σ_n) on a fracture plane or axial stress (σ_1) versus confining pressure ($\sigma_2 = \sigma_3$) (Adams and Bancroft,

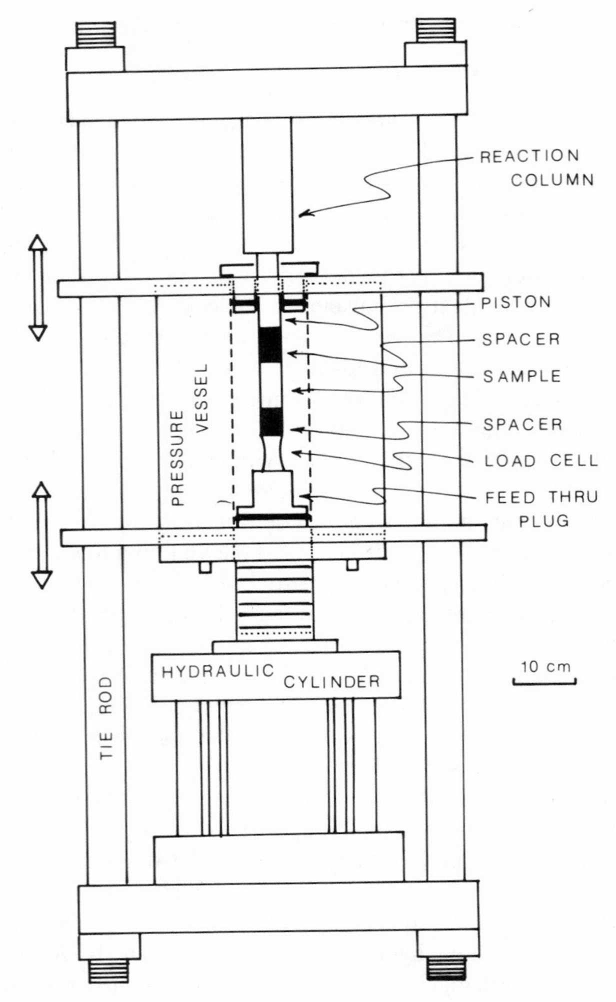

Fig. 3–1. A cross section of The Pennsylvania State University's large sample triaxial rock deformation apparatus. The inside bore of the pressure vessel is 14 cm. The unconventional aspect of this apparatus is that the pressure vessel and feed-thru plug are tied to the hydraulic cylinder. During a compression test the pressure vessel and feed-thru plug move upward relative to a fixed upper piston and reaction column. This configuration permits large samples to be assembled on the feed-thru plug without the necessity of moving the entire assembly (> 50 kg) from a lab bench. The pressure vessel is raised and lowered over the sample assembly by hydraulic cylinders not shown in this cross section.

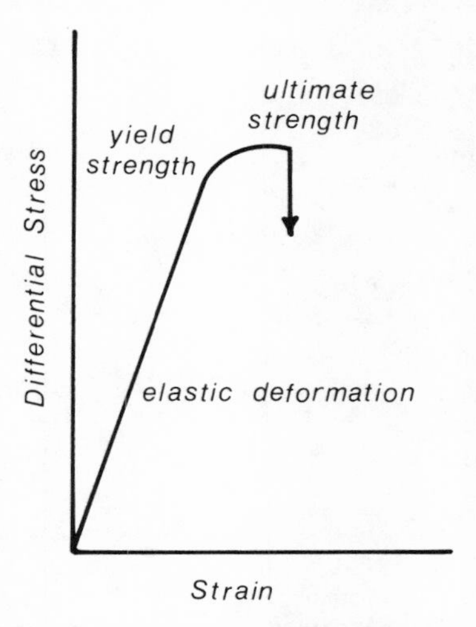

Fig. 3–2. A typical σ_d-axial strain curve for a brittle rock-strength experiment. Brittle failure is represented by the instantaneous loss of strength (downward arrow).

1917; Griggs and Handin, 1960). The graph of a failure criterion shows a line marking the boundary between stability at lower σ_d and shear rupture at higher σ_d where the failure criterion is

$$\sigma_1 = f(\sigma_2, \sigma_3). \tag{3–1}$$

Such a failure criterion describes the macroscopic shear strength of the rock without regard to the microscopic deformation mechanisms. Although crack propagation takes place during shear rupture, failure in the shear-rupture regime is characterized by some mode II or III crack propagation (e.g., Cox and Scholz, 1988), whereas failure in the crack-propagation regime is mainly by mode I crack propagation.

The effect of burial on rock strength in the schizosphere is expressed by a number of failure criteria. The simplest of these criteria was developed during the late eighteenth century when a French naturalist, Charles Augustin de Coulomb, observed that shear stress, τ, necessary to cause brittle failure is resisted by the cohesion of the material, τ_o, and by a constant, μ^*, the coefficient of internal friction, multiplied by the normal stress, σ_n, across the failure plane so that

$$\tau = \frac{\sigma_d}{2} = \tau_0 + \mu^* \sigma_n. \tag{3–2}$$

where μ^* is an empirically derived expression of rock strength and not to be mistaken for μ, the coefficient of sliding friction, which is discussed later in

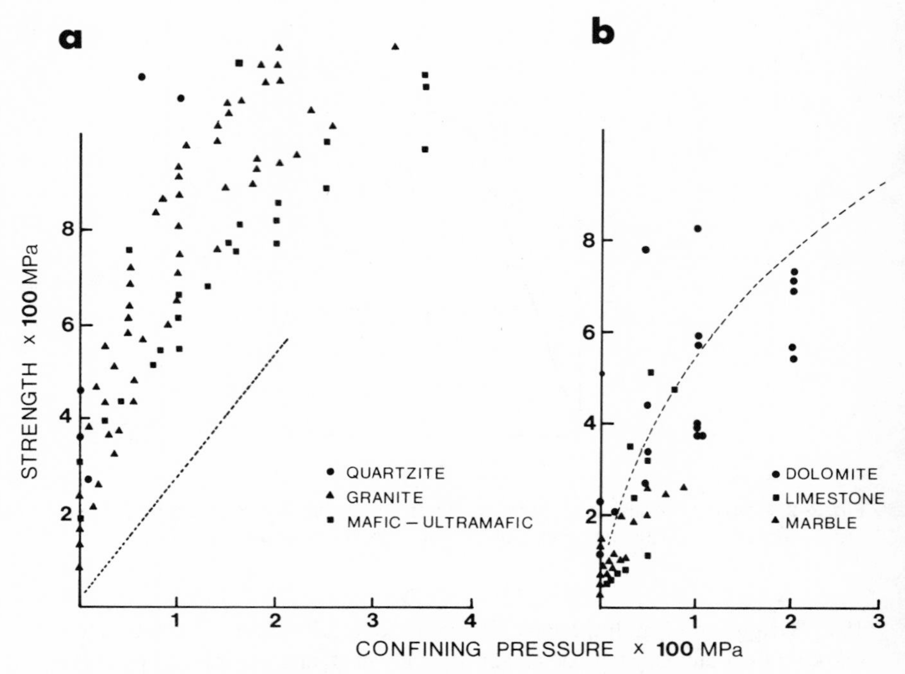

Fig. 3–3. (a) Strength (σ_d) at shear failure for a wide range of crystalline rocks. (b) Strength (σ_d) at shear failure for a wide range of carbonates. The dashed lines, which indicate the lower limit to the strength of rocks in the adjacent plot, show that crystalline rocks are stronger than carbonates. Assuming hydrostatic P_p, 100 MPa (confining pressure) equals a depth of approximately 8 km. Data compiled by Ohnaka (1973).

this chapter. Because the plane of failure is subject to a significant shear stress, the rupture plane is called a shear fracture. This criterion is also known as the Coulomb-Navier criterion (Anderson, 1951).

Some experiments on rock at high confining pressure confirm the general prediction of the Coulomb-Navier criterion that rock strength is modeled by a linear relation between τ and σ_n. A linear behavior is illustrated by the Oil Creek Sandstone (a massive, very fine-grained, well-sorted, well-cemented Ordovician sandstone from Grayson County, Texas) which increases in strength by 200 MPa for every 4 km of burial assuming P_p = hydrostatic (Handin and Hager, 1957). σ_d must exceed 600 MPa for Oil Creek Sandstone to rupture in shear at 8 km in the crust (P_p = hydrostatic). With a strength equivalent to crystalline rocks (fig. 3–3) Oil Creek Sandstone is atypically strong for a sedimentary rock.

In 1900 Mohr introduced a general form of equation 3–2. Mohr's (1900) failure criterion states that the normal stress, σ_n, and shear stress, τ, are related by a function

$$\tau = f(\sigma_n). \qquad (3\text{–}3)$$

which is not necessarily linear. To graphically depict the strength of materials, Mohr (1900) introduced the Mohr circle diagram, an extremely valuable tool for illustrating state of stress in the earth (fig. 3–4). State of stress in two dimensions is represented by a Mohr circle, plotted on Cartesian axes: σ_n (x-axis) versus τ (y-axis) with the diameter of the Mohr circle representing σ_d. The coordinates of each point on the circle represent the values of the σ_n and τ acting across a plane of an orientation specified by θ[2], which is measured between the pole of that plane and σ_1.

It is convenient to have expressions for σ_n and τ in terms of the principal stresses, σ_1 and σ_3. The following equations emerge from a force-balance analysis (e.g., Jaeger and Cook, 1969; section 2.3)

$$\sigma_n = \tfrac{1}{2}(\sigma_1 + \sigma_3) + \tfrac{1}{2}(\sigma_1 - \sigma_3)\cos 2\theta \qquad (3\text{–}4)$$

and

$$\tau = \tfrac{1}{2}(\sigma_1 - \sigma_3)\sin 2\theta. \qquad (3\text{–}5)$$

The Mohr circle is the locus of all points that satisfy equations 3–4 and 3–5 (fig. 3–4). The Mohr failure envelope for a particular rock is obtained by plotting the tangent line touching several Mohr circles, each corresponding to one experiment to determine σ_d at failure under a particular confining pressure (i.e., σ_3) (fig. 3–4a). Once the Mohr failure envelope is constructed, the envelope can be used to constrain the size of the Mohr circle and, hence, predict a limit to the magnitude of σ_d at any confining pressure (fig. 3–4b). Figure 3–5 is a Mohr diagram showing a curved failure envelope for crystalline rocks using data taken from figure 3–3. Assuming that granite is the strongest rock in the lithosphere, this Mohr diagram shows that σ_d is limited to about 470 MPa at an effective confining pressure of 50 MPa. If P_p is hydrostatic this effective confining pressure is found at a depth of about 4 km.

Data from laboratory compression experiments using rock cylinders where $\sigma_2 = \sigma_3$ are most conveniently represented by two-dimensional Mohr circle diagrams. Such two-dimensional stress analyses are generally appropriate be-

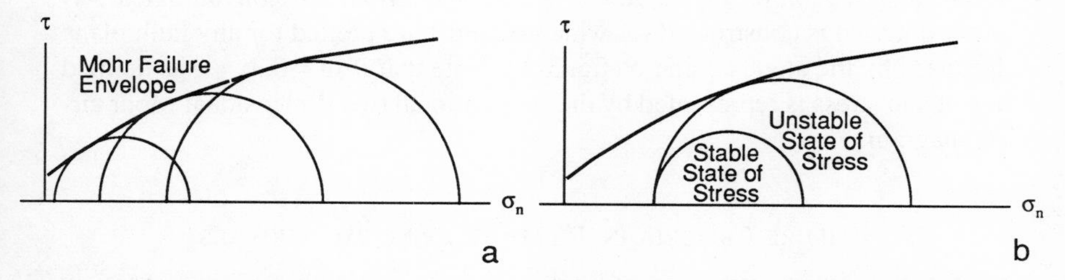

Fig. 3–4. Mohr circle diagrams. (a) The compilation of three Mohr circles to determine the Mohr Failure Envelope. (b) Mohr circles representing stable and unstable states of stress within an intact rock. σ_d is represented by the diameter of the Mohr circles.

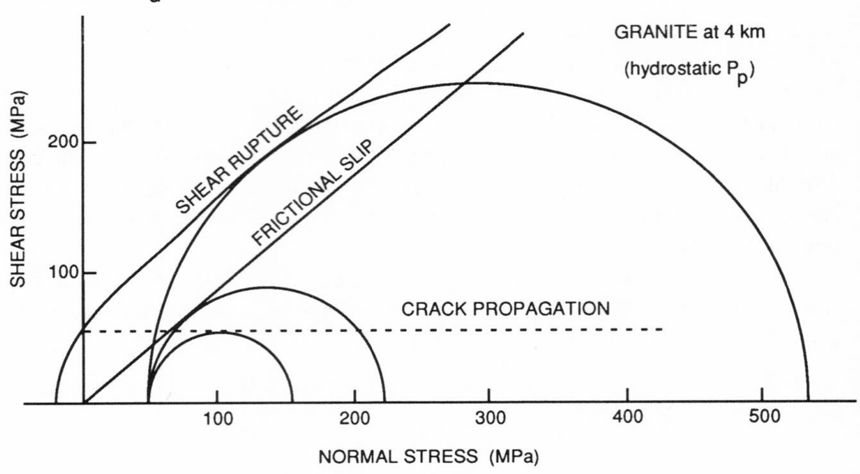

Fig. 3–5. Limits to σ_d in three stress regimes of the lithosphere at an effective confining pressure of 50 MPa. Plots include: (1) the Mohr Failure Envelope for data on crystalline rocks based on an average of Ohnaka's (1973) compilation (the shear-rupture regime); (2) a failure curve for frictional slip based on Byerlee's (1978) compilation (the frictional-slip regime); and (3) a curve showing the limit of stress which would favor crack propagation (the crack-propagation regime). σ_d is shown as the diameter of the Mohr circles on a Mohr diagram of τ versus σ_n. A Mohr circle for each stress regime is drawn assuming thrust faulting where $S_v = \sigma_3$ and assuming P_p is hydrostatic. For this case each circle represents the limit to σ_d at a depth of approximately 4 km.

cause the magnitude of σ_2 between σ_1 and σ_3 is a second-order effect on rock strength (i.e., Handin et al., 1967). However, because the condition where $\sigma_2 = \sigma_3$ is rare in the earth, a Mohr's representation of stress in three dimensions is more appropriate for some analyses (Jaeger and Cook, 1969; section 2.6). The stress state for three dimensions consists of three Mohr's circles as shown in figure 3–6 for arbitrary principal stresses. We can specify the normal, **n**, to a fault plane using three direction cosines: cos ϕ = l, cos ω = m, and cos χ = n where ϕ is $\sigma_1 \angle$ **n**, ω is $\sigma_2 \angle$ **n**, and χ is $\sigma_3 \angle$ **n**. With this convention, a 3-D Mohr diagram is constructed for which σ_n and τ are plotted for any fault plane specified by the angles χ and ϕ (fig. 3–6). Note that if m = 0, $\phi = \theta$ as defined above and stress is represented by the conventional two-dimensional Mohr circle diagram.

Failure Criteria in Terms of Principal Stresses

During the analysis of stress regimes in the lithosphere it is more convenient to write failure criteria in terms of principal stresses. Substituting equations 3–4 and 3–5 into 3–2, the Coulomb-Navier failure criterion becomes

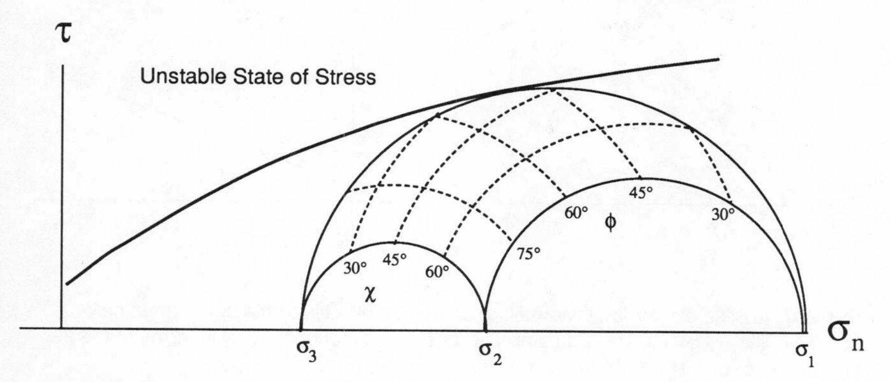

Fig. 3–6. Representation of stress in three dimensions using Mohr's circles.

$$\tau - \mu^*\sigma_n = \tfrac{1}{2}(\sigma_1 - \sigma_3)[\sin 2\theta - \mu^*\cos 2\theta] - \tfrac{1}{2}\mu^*(\sigma_1 + \sigma_3). \qquad (3\text{–}6)$$

Jaeger and Cook (1969, section 4.6) simplify equation 3–6 using the hypothesis that shear rupture follows the plane along which the ratio of τ to σ_n is a maximum. A qualitative justification for this hypothesis is that, prior to shear rupture, σ_n acts to inhibit shear rupture whereas τ acts to drive shear rupture. The fault plane, on which τ/σ_n is maximum, is found by taking the first derivative of equation 3–6 with respect to θ and solving for the critical point. At the maximum

$$\tan 2\theta = -\frac{1}{\mu^*} \qquad (3\text{–}7)$$

so 2θ lies between 90° and 180°. Multiplying the right side of equation 3–7 by $\{(\mu^{*2} + 1)^{-1/2}\}/\{(\mu^{*2} + 1)^{-1/2}\}$ gives the result that

$$\sin 2\theta = \left(\mu^{*2} + 1\right)^{-\frac{1}{2}} \qquad (3\text{–}8)$$

and

$$\cos 2\theta = -\mu^*\left(\mu^{*2} + 1\right)^{-\frac{1}{2}} \qquad (3\text{–}9)$$

Substituting equations 3–8 and 3–9 into equation 3–6 for the Coulomb-Navier failure criterion we derive the equation for the maximum value of $\tau - \mu^*\sigma_n$ in terms of the principal stresses

$$\tau - \mu^*\sigma_n = \tfrac{1}{2}(\sigma_1 - \sigma_3)[\mu^{*2} + 1]^{\frac{1}{2}} - \tfrac{1}{2}\mu^*(\sigma_1 + \sigma_3). \qquad (3\text{–}10)$$

Rearranging equation 3–10, shear rupture occurs when

$$2\tau_o = \sigma_1[(\mu^{*2} + 1)^{\frac{1}{2}} - \mu^*] - \sigma_3[(\mu^{*2} + 1)^{\frac{1}{2}} + \mu^*]. \qquad (3\text{–}11)$$

This equation is later used to derive the failure line for frictional sliding in the schizosphere. Figure 3–7a is a plot of the Coulomb-Navier failure criterion in

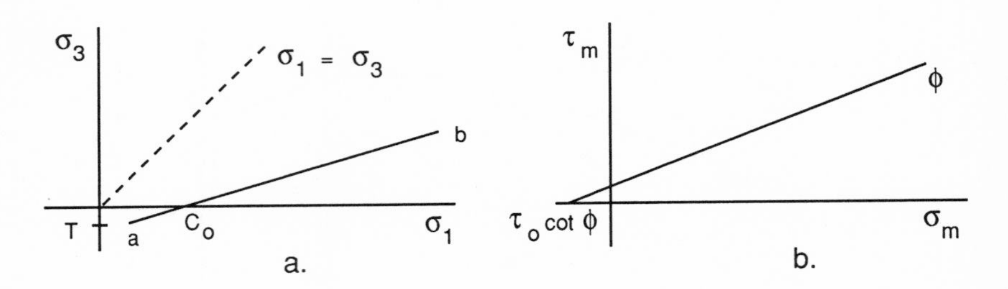

Fig. 3–7. (a) The Coulomb-Navier failure criterion plotted in terms of the principal stresses (line a - b). (b) The Coulomb-Navier failure criterion plotted in terms of the average stress (σ_m) and the maximum shear stress (τ_m).

the σ_1 - σ_3 plane. If the uniaxial compressive strength, C_o, is defined as the point where the Coulomb-Navier failure criterion crosses the abscissa ($\sigma_3 = 0$), then

$$C_o = 2\tau_o\,[(\mu^{*2} + 1\,)^{1/2} - \mu^*]^{-1}. \tag{3–12}$$

We may also plot the Coulomb-Navier failure criterion in terms of average stress σ_m and maximum shear stress τ_m where

$$\sigma_m = 1/2(\sigma_1 + \sigma_3) \text{ and } \tau_m = 1/2(\sigma_1 - \sigma_3) \tag{3–13}$$

Note that τ_m is the radius of the Mohr circle, σ_m is the center of the Mohr circle, and σ_d is the diameter of the Mohr circle. Substituting into equation 3–10, the Coulomb-Navier failure criterion becomes

$$(\mu^{*2} + 1)^{1/2}\tau_m - \mu^*\sigma_m = \tau_o \tag{3–14}$$

By setting $\mu^* = \tan\phi$ where ϕ is the angle of internal friction, Jaeger and Cook (1969) derive the equation,

$$\tau_m = \sigma_m \sin\phi + \tau_o\cos\phi, \tag{3–15}$$

which is a line in τ_m - σ_m space with inclination $\tan^{-1}(\sin\phi)$ and intercept $\tau_o\cos\phi$ on the τ_m-axis (fig. 3–7b). This result is the Coulomb-Navier criterion represented by a straight line of slope $\mu^* = \tan\phi$ and intercept τ_o on the $|\tau|$ axis.

Many suites of fracture experiments show that a linear Coulomb-Navier failure criterion is an oversimplification (Murrell, 1971). A nonlinear shape to the Mohr failure envelope is illustrated by the suite of data on the strength of granite (fig. 3–3). Two nonlinear forms of the failure criterion are

$$\tau^2 = h_0 + h_1\sigma_n \tag{3–16}$$

and

$$\tau = h_0 + h_1\sigma^a \tag{3–17}$$

where h_0, h_1, and a are material constants. Mogi (1972) proposes similar forms where the equation for Coulomb-Navier failure criterion is written in terms of principal stresses:

$$\sigma_1 = h_0 + h_1\sigma_3^a \tag{3–18}$$

$$\sigma_1 = h_0 + \sigma_3 + h_1\sigma_3^a \tag{3–19}$$

$$\sigma_1 = \sigma_3 + h_0 (\sigma_1 + \sigma_3)^a. \tag{3–20}$$

The constants for these failure criteria are determined in the laboratory. With constants for appropriate rock types we are able to place an upper limit on the strength of the schizosphere and, hence, give an indication of the theoretical limit to the magnitude of σ_d in the earth. That limit is on the order of 1000 MPa to 1500 MPa.

Microscopic Mechanisms

Other failure criterion are based on microscopic deformation mechanisms. The most successful is the Griffith (1924) criterion based on the surface energy to generate new fracture surfaces (eq. 2–11). Griffith assumed that failure would initiate from the longest, most critically oriented crack in the rock. McClintock and Walsh (1962) write a comprehensive failure criterion for both shear-rupture and crack-propagation in terms of σ_1 and σ_3 as

$$(\sigma_1 - \sigma_3)^2 - 8T_0 (\sigma_1 + \sigma_3) = 0 \text{ if } \sigma_1 > -3\sigma_3 \tag{3–21}$$

$$\sigma_3 = -T_0 \text{ if } \sigma_1 < -3\sigma_3 \tag{3–22}$$

Note that equation 3–22 is the same as equation 2–11. For the first time we see that σ_d in the crack-propagation regime is quite restricted compared with σ_d in the shear-rupture regime. Equation 3–22 indicates that the crack-propagation regime is limited to conditions where $\sigma_d < 3T_0$. Otherwise, the rock will fail by shear rupture long before a tensile effective stress develops.

Factors Controlling Rock Strength

Failure by shear rupture or crack propagation takes place at the extreme boundaries of the shear-rupture and crack-propagation regimes. If the principal stresses are compressive and σ_d is low, the rock may be stable. There are, however, several phenomena that may lead to shear rupture without necessarily causing a change in σ_d.

First, pore pressure has a strong influence on rock strength. Equation 2–25 indicates that crack propagation takes place when pore pressure within the rock exceeds a combination of the tensile strength and applied stress. Pore pressure also strongly influences the shear strength of rock as indicated by rewriting the Coulomb-Navier criterion to incorporate effective stress (eq. 1–38)

$$\tau = \frac{(\sigma_1 - P_p) - (\sigma_3 - P_p)}{2} = \tau_0 + \mu^*(\sigma_n - P_p). \qquad (3\text{–}23)$$

The Mohr diagram is useful for comparing stress conditions favoring either crack propagation or shear rupture. An increase in pore pressure causes a decrease in effective normal stress without affecting σ_d (Handin et al., 1963). Any increase in P_p is reflected on the Mohr diagram with the Mohr circle shifting to the left by the value of ΔP_p without changing its diameter. The limit to shear stress which favors crack propagation is the dashed line in figure 3–5. This line is drawn to indicate that once σ_d is greater than three times the tensile strength the rock will fail by shear rupture under increasing P_p before effective tensile stresses develop. When $\sigma_3 = S_v$, crack propagation will take place only if $P_p > T_o + \sigma_3$. This condition is met only when P_p increases well above hydrostatic pressures. From the law of effective stress, we conclude that a sedimentary rock with hydrostatic pore pressure at a depth of 9 km has about the same strength as the same rock buried to a depth of 4 km under dry conditions. Provided σ_d is relatively high, an increase in P_p will lead to shear rupture in an otherwise stable rock, and, so, P_p is a very important parameter for limiting the magnitude of σ_d in the schizosphere.

During triaxial compression tests, samples of rock start to exhibit significant inelastic deformation at σ_d of about half the fracture strength of the rock. This inelastic deformation is a manifestation of the propagation of many microcracks within the sample. The increased number of microcracks causes an increased porosity, a phenomenon called *dilatancy* (Brace et al., 1966). Upon dilatancy, pore fluid pressures will drop, thereby making the effective stress within the sample much higher, and hence, the sample will become relatively stronger, a phenomenon called dilatancy hardening (Brace and Martin, 1968). Dilatancy hardening can, at least momentarily, stabilize a rock so that a much higher σ_d is required to cause shear rupture.

Water may act as a chemical agent to weaken the rock through stress corrosion cracking (e.g., Dunning et al., 1984) or through enhanced plasticity by hydrolytic weakening (e.g., Griggs and Blacic, 1965). These two processes will limit σ_d in the lithosphere. Laboratory experiments suggest that scale affects rock strength with larger samples being weaker (e.g., Pratt et al., 1972). The idea here is that the strength of the sample is controlled by the largest flaw. Although the fracture toughness, K_{Ic}, of the sample is scale-independent, longer flaws cause failure at lower effective tensile stresses (eq. 2–18).

In summary, laboratory data shows that intact rock is relatively strong. As a consequence, the stress field covered by the shear-rupture regime is large. In contrast, crack propagation takes place only if $\sigma_d < 3T_0$, so the stress field covered by the crack propagation regime is limited to a small portion of the stress field in the shear-rupture regime (fig. 3–5).

Stress within the Frictional-Slip Regime

The earth's crust is pervaded by discontinuities, including joints, shear fractures, and their reactivated counterparts, faults. Although the generation of shear fractures by rupture is a local phenomenon occurring near local faults and folds, reactivation of planar discontinuities is common. Some of these discontinuities have evolved in size to the point where they are now faults bounding lithospheric plates. Because of the pervasive nature of these discontinuities, a buildup in earth stress is generally relieved by frictional slip long before σ_d becomes high enough to shear intact rocks. It is stress relief by frictional slip that makes shear rupture of intact rock such a local phenomenon. In-situ stress measurements confirm that σ_d in the upper crust rarely exceeds the frictional strength of rocks (Zoback and Healy, 1984). Thus the frictional properties of rocks are an extremely important governor for σ_d in the lithosphere.

Information on the frictional strength of rocks comes from laboratory tests using a variety of configurations shown in figure 3–8. Force-displacement curves recorded during friction tests reflect two general types of slip: stable sliding and stick-slip (fig. 3–9). If displacement is measured at some distance from the sliding surface, a common practice in polyaxial friction tests, elastic deformation of the sample prior to slip is recorded as part of the displacement. An ideal curve for *stable sliding* shows force increasing until the rock slips at the frictional yield point, but once slip is initiated it will continue without interruption. After the initiation of stable sliding, some force-displacement curves show a gradual increase in force caused by slip hardening. Slip hardening, which arises from damage to the sliding surfaces during slip, will continue at a diminishing rate. During *stick-slip sliding,* force builds and then suddenly drops with the initiation of slip. After the force drop, slip stops. This force-displacement cycle may repeat many times throughout an experiment with each cycle characterized by slip accompanying a major drop in frictional resistance. Such behavior is an analogue for earthquakes along fault zones, large and small (Brace and Byerlee, 1970). Regardless of whether slip is stable or by stick slip, σ_d in the frictional-slip regime is governed by the frictional strength defined at the onset of slip.

The Friction Line

Laboratory studies provide a phenomenological framework in which to understand the physical parameters controlling slip along fault zones and, ultimately, to predict the strength of a fractured schizosphere. The basic parameter mea-

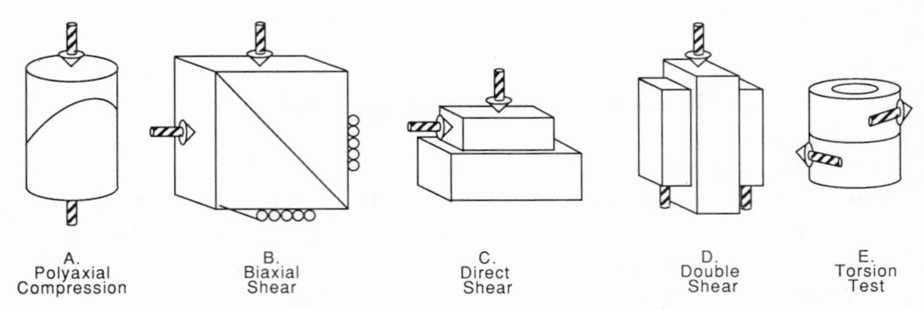

Fig. 3–8. Five different sample configurations for evaluating the frictional properties of rock. Sample A represents a cylinder cut for polyaxial testing (Byerlee and Brace, 1968). B is the configuration for biaxial tests where the samples are rectangular solids (Scholz et al., 1972). Loading is not orthogonal to the sliding surface for configurations A and B; therefore, normal stress across the sliding surface increases in proportion to the axial load. C is a conventional biaxial shear test without confining pressure but with a constant normal stress indicated by the vertical arrow (Barton, 1976). D is a double shear test with one plate sliding between two other plates (Dieterich, 1972). E is a torsion test (Tullis and Weeks, 1986). σ_n and τ are measured directly in tests C and D, whereas they are calculated from the principal stresses (eqs. 3–4 and 3–5) in the tests of configurations A and B. For the torsion test τ must be integrated over the surface of the cylinder. The advantage of the torsion test over the other four is that it allows very large displacements which permit the evolution of steady state conditions along the sliding surface.

sured during these studies is the shear stress, τ, necessary to initiate or reactivate slip along a joint or fault within the schizosphere. In this context, τ is controlled by the static friction of rock which for a single experiment is defined by the coefficient of static friction

$$\mu_s = \frac{\tau}{\sigma_n}. \tag{3–24}$$

If τ versus σ_n for several experiments are plotted, the line connecting the data points, a frictional strength line, may extrapolate to the origin. In this case, μ_s is the slope of the line, and friction has the same meaning as for the single experiment. But if the line does not converge to the origin, it will cut the τ axis at the value τ_0 which represents the cohesive strength of the fault zone. Here again μ_s is the slope of the line but

$$\mu_s = \frac{\tau - \tau_0}{\sigma_n}. \tag{3–25}$$

Some startling conclusions emerged from a compilation of static friction data by Byerlee (1978). At very low σ_n ($\approx$2.5 MPa), μ_s can vary widely because of

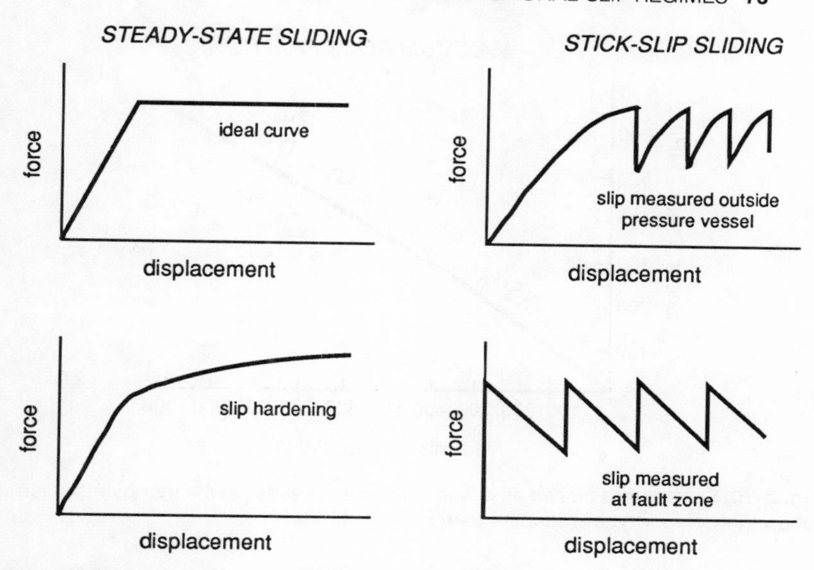

Fig. 3–9. Stress-displacement curves during frictional sliding during polyaxial compression tests.

the strong dependence on surface roughness, a result largely applicable to engineering problems. However, at intermediate to high σ_n, μ_s is less dependent on surface roughness; furthermore, a simple friction law applies to most rocks except those containing large quantities of clay (fig. 3–10). Byerlee's compilation suggests that at σ_n up to 200 MPa, τ required to initiate sliding is

$$\tau = 0.85\sigma_n \tag{3–26}$$

and at σ_n above 200 MPa, the τ required to initiate sliding is

$$\tau = 0.5 + 0.6\sigma_n. \tag{3–27}$$

These equations, first published by Byerlee (1978), became known as the general friction law for rocks within the lithosphere. Frictional strength is affected by such variables as roughness, hardness, temperature, and ductility as reviewed by Scholz (1990). Because the real contact area of sliding surfaces is small, the simple law of effective stress applies to the problem of determining frictional strength in the schizosphere

$$\tau = \tau_0 + \mu_s(\sigma_n - P_p). \tag{3–28}$$

Many studies of in situ stress have shown that general friction laws constrain σ_d in the schizosphere (e.g., Zoback and Healy, 1984; Hickman, 1991). When the envelope for frictional slip is plotted on a Mohr diagram, it is clear that fric-

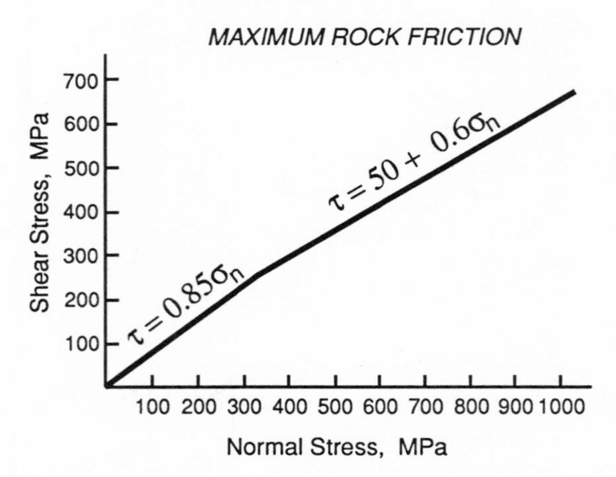

Fig. 3–10. Shear stress plotted as a function of normal stress at the maximum friction for a variety of rock types (adapted from Byerlee, 1978).

tional slip will take place at much lower σ_d than is required for shear rupture (fig. 3–5).

With knowledge of a general law for frictional slip on faults and joints, an upper bound on σ_1 in the thrust-fault regime and a lower bound on σ_3 in the normal-fault regime are defined for a crust pervaded by discontinuities. If we assume that all discontinuities are cohesionless (i.e., $\tau_0 = 0$), such bounds are found using equation 3–11 and replacing μ^* with μ_s, so that

$$\sigma_1\{(\mu_s^2 + 1)^{1/2} - \mu_s\} - \sigma_3\{(\mu_s^2 + 1)^{1/2} + \mu_s\} = 0. \qquad (3\text{–}29)$$

On multiplying by $\{(\mu_s{}^2 + 1)^{1/2} + \mu_s\}$ and rearranging, an equation for the relationship between σ_1 and σ_3 emerges for thrust faulting

$$\sigma_1 = \sigma_3\{(\mu_s^2 + 1)^{1/2} + \mu_s\}^2 \qquad (3\text{–}30)$$

and for normal faulting

$$\sigma_3 = \sigma_1\{(\mu_s^2 + 1)^{1/2} + \mu_s\}^{-2}. \qquad (3\text{–}31)$$

If pore fluids are present at depth, the effective stress law applies and P_p is subtracted from σ_1 and σ_3. These are the friction lines which appear in the literature on many plots of in situ stress versus depth.

Stress measurements confirm that the general law of friction as determined in the laboratory applies in the schizosphere. This point is illustrated using data from Yucca Mountain, Nevada, where hydraulic-fracture stress measurements show a normal faulting regime with $\sigma_1 = S_v$ and $\sigma_3 = S_h$ (Stock et al., 1985). At Yucca Mountain the rock is dry down to about 590 m in two wells, G-1 and G-2, below which the effective stress law applies. Assuming that $\mu = 0.6$, equation 3–31 may be used to construct a friction line for normal faulting (fig. 3–11). The friction line identifies the values of σ_3 below which slip will occur on favorably oriented fractures. Because data points for S_h at Yucca Mountain fall along the friction line, contemporaneous normal faulting is, indeed, likely in the vicinity of Yucca Mountain. σ_d has reached its limit in a rock that is pervaded by fractures because fault reactivation will prevent σ_d from growing larger. This is consistent with the interpretation that the Basin and Range, the site of Yucca Mountain, is an active tectonic province with σ_d at the level necessary for continued faulting and earthquake activity.

Limits to σ_d in the Schizosphere

Boundaries of the stress regimes in the lithosphere are defined in terms of the maximum σ_d which is permitted by the three general classes of brittle failure: crack propagation, shear rupture, and frictional slip. With an understanding of the nature of brittle failure, one can illustrate the limits within which σ_d must lie in the schizosphere. For example, the Mohr diagram gives an indication of the relative limits of σ_d, the diameter of the Mohr circle, for each of the three brittle-stress regimes (fig. 3–5). Each of the three stress regimes is defined by a set of some, but not all, Mohr circles that fit below the shear rupture, frictional slip,

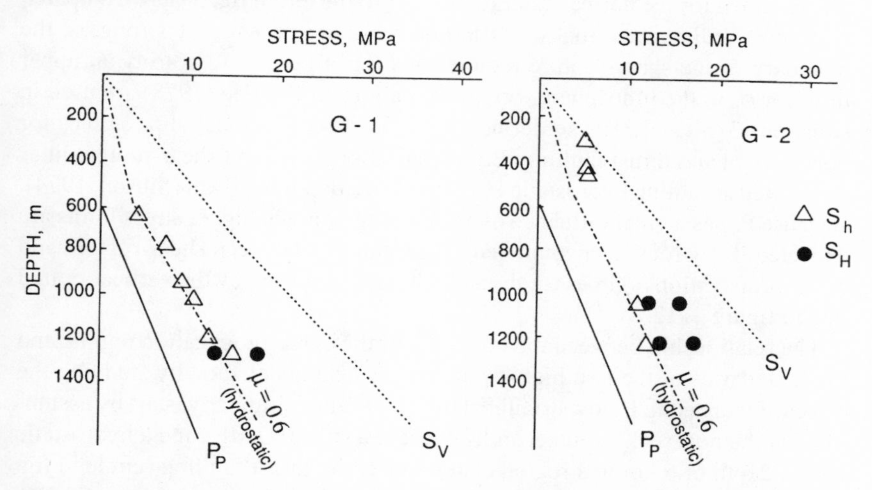

Fig. 3–11. Stress measurements at Yucca Mountain, Nevada (adapted from Stock et al., 1985). Stress (S_h) data falling to the left of the dashed lines signify stress states for which slip is expected to occur on preexisting favorably oriented normal faults, for values of $\mu_s = 0.6$. Different assumptions lead to two estimates of S_H.

and crack propagation lines, respectively. The set of σ_d falling within the frictional-slip regime overlaps with the set of σ_d falling within the shear-rupture regime. Note that a significant increase in P_p can cause a certain set of σ_d to move out of the frictional-slip regime while staying within the limits of the shear-rupture regime. This will occur only if there are no favorably oriented fractures for frictional slip. In contrast, the set of σ_d falling within the crack-propagation regime does not overlap with the set of σ_d falling within the shear-rupture regime. The dashed line in figure 3–5 is a lower boundary for the shear-rupture regime. Furthermore, the lower left end of the Mohr envelope in figure 3–5 is a second boundary for the crack-propagation regime which does not extend very far into negative effective stress space (i.e., the negative direction on the normal stress axis of the Mohr diagram).

Another means of illustrating the limits of the three brittle-stress regimes is through a plot of σ_d versus vertical effective stress (i.e., depth) for both the thrust-faulting and normal-faulting regimes. The following is Brace's and Kohlstedt's (1980) approach which assumes that the effective stress principle operates for both shear rupture and frictional slip. Using Ohnaka's data (1973) for the shear strength of crystalline rock (fig. 3–3) a dashed line is drawn to represent the limit of σ_d in the shear-rupture regime for thrust faulting (σ_d to the right in figure 3–12). Theoretically, crystalline rocks are strong enough (a uniaxial compressive strength of 200 MPa) to support a free-standing column of rock greater than 5 km in height. So, σ_d in the normal-fault regime does not reach the limit for shear rupture at least down to 5 km and, consequently, no line is drawn for the normal-fault regime (σ_d to the left in fig. 3–12). Compared with most sedimentary rocks, crystalline rocks are excessively strong, so the boundary of the shear-rupture regime shown in figure 3–12 is truly an upper limit for σ_d in the lithosphere. Frictional data from Byerlee (1978) are used in equations 3–30 and 3–31 to define the limits of the frictional-slip regimes for both normal and thrust faulting. Both the frictional-slip and shear-rupture lines are plotted assuming hydrostatic P_p is found to a depth of 10 km (Sibson, 1974). Because P_p has a great influence on rock strength, much higher stress limits are expected if dry rock is encountered. The boundary between shear-rupture and crack-propagation regimes for thrust faulting is indicated by the vertical-dotted line in figure 3–12.

The relationship between the boundary of the crack-propagation regime and the boundaries of the two high-σ_d regimes is also understood by studying the effect of increasing P_p toward lithostatic values. For simplicity, start by assuming that the host rock[3] is intact and subject to a $\sigma_d = 100$ MPa and a hydrostatic P_p at a depth of 6 km in a region characterized by thrust faulting (circle in fig. 3–12). An increase in P_p is seen as a decrease in effective stress (solid line leaving circle in a vertical direction in fig. 12). Although this will seem confusing because depth and effective stress are plotted on the same axis, the decrease in effective stress accompanying an increase in P_p does not mean that there has

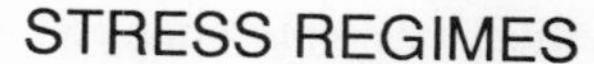

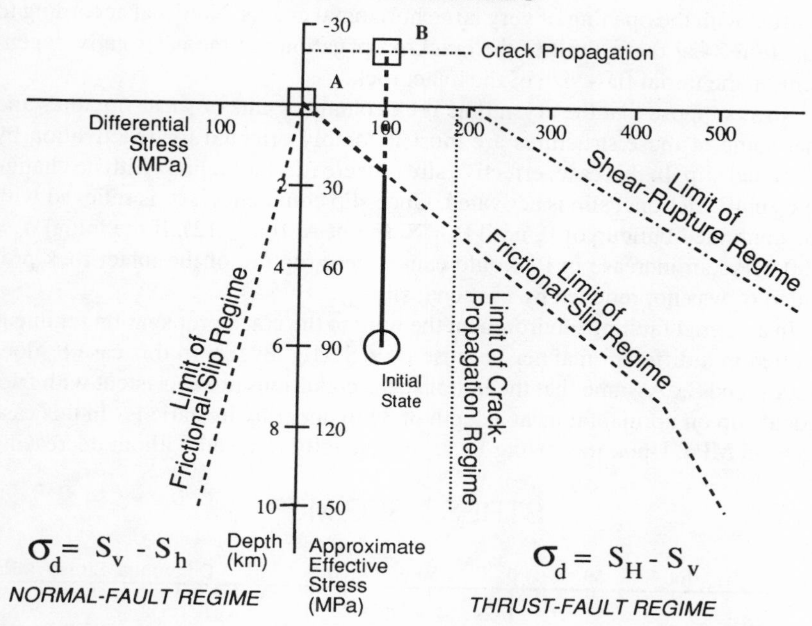

Fig. 3–12. The limits of differential stress (σ_d) in the upper crust with an example from the thrust-fault regime. σ_d in the shear-rupture regime is limited by the fracture strength of an "average" crystalline rock. The dashed line representing this limit is taken from Ohnaka's (1973) compilation. σ_d in the frictional-slip regime is limited by the frictional strength of non-clay-bearing rocks. The dashed line representing this limit is taken from Byerlee's (1978) compilation. σ_d in the crack-propagation regime is limited approximately by the uniaxial compressive strength of "average" crystalline rock (dotted vertical line). At σ_d much less than this value, crystalline rock will not fail in compression even when $\bar{\sigma} < 0$. There are limits to negative effective stress in the crack-propagation regime. The horizontal dashed line shows this limit based on the assumption that fracture toughness of crystalline rock is about 2 MPa · m$^{1/2}$ and the intact rock has flaws on the order of several mm in diameter. The effect of increasing P_p is shown where the initial σ_d is identified by a circle indicating that $S_H > S_v$. The solid line from the initial stress state indicates no change in σ_d if P_p increases in intact rock until cracks propagate at $\bar{\sigma} < 0$. However, if a crystalline rock with favorably oriented slip surfaces is subject to an increase in P_p, then slip acts to reduce σ_d until high fluid pressures act to open existing cracks.

been a concomitant decrease in depth. A σ_d of 100 MPa is not large enough to fracture a strong crystalline rock even when unsupported (i.e., $\bar{\sigma}_3 = 0$). Assuming the rock has initial flaws on the order of 1 cm, P_p will continue to build until crack propagation commences at a P_p (see eq. 2–27) on the order of 170 MPa or approximately 20 MPa above $\sigma_3 = S_v \approx 150$ MPa (point B, fig. 3–12). This, then, is the limit of the crack-propagation regime in terms of effective stress where $\bar{\sigma}$ is plotted as the vertical axis representing $\bar{S}_v$. At this point overburden is lifted with the opening of very large horizontal cracks. Note that according to equation 2–27 the $\bar{\sigma}$ limit to the crack-propagation regime is strongly dependent on the initial flaw size of the intact rock.

Now, suppose that the crystalline rock contains joints or shear fractures and that some of these structures are most favorably oriented for reactivation by frictional slip. In this case, effective stress decreases (solid line) with no change in σ_d until frictional slip is activated. Once slip commences, σ_d is relieved with the continued buildup of P_p until $P_p = S_v$ (point A; fig. 3–12). If the initial $\sigma_d >$ 200 MPa, an increase in P_p would cause shear rupture of the intact rock provided σ_d was not relieved by frictional slip.

In a normal faulting environment the limit to the crack propagation regime is plotted in a different manner because $\sigma_3 \neq S_v$ (fig. 3–13). In this case P_p does not exceed S_v. Assume that the initial stress conditions are consistent with frictional slip on normal faults at a depth of 3 km under hydrostatic P_p. In this case $\sigma_d \approx 35$ MPa. Upon increasing P_p, σ_d is gradually relieved without decreasing

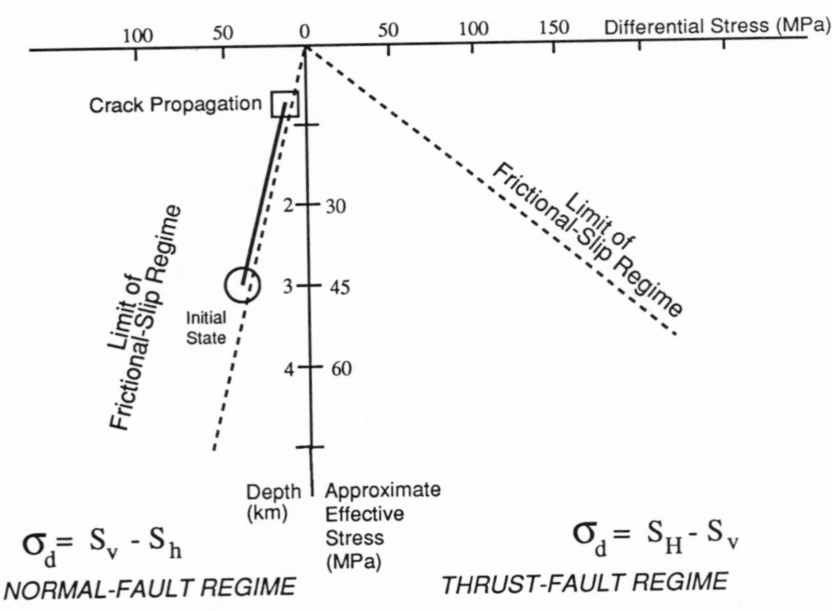

Fig. 3–13. The limits of differential stress (σ_d) in the upper crust with an example from the normal-fault regime. For this plot the horizontal axis is expanded relative to figure 3–12. Here initial σ_d is reduced by slip on favorably oriented normal faults until vertical cracks propagate under a horizontal effective stress which is less than the vertical stress.

depth of burial. Follow a path from the initial conditions to the present stress conditions found at a depth of 3 km in the Gulf of Mexico where $S_h = 0.85\ S_v$ and $\overline{S}_h = \overline{\sigma}_3 = 0$ according to arguments in chapter 2 (fig. 2–7). Under present conditions in the Gulf of Mexico, the total stress, $\sigma_1 = S_v = 75$ MPa, and the total stress, $S_h = 64$ MPa, so $\sigma_d = 11$ MPa, even through $\overline{S}_h = \overline{\sigma}_3 = 0$ (square in fig. 3–13). In a region of normal faulting crack propagation will take place at a low but positive vertical effective stress as shown in figure 3–13. An effective-stress boundary for the crack-propagation regime on a σ_d-effective stress diagram is dependent on whether or not σ_3 is plotted as the vertical axis.

Note that the average maximum shear strength for faults in the continental schizosphere under a typical continental geotherm and strain rate is about 35 MPa and 150 MPa, in normal and thrust fault regimes, respectively (Hickman, 1991). The reason for calling these values a maximum is that many faults in the schizosphere exhibit a ductile behavior and undoubtedly slip at much lower τ (chap. 4).

Regional Stress Patterns from Fault-Slip Data

Tectonically active regions may contain dozens if not hundreds of outcrops with exposed fault surfaces including primary shear fractures plus reactivated joints and bedding planes. It is reasonable to hypothesize that slip on many, if not all, of these faults occurred in response to stress from one or more tectonically significant events. If so, then the orientation of structures on these fault surfaces contains information on the orientation and relative magnitude of the regional tectonic stress accompanying these tectonic events. The slip direction on many of these fault surfaces is indicated by lineations which are structures such as rough and smooth facets, carrot-shaped wear grooves, tension gashes, and fibrous minerals (Petit, 1987). Commonly, the sense of shear is also apparent on these structures. Data available from field measurements on fault zones includes the dip direction (θ), the dip (δ), the rake of the slip lineations (λ), and the sense of slip (fig. 3–14). From these data come two unit vectors which are important in deriving information about the orientation and relative magnitude of the regional stress causing slip on these faults. These unit vectors are the normal to the fault surface, **n**, and the slip lineation vector, **s**. To understand how these unit vectors relate to the regional tectonic stress, we must examine the nature of the stress tensor (**S**).

The Stress Tensor

Although the complete stress tensor (**S**) is characterized by six parameters, there is more than one combination of parameters which can represent the com-

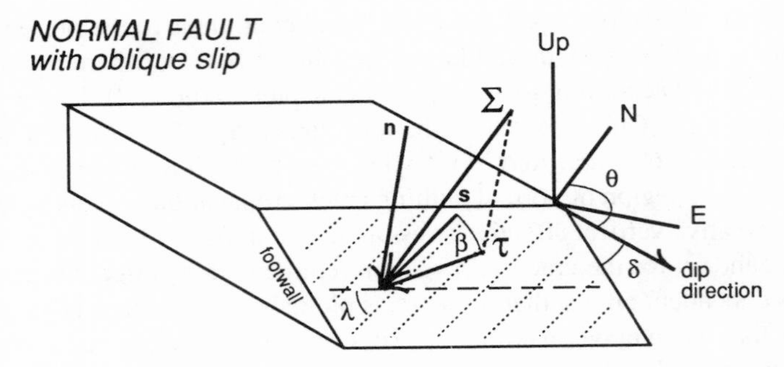

Fig. 3–14. A block diagram of a normal fault with oblique slip. The dip direction, dip, and rake are θ, δ, and λ, respectively. **n** and **s** are the outward normal to the foot wall block and the slip vector within the fault plane, respectively. σ is the stress vector on the fault plane and τ is the shear stress traction on the fault plane. β is the angle between **s** and τ.

plete tensor. Given an arbitrary coordinate system, six stress components are adequate such that

$$\mathbf{S} = \begin{bmatrix} \sigma_{11}\sigma_{12}\sigma_{13} \\ \sigma_{21}\sigma_{22}\sigma_{23} \\ \sigma_{31}\sigma_{32}\sigma_{33} \end{bmatrix} \tag{3–32}$$

where $\sigma_{12} = \sigma_{21}$, $\sigma_{13} = \sigma_{31}$, $\sigma_{23} = \sigma_{32}$. There exists another coordinate system in which the off diagonal components of the stress tensor go to zero and the diagonal components become principal stresses. In this new coordinate system, the coordinate axes are parallel to the direction of the principal stresses. The principal stresses are the eigenvalues of the matrix in equation 3–32 and the orientation of the principal stress is given by the eigenvectors of the matrix so that

$$\mathbf{S} = \begin{bmatrix} \sigma_1 & 0 & 0 \\ 0 & \sigma_2 & 0 \\ 0 & 0 & \sigma_3 \end{bmatrix}. \tag{3–33}$$

In this latter case the three additional parameters which specify the stress tensor are the eigenvectors of the matrix in equation 3–33 which are the directions of the principal stresses. The matrix representing the stress tensor is the sum of two matrices: the deviatoric stress and the mean stress

$$\mathbf{S} = \begin{bmatrix} \sigma_1 - \sigma_m & 0 & 0 \\ 0 & \sigma_2 - \sigma_m & 0 \\ 0 & 0 & \sigma_3 - \sigma_m \end{bmatrix} + \begin{bmatrix} \sigma_m & 0 & 0 \\ 0 & \sigma_m & 0 \\ 0 & 0 & \sigma_m \end{bmatrix}. \tag{3–34}$$

Etchecopar et al. (1981) call the latter part of equation 3–34 the isotropic pressure part of the stress tensor. If the isotropic pressure part of the stress tensor is just a unit tensor, **I**, multiplied by a constant, b, and if the deviatoric part is a normalized deviatoric tensor, $\mathbf{D}_o$, multiplied by a constant, a, then

$$\mathbf{S} = a\mathbf{D}_o + b\mathbf{I}. \tag{3–35}$$

The most common method for normalizing the deviatoric tensor is to set $\sigma_1 - \sigma_3 = 1$. This allows the components of the normalized deviatoric stress tensor to vary from zero to one. Thus, given the orientation of the principal stresses, the stress tensor can be represented by three parameters other than the principal stresses. These are the mean stress (the isotropic pressure), the differential stress, and a stress ratio which is defined as

$$R = \frac{\sigma_2 - \sigma_3}{\sigma_1 - \sigma_3}. \tag{3–36}$$

The Stress Ratio, R

The stress ratio, R, is important in reducing fault-slip data for the following reason. Because fault-slip data contain four parameters (i.e., θ, δ, λ, and sense of slip), there is not enough information to constrain all six parameters in the stress tensor. Because the four fault-slip parameters carry no information on the absolute magnitude of the stress components, we cannot solve for either differential stress or the isotropic pressure component. This leaves four parameters including the orientation of the principal stresses and the stress parameter, R, which describes the ratio of principal stresses. These four parameters are found using one of a number of inversion techniques. Theoretically, four fault planes are required to determine these four parameters. Most inversion techniques use many more than four sets of fault-plane data to determine a minimum misfit solution to this overdetermined system. Before discussing inversion techniques, it is appropriate to explain the characteristics of R.

Many joints, shear fractures, and faults are reactivated to slip within a stress field which was not responsible for their rupture. Thus, for any given joint, fracture or fault, slip may occur in any of the three Andersonian fault-stress states (i.e., $\sigma_1 = S_v$, $\sigma_2 = S_v$, or $\sigma_3 = S_v$) or a stress state with principal axes which are not orthogonal to the earth's surface. For a fault in a particular orientation, each of the three Andersonian fault-stress states is represented by a range of pitches for fault-slip lineation depending on the relative magnitudes of the three principal stresses (fig. 3–15). In figure 3–15, the principal stresses, σ_1 and σ_3, are projected on the fault plane at t_0 and t_1 by drawing a plane through the normal to the fault plane and either σ_1 or σ_3 (Angelier, 1979). Note that the range of pitches permissible for each stress state is bounded by the projection of σ_1 and σ_3 on the fault plane (heavy arrows in fig. 3–15). However, for a fault in any arbitrary orientation, any pitch for lineation from 0° to 90° is possible, depending only on the orientation of the three principal stresses. If the pitch of

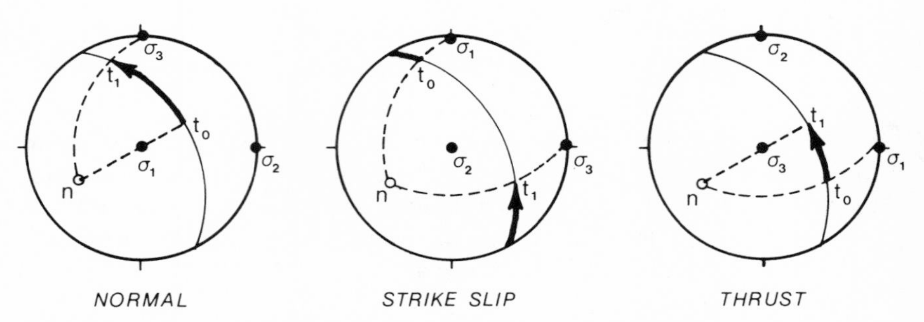

Fig. 3–15. Angular variation of shear stress with R when stress axes σ_1, σ_2, and σ_3 are fixed. n = perpendicular to fault plane F; t_o = projection of σ_1 on F; t_1 = projection of σ_3 on F. When R goes from 0 to 1, the shear stress goes from τ_o to τ_1, traveling the angle δ (arrows), given by $\cos\delta = \tan\alpha_1 \times \tan\alpha_3$, where α_1 and α_3 are the angles of F with σ_1 and σ_3, respectively. Fault types are normal, strike-slip (dextral), and thrust (adapted from Angelier, 1979).

the lineation points near the projection of σ_1 on the fault plane, this means that the magnitude of σ_2 is close to σ_1, in which case R is close to 1. On the other hand, if the pitch of the lineation points near the projection of σ_3 on the fault plane, this means that the magnitude of σ_2 is close to σ_3, in which case R is close to 0. Thus, provided their orientation is known, the relative magnitude of σ_1, σ_2, and σ_3 can be estimated based on where the slip lineation projects on the fault plane relative to the projection of σ_1 and σ_3.

Gephart and Forsyth (1984) derive the stress ratio, R′, using the coordinate system where the pole to the fault plane is x'_1, the normal to the lineations in the fault plane is x'_2, and the direction of the fault lineations is x'_3. The directions of principal stresses are designated as x_1, x_2, and x_3, respectively. If the direction cosine between the fault plane coordinate system x'_i and the orientation of principal stresses x_j is β_{ij}, then

$$\sigma'_{ij} = \beta_{ik}\beta_{jl}\beta_{kl} \tag{3–37}$$

where σ'_{ij} are the stresses on the fault plane and σ_{kl} are the regional stresses. σ'_{11} is the normal traction on the fault plane, σ'_{12} is the shear traction perpendicular to the striae which is zero, and σ'_{13} is the shear traction parallel to the striae. From the assumption that slip is in the direction of resolved shear stress (i.e., $\sigma'_{12} = 0$) which means that

$$\sigma'_{12} = 0 = \sigma_1\beta_{11}\beta_{12} + \sigma_2\beta_{12}\beta_{22} + \sigma_3\beta_{13}\beta_{23} \tag{3–38}$$

and the orthogonality of x'_1 and x'_2 which means that

$$0 = \beta_{11}\beta_{12} + \beta_{12}\beta_{22} + \beta_{13}\beta_{23}, \tag{3–39}$$

Gephart and Forsyth (1984) derive the stress parameter

$$R' = \frac{\sigma_2 - \sigma_1}{\sigma_3 - \sigma_1} = -\frac{\beta_{13}\beta_{23}}{\beta_{12}\beta_{22}}. \qquad (3\text{–}40)$$

This expression is equivalent to one derived by McKenzie (1969a) and is the inverse of R as identified by Etchecopar et al. (1981).

Inversion Techniques

From a large set of fault-slip data, inversion techniques allow the calculation of the orientation of the principal stresses affecting that tectonically active area (Carey and Brunier, 1974; Angelier, 1979). Inversion techniques are either graphical (e.g., Angelier and Mechler, 1977; Aleksandrowski, 1985; Lisle, 1987) or numerical (e.g., Armijo et al., 1982; Gephart, 1988; Huang, 1988). The inversion of fault-slip data for stress determination requires a number of assumptions. First, all the faults within a region are assumed to slip independently (Angelier, 1979; Reches, 1987). Second, faults are assumed to slip in response to the same stress field regardless of the age of the faults. In other words, the last motion on all faults took place synchronously and independently of the time of formation of the faults. For this reason the present orientation of the faults gives no information on the stress field during the last motion on the faults. Third, the shear force applied on a given fault plane causes a slip in the direction and orientation of that shear force. The fourth assumption is that local stress variations cancel so that the mean stress deviator is representative of a regional stress (Angelier, 1979). Zoback (1983) argues that the best justification for these constricting assumptions is that the method seems to work; using a well-constrained data set from the Wasatch Front, Utah, she showed that the observed and predicted slip directions agree well. In another example, stress orientation determined from fault-slip data near the Nevada Test Site show that S_h is N60°W and consistent with that determined from earthquake focal mechanisms, wellbore breakouts, and hydraulic fracturing measurements (Frizzell and Zoback, 1987).

Inversion of fault-slip data using graphical techniques is similar to producing a composite earthquake fault-plane solution (fig. 3–16a) for which each fault can slip without being in the orientation of maximum shear stress (Bott, 1959; McKenzie, 1969a). However, it is understood that fault slip is in the direction of maximum resolved shear stress on the fault plane. The fault-slip inversion is more powerful than the earthquake fault-plane solution in that the active fault plane is unambiguous, whereas in the fault-plane solution two nodal planes are documented but the actual plane of slip must be inferred from other evidence. The use of fault-slip data is best visualized in a stereographic (lower hemisphere) projection of the fault plane, the normal to the fault plane (**n**), and the

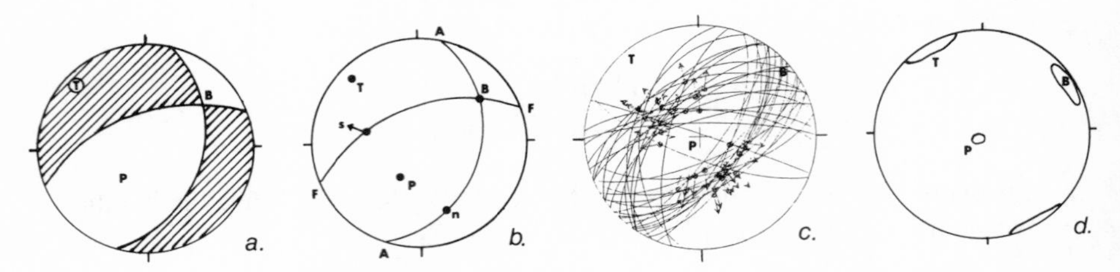

Fig. 3–16. (a) Earthquake fault-plane solution with the compressional first motion quadrant (T-axis) in black. The fault motion is normal-sinistral (adapted from Angelier, 1979); (b) Fault-slip mechanism with tectonic and seismological nomenclatures presented for each axis or vector. F - fault plane; A - auxiliary plane; n is the pole to F, s is the slickenside lineation; B is perpendicular to A and C. P and T are compression and tension stress axes of focal mechanism of earthquakes, making 45° angles with F and A; (c) Neogene fault slip data from central Crete, Greece. Fault plane and slickenside lineation are plotted; (d) The 95% confidence areas for P, T, and B directions for Angelier's (1979) data from central Crete, Greece.

pitch of the lineation along the fault plane (**s**) (fig. 3–16b). In stereographic projection, it is assumed that the fault plane has a conjugate or auxiliary plane at 90° so that the auxiliary contains the normal to the fault plane. The auxiliary plane is drawn so that its great circle intersects the plot to the fault plane and a point 90° from the lineation direction along the fault plane. For earthquake fault-plane solutions, the intersection line between the fault and its auxiliary is in the vicinity of the intermediate principal stress. The orientation of "compression" is on the great circle drawn through the normal to the fault plane and the lineation on the fault. Compression is placed on the same great circle at 45° from the lineation so that the sense of slip on the fault plane is correct as indicated by the lineation structure. For an earthquake fault-plane solution, this compression is taken as being in the vicinity of σ_1 but not in its exact orientation. A single fault plane with lineations will give an approximate estimate of the orientation of the three principal stresses. The compilation of several dozen fault planes gives a better indication of the orientation of the principal stresses which plot as a cluster on the fault-slip mechanism diagram (fig. 3–16c). Following a convention used in seismology, $\sigma_1 \approx$ P axis, $\sigma_2 \approx$ B axis, and $\sigma_3 \approx$ T axis. The location of the T, P, and B axes may be found by visual inspection of this cluster. This graphical exercise gives an approximate orientation for the three principal stresses and these are just half the parameters required for a complete stress tensor.

There are a number of numerical techniques for the direct inversion of fault-slip data to obtain the orientation of regional principal stresses. Angelier and Goguel (1979) devised a method based on a least squares minimization of the component of tangential stress perpendicular to the measured fault lineation. The idea is to find the orientation of the principal stresses which maximizes the resolved shear stresses on a given suite of faults with known slip lineations.

To help understand the inversion technique the concept of a *stress vector* is important (Nye, 1957; Means, 1976). A stress vector at a point on any plane such as a fault within a solid is nothing more than the force or *surface traction* which would act on the surface of the same orientation if half the body were removed along the plane in question. Because the term stress vector is inherently confusing for the student who is led to think that it is somehow a stress, its use should be discontinued in favor of the term surface traction. Commonly, geologists recognize the normal traction and shear traction as two components of the surface traction acting on a fault plane. The link between these two components of force acting on a surface such as a fault zone and the regional stress tensor is the surface traction. The force acting across a fault plane, specified by the unit normal vector, **n**, is a surface traction skewed at an odd angle depending on both the regional stress tensor, **S**, and the orientation of the unit normal vector to the fault. This surface traction, $\mathbf{\Sigma}$, on the fault plane is found by multiplying the stress tensor matrix by the unit normal vector

$$\mathbf{\Sigma} = \mathbf{S} \cdot \mathbf{n} \qquad (3\text{–}41a)$$

or

$$\Sigma_i = \sigma_{ij} \cdot n_j \qquad (3\text{–}41b)$$

Equation 3–41b is the definition of stress in its most rigorous form. This equation says that stress, for which there are nine components, is the reaction of the interior of a solid body to the application of a force (specified by three components) on a surface of the body (specified by the three components of its unit normal vector). The magnitude of the stress and force plus the orientation of the normal to the surface remain unchanged by a coordinate transformation.

The surface traction, $\mathbf{\Sigma}$, on the fault plane has a tangential component which is called a shear traction, τ (fig. 3–14). The shear traction on the fault plane is found by subtracting the normal traction from the surface traction

$$\tau = \Sigma - \{\Sigma \cdot \mathbf{n}\,]\,\mathbf{n} \qquad (3\text{–}42)$$

Note that the term inside the brackets of equation 3–42 is a scalar, which when multiplied by the unit normal vector becomes a vector which is the traction normal to the fault surface. The basis for the inversion of fault-slip data is that, under ideal circumstances, the direction of the unit shear traction vector, **t**, should be parallel to the unit slip vector, **s**, for all fault planes (fig. 3–14). This falls out of Bott's (1959) assumption that the slip direction on faults is in the direction of the resolved shear stress. The unit shear traction, a vector, is found by normalizing τ as

$$\mathbf{t} = \frac{\vec{\tau}}{\|\vec{\tau}\|}. \qquad (3\text{–}43)$$

The idea of the inversion technique is to search for an $\mathbf{S}_R$ which minimizes the angle, β, between τ and $\mathbf{s}$ (fig. 3–14). $\mathbf{S}_R$ is a reduced stress tensor consisting of four of the six parameters required for the complete stress tensor, $\mathbf{S}$. In this search various $\mathbf{S}_R$ are specified as

$$\mathbf{S}_R = \begin{bmatrix} x_1 x_2 x_3 \\ y_1 y_2 y_3 \\ z_1 z_2 z_3 \end{bmatrix} \begin{bmatrix} 100 \\ 0R0 \\ 000 \end{bmatrix} \begin{bmatrix} x_1 y_1 z_1 \\ x_2 y_2 z_2 \\ x_3 y_3 z_3 \end{bmatrix} \qquad (3\text{–}44)$$

where x_i, y_i, and z_i are direction cosines of any possible regional principal stress axes σ_i. Note that $\mathbf{S}$ is an affine function of $\mathbf{S}_R$ which means that the principal axes of both tensors have the same orientation.

In practice, the direction of the unit shear traction vector, $\mathbf{t}$, is rarely parallel to the unit slip vector, $\mathbf{s}$, for all fault planes. For many if not all of the fault planes in a regional survey of fault-slip data the difference between $\mathbf{t}$ and $\mathbf{s}$ is not zero. For this reason, inversion techniques search for a reduced stress tensor which minimizes the function

$$F = \sum_0^k |\mathbf{s}_k - \mathbf{t}_k|^2. \qquad (3\text{–}45)$$

There are a number of functions which can be used in the minimization procedure (Angelier, 1984).

Using an inversion technique for Angelier's (1979) data from Crete, Michael (1984) was able to determine the orientation of the 95% confidence areas for σ_1, σ_2, and σ_3 directions (fig. 3–16d). Modifications of this technique include: an algorithm to separate tectonic phases (Etchecopar et al., 1981); a grid search for the analysis of earthquake fault-plane data (Gephart and Forsyth, 1984); an inversion technique using the constraint that one principal stress is vertical (Michael, 1984); an inversion technique which incorporates the Coulomb yield criterion (Reches, 1987) or friction law (Céléier, 1988); and a define-and-select technique for the separation of mixed data sets (Hardcastle, 1989).

There are many examples of the successful inversion of fault-slip data to determine stress orientation. Zoback (1983) examined fault slip data from northern Utah to evaluate the kinematics of recent faulting along the Wasatch Front in Utah. Her analysis used a modification of the Angelier (1979) method for stress determination by minimizing the component of shear stress acting on the fault plane in a direction perpendicular to the observed slip vector. In the Utah study the principal stresses were constrained to lie in the vertical and horizontal planes for both earthquake focal mechanisms and fault-slip data (e.g., Michael, 1984). The orientation of P and T axes for all northern Utah focal mechanisms shown in figure 3–17 indicates an average S_h of N86°E (Zoback, 1983). The fault-slip data give an S_h of N78°E. Because both of these estimates have an uncertainty of ± 20° they are assumed to be indicative of the same

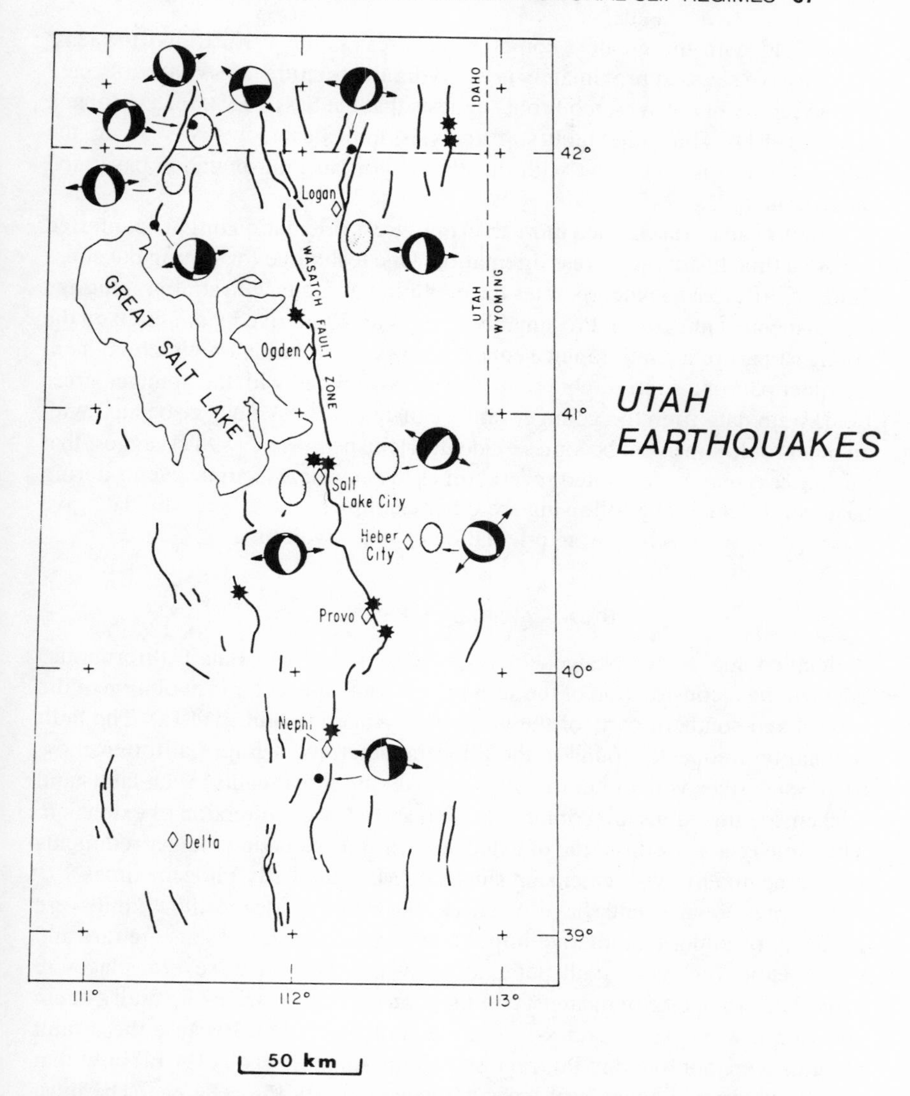

Fig. 3–17. Lower hemisphere equal-area projections of fault-plane solutions from earthquakes in the Wasatch Front area, Utah. Compressional first motion quadrants are shown in black. Focal mechanisms compiled by Zoback (1983) after Arabasz et al. (1979). Large dots show location of single-event solutions; lined zones show sample area for composite solutions. Fault-slip data gathered at starred localities (adapted from Zoback, 1983).

stress field with the greatest compressive stress being vertical and the least compressive stress approximately E-W. Hydraulic fracture stress data collected in the vicinity of the Wasatch Front suggests that the S_h is N73°E ± 15° (Zoback et al., 1981). Thus, the fault-slip analysis indicates that stress along the Wasatch Front is consistent with the classic normal fault-bounded basin and range structure.

If faults can be dated, then more than one stress orientation might be inferred to give a time history for stress orientation. One technique for sorting out stress fields of different age incorporates the evolution of faults with stratigraphic age (Kleinspehn et al., 1989). Presumably a stress tensor derived from faults in the youngest part of a stratigraphic sequence gives information on which faults in the older part of a stratigraphic sequence are associated with the younger stress field. With data from the younger faults isolated, a stress field associated with an older fault-slip event becomes evident. Kleinspehn et al. (1989) suggest that this process may be reiterated several times to determine tectonic events during basin subsidence. The following are other examples where fault-slip data give information on the changes in orientation of stress with time.

Baja California Peninsula

A detailed analysis of Neogene and Quaternary faults in Baja California has enabled the reconstruction of the stress pattern and the tectonic evolution of the central and southern parts of the peninsula (Angelier et al., 1981a). The fault population frequently found in the Pliocene outcrops of Baja California show NNW-SSE strikes with either dip-slip or strike-slip movements. Such faults and slickenside lineations are compatible with an ENE-WSW direction of extension. The same ENE-WSW direction of extension was measured in younger sediments indicating an ENE-WSW extension since late Miocene-early Pliocene times.

Compressional events have occurred since late Neogene times, but were probably of minor quantitative importance because reverse faults are rare and small. However, older fault populations were observed in several places in Baja California. One of the most common is a conjugate strike-slip fault system showing E-W extension and N-S compression (fig. 3–18). Because these fault systems were not found in Pliocene sediments Angelier et al. (1981a) infer that the compressional event is of latest Miocene to early Pliocene age. The most common faults in Baja California show show dextral strike-slip motion along NNW-SSE planes. These are interpreted to be Riedel shears related to dextral strike-slip of the NW-SE Gulf of California fault zone.

Southwestern Anatolia

Even more complex than Baja California is the area along the Anatolian fault in central Turkey. The fault-slip technique has been used to deduce the follow-

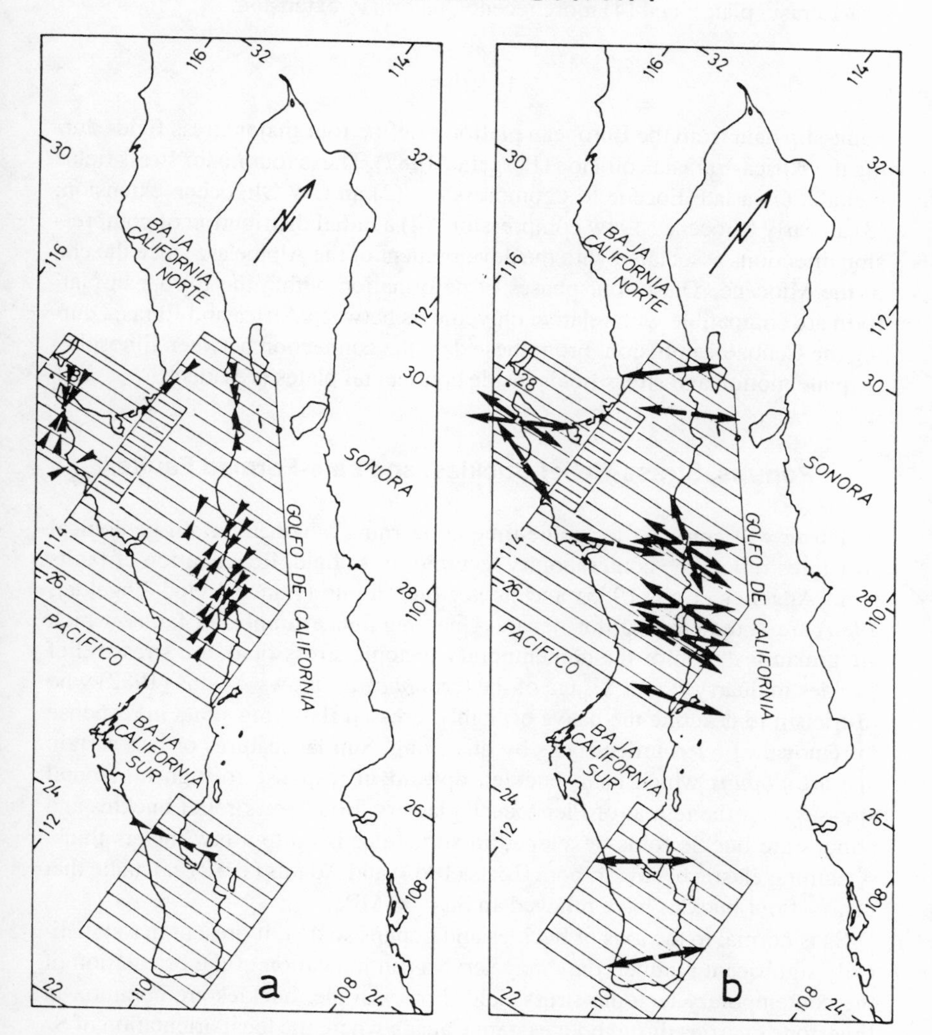

Fig. 3–18. Regional pattern of compression and extension stresses, as inferred from field analyses of faulting in the Baja California Peninsula (adapted from Angelier et al., 1981a).

ing sequence of events in the southwestern Anatolia region (Angelier et al., 1981b): (1) Miocene compression of three ages (23–20 m.y., 15 m.y., 6–4 m.y.) (This event relates to the changes in relative plate motion between Africa, Eurasia, Arabia, and Anatolia); (2) Large Pliocene and early Quaternary extension (common for the larger Mediterranean region); (3) Strike-slip event of probable early Quaternary age (compression due to minor changes in the Africa-Eurasia plate); and (4) more recent Quaternary extension.

EUROPE

Fault-slip data from the European platform define four major stress fields during the Africa-Eurasia collision (Bergerat, 1987). These four major stress fields include: (1) a late Eocene N-S compression; (2) an E-W Oligocene extension; (3) an early Miocene NE-SW compression; (4) a radial distribution of compression directions associated with the development of the Alpine arc since the end of the Miocene. These four phases of deformation within the European platform are compatible with relative movements between Africa and Eurasia during the Cenozoic collision. From these data the connection between lithospheric plate motion and stress fields inside continental plates is evident.

Popups, Stress-Relief Buckles, and Late-Formed Faults

A debate will continue for some time concerning whether certain geological structures reflect the contemporary tectonic stress field. Reactivation of faults (e.g., Angelier et al., 1985) and neotectonic joints (Hancock and Engelder, 1989) are at the heart of this debate. Unambiguous examples of faults, which are a manifestation of the contemporary tectonic stress field, are stress-relief buckles in quarry floors. Usage of the term *buckle* follows Adams (1982) who restricts it to describe the heave of highly stressed flat-lying strata in response to removal of overburden rocks by quarrying. Similar features of less certain age are *popups* which have buckled upward in response to high horizontal stresses after the retreat of Pleistocene glaciers. Both stress-relief buckles and popups are buckle folds developed in surface beds up to a few meters thick. Assuming elastic behavior, both Coats (1964) and Adams (1982) calculate that quarry-floor buckles have relieved an $S_H = 30$ MPa.

S_H is normal to the axis of buckles and popups so that, if present in a statistically significant number, they may serve as an indication of the orientation of the contemporary tectonic stress field. For example, buckles are common in limestone quarries throughout eastern Canada where the local orientation of S_H is 050° (Port Colborne; Williams et al., 1985), 046° (Ottawa; Adams, 1982), 040° (Montreal; Saull and Williams, 1974), and 045° (Milton; White et al., 1973). These data are consistent with other measures of the orientation of S_H from the Canadian Shield (Hasegawa et al., 1985). Post-glacial popups are mapped in the same manner as buckles. By mapping popups White and Russell (1982) argue that S_H is about 040° throughout southern Ontario but Hasegawa et al. (1985) suggest that the orientation of popups in Ontario is too variable to draw any conclusions. Many popups in sandstones in the vicinity of Alexandria Bay, New York, also suggest that S_H is ENE (Engelder and Sbar, 1977).

Other neotectonic structures may not reflect the contemporary tectonic stress field. For example, post-glacial thrust faults[4] in eastern Canada reflect a N-S S_H (Adams, 1982). Hasegawa et al. (1985) suggest that "these structures reflect a transient stress field that was generated during glaciation prior to 10,000 years ago and partially released soon after the ice retreated."

In the coal mining district of southern Illinois, north-south trending thrust faults cut the coals down to depths of at least 80 m (Nelson and Bauer, 1987). These faults extend for several hundred meters along strike in underground coal mines. Displacements range from less than a couple of cm to about 3 m. Most faults are fracture zones composed of parallel minor faults which overlap or branch along strike. The strike of the fault zones is remarkably consistent; in Saline and Williamson counties, Illinois, the strike is within a few degrees of true north, whereas in western Kentucky the same faults strike N 20° to 25° E. In southern Indiana faults of the same orientation are known (Ault et al., 1985). Nelson and Bauer (1987) argue that these north-south trending thrust faults are a response of the upper hundred meters of the Illinois Basin to the contemporary tectonic stress field. This argument is based in part on the correlation in orientation of S_H among hydraulic fracture measurements, overcoring, and earthquake fault plane solutions. All three types of measurements show that S_H is roughly east-west and, thus, compatible with north-south trending thrust faults.

The Near-Surface Horizontal-Stress Paradox

Compilation of data in the upper 2.5 km of the earth's crust shows that horizontal stresses become significantly larger than vertical ones (Brown and Hoek, 1978) (fig. 3–19). Popups, stress-relief buckles, and late-formed faults are apparently a manifestation of the relief of such large near-surface horizontal stresses. Yet, this characteristic distribution of earth stress in the upper crust is a paradox because it is difficult to reconcile with other aspects of the brittle behavior of the upper crust. Field observations suggest that recent joints are common and can be induced with just a few hundred meters of uplift and erosion (Hancock and Engelder, 1989). These joints form near the earth's surface and apparently propagate in response to a low effective S_h. The observation that many outcrops contain joints (i.e., unloading joints in Engelder's [1985] classification), which are not found in deep cores of the same formation (e.g., Cliffs Minerals, Inc., 1982), leaves little doubt that low effective stresses have been common in both time and space in the near surface. If stresses of the magnitude shown in figure 3–19 were present during joint propagation, an un-

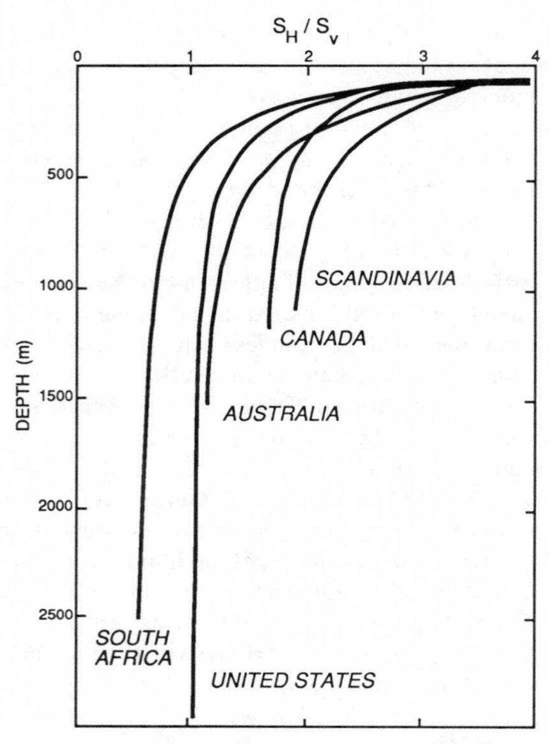

Fig. 3–19. Curves defining the ratio of the maximum horizontal to vertical stress (S_H/S_V) as a function of depth in the earth's crust from various continents (adapted from Bieniawski, 1984).

usual drive mechanism for joint propagation was responsible. Abnormally high pore pressure is so rare within a few hundred meters of the surface that a water drive mechanism seems unlikely.

In the laboratory axial splitting fractures are driven parallel to σ_1 under high compressive loads when confining pressure is near zero. Lorenz et al. (1991) suggest that such a fracture mechanism may lead to extension fracturing throughout regions of considerable extent provided that the least effective stress is near zero. While this seems to be a reasonable theory for the abundance of near-surface joints, a complete explanation still requires a mechanism for reducing S_h in light of the large quantity of data supporting a marked increase in horizontal stress in the near surface.

There are trends in certain suites of stress data which point to the development of effective tensile stresses in the upper km of the crust. One example is the extrapolation of S_h data from hydraulic fracture tests. S_h from the Auburn, New York, well (Hickman et al., 1984) extrapolated up to the tensile stress

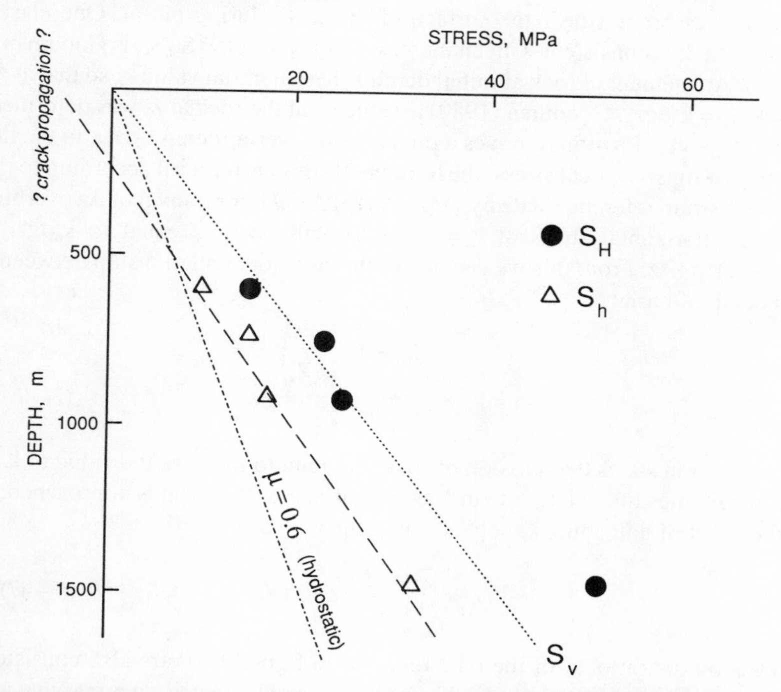

Fig. 3–20. A plot of horizontal stress versus depth for the hydraulic fracture measurements taken at the Auburn, New York, geothermal well (adapted from Hickman et al., 1985). An extrapolation of the least principal stress back toward the surface suggests that the tensile strength of the local rocks should be exceeded when rocks are uplifted to a depth of about 200 m.

regime at 200 m below the surface (fig. 3–20). But even in the vicinity of Auburn, stress measurements within the top 30 m of the crust at Oswego, New York (Dames and Moore, 1978) show compressive horizontal stresses as the Brown and Hoek (1978) compilation predicts.

The behavior of earth stress in the upper crust is still uncertain because there is still no unifying model for crustal stresses developed during uplift nor a clear understanding of the nature of the reference state of stress during erosion. The initial stages of the deposition of sediments impose no lateral loads, therefore the starting point for horizontal stress during burial is 0 MPa. If rocks behave elastically during burial and uplift, if the uniaxial-strain reference state applies, and if no change in elastic properties takes place, then there should be no net horizontal stress on return to the surface, but changes in rock properties do take place during burial and uplift and probably those changes contribute to the net horizontal stresses (Karig and Hou, 1992).

After burial and uplift, sedimentary rocks do not return to the same stress state as was present near the surface of the earth during burial. One elastic model which is consistent with an increase in the ratio, $k = S_h/S_v$, is Goodman's (1980). An element of rock at initial depth z_o has an initial value k_o so that at z_o, $S_h = k_o S_v = k_o \rho g z_o$. Goodman (1980) assumes that the rock at z_o was deformed, $k_o \neq \{\nu/(1 - \nu)\}$. Erosion removes a thickness of overburden Δz. Due to the unloading of $\rho g \Delta z$ vertical stress, the horizontal stress is reduced according to the uniaxial-strain reference state by $\{\nu/(1 - \nu)\}\rho g \Delta z$. After removal of Δz overburden, the horizontal stress at $z = z_o - \Delta z$ will become equal to $k_o \rho g z_o - \{\nu/(1 - \nu)\}\rho g \Delta z$. From this we can derive the functional relationship between k and depth of burial

$$k(z) = k_0 + \frac{\left[k_0 - \left(\frac{\nu}{1-\nu}\right)\right]\Delta z}{z} \qquad (3\text{–}46)$$

This equation states that erosion of rock will tend to increase the value of k so that it becomes much larger than 1 as the surface of the earth is approached. If at the onset of unloading $k_o = \{\nu/(1 - \nu)\}$, then

$$k(z) = \frac{\nu}{1-\nu} = \text{const.} \qquad (3\text{–}47)$$

Data on the ratio, k, of the type reported in figure 3–19 are also consistent with a model of the earth as a self-gravitating shell situated on a massive and unyielding interior (McCutchen, 1982). The ratio, k, also increases in the near surface in models using a constant-horizontal-stress reference state (McGarr, 1988).

The horizontal stress paradox leads to the suggestion that a "buffering" process prevents the horizontal tensile stresses from becoming larger once the stresses drop below a moderately compressive value (Bruner, 1984). The development of joints and normal faults can prevent the generation of large horizontal tensile stresses but their formation does not generate compressive stresses. The components of total earth stress are never tensile. Ground water at hydrostatic pressures acts as a water drive for crack propagation when effective tensile stresses exceed the tensile strength of the rock. Possible mechanisms for maintaining large near-surface compression as observed by stress measurements include the absorption of water by clay near the surface of the earth. In shales the absorption of water appears to cause the time-dependent expansion of unconfined cores (Harper et al., 1979). Water absorption is important in rocks with a high clay content, but crystalline rocks are more likely to be influenced by the relaxation of internal stresses which are large on the grain scale and are induced by large temperature changes in the rock. Bruner's (1984) calculations show that "crack growth leads to compressive changes in macroscopic stress in increasing the elastic compliance of the rock and by decreasing its linear thermal expansion coefficient. The growth and opening of cracks also produces an expansion which must be canceled by increased horizontal compression in order to satisfy the lateral constraints imposed by the presence of adjacent rocks."

Concluding Remarks

The boundaries for the three brittle-stress regimes are defined in terms of increasing σ_d from the crack-propagation to the shear-rupture regime. Inferences about the occurrence of these three stress regimes come from mapping the regional distribution of joints and primary shear fractures. The regional distribution of joints suggests that lower σ_d is relatively common and that reactivation by frictional-slip may moderate σ_d within the lithosphere. The local occurrence of shear fractures near faults and folds suggests that tectonic stress has to be concentrated above a "background" level controlled by frictional slip in order for entry into the shear-rupture regime. Modern examples of the shear-rupture regime are found in the vicinity of the tips of active but blind thrusts such as those located under the Coalingua and Kettleman Hills anticlines of California (Stein and Yeats, 1989). Folds over these blind thrusts are the host of a diffuse pattern of earthquakes, suggesting the rupture of intact rock as the folds grow.

4

Stress in the Ductile-Flow Regime

All the petrographic data, and especially those relating to internal rotation, can be explained on the orthodox assumption that the essence of plastic deformation in ionic crystals is gliding in one or more glide systems for which the applied resolved shear stress exceeds the critical value.

—Francis Turner, David Griggs, and Hugh Heard (1954) after observing the effect of stress on single crystals of calcite

Constraints on lithospheric stress are based on ductile as well as brittle rock strength. Temperature and confining pressure have fundamentally different effects on brittle and ductile rock strength. Ductile strength is almost unaffected by confining pressure, whereas the brittle strength of rock increases markedly at higher confining pressure (Heard, 1960). In contrast, temperature has a large effect on ductile processes such as dislocation motion and diffusion-assisted deformation, whereas brittle strength shows little dependence on temperature (Heard, 1963). In laboratory experiments, rocks are markedly weaker during ductile deformation at high temperature which means that σ_d is much lower in the warmer portions of the plastosphere and in regions of the schizosphere dominated by ductile deformation. Semibrittle behavior occurs in the transition from pressure-dominated brittle deformation to temperature-dominated ductile deformation (Tullis and Yund, 1977). Semibrittle behavior is found in a transition zone of the lithosphere where σ_d gradually drops from a maximum at depths of 10–15 km to smaller values at depths near the Moho. This brittle-ductile transition zone is the boundary between the schizosphere and plastosphere. The Moho apparently represents a mechanical discontinuity within the lithosphere where the underlying peridotite is stronger than rocks of the lower crust (Chen and Molnar, 1983; Kirby, 1985). As is the case with the lower crust, strength of the mantle drops as temperature increases with depth.

Chapter 4 deals with the influence of ductile rock deformation on stress in the lithosphere. Although the ductile flow is found throughout the lithosphere, its role as a governor of lithospheric stress is best understood by subdividing the ductile-flow regime into three tiers. The upper tier is demarcated by low-temperature ductile deformation including diffusion-mass transfer which is found throughout the schizosphere (Durney, 1972; 1978). Other low-temperature flow processes are work-hardening with rate-controlling mechanisms being restricted to dislocation glide (slip, twinning) (Carter and Tsenn, 1987). The middle tier is found in the brittle-ductile transition at the boundary between the schizosphere and plastosphere. In this tier, dislocation and diffusion creep plus other thermally activated processes operate in conjunction with brittle mechanisms to contribute to creep strain (Carter and Kirby, 1978). The lower tier of the ductile-flow regime is found in the plastosphere where ductile flow is the dominant deformation process.

To introduce a three-tier ductile-flow regime, rudimentary elements of crystal plasticity are discussed. From a knowledge of crystal plasticity there arises considerable information on stress orientation and magnitude at the time of deformation in all three tiers of the ductile-flow regime. Finally, the relationship between the ductile-flow regime and the three brittle regimes is explained in terms of σ_d-depth profiles.

Crystal Plasticity

After World War II David Griggs and his associates at UCLA started a systematic investigation of the strength of rocks, focusing largely on ductile behavior. A critical step occurred when F. J. Turner (1948) hypothesized that twin lamellae in calcite or deformation lamellae in quartz originated by plastic flow and formed when the slip systems were in orientations of high shear stress. By 1950 Griggs's group narrowed their effort to focus on the experimental deformation of marble largely because calcite is ductile under a wide range of experimental conditions. Yule marble, a relatively pure and homogeneous, if not isotropic, carbonate, was selected for mechanical tests. Small blocks of unweathered Yule marble were readily available as scrap from the building of such famous structures as the Lincoln Memorial and the Tomb of the Unknown Soldier in Washington, D.C. Through experiments on Yule marble the validity of the Turner hypothesis became clear; that is, for small strains both translation and twin gliding favor those intracrystalline planes lying in orientations of high shear stress. Further confirmation came with Turner's (1953) study of the orientations of calcite twin lamellae in three naturally deformed marbles: a banded marble from Moray Firth, Scotland; the Yule marble from Colorado; and a foliated marble from Sonora, California. The lamellae in all three marbles formed in an orientation of high shear stress which orientation was indicated by local faults and folds. From 1950 until the development and routine use of the high voltage electron microscope around 1970, intragranular flow was studied largely through petrographic observations of deformation lamellae. These petrographic analyses of intragranular flow mechanisms contributed greatly to an understanding of the orientation and magnitude of lithospheric stress in the ductile flow regime.

Two distinct intracrystalline slip mechanisms are evident in petrographic thin sections of deformed rocks. These mechanisms, translation gliding and twin gliding, are controlled by crystallographic structures. Both mechanisms obey *Schmid's law* which states that an intracrystalline slip mechanism will operate only when the shear stress resolved along the slip direction in the slip plane has reached a certain critical value (τ_c). Although τ_c is a function of temperature and strain rate, it is independent of the stress normal to the active slip plane. Therefore, the presence of deformation lamellae indicate that τ_c in the slip direction has been reached regardless of the confining pressure. Hence, deformation lamellae serve as convenient stress gauges.

Translation gliding is the macroscopic manifestation of edge dislocation motion along a slip plane (Poirier, 1985). The large-scale effect of translation gliding is the deformation of the host grain in simple shear with each atomic plane displaced an integral atomic distance relative to the plane below (fig. 4–1a). Displacement along any slip plane is limited by crystallographic symmetry to one or a few slip directions with the slip plane and slip direction comprising a *slip system*. Crystals may have more than one slip system, each with a unique τ_c. The slip system with the lowest τ_c is the most active system. The resolved shear stress (τ_r) along a given slip direction is given by

$$\tau_r = (\sigma_1 - \sigma_3)\Omega_o \quad \text{and} \quad \Omega_o = \cos\lambda\cos\phi, \qquad (4\text{–}1)$$

where λ is the angle between σ_1 and the normal to the slip plane and ϕ is the angle between σ_1 and the slip direction (fig. 4–2a). The maximum τ_r occurs on a slip system where $\lambda = \phi = 45°$ so that $\Omega_o = 0.5$. Slip will occur only if $\tau_r \geq \tau_c$. However, for homogeneous deformation of an aggregate, five slip systems must operate, an observation called the von Mises criterion.

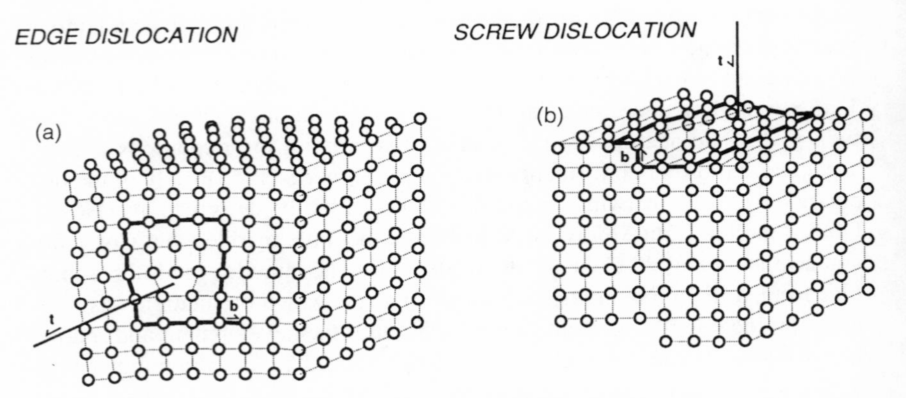

Fig. 4–1. (a) Schematic block diagram of an edge dislocation in a cubic lattice. The Burger's vector, **b,** is defined as the closure vector for the loop counted around the dislocation. The tangent vector, **t,** is defined to lie parallel to the dislocation line. (b) Schematic block diagram of a screw dislocation in a cubic lattice. The Burger's and tangent vectors are perpendicular for an edge dislocation; in contrast, they are parallel for a screw dislocation.

GLIDE SYSTEM -- calcite e-twins

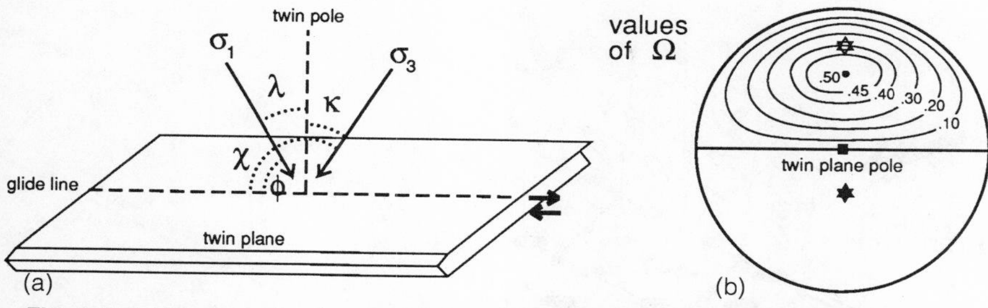

Fig. 4–2. (a) Diagram showing the nature of the gliding system for calcite *e*-twins relative to principal stress axes and angles λ, ϕ, κ, and χ. (b) Contoured values of Ω_0 for a horizontal twin plane. Twin plane corresponds to the primitive circle of the lower hemisphere equal-area projection. Locations of the c axis for calcite (solid star) and dolomite (open star) are shown (adapted from Jamison and Spang, 1976).

Although *twin gliding* is also regarded as a simple shear of the crystalline lattice along the slip plane, it differs from translation gliding in two respects. First, it is homogeneous which means that each lattice plane is displaced the same amount relative to the plane below. There are no such constraints on the motion of each lattice plane during translation gliding. Secondly, the twinned portion of a crystal is deformed into a mirror image of the undeformed crystalline lattice across the twin plane. This mirror image is the manifestation of a screw dislocation moving through the crystalline lattice. After deformation each atomic plane has translated some fixed fraction of an integral atomic spacing depending on the crystal structure (fig. 4–1b). Because this fixed translation is homogeneous throughout the twinned portion of a crystal, the shear strain for a given twin law is constant.

CALCITE

During the 1950s Griggs's lab made several discoveries concerning the mechanical properties of calcite in samples of Yule marble (Turner et al., 1954). The mechanism for plastic deformation of calcite in the upper crust, where temperatures were less than 600°C, was twin-gliding parallel to the direction $[e_1{:}r_2]$ on one or more of the $e\{01\bar{1}2\}$ planes. Calcite twinning is one of the key deformation mechanisms used in the study of stress during ductile flow in the schizosphere. For twinning on the *e*-planes, the atomic layers above the twin plane move toward the *c*-axis in the host crystal with respect to the atomic layers below the twin plane (fig. 4–3). This sense of shear is defined as positive. Twin-gliding with a negative sense of shear does not occur along the *e*-plane. The *c*-axis in the twinned crystal rotates 52.5° with a negative sense of

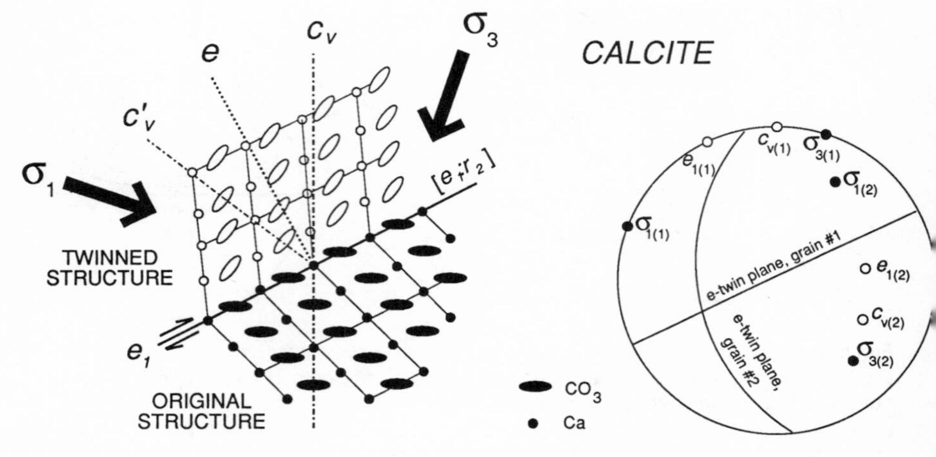

Fig. 4–3. Twin-gliding in calcite. Diagrammatic projection of the twinned calcite structure parallel to the plane containing the *c*-axis and the normal to e_1, showing the twinning elements (adapted from Friedman, 1964). Equal-area projection showing the graphic method for constructing principal stress axes best oriented to produce twin-gliding in calcite for two *e*-planes (adapted from Carter and Raleigh, 1969).

shear. Although Griggs's lab found other flow mechanisms, twin gliding on *e*-plane had the lowest τ_c. With petrographic examination of the deformed Yule marble, Turner (1953) recognized that a statistical σ_1 and σ_3 are identified by measuring the orientation of both the *e*-twin plane and the original *c*-axis in more than 50 grains. One data point for σ_1 and σ_3 is obtained for each grain by plotting in stereographic projection the pole to the *e*-twin plane and the *c*-axis for that grain (fig. 4–3). There is an uncertainty of 2° in locating both the pole to the *e*-twin plane and the *c*-axis using a universal-stage petrographic microscope. If the pole and *c*-axis are located correctly they will be 26°15′ apart, although in general practice an angle between 24° and 28° is acceptable (fig. 4–4). σ_1 and σ_3 are located in the e_1-c plane with σ_1 found by tracing 45° along the great circle starting at the pole to the *e*-twin plane and moving away from the the *c*-axis. σ_3 which plots relatively close to the *c*-axis and is found by tracing 45° along the great circle in the opposite direction starting at the pole to the *e*-twin plane. Note that σ_1 is plotted to place the *e*-twin in the plane of maximum shear stress.

In early work on calcite deformation, a distinction was made between calcite twins and calcite twin lamellae (Turner and Ch'ih, 1951; Borg and Turner, 1953; Friedman, 1963). In both experimentally and naturally deformed limestones, *twin lamellae* are nonvisibly twinned, which is to say they are too narrow to permit observation of the layers in the twinned orientation. Twin lamellae are particularly common in lightly deformed natural rocks of the schizosphere (Groshong, 1972).

CALCITE TWIN LAMELLAE

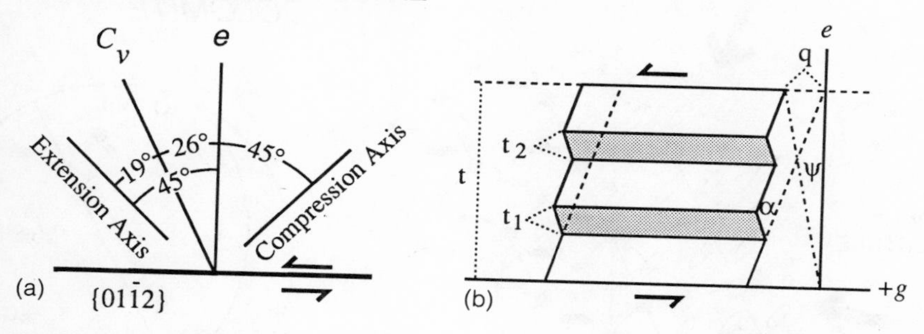

Fig. 4–4. (a) Position of compression and extension axes that would most favor development of calcite twin lamellae where plane of diagram is normal to *e*-twin plane and contains glide direction $[e_1{:}r_2]$. (b) Shear strain in a partially twinned calcite grain (adapted from Groshong, 1972). The values t_1 and t_2 are the widths of twins and t is the width of the host grain perpendicular to the twin planes. α (= 38°17′) is the angle of rotation of the grain edge from the untwinned to the twinned position (Handin and Griggs, 1951). ψ is the angular change of a line originally at right angles to the twin plane.

DOLOMITE

The other common carbonate in the crust of the earth, dolomite, proved much stronger and more brittle than Yule marble under the same conditions. Elevated temperatures and stresses were necessary to activate twinning in dolomite rocks, whereas marble developed dense twinning under modest confining pressures at room temperature. At 300° to 400°C active twin gliding occurs on the *f* {02$\bar{2}$1} plane parallel to $[f_2{:}a_3]$ with a negative sense of shear (Handin and Fairbairn, 1953). Unlike calcite, τ_c for *f* twin-gliding decreases as a function of increasing temperature. The dynamic analysis developed for calcite also applies to dolomite except the sense of shear is negative and the *c*-axis rotates 54° in a positive sense (fig. 4–5). Because the angle *c* to *f* is 62.5°, σ_1 is located 45° from f_1 but between *c* and *f*, due to the negative sense of shear (σ_3 is 45° from *f* in the opposite direction).

QUARTZ

One slip system of many, which activated during the experimental deformation of quartz, is slip on the basal plane {0001} (Carter et al., 1964). Slip on the basal plane leads to the buildup of dislocations in zones forming deformation lamellae, which are visible because the refractive index of the quartz crystal changes around dislocation clusters that elastically strain the crystal. In natural tectonites, lamellae are commonly inclined at 10° to 30° to {0001}. Such subbasal lamellae were produced experimentally by Heard and Carter (1968) who

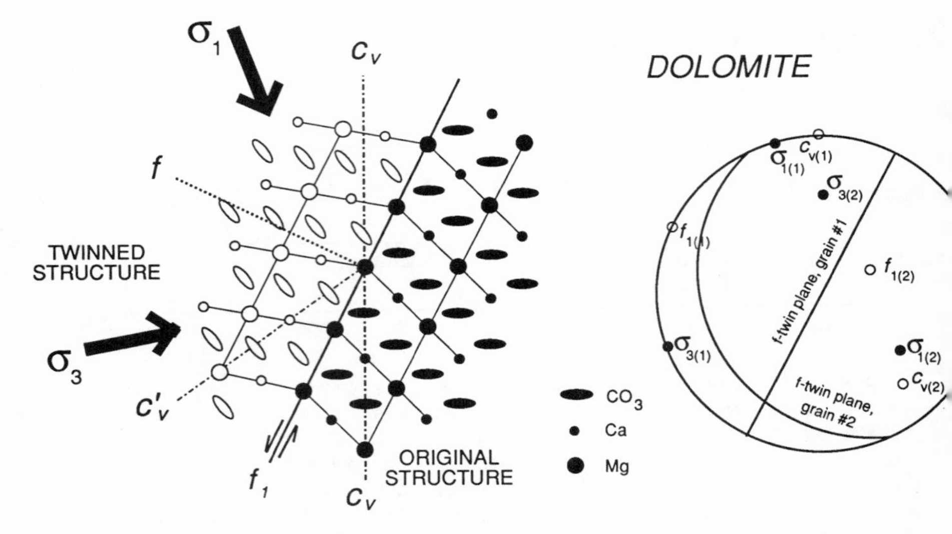

Fig. 4–5. Twin-gliding for dolomite. Diagrammatic projection of the dolomite structure parallel to {11$\bar{2}$0}, showing elements of twinning of *f* {02$\bar{2}$1} planes and translation gliding on the basal plane (adapted from Friedman, 1964). Equal area projection showing the graphic method for constructing principal stress axes best oriented to produce twin-gliding (adapted from Carter and Raleigh, 1969).

showed that they are largely irrational, which is to say they are not associated with any particular slip system. There are three techniques for locating the orientation of σ_1 using the orientation of quartz deformation lamellae (Heard and Carter, 1968). First, a plot of the poles to the deformation lamellae should yield two small circles in stereographic projection for which the poles to these small circles define σ_1 (fig. 4–6a,d). Second, in areas where the density of lamellae is high, the crystal structure is externally rotated such that the *c*-axis in the rotated zone lies closer to the σ_1 than in adjacent unrotated zones. A cluster of data from several crystals with external rotation portions points the direction of σ_1. Third, the *c*-axes in the undeformed parts of one crystal and the poles to deformation lamellae are nearly cozonal, so that an arrow from the *c*-axes to the poles of the lamellae will point away from σ_1 and toward σ_3. Figure 4–6 shows the small circles for quartz lamellae, the *c*-axes in the deformed and undeformed quartz grains, and the poles to lamellae (points of arrows) and *c*-axes (ends of arrows) in lower hemispheric projection.

Olivine, Mica, Diopside, Plagioclase

Other crystals are also used to infer the orientation of principal stresses although their use is far less common than that of quartz, calcite, and dolomite

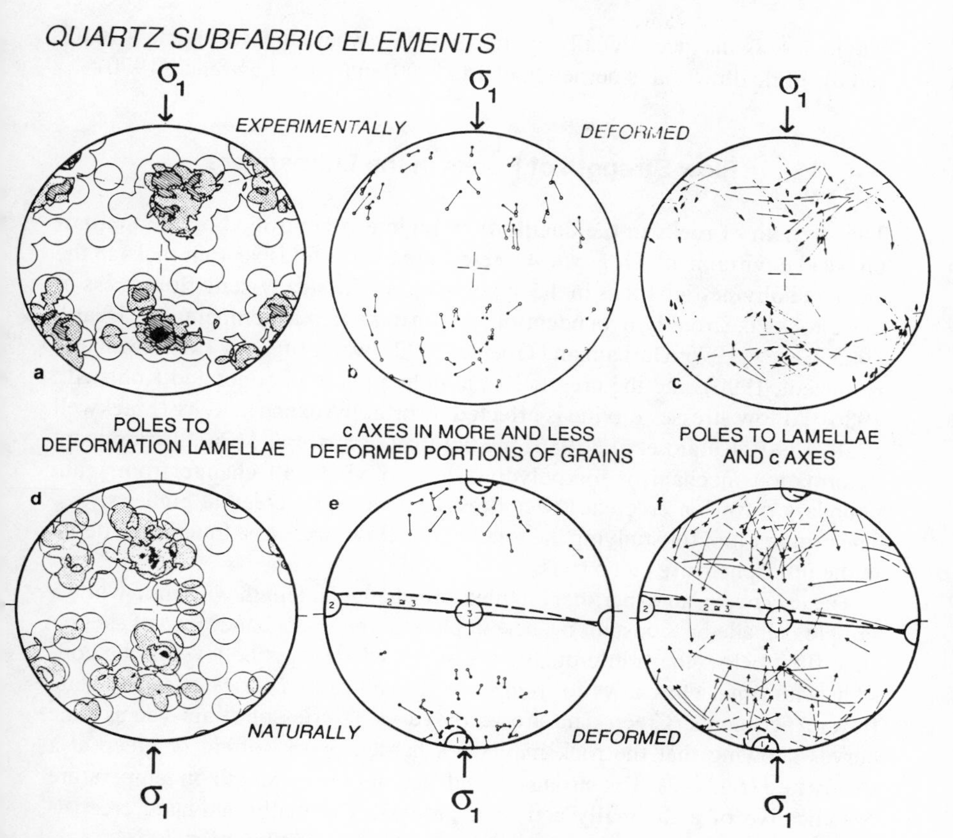

Fig. 4–6. Orientation of subfabric elements in experimentally (Heard and Carter, 1968) and naturally (Carter and Friedman, 1965) deformed quartzites (adapted from Carter and Raleigh, 1969). a & d. Poles to 128 sets of deformation lamellae. Contour intervals are 7, 5, 3, 1 (a) and 6.7, 4.5, 2.2, 1.1 (d) percent per 1 percent area. b & e. c-axes in more (solid circles: c_1) and less (open circles: c_0) deformed parts of individual grains. c & f. Poles to lamellae (points of arrows) and c-axes (ends of arrows) in grains containing lamellae.

(Carter and Raleigh, 1969). Several slip systems are active in olivine (Raleigh, 1968; Durham and Goetze, 1977), but, at low strains olivine is not useful for inferring stress orientations. However, at high strains, the preferred orientation developed by slip on (010)[100] results in a fabric parallel with mantle seismic anisotropy (Carter and Avé Lallement, 1970). Mica in foliated rock will kink when the foliation plane and, hence, most mica {001} planes are parallel to σ_1. Kinking is also common in orthopyroxenes. In diopside, a clinopyroxene, slip occurs on (100)[001] to produce mechanical twins (Raleigh and Talbot, 1967).

Plagioclase twins mechanically by the albite law where the twin plane is (010) and the glide direction is perpendicular to [100] in (010) (Lawrence, 1970).

Flow Strength of Rocks in the Lithosphere

The strength of rocks in the ductile-flow regime is controlled by the thermophysical environment. Of greatest interest are quartz-feldspar-rich rocks in the crust and olivine-rich rocks in the upper mantle. The steady-state flow stress of these rocks is strongly dependent on such parameters as temperature (Heard, 1963; Carter, 1976; Durham and Goetze, 1977), water fugacity (Mackwell and Kohlstedt, 1990), and the presence of a molten phase (Cooper and Kohlstedt, 1986). At low stresses olivine is affected by orthopyroxene activity (Mackwell et al., 1990). Karato et al. (1986) predict that, at a $\sigma_d = 1$ MPa, the dominant deformation mechanism for polycrystalline olivine will change from grain boundary diffusion creep at lower stress to dislocation creep at higher stress. The primary tool for studying the steady-state flow stress as a function of depth in the lithosphere are creep tests.

The *creep test,* an experiment to measure change in length with time while a sample is loaded at constant σ_d, best duplicates the stress conditions of steady-state flow below the brittle-ductile transition where σ_d is believed to be constant with time. Flow laws for rocks are derived from a series of steady-state flow experiments. Creep data are recorded as a series of strain-versus-time curves showing that the rock may work-harden, work-soften, or creep at a steady rate (fig. 4–7). The strong dependence of creep strength on temperature is indicative of a thermally activated process. Thermally activated creep is modeled by giving a steady-state creep rate, $\dot{\varepsilon}_s$, as a function of σ_d,

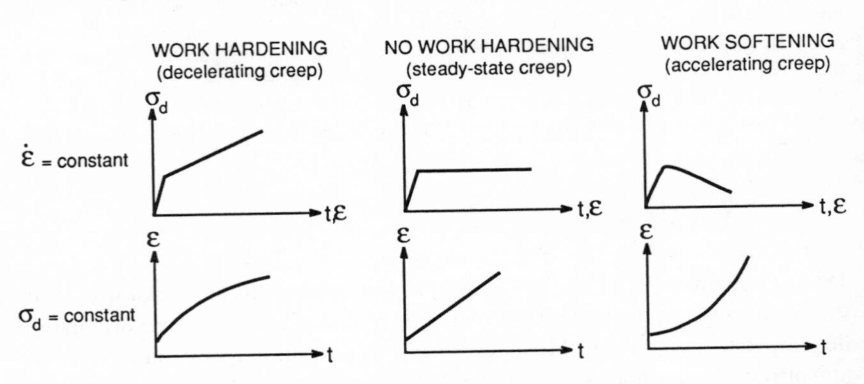

Fig. 4–7. Comparison of stress-strain curves (constant strain rate) and creep curves (stress constant) for materials with various properties: work-hardening, decelerating creep; no work-hardening, steady-state creep; and work-softening, accelerating creep-rate (adapted from Poirier, 1985).

$$\dot{\varepsilon}_s = \dot{\varepsilon}_0 \exp\left(-\frac{H^*}{nRT}\right) f(\sigma_d) \qquad (4\text{–}2)$$

where $\dot{\varepsilon}_o$ is a material-dependent constant, H* is the activation enthalpy, R is the gas constant, and T is the absolute temperature. Two common flow equations are the exponential law

$$\dot{\varepsilon}_s = A \exp\left(\frac{\sigma_d}{\sigma_0}\right) \exp\left(-\frac{H^*}{nRT}\right) \qquad (4\text{–}3)$$

and the power law

$$\dot{\varepsilon}_s = A(\sigma_d)^n \exp\left(-\frac{H^*}{nRT}\right) \qquad (4\text{–}4)$$

where A, σ_o, and n are constants. The power-law equation describes the creep of many rocks in the stress range found in the lithosphere. Yet, no single flow law represents rock deformation at all temperatures, $\dot{\varepsilon}$, and σ_d, largely because the rate-controlling deformation mechanisms also vary as a function of these parameters. Different flow laws describe the behavior of rocks when different deformation mechanisms are rate controlling. Various flow regimes are identified with subdivisions of a temperature-$\dot{\varepsilon}$-σ_d plot known as a *deformation mechanism map* as shown in figure 4–8 (Stocker and Ashby, 1973; Frost and Ashby, 1982). In the upper tier of the ductile-flow regime, grain-size-sensitive creep mechanisms are important. Three grain-size mechanisms include diffusional flow, stress solution,[1] and superplastic flow (e.g., Durney, 1972; Schmid, 1983). The middle tier of the ductile-flow regime is characterized by low-temperature creep for which flow stress is relatively insensitive to temperature and strain-rate changes (Kirby, 1980). Here dislocation glide dominates and dislocation pile-up leads to failure which may be brittle. In the high-temperature creep (*power-law creep*) regime, relatively small temperature changes have a marked effect on the steady-state flow stress, σ_d^s (Kirby, 1983). We know that this behavior is found in the lower tier of the ductile-flow regime because deformation structures developed experimentally at high temperatures are virtually identical to ones that occur in naturally deformed olivine of mantle origin (e.g., Kirby and Raleigh, 1973; Gueguen, and Nicolas, 1980).

The transition from high-temperature to low-temperature creep is commonly referred to as *power-law breakdown* and is explained by considering dislocation motion which is retarded by defects or obstacles within the crystal lattice. If the obstacles are overcome by thermal agitation, the creep is glide controlled whereas, if the obstacles are surmounted by cross-slip or diffusion-controlled climb, the creep is recovery controlled (Poirier, 1985). If dislocation glide is the rate-controlling step during deformation, then an exponential stress dependence of $\dot{\varepsilon}$ is observed as is characteristic of the low-temperature creep regime (fig. 4–8). As temperature increases, diffusion-controlled climb is the rate-

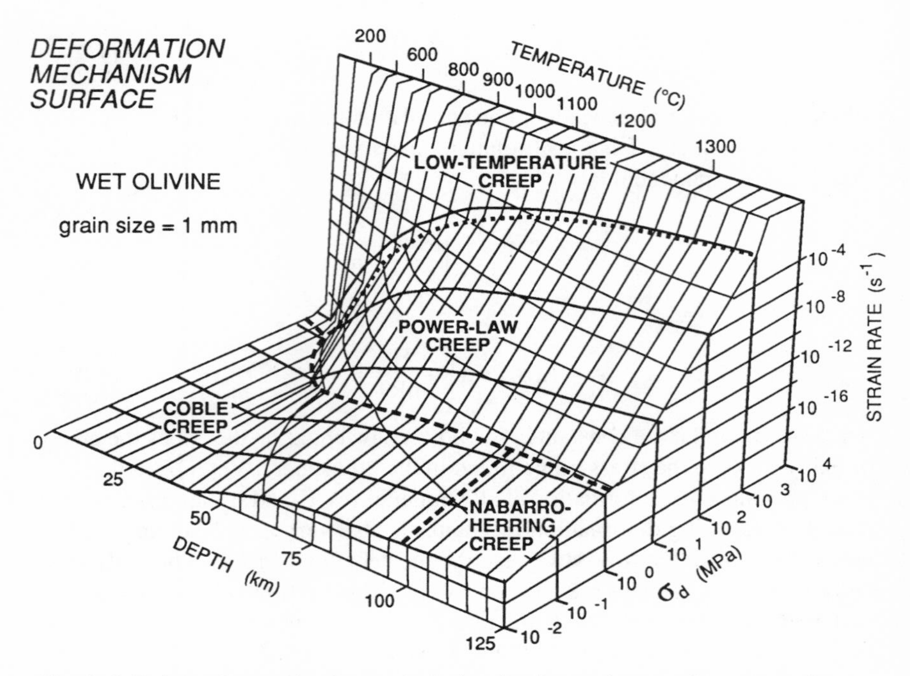

Fig. 4–8. Deformation mechanism surface showing flow regimes and processes for wet olivine (adapted from Tsenn and Carter, 1987).

controlling step which leads to the arrangement of dislocations in subgrain walls leaving areas of the grain with relatively few dislocations in a process called *polygonization.* Extensive dislocation climb is represented by the power-law creep field (fig. 4–8). Grain boundaries eventually migrate through the crystalline lattice to sweep out dislocations in a process called recrystallization. The driving force for grain-boundary migration arises from the difference in strain energy between deformed grains and dislocation-free nuclei. The creep process is characterized by dislocation motion which produces strain, whereas the recovery/recrystallization process does not produce strain.

If the various mechanisms of deformation and their flow-laws are known, then we can constrain the magnitude of σ_d in the three tiers of the ductile-flow regime. These constraints come from a consideration of the steady-state flow stress as a function of depth in the lithosphere (Kirby, 1985). Rearranging equation 4–4 gives the steady-state flow stress, σ_d^s, for the ductile-flow regime

$$\sigma_d^s = \left(\frac{\dot{\varepsilon}_s}{A}\right)^{-n} \exp\left(\frac{H^*}{nRT}\right). \tag{4–5}$$

Data from creep experiments on various felsic, mafic, and ultramafic rocks show that σ_d^s varies as an exponential function of $^1/_T$ at a fixed strain rate. Stress-depth profiles such as those shown in figure 4–9 are generated using Mercier's (1980a) continental and oceanic geotherm assuming $\dot{\varepsilon} = 10^{-15}s^{-1}$ for midcontinent regions and $\dot{\varepsilon} = 10^{-14}s^{-1}$ for orogenic provinces. Such profiles suggest that at a depth of 10 km and $\dot{\varepsilon} = 10^{-15}s^{-1}$ the strength of wet granite has dropped to 10 MPa (Hansen and Carter, 1983), a huge stress drop from a maximum $\sigma_d >$ 200 MPa for thrust faulting in the frictional slip regime of the schizosphere.

Paleopiezometry

Paleopiezometry is any of a number of petrofabric techniques for inferring the σ_d to which a tectonite, a deformed rock, was last subjected during deformation in the ductile-flow regime. These techniques fall into several groups based on microstructural elements, including twin density, dislocation density, subgrain size, and recrystallization grain size. Each technique is based on the correlation

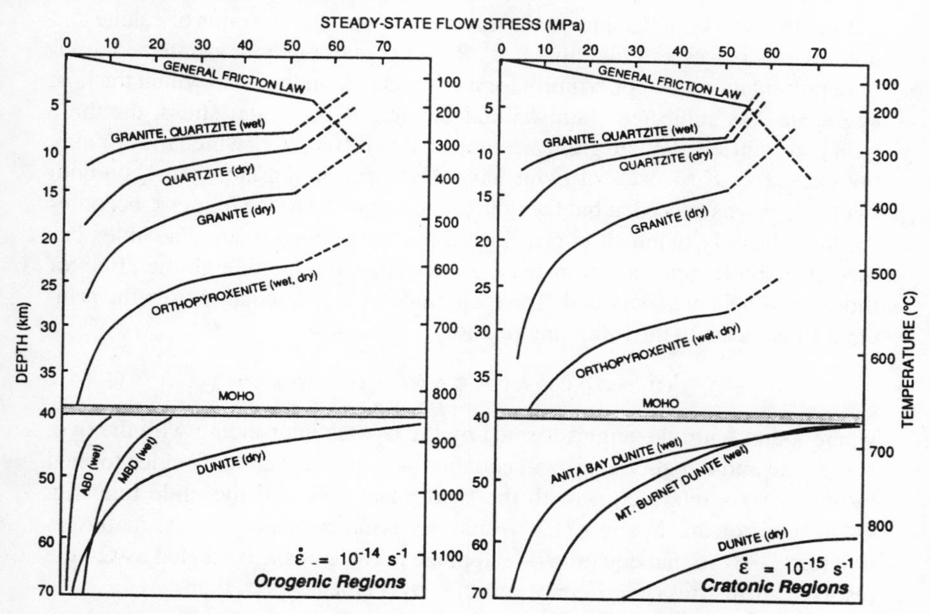

Fig. 4–9. σ_d^s-depth profiles which give the limit to stress in the lithosphere based on the average steady-state flow stress data from a number of rocks (adapted from Carter and Tsenn, 1987). Data includes Anita Bay Dunite (Chopra and Paterson, 1981); Mt. Burnet Dunite (Post, 1973); Websterite (WEB-Avé Lallemant, 1978); Aheim Dunite (Chopra and Paterson, 1981); Dunite (dry-Chopra and Paterson, 1984); Orthopyroxenite (Ross and Nielsen, 1978); Aplite (Shelton, 1981); Quartzite (dry-Koch, 1983); Granite (wet-Hansen and Carter, 1983).

between steady-state flow stress and the microstructural element (Bird et al., 1969; Luton and Sellars, 1969; and Glover and Sellars, 1973). The latter three microstructural elements vary monotonically with σ_d and operate independently of temperature and total strain.

Twin Density

Laboratory experiments indicate that the number of twin lamellae (i.e., the *twin lamellae index* or the number of lamellae per mm of crystal normal to the *e*-plane) in low porosity limestones increases as a function of both σ_d and duration of the load (Friedman and Heard, 1974). Remember that twin lamellae are so narrow that twinned crystal cannot be observed. For creep tests, the twin lamellae index was found to increase at the rate of 10 per 3.5 MPa of σ_d up to a $\sigma_d = 40$ MPa (Friedman and Heard, 1974). If the twin lamellae index is properly calibrated as a function of σ_d and duration of load, then the index in naturally deformed rocks is a measure of the maximum σ_d to which the rock was subjected.

Like the twin lamellae index, the number of twin sets in grains of calcite and dolomite increases as a function of σ_d. Rocks containing both calcite and dolomite present a unique opportunity for a cross check on the σ_d to which the host aggregate was subjected (Jamison and Spang, 1976). Under stress, the three twin planes in a carbonate grain are subjected to different τ_r, which may or may not exceed τ_c. If two sets of twins have developed in a grain, then τ_c on both twin planes was exceeded but the twin plane subjected to the higher τ_r becomes the more heavily twinned. In principle, an increase in twin lamellae index has the same significance as an increase in twin thickness, although the effect of time is still poorly understood. The magnitude of τ_r is calculated from the principal stresses using the following relationship:

$$\tau_r = (\sigma_1 - \sigma_3) \cos \lambda \cos \phi + (\sigma_2 - \sigma_3) \cos \kappa \cos \chi \qquad (4\text{–}6)$$

where λ and ϕ are the angles formed by the σ_1 axis intersecting with the twin plane pole and the glide line as in equation 4–1. κ and χ are the angles formed by the σ_3 axis intersecting with the twin plane pole and the glide line (fig. 4–2a). Jamison and Spang's (1976) analysis assumes that $\sigma_2 - \sigma_3$ is small relative to $\sigma_1 - \sigma_3$ so that equation 4–1 applies. If $\cos \phi \cos \lambda$ is labeled as Ω_0, the resolved shear stress coefficient, and $\sigma_1 - \sigma_3$ is labeled as σ_d then

$$\sigma_d = \frac{\tau_c}{\Omega_0}. \qquad (4\text{–}7)$$

Ω_0 varies relative to the pole to the twin plane and the *c*-axis according to figure 4–2b. In carbonates, Ω_0 is evaluated by counting the percentage of twinned calcite and dolomite grains containing one, two, and three twin sets. From this

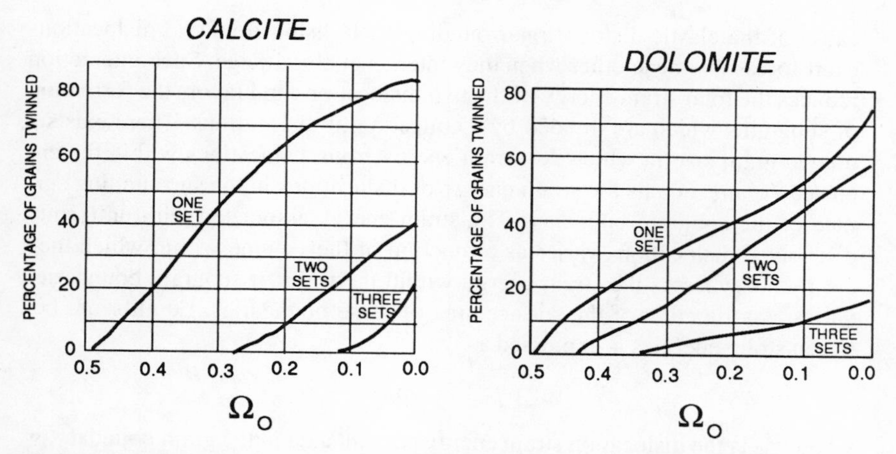

Fig. 4–10. Plots of twin-set development versus Ω_0 for calcite and dolomite. Curves are cumulative (that is, one twin-set curve includes all grains that also contain two and three sets) (adapted from Jamison and Spang, 1976).

analysis Ω_0 is determined using the plot in figure 4–10. If τ_c is known, then σ_d is calculated using equation 4–7.

A proper assessment of σ_d is critically dependent on the correct assessment of τ_c. The experimentally determined value for τ_c in calcite is given as 10 MPa (Turner et al., 1954), whereas in dolomite τ_c is 60 MPa (Higgs and Handin, 1959). Tullis's (1980) compilation of data on dolomite suggests that τ_c is as high as 100 MPa. In the analysis of natural samples, Laurent et al. (1990) suggest that τ_c for calcite is lower than 10 MPa. This question is still without a resolution. A correct assessment of τ_c is further complicated by the observation that in forelands such as the Appalachian Plateau large calcite grains are more likely to contain twins than small grains. In the laboratory Spiers and Rutter (1984) have confirmed that grains in the coarse Carrara marble twin at lower σ_d than grains in the fine Solnhofen limestone. Although the τ_c-rule still holds, grain boundaries impede twinning, so the probability that a grain of a certain size will have twins increases as a function of σ_d.

Dislocation Density

The theory for the correlation between the microstructural elements (dislocation density, subgrain size, and recrystallization grain size) and flow stress follows Twiss's (1977) analysis. During hot creep in rocks, dislocations are generated at a certain rate, depending on σ_d and composition of the host grain. As the velocity of the dislocation is largely independent of stress, higher stresses lead to a higher dislocation density, ρ_{dis}, because of a higher nucleation rate. The strain energy of the host increases with the density of dislocations. Because of the elastic distortion surrounding each dislocation, two dislocations exert forces on one another when they interact at close range. Such interaction reduces the total strain energy of the two dislocations and favors the formation of subgrains which are divided by a collection of dislocations. There exists a unique grain size in which the strain energy from dislocations within the enclosed volume equals the strain energy of dislocations in the surrounding surface of the subgrain boundary. The strain energy associated with uniformly distributed dislocations varies as a function of the volume within which they are found, whereas the strain energy within the walls of subgrain boundaries varies as a function of the surface area of those boundaries. Equilibrium between strain energies is expressed as

$$6\gamma_{dis}d^2 = w\rho_{dis}d^3 \tag{4–8}$$

where γ_{dis} is the dislocation strain energy per unit area in the grain boundary, w is the dislocation strain energy per unit length in the grain volume, d is the grain diameter, and ρ_{dis} is the steady-state dislocation density in the grain volume (Twiss, 1977).

The internal stress within the host (σ_{int}) varies directly with dislocation density and is proportional to the steady-state flow stress ($\sigma_d^s = \sigma_1 - \sigma_3$), so that

$$\sigma_d^s \propto \sigma_{int}\,. \tag{4–9}$$

Internal stress (σ_{int}) is equivalent to the elastic strain energy within the host. Dislocations impart a strain energy within their host by elastically distorting the crystal lattice near the dislocation. Internal stress felt by a nearest neighbor dislocation is

$$\sigma_{int} \propto \frac{\zeta \mathbf{b}}{r} \tag{4–10}$$

where r is the distance between dislocations, **b** is the Burgers vector, and ζ is the shear modulus of the host. The geometric relationship between dislocation spacing and density is

$$\rho_{dis} = 1/r^2. \tag{4–11}$$

Combining equations 4–9, 4–10, and 4–11,

$$\sigma_d^s = \alpha\zeta\mathbf{b}(\rho_{dis})^{1/2} \tag{4–12}$$

where α is an empirical parameter on the order of 1. This theory assumes that, at steady-state creep, the back stress on a dislocation caused by its interaction with other nearby dislocations must equal σ_{int}. The σ_d^s necessary to produce dislocation motion in a uniform array of edge dislocations is then

$$\sigma_d^s > \alpha\zeta\mathbf{b}(\rho_{dis})^{1/2}. \tag{4–13}$$

Using equation 4–13, a minimum estimate of creep stress, σ_d^s, is available, provided dislocation density is known.

SUBGRAIN AND RECRYSTALLIZATION GRAIN SIZE

Dislocations will migrate within the host grain into planar walls which divide the host into subgrains. After modest strain the subgrain walls and subgrain size reach a steady-state configuration which depends on the ρ_{dis} within the host grain. The size of the subgrains within the host grains is a function of ρ_{dis} where higher ρ_{dis} leads to smaller subgrain size. The host grains may then undergo a dynamic recrystallization to a mean grain size which is also a function of the steady-state flow stress. Dynamic recrystallization will result in nucleation of strain-free grains on grain boundaries of strained grains. As the new grains grow, they become strained and new nuclei form on their boundaries. At even higher σ_d old grains do not grow as large before they are consumed by new grains.

Although measuring dislocation density, subgrain size, and recrystallization grain size is straightforward, there are some limitations to the use of microstructural paleopiezometers (Twiss and Sellars, 1978). Overprinting by a later low T tectonic event or a later thermal pulse may obliterate an earlier signal by resetting the microstructures associated with the deformational event of interest. Usually there is no clear indication which stage in the tectonic history of the rock is recorded by the microstructural signal. Recent work suggests that piezometers show a sensitivity of substructure to stress that decreases with decreasing stress (Twiss, 1986). To compensate for this behavior, piezometer equations relating stress to dislocation density and subgrain size should be written as the sum of the σ_d^s plus an empirical constant, rather than simply σ_d^s as is the case for equation 4–12. Experimental work shows that ρ_{dis}, subgrain size, and recrystallization grain size reequilibrate at different rates after changes in stress and environment. Grain size is most stable under changing stress conditions, whereas the dislocation density is the most easily altered. Such limitations are illustrated by Goetze's (1975) analysis of the deformational history of olivine-bearing kimberlite nodules using recrystallizated grain size, as well as ρ_{dis}. In these nodules the grain size distribution is bimodal where the large grains indicated $\sigma_d^s = 10$ MPa, whereas the smaller second generation grains indicated $\sigma_d^s = 300$ MPa. Apparently, the nodules were subjected to high stresses during a rapid ascent into the upper crust and the low stresses are more likely representative of upper mantle conditions. ρ_{dis} of both the large and small grains record a $\sigma_d^s = 300$ MPa, and thus indicate that ρ_{dis} is more sensitive to changes in stress state than recrystallizated grain size. The sensitivity of ρ_{dis} to a secondary stress state is also demonstrated in static annealing experiments (Kohlstedt and Weathers, 1980). A high ρ_{dis} relative to recrystallization grain size is generally believed to indicate a late stage high-σ_d event as is found in the Barre granite, Vermont (Schedl et al., 1986) and along the Glarus Thrust Zone, Switzerland (Briegel and Goetze, 1978).

Successful applications of ρ_{dis}, subgrain size, and recrystallization grain size paleopiezometry include the determination of σ_d^s in the lower crust and mantle based on the microstructures found in mantle peridotites, along exhumed fault zones, in folds, and in salt domes plus décollements through salt. Two steps are involved: (1) several laboratory creep experiments to calibrate the relationship between σ_d^s and ρ_{dis}, subgrain size, and recrystallization grain size; and (2) measurement of the microstructures in natural samples to match the empirical curves and estimate flow stress. An example of a laboratory calibration for quartz and olivine is shown in figure 4–11 in which normalized dislocation density (ρ_{dis}) is plotted against normalized flow stress (σ_d).

Stress in the Upper Tier of the Ductile-Flow Regime

Dynamic Analysis

Constraints on stress in the upper tier of the ductile flow regime (i.e., ductile deformation in the schizosphere) come from the behavior of calcite. Turner's (1953) technique for identifying the orientation of stress axes based on the measurement of twins in calcite is called *dynamic analysis* (Friedman, 1964).

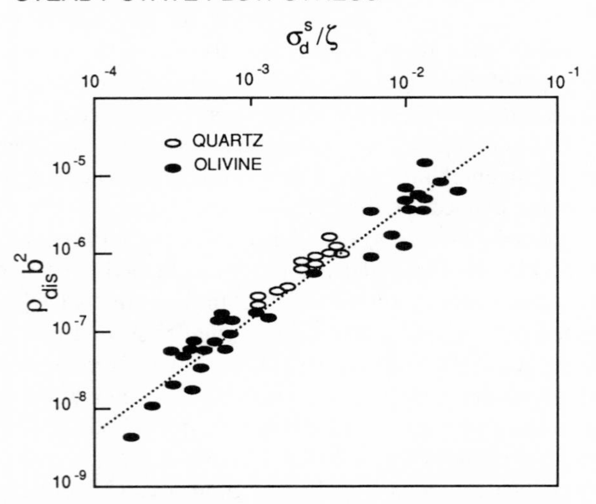

Fig. 4–11. Relationship between steady-state flow stress and dislocation density in quartz and olivine (adapted from Kohlstedt and Weathers 1980).

Dynamic analysis takes advantage of the fact that calcite grains with *c*-axes subnormal to the compression direction (i.e., σ_1) are most likely to have well-developed twin sets. Crystals so oriented have *e*-planes in an orientation of high shear stress for a positive sense of shear, whereas calcite does not develop *e*-twins with a negative sense of shear which is the sense of shear for crystals with *c*-axes oriented parallel to σ_1 (fig. 4–3). The correlation between σ_1 as deduced from the dynamic analysis of calcite and σ_1 inferred from other geological indicators gives confidence that dynamic analysis is applicable in natural situations (e.g., Nickelsen and Gross, 1959; Friedman and Conger, 1964; Carter and Friedman, 1965). A compelling example of the use of calcite for dynamic analyses is Friedman and Conger's (1964) study of a gastropod which was taken from a series of north-south trending folds in the Lower Cretaceous Kootenai Formation near Drummond, Montana. Gastropods consist of multicrystalline shells in which calcite is precipitated so that *c*-axes form a radial pattern in a cross section through the wall of the shell. Because *e*-twins grow with a positive sense of shear, only part of the radially distributed calcite crystals within the wall of the gastropod are in the correct orientation for shear with a positive sense in response to a compression across any diameter of the gastropod. Twins within grains of the naturally deformed gastropod show a markedly uneven development with grains along the north and south walls more densely twinned than grains along the east and west walls (fig. 4–12). Such a distribution of twinned grains is compatible with an east-west compression, the same responsible for folds in the area.

The first numerical technique for dynamic analysis based on twin gliding was formulated by Spang (1972). Compression and tension axes for maximum τ_r on each twin set are treated as components of a tensor with axes of unit length. Each set of axes is transformed into a common coordinate system and added to a composite 3×3 matrix which is then solved to find the principal "stress" axes by taking the eigenvalues and eigenvectors of the matrix. The solution to the eigenvalue problem yields the principal stresses acting to deform the twinned aggregate. Numerical dynamic analysis has also been applied to quartz deformation lamellae and dolomite twin lamellae (Spang and Van der Lee, 1975). A variation of this technique is to maximize the shear stress on twinned planes and minimize the shear stress on untwinned planes (Laurent et al., 1981). Other variations on these techniques have been suggested by Nissen (1964) and Deitrich and Song (1984).

A major rationale for the use of calcite twinning as a tool for dynamic analysis was the correlation of principal stress axes inferred from calcite twinning with known stress directions based on laboratory experiments (Friedman, 1963). However, the application of dynamic analysis to natural samples is somewhat problematical for samples with a large strain or complex stress history (Friedman and Sowers, 1970). In the laboratory experiments, the principal stress and strain directions tend to be coaxial, particularly if strain is limited to

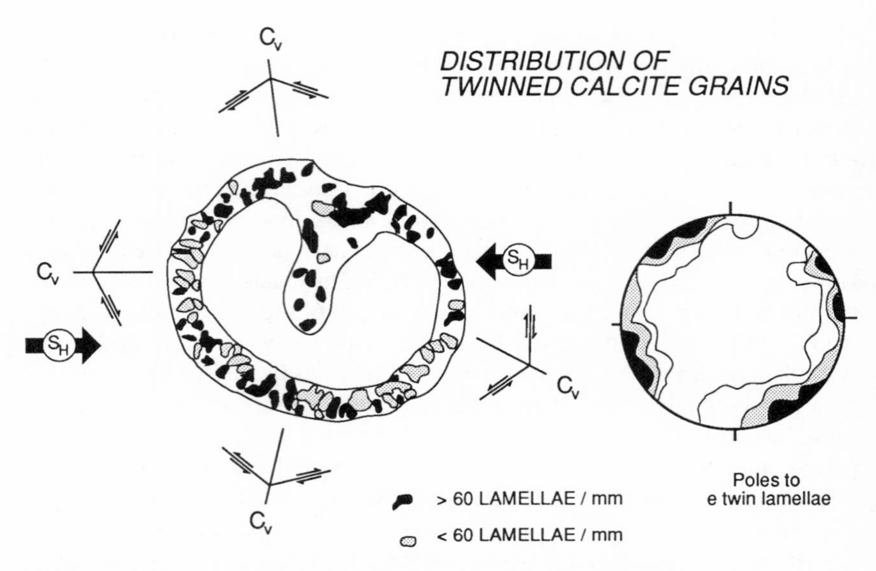

Fig. 4–12. Dynamic analysis of a gastropod from Montana (adapted from Friedman and Conger, 1964). Transverse section through a gastropod (*left*). Gastropod shell composed of calcite with *c*-axes oriented radially. Blackened grains are those most heavily twinned suggesting an east-west orientation of σ_1. Distribution of normals to 100 sets of e_1 lamellae showing slip on those planes favorably oriented for an east-west σ_1 (*right*). Also shown are the approximate orientations of two *e*-twin planes and the *c*-axis in calcite crystals at four points within the gastropod shell. With respect to an assumed general east-west σ_1, grains along the north and south sides of the shell are favorably oriented for twinning because of the sense of shear on the twin planes, whereas those on the east and west sides are unfavorably oriented for twinning.

a few percent. In the field, there are many examples where the orientation of the principal axes of progressive strain, which is well constrained on a regional scale, is not coaxial with stress axes based on the orientation of joint sets (e.g., the Appalachian Plateau [Engelder and Geiser, 1980]). When heavily twinned calcite is used for a numeric dynamic analysis, derived stress directions are only approximate.

The problem of sorting out multiple paleostress orientations from complexly twinned calcite is resolved using the inverse technique introduced for the analysis of fault-slip data as described in chapter 3 (Tourneret and Laurent, 1990). Like the analysis of fault-slip data, the inverse technique applied to calcite twinning uses four parameters for calculation of a stress tensor defining the orientation of the principal stresses and the value of R, the ratio of the magnitude of the principal stresses. The idea behind the inverse technique is to find the stress tensor which best groups twinned and untwinned planes into regions of high and low resolved shear stress. This is done by applying between 100 and 1000 random tensors to identify the tensor with the best fit. There will always be some data which are not compatible even with the best fit tensor. For the determination of stress orientations in terrane subject to polyphase deformation, a best-fit stress tensor is determined based on the orientation of untwinned *e*-planes. In this case only a modest 50% of the twinned planes are compatible with a stress tensor solution. These data are removed and a second best-fit stress tensor is calculated using the remainder of the data. In the Pyrenean foreland Tourneret and Laurent (1990) observed two stress tensors based on calcite twins: a N-S compression and an extension with σ_1 vertical.

Heavily twinned calcite is appropriately used for strain analysis (Conel, 1962; Groshong, 1972). Shear strain is readily apparent in a heavily twinned calcite grain. Engineering shear strain for a heavily twinned calcite crystal, γ_τ, is tan ψ or $\gamma_\tau = q/t$ with q and t defined in figure 4–4b. The displacement of the *e*-plane, p_i, is equal to $2\tau_1 \tan(\alpha/2)$. The length of q is the sum of all p_i within a grain. The shear strain in a partially twinned grain is thus

$$\gamma_\tau = \Sigma p_i / t. \tag{4–14}$$

A minimum of five twin sets with known shear strains and orientations are necessary to calculate the strain in an aggregate of calcite crystals (Groshong, 1972). For the most well-constrained solutions, measurements in 30–50 grains are required. The calcite strain-gauge technique has been used quite effectively for a number of studies in naturally strained rocks (Engelder, 1979; Groshong et al., 1984).

As long as it is appreciated that twinned calcite may better serve as a strain gauge than a stress gauge because the effect of load duration is poorly understood, dynamic analysis works for such applications as the study of outcrop scale buckle folds. Elastic theory for buckling suggests that, once folding is initiated, folds might contain a neutral fiber dividing the fold into interior and exterior parts. Interior to the neutral fiber σ_1 is parallel to the folded bed throughout the history of the fold. Exterior to the neutral fiber σ_1 rotates into a position normal to the bedding plane so that outer arc extension takes place parallel to bedding. While most field studies fail to find this outer arc extension (e.g., Friedman and Stearns, 1971; Chapple and Spang, 1974; Groshong, 1975), Conel (1962) was successful during the study of a folded layer of limestone taken from the Silurian McKenzie limestone near Hancock, Maryland. Folds in the McKenzie limestone show a layer-parallel extension which was recorded in the outer arc of the fold mainly in calcite-filled veins. Analysis of this fold, based on calcite twinning in late-formed veins, show that exterior to the neutral fiber σ_1 was perpendicular to both the layer and the fold axes (Spang and Groshong, 1981).

Quartz deformation lamellae also serve for dynamic analyses of natural samples. One of the first field tests was a study of the Devonian Oriskany Sandstone in the Appalachian Valley and Ridge of Pennsylvania where the orientation of principal stress axes deduced from an analysis of quartz deformation lamellae is nearly the same as those deduced from an analysis of calcite twins in the matrix of the sandstone (Hanson and Borg, 1962). For an analysis of stress during folding of the Dry Creek Ridge anticline, Carter and Friedman (1965) used both quartz-deformation and calcite-twin lamellae within the Jurassic Swift Formation, a calcite-cemented sandstone. This anticline is located in the structurally complex mountain-front zone at the northwest corner of the Beartooth Mountains, Montana. Dynamic analyses show that in the fold limbs σ_1 was inclined at low angles to the dip and σ_3 was normal to the bedding (fig. 4–13). Although dry quartz is very strong relative to calcite, Carter and Friedman (1965) also concluded that under the ambient conditions of deformation (about 100 MPa overburden pressure, 100° to 150° C), the σ_d for slip in quartz was probably as low as that for twinning of the calcite cement.

Differential Stress

For laboratory creep tests, the twin lamellae index was found to increase at the rate of 10 per 3.5 MPa of σ_d up to a $\sigma_d = 40$ MPa (Friedman and Heard, 1974). This calibration was used to measure σ_d in the Texas Gulf Coast where $S_v > S_H > S_h$. Samples of limestone core from depths up to 6.5 km show an increase in twin lamellae index of 10 per 1800 m of depth. Using their laboratory calibration, Friedman and Heard (1974) conclude that at a depth of 6.5 km, σ_d is approximately 12.6 MPa, so that $S_h/S_v = .93$. This result is interesting when considered in conjunction with the observation that pore pressure in the Gulf of Mexico rises to about 85% to 90% of overburden at depths below 3 km (fig. 2–7). Recall from arguments in chapter 2 that P_p is held at 85% to 90% of overburden because at this level P_p and S_h were balanced, and thus, equal in the clastic wedge of the Gulf of Mexico. Arguing that S_h is 85% to 90% of overburden based on P_p data is consistent with Friedman's and Heard's (1974) measure of horizontal stress.

Another estimate of σ_d is based on the presence of both dolomite and calcite in the same aggregate. Jamison and Spang (1976) measured the number of mechanical twins in dolomite and calcite grains in four samples from the McConnell thrust sheet of the Front Ranges of the Canadian Rocky Mountains. They then applied equations 4–6 and 4–7 to calculate σ_d during the emplacement of the McConnell thrust sheet and found that σ_d reached a fairly uniform maximum value of about 125 MPa. This is well below the σ_d predicted by the general law of friction for thrust faulting at burial depths of 5 km and hydrostatic P_p but is consistent with frictional stresses if P_p was about 75% of overburden (fig. 4–14). Note that σ_d is nearly an order of magnitude higher for thrusting in the Canadian Rocky Mountains than was found for normal faulting in the Texas Gulf Coast at comparable depths of burial (i.e., 5–7 km). Data from the Pyrenean foreland, a setting forward of major thrusting, suggest that σ_d was about 80 MPa and, hence, less than that found during thrusting in the

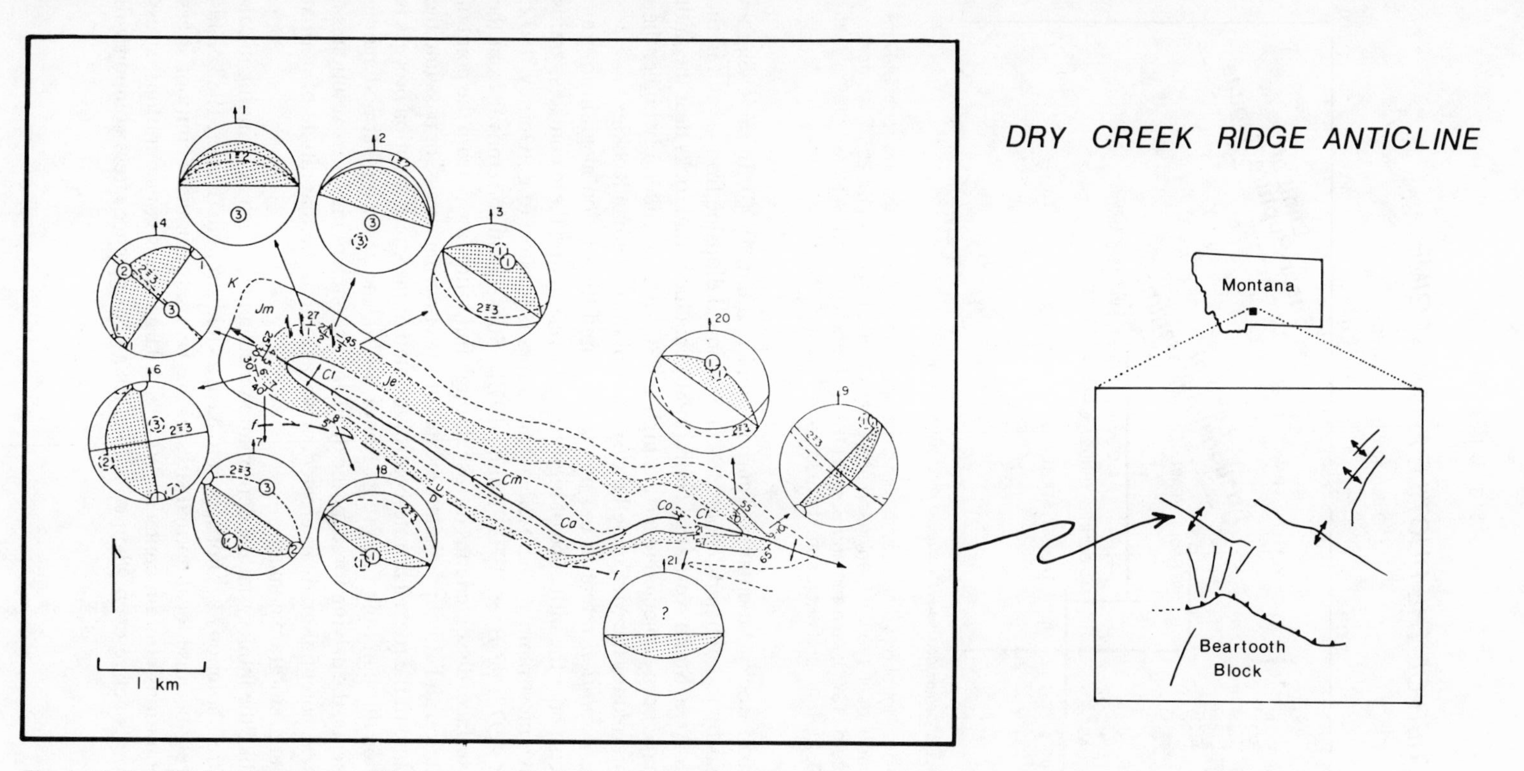

Fig. 4–13. Dry Creek Ridge anticline. Orientations of principal stresses around a large fold deduced from twinning in calcite (dashed) and intragranular flow of quartz (solid). Numbers refer to subscript of σ_1, σ_2, and σ_3. Stippled planes show orientation of bedding (adapted from Carter and Friedman, 1965).

UPPER TIER OF THE DUCTILE-FLOW REGIME

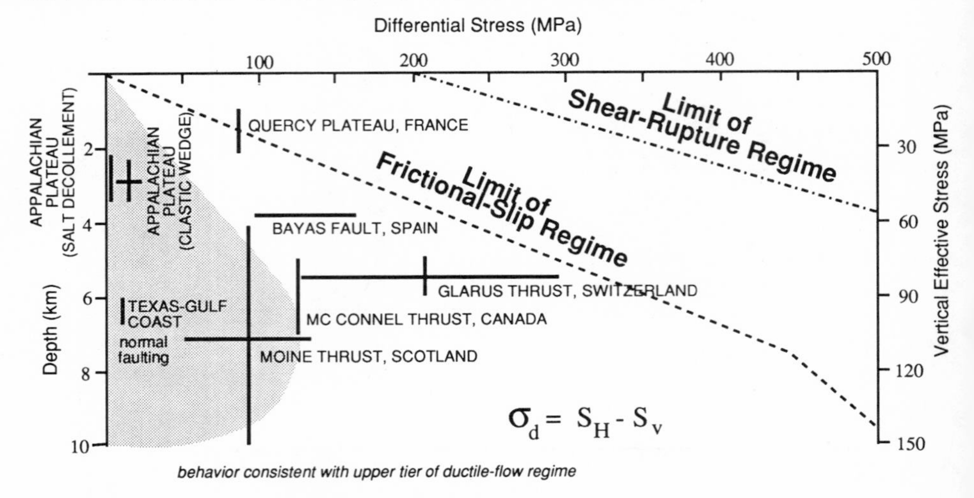

Fig. 4–14. Plot of σ_d versus depth within the upper tier of the ductile-flow regime. Data from Carter et al. (1982); Blenkinsop and Drury (1990); Engelder (1982c); Friedman and Heard (1974); Jamison and Spang (1976); Kappmeyer and Wiltschko (1984); Pfiffner (1982); Tourneret and Laurent (1990); and Twiss (1977).

Canadian Rocky Mountains (Tourneret and Laurent, 1990). In reviewing an analysis by Raleigh and Talbot (1967) of twinned diopside from a fault in the Palmer Area, South Australia, Tullis (1980) concludes that rocks near the fault may have been subjected to a σ_d as high as 280 MPa, considerably higher than that inferred for the McConnell thrust sheet of the Canadian Rockies.

Stress solution refers to the process of rapid dissolution at grain-to-grain contacts under the influence of high normal stress across these contacts; grains dissolve most rapidly where the normal stress is greatest (e.g., Durney, 1972; Elliott, 1973; de Boer, 1977; Robin, 1978). Subsequent to removal from the grain surface, dissolved ions diffuse through fluid films away from the point of dissolution and either precipitate at sites of lower normal stress, or pass into the pore fluid of the rock. The rate-controlling step of the stress-solution process is diffusion through fluid films (e.g., Rutter, 1976), and based on thermodynamic arguments, the driving force for the diffusion accompanying stress solution is a chemical potential gradient created by variations in the magnitude of normal stress at grain-grain contacts.

Unlike the intracrystalline deformation mechanisms such as twinning in calcite, stress solution does not appear to have a yield strength (Rutter, 1983). This is largely because local normal stresses are very large in the vicinity of stress concentrations at grain contacts. With such large stress concentrations, stress solution is active even when σ_d is low. As a consequence stress solution will operate at σ_d less than necessary to activate crystal-plastic slip systems. Evidence includes solution-pitted pebbles never buried more than 30–40 m (Trurnit, 1968) and sutured stylolites at burial depths of 90 m (Schlanger, 1964). In foreland fold-thrust belts, the presence of stress solution as the active deformation mechanism in the absence of calcite twinning further demonstrates that stress solution operates at low σ_d (Alvarez et al., 1976). These observations led to the conclusion that deformation in foreland fold-thrust belts occurs at considerably lower σ_d than necessary to initiate slip according to the general friction law. Although it is difficult to constrain a lower limit for σ_d during active stress-solution creep, Engelder (1982c) uses the stress concentration at the axial canal of crinoid columnals to argue that locally $\sigma_d < 6.2$ MPa for deformation of the Appalachian Plateau during the Alleghanian Orogeny.

If σ_d was too low for brittle frictional slip during certain phases of mountain building, what is the explanation of the pervasive faults throughout forelands? Part of the explanation is tied to the nature of faulting which is characterized by slickensides coated with secondary mineral growth. The *slickenlines* on these surfaces are not brittle wear grooves indicative of high frictional stresses but rather fibrous minerals indicative of local dissolution and mineral growth. Technically, slip by a ductile mechanism is also frictional slip, however in the context of this book the frictional-slip regime refers to σ_d developed as a consequence of brittle wear. The Umbrian Apennine fold-thrust belt contains an array of faults displaying slickenlines. Such faults have a preferred orientation and fall into one of two clusters forming conjugate sets with a 90° dihedral angle (Marshak et al., 1982). A conjugate set of faults forming at a 90° dihedral angle is unusual for brittle behavior. A 90° dihedral angle suggests that the faults slipped at the maximum τ, a situation indicative of a ductile creep mechanism (i.e., stress solution) rather than a brittle friction. If so, these fault surfaces slipped under relatively low τ in the upper tier of the ductile-flow regime. In other mountain belts σ_d is sufficiently low so that the crack propagation under high pore pressure is the favored brittle deformation mechanism. During the Alleghanian Orogeny, not only was brittle deformation of clastic sediments of the Appalachian Plateau restricted to crack propagation (e.g., Bahat and Engelder, 1984), but also the Alleghanian joints were not reactivated in frictional slip after the stress field rotation to subject early joints to a τ (Geiser and Engelder, 1983). This further suggests that σ_d remained relatively low[2] throughout large portions of the Appalachian Plateau.

Stress in Deformed Salt

Estimates of σ_d within the salt décollement of the Appalachian Plateau are consistent with the low stresses found in most of the clastic section above the décollement. Carter et al. (1982) illustrate the use of paleopiezometry to estimate the σ_d in natural rock salt, one of the weakest of rocks in the crust of the earth. Figure 4–15 shows a plot of subgrain size (μm) versus log σ_d (MPa) for several creep experiments using Avery Island domal salt. The σ_d versus subgrain-size curves suggest the relation

$$d\ (\mu m) = 190\ \sigma_d^{-1}\ \text{MPa}. \quad (4\text{–}15)$$

These curves are then used to plot subgrain sizes for naturally occurring rocksalt (fig. 4–15). For the deformation of salt σ_d was found to vary between 0.5 and 2.5 MPa, values significantly lower than found in other geological settings in the upper crust. The highest value of σ_d (= 2.5 MPa) was found in salt from the Appalachian Plateau décollement at Lansing, New York. From this study Carter et al. (1982) concluded that bedded (i.e., undeformed) salt was subject to lower σ_d than naturally deformed salt, either domal salt or salt from bedding plane décollements.

A compilation of data from the upper tier of the ductile-flow regime reveals that while σ_d near fault zones approaches the frictional strength of the schizosphere, there is plenty of evidence indicating that the heart of foreland fold-thrust belts was generally at σ_d, much less than required for frictional slip (fig. 4–14). Data from the Glarus Thrust in Switzerland indicate that σ_d was near the frictional strength of the upper crust (Briegel and Goetze, 1978). Within the Marquette Synclinorium, Michigan, σ_d varied from 1 to 36 MPa (Kappmeyer and Wiltschko, 1984). σ_d near other major faults was lower, ranging from 125 MPa near the McConnell Thrust of the Canadian Rockies to 2.5 MPa along the Appalachian Plateau décollement. Fold-thrust belts riding atop a salt décollement (e.g., the Appalachian Plateau, the Franklin Mountains of Canada, the Jura of the Alps) are extremely wide and characterized by a small cross-sectional taper angle (Davis and Engelder, 1985). The width of these fold-thrust belts is attributed to the extremely weak nature of the salt décollement and a relatively low σ_d within the clastic wedges. All of this means that earthquakes, a frictional phenomenon responding to higher σ_d, are not common within the clastic wedges of fold-thrust belts in the schizosphere and that ductile flow at relatively low σ_d plays a major role in foreland deformation (Groshong, 1988).

Stress in the Middle Tier of the Ductile-Flow Regime

Although ductile flow is commonly encountered in foreland settings where clastic rocks, shales, siltstones, carbonates, and salt are the dominant lithology, crystalline basement in the upper 10 km of the crust is strong enough to favor frictional slip. As temperature increases in the lithosphere some minerals in crystalline basement deform in a ductile manner and the bulk rock behaves as a semibrittle material. This is because at crustal depths in excess of 10 km, dislocation creep and other thermally activated processes begin to affect the

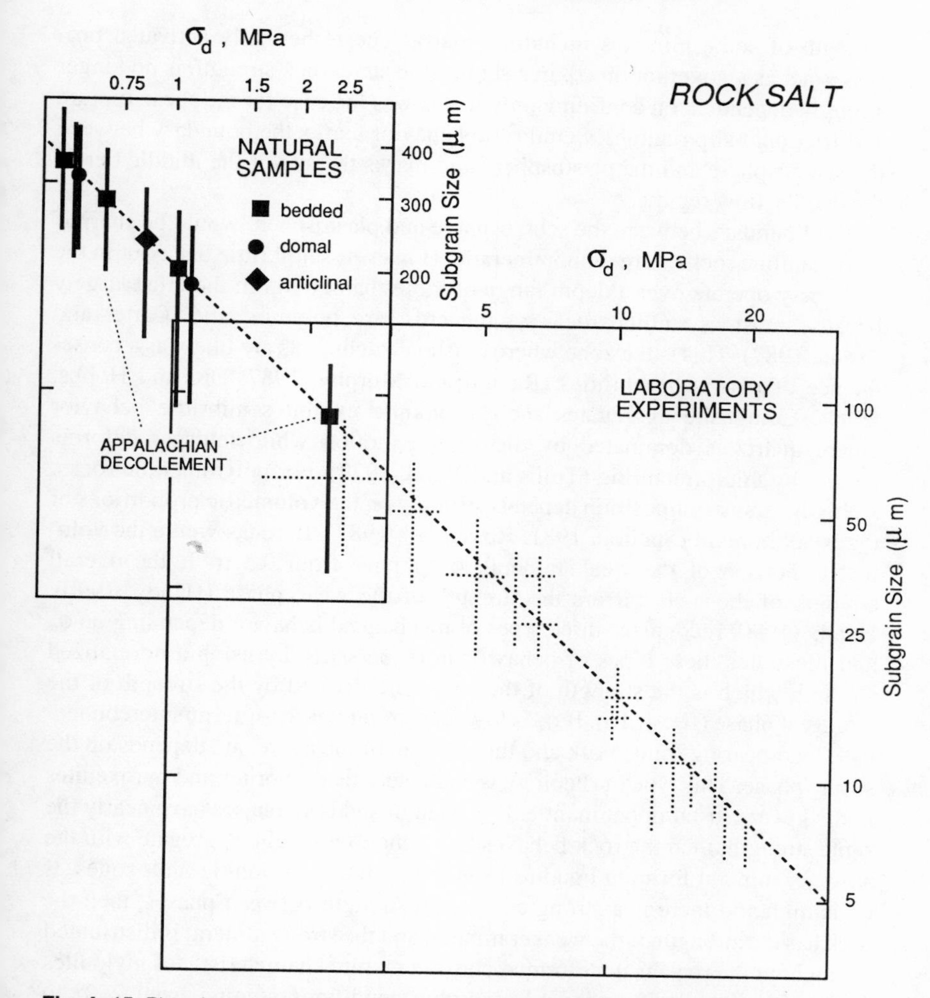

Fig. 4–15. Plot of log subgrain size (μm) versus log σ_d (MPa) for creep experiments on Avery Island domal salt. Horizontal bars show maximum and minimum stress levels recorded during experiments and vertical bars show one standard deviation from mean subgrain diameter. Data fit the dashed line with a correlation coefficient of 0.9775. Also shown are mean subgrain sizes and corresponding σ_d for seven naturally bedded, domal, and anticlinal rock-salts. Vertical bars show one standard deviation from mean subgrain diameter (adapted from Carter et al., 1982).

strength of some minerals including quartz. These thermally activated processes act as a governor on σ_d. In a semibrittle state, rock strength is no longer strongly dependent on confining pressure as was the case for the shear-rupture and frictional-slip regimes. Semibrittle behavior marks the boundary between the schizosphere and the plastosphere and marks the site of the middle tier of the ductile-flow regime.

The boundary between the schizosphere and plastosphere would be sharp if all crystalline rocks were monomineralic. However, semibrittle behavior in the crust may operate over a depth range of more than 20 km in the crust largely because most crystalline rocks of the crust are polymineralic (Carter and Tsenn, 1987). This is the zone where crustal detachments are likely as a consequence of plastic instabilities (Ranalli and Murphy, 1987; Ord and Hobbs, 1989). Composite aggregates such as granite exhibit semibrittle behavior where quartz is dominated by dislocation motion while feldspar deforms largely by microfracturing (Tullis and Yund, 1977). For polymineralic rocks, strength versus composition depends strongly on the volumetric proportions of the weak mineral (Shelton, 1981; Ross et al., 1987). In rocks where the volumetric portion of the weak mineral, ϕ_w, ranges from 0.3 to 1, the overall strength of the rock mirrors the strength of the weak phase (Handy, 1990). Handy (1990) recognizes three types of mechanical behavior depending on ϕ_w and illustrates these types of behavior in σ_d-ϕ_w space by using a normalized strength which is the strength of the aggregate divided by the strength of the strongest phase (fig. 4–16). If ϕ_w is low, the strong phase forms an interconnective, load bearing framework and the strength of the aggregate depends on the strong phase. This type of rheology is characteristic of diorites and pyroxenites in the lower crustal/upper mantle. If ϕ_w is high and both phases have nearly the same strength, then the rock behaves like a monomineralic aggregate with the stronger mineral forming boudins as is found in ultramylonitic fault zones. If ϕ_w is high and there is a strong contrast in strength between phases, then the rock has the strength of the weaker mineral and the strong mineral is distributed throughout the rock as undeformed clasts, a texture characteristic of mylonites and quartz-rich rocks under metamorphic conditions ranging from 250° to 800°C. Handy (1990) applies a grain-size paleopiezometer to show that the quartz matrix in a mylonitized granodiorite was subjected to stress contrasts of $\approx$ 300% over a distance of about a mm. Handy concludes that "the rate of stress drop at the end of deformation must have far exceeded the rate at which dynamically recrystallized grain size could equilibrate with stress."

The Architecture of Fault Zones

Faulting is affected by changes in mechanism from brittle to ductile behavior as evidenced by exhumed fault zones which were once active at great depths. Some famous examples include the Moine Thrust of Scotland, the Alpine Fault

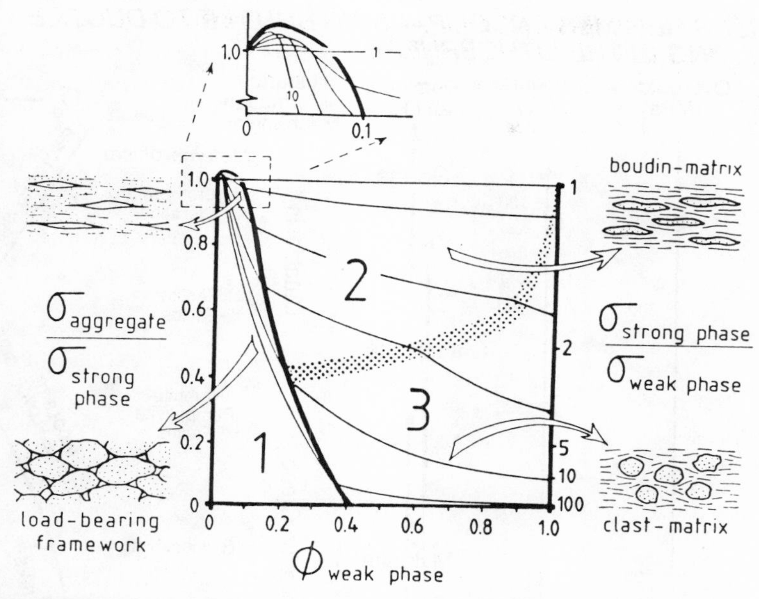

Fig. 4–16. Hypothetical diagram of normalized aggregate strength versus volume proportion of the weak phase for the two-phase aggregate (adapted from Handy, 1990). Right-hand axis is the relative strength of the strong and weak phases. Heavy line is the boundary between framework-supported rheologies (domain 1) and matrix-controlled rheologies (domains 2 and 3). Stippled curve indicates the transition from a boudin-matrix rheology (domain 2) to a clast-matrix rheology (domain 3). Thin curves are the contours of normalized rock strength at various mineral strength ratios that are indicated on the right-hand axis.

of New Zealand, the Outer Hebrides Thrusts of Scotland, the Greenland Shear Zones, and the mantled gneiss domes of the southwestern United States. Figure 4–17 indicates the correlation between depth of burial and several parameters including temperature, confining pressure, metamorphic grade, and general type of fault rock (Sibson, 1977a, 1986). Fault rock is characterized by grain reduction through either brittle or ductile mechanisms with the three main types of fault rock being gouge, cataclasite, and mylonite. *Gouge*, a product of frictional slip, is a cataclastic material without fabric or cohesion; grain-size reduction is dominated by brittle fracture. A *cataclasite* is a nonfoliated fault-zone material with cohesion, whereas a *mylonite* is a foliated fault-zone material with cohesion. Both cataclasites and mylonites are the product of semibrittle deformation with differences depending, in part, on strain rate versus recovery rate of grains (fig. 4–18). As a general rule, higher σ_d is associated with higher strain rate.

Strike-slip faults differ somewhat from thrust faults with depth of burial. One good exposure of a strike-slip fault at depth is the Nordre Stromfjord Shear

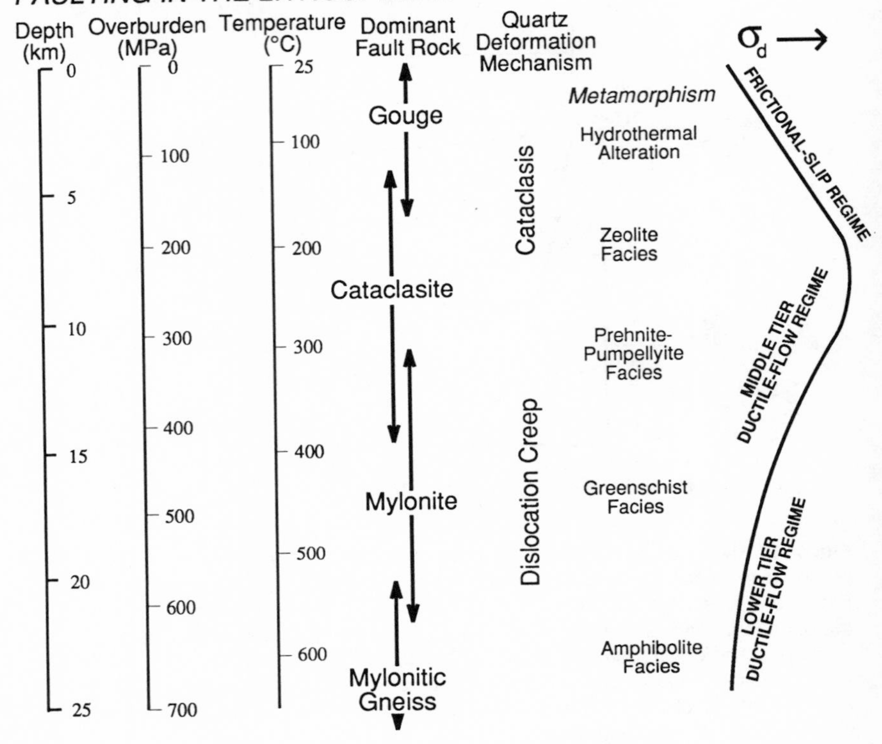

Fig. 4–17. A conceptual diagram for a major fault zone encompassing both brittle and ductile rocks in the continental crust. This diagram schematically relates different stress regimes to likely metamorphic environment, dominant quartz steady-state deformation mechanisms, and associated fault rocks. Profile of σ_d is in arbitrary units (adapted from Sibson, 1986).

Zone in western Greenland where the crystalline terrain dips to the east. The shear zone, exposed in amphibolite and granulite facies rocks, becomes wider with depth in the crust of the earth. A cross section through the San Andreas fault might look like the section mapped in western Greenland where the shear zone widens with depth through amphibolite and into granulite terrain. Major earthquakes are located in the upper portion of large strike-slip faults. Along the San Andreas fault, for example, the majority of earthquakes originate in the top 15 km of the crust where the fault zone is relatively narrow (fig. 4–17). Below the zone of seismicity, fault zones become ductile and much wider. This transition in seismic behavior correlates with the middle tier of the ductile-flow regime.

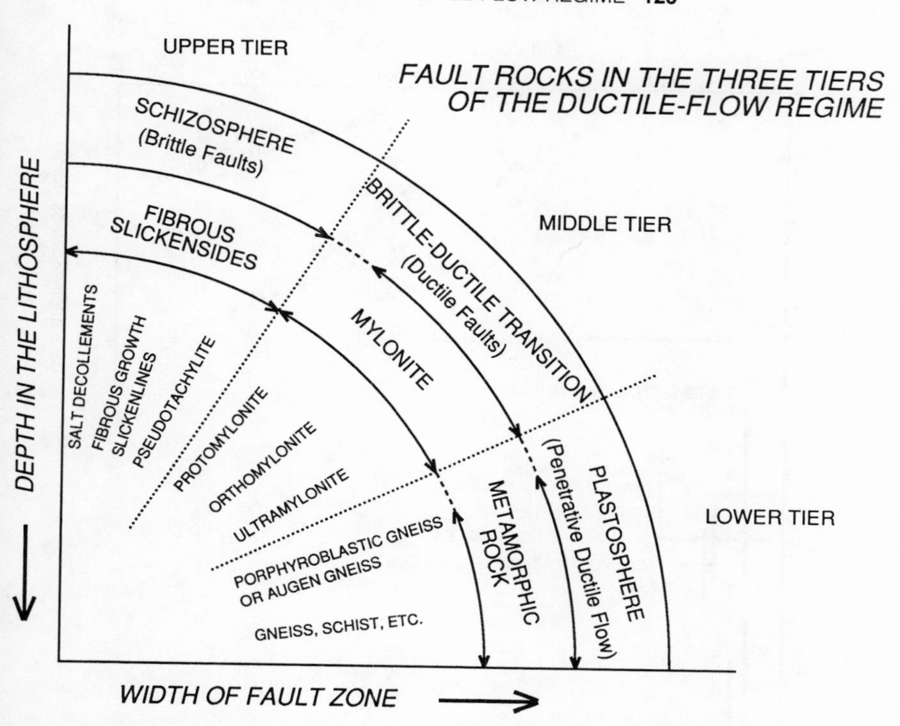

Fig. 4–18. Terminology of fault-related rocks in the three tiers of the ductile-flow regime. Horizontal and vertical scales are variable depending on composition, grain size, and fluids of the fault rocks (adapted from Wise et al., 1984).

Stress within Fault Zones

Paleopiezometry yields a variety of estimates for σ_d along fault zones and within other crustal structures. Ductile shear zones within metamorphic core complexes represent examples of midcrustal faults with large amounts of slip (Wernicke, 1981). Data from quartz paleopiezometry suggest that during mylonitization at 630°C, σ_d was between 38 and 64 MPa for ductile shear zones within the Ruby Mountains (Hacker et al., 1990). During mylonitization in the Ruby Mountains S_v was 400 ± 100 MPa according to a garnet-biotite-muscovite-plagioclase thermobarometer (Hurlow, 1988). Extrapolation of quartzite flow laws indicates that the mylonitization occurred within an order of magnitude of $\dot{\varepsilon} = 10^{-12}s^{-1}$.

In fault zones from quartz-bearing rocks σ_d ranges from as low as 20–40 MPa along the Ikertoq shear zone in Greenland (Kohlstedt et al., 1979) to as high as 130 MPa in the Massif Central of France (Burg and Laurent, 1978) (fig. 4–19). Some of the fault zones come from depths as great as 30 km where temperatures range up to 800°C. Work on calcite from the Glarus mylonite and

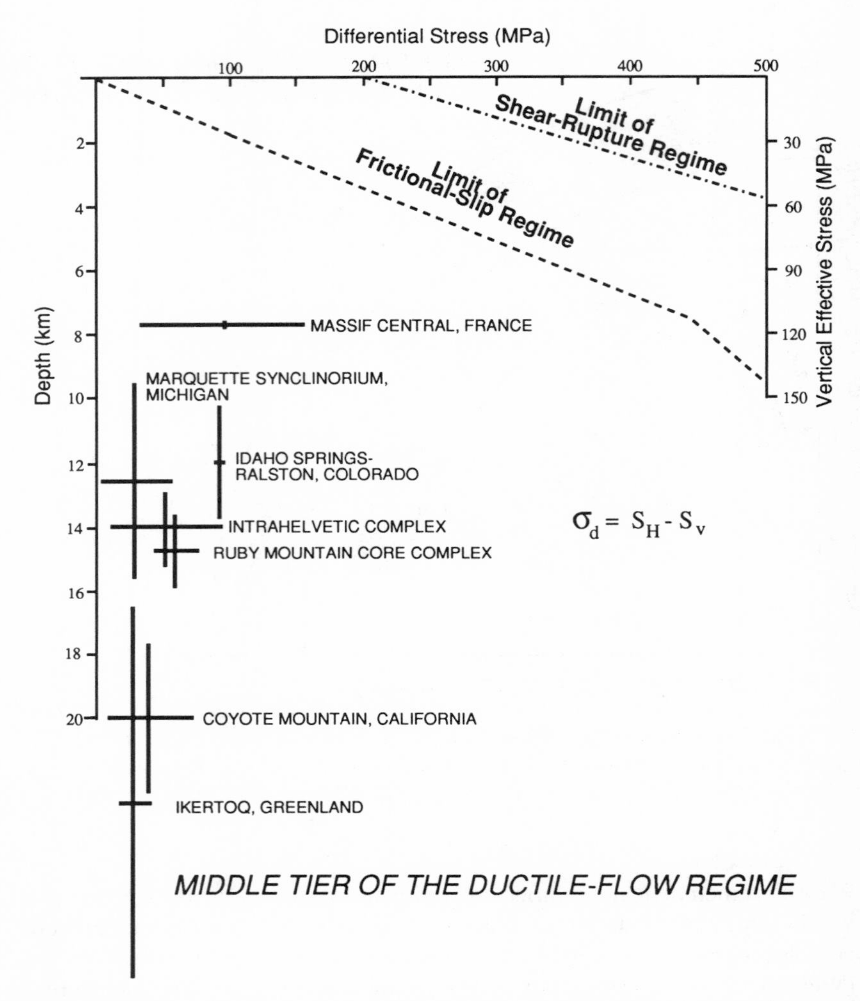

Fig. 4–19. Plot of σ_d versus depth from fault zones within the middle tier of the ductile-flow regime using only recrystallized grain size data. Data from Burg and Laurent (1978); Christie and Ord (1980); Hacker et al. (1990); Kohlstedt et al. (1979); Kohlstedt and Weathers (1980); and Pfiffner (1982).

within the Intrahelvetic complex yield σ_d estimates of 25–260 MPa and 42–100, respectively (Pfiffner, 1982). Although stresses as high as 200 MPa are reported for the Mullen Creek-Nash Fork shear zone in Wyoming (e.g., Weathers et al., 1979), these data are based on free dislocation density which is notoriously sensitive to late-stage stress changes. In some fault-zone samples, a low ρ_{dis} or large recrystallization grain size may reflect a statically annealed state associated with stresses generated during uplift, erosion, and cooling (Kohlstedt and Weathers, 1980). Samples from the Ikertoq shear zone, for example, indicate a relatively low σ_d which, Kohlstedt and Weathers (1980) suggest, does not represent the major episode of deformation within the fault zone but rather some other point in a complicated history during uplift and erosion. Yet, the low σ_d is consistent with data on the creep strength of crustal rocks (e.g., fig. 4–9) and probably is representative of stresses at 20–30 km. The Woodruffe Thrust in Australia (Bell and Etheridge, 1973) and the Idaho Springs-Ralston shear zone in Colorado (Kohlstedt and Weathers, 1980) display an increase in deformation intensity from the boundaries inward. At the fault zone edges quartz grains show undulatory extinction and low-angle boundaries, whereas in the center of the fault zones the quartz is completely recrystallized. This is compelling evidence that σ_d during faulting was recorded by the fault-zone samples. In general, σ_d in the middle tier of the ductile-flow regime decrease with depth in a manner consistent with laboratory creep data for quartzite and granitic rock (fig. 4–19).

Stress in the Lower Tier of the Ductile-Flow Regime

A direct measurement of stress in the mantle would be very expensive and maybe beyond present technical capabilities. Therefore, indirect measurements of σ_d using paleopiezometry must suffice for some time to come (Avé Lallement et al., 1980). Olivine-bearing xenoliths within intrusions and ultramafic rocks in obducted ophiolites give an indication of σ_d^s in the upper mantle (Post, 1973; Mercier et al., 1977). One estimate of the range in mantle stresses comes from measuring the recrystallization grain size of nodules sampled in the Basin and Range and in Southern Africa (Mercier, 1980b). The interest in comparing xenoliths from these two locations is that the Basin and Range samples represent the shallow mantle under a continental extension zone, whereas those from Southern Africa represent deeper intracratonic mantle. By assuming one mechanism for recrystallization (e.g., Ross et al., 1980) the nodules from both localities seemed to reflect a discontinuity in the σ_d^s somewhere below 60 km. Origin depths of the nodules were estimated using Mercier's (1980a) single-pyroxene thermobarometer. Mercier (1980b) assumed that this discontinuity was indicative of a change in recrystallization mechanism in which case the σ_d^s-versus-depth profile was a continuous curve. Avé Lallemant's (1985) experiments suggested that there was no basis for assuming a change in recrystallization mechanism in which case there was a discontinuity in flow stress. If so, then as upward flow of mantle rocks crossed the lithosphere-asthenosphere boundary at a depth of 65 km, the flow diverged causing flow velocities to decrease abruptly. According to equation 4–5, a consequence of the decrease in strain rate by an order of magnitude is that σ_d^s drops by a factor of two. σ_d^s in the uppermost mantle under a continental extension zone is as high as 45 MPa, whereas σ_d^s in the intracratonic mantle is less than 20 MPa; σ_d^s decreases to less than 10 MPa at depths below 100 km in intracratonic mantle (Mercier, 1980b) (fig. 4–20).

Stress-Depth Profiles

Laboratory data on rock strength in both the brittle and ductile regime constrains the stress distribution in the lithosphere (e.g., Goetze and Evans, 1979; Brace and Kohlstedt, 1980; Kirby, 1980). If both brittle and ductile deformation mechanisms are active as is the case for semibrittle behavior, σ_d is defined by the weaker active mechanism (e.g., Handy, 1990). In the schizosphere, fric-

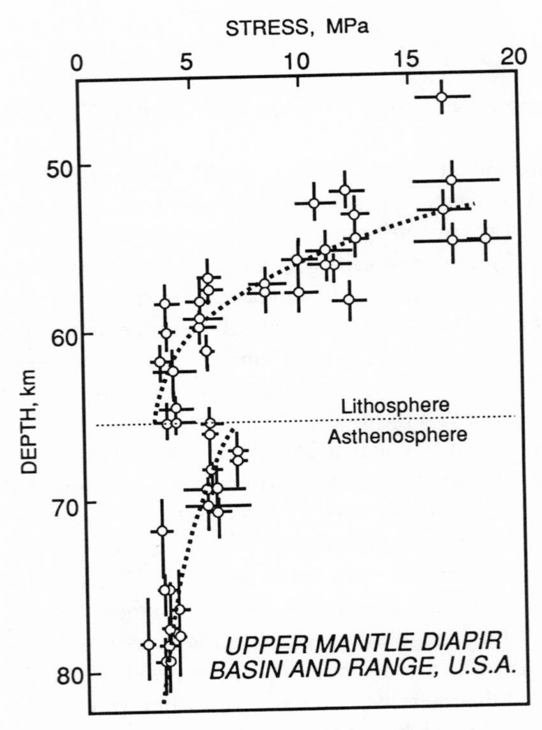

Fig. 4–20. Estimates for σ_d in the Basin and Range of the United States using recrystallization grain size (adapted from Mercier, 1980b). Estimates are obtained by using Ross's et al. (1980) grain boundary migration piezometer. These data are characteristic of the stress-depth trend for the lower tier of the ductile-flow regime.

tional slip, a common mechanism as indicated by the distribution of earthquakes, is not the only active mechanism. Low-temperature ductile-flow mechanisms often limit σ_d to well below that necessary for frictional slip and thereby necessitate the postulate that the schizosphere is also the upper tier to the ductile-flow regime. Semibrittle and ductile deformation, representing the middle and lower tiers of the ductile-flow regime, become increasingly important below the level of the crust where earthquakes are common. Below the bottom of the seismic zone, creep relieves stresses at levels below the frictional strengths.

Plots of σ_d-depth relationships are constructed from laboratory data as in figure 4–9. These plots give the maximum σ_d that the rock can tolerate, and thus are not meant to indicate that σ_d actually gets this high throughout the lithosphere. Because the type of rock or moisture content varies within the crust, there may be several weak zones as indicated by the change in strength of wet granite versus dry granite or a change in rock type. For example, the Moho is also believed to be a strength discontinuity and, hence, discontinuity in σ_d, as rock composition changes from a felsic to mafic.

The boundary between each tier of the ductile-flow regime is arbitrarily picked in much the same manner that Kirby (1985) defines the base of the lithosphere as the depth to the critical isotherm above which strength of olivine drops below 100 MPa at $\dot{\varepsilon} = 10^{-15}s^{-1}$. The base of the upper tier of the ductile-flow regime correlates with the base of the seismogenic zone (Meissner and Strehlau, 1982). The base of the middle tier of the ductile-flow regime is found where polymineralic rocks with significant quartz give way to mafic rocks. By the base of the middle tier of the ductile-flow regime σ_d is less than 100 MPa (fig. 4–9).

The σ_d in the schizosphere (i.e., the brittle regime and the upper tier of the ductile-flow regime), is confirmed using several direct sampling techniques (chaps. 5–8). However, σ_d in the plastosphere is known only through laboratory measurements and a series of indirect measurements and calculations (chaps. 11 and 12).

Concluding Remarks

The ductile-flow regime within the lithosphere has three tiers which correspond to the schizosphere, the brittle-ductile transition, and the plastosphere. Within all three tiers ductile flow serves to modulate σ_d which otherwise would climb until the point of inducing either frictional slip or shear rupture. From the upper to lower tiers, the dominant deformation mechanisms are diffusion mass transfer, restricted dislocation glide, and dislocation creep, respectively. The upper tier of the ductile-flow regime is found in that portion of the lithosphere characterized by the three brittle-stress regimes. In foreland settings there is apparently a symbiosis between the upper tier of the ductile-flow regime and the crack-propagation regime such that ductile flow suppresses σ_d to favor crack propagation upon increase in P_p above hydrostatic pressure. In a sense, the upper tier of the ductile-flow regime can occupy the same space as the frictional-slip or shear-rupture regimes, but the ductile-flow and high-stress regimes do not occupy that space concurrently. In the middle and lower tiers of the ductile-flow regime, brittle deformation is suppressed by high confining pressure in favor of dislocation glide (slip, twinning), dislocation creep (climb, cross-slip), and other high-temperature diffusion mechanisms.

5

Hydraulic Fracture

Hydraulically induced fractures should be formed approximately perpendicular to the least principal stress. Therefore, in tectonically relaxed areas they should be vertical, while in tectonically compressed areas they should be horizontal.

—King Hubbert and David Willis (1957) during a discussion of the mechanics of hydraulic fracturing.

The evolution of hydraulic fracture stress measurements starts shortly after 1859 when Edwin L. Drake bored a 23 m hole into an oil reservoir near Titusville, Pennsylvania, and thereby completed Pennsylvania's first oil well. One of the first problems encountered in Pennsylvania oil fields was that wells clogged with paraffin shortly after production started. Henry Dennis discovered that he could relieve paraffin clogging by setting off explosive shots in oil wells (Herrick, 1949). Dennis used a rifle powder mixture of niter, charcoal, and sulphur in a technique which was later called torpedoing. By 1865, Alfred Nobel's blasting oil, nitroglycerin, was introduced in the United States, and later that same year Col. E.A.L. Roberts took out a patent to replace rifle powder with nitroglycerin in oil-well torpedoes. Thus, starting in the 1860s, oil recovery was routinely enhanced by fracturing of rock adjacent to the wellbore. However, by 1900 the flow of oil declined in the major oil fields of the Appalachian Basin.

The 1900 production decline was arrested with the introduction of water flooding, a practice during which water is driven through the pay zone from injection wells to sweep oil toward producing wells. Hydraulic fracturing evolved from a practice of pumping water into an injection well at a rate necessary to maintain a gradually increasing well-head pressure. After a certain pressure increase, an unexpectedly large volume of water was required to support the next increment in well-head pressure (Dickey and Andersen, 1950). Simultaneously, producing wells started to yield water at an increased rate. One explanation for this behavior was that the formation parted or split under high water pressure, a phenomenon we now recognize as hydraulic fracturing. Often the injection pressure was near the weight of the overburden when flow rate increased, so some suggested that the formation had parted parallel to bedding and the water flood had lifted overburden (Yuster and Calhoun, 1945).

Petroleum engineers struggled with an explanation for formation parting because in some wells the *lifting factor*[1] (ratio of injection pressure to weight of overburden) was significantly less than one and the lifting factor actually decreased with depth. Pressure parting data from the Gulf of Mexico is noteworthy for low lifting factors (Howard and Fast, 1970). Some interpreted the decrease in lifting factor as a manifestation of overburden partially supported by bridging, a process by which high horizontal stress causes upward buckling of beds (Yuster and Calhoun, 1945). By pointing out that S_h/S_v decreased from the surface to depths of a few km in the earth's crust, Yuster and Calhoun (1945) were among the first to explicitly discuss data on behavior of earth stress in a vertical profile.

By the 1950s engineers entertained the possibility that formation parting followed a vertical plane. In their theoretical treatment, Hubbert and Willis (1957) pointed out that hydraulic fractures near the well bore should open up against σ_3 and, hence, propagate in a vertical plane, if stress at depth is anisotropic and if $S_h < S_v$. Concurrently, Ed Heck (1988, personal communication) noticed that hydraulically fractured wells near Bradford, Pennsylvania, shut in at pressures less than the weight of the overburden. Like Hubbert and Willis (1957), Heck reasoned that this would occur if the fractures were vertical and $S_h < S_v$. The debate over fracture attitude was solved in the oilfields of the Appalachian Basin where Jim Bird, founder of Birdwell Logging Company in Bradford, Pennsylvania, realized that if he pumped radioactive sand into a hydraulic fracture, he could infer its attitude (i.e., vertical versus horizontal position) based on the signal from a gamma-ray log. A radioactive signal along several meters of borehole led Bird to conclude that many hydraulic fractures were, indeed, vertical (Heck, 1960). To look for fracture orientation in the wall of open boreholes, Heck (1984, personal communication) used a rubber impression packer designed to check for casing leaks.[2] The soft rubber would flow into the crack and, in doing so, leave an impression in the packer when it was recovered from the borehole. Heck observed vertical fractures and thereby confirmed Bird's interpretation of gamma-ray logs.

Once it was clear that water floods were more efficient using wells treated by hydraulic fractures rather than torpedoes, hydraulic fracturing became popular as a means of connecting oil wells for the secondary recovery of oil in the Bradford fields. Still, a major problem remained. Even if symmetrical, vertical hydraulic fractures were driven out from each side of the wellbore, there was no guarantee that the fractures would propagate in the desired direction which was toward adjacent injection wells. At first, the general practice was to align wells on some convenient grid such as lines of latitude and longitude. However, impression packer tests in the Bradford oil fields showed that most fractures were leaving the wellbore at an azimuth of about ENE-WSW (e.g., Heck, 1960; Anderson and Stahl, 1967). Karney Cochran (1982, personal communication) of Bradley Producing Company in Wellsville, New York, was aware of Heck's work in the vicinity of Bradford, Pennsylvania. Without understanding why hydraulic fractures should propagate ENE-WSW, Cochran developed a property for Quaker State near Alma, New York, by drilling along a grid aligned with the ENE trend discovered by Heck (fig. 5–1). In this alignment, hydraulic fractures actually hit adjacent wells to make a very effective water flood.

Obert (1962) was the first to suggest that S_H aligned to the ENE reflected tectonic stress in the North American crust. A decade later, Sbar and Sykes (1973) proposed that the ENE direction of S_H associated with the contemporary tectonic stress field was found throughout the eastern United States. Unwittingly, with their use of hydraulic fractures for enhanced oil recovery, Heck and Cochran were among the first to map the orientation of the contemporary tectonic stress field affecting a large portion of the North American continent.

Borehole Stress Concentration

The practical basis for hydraulic fracture stress measurements is the assumption that the propagation of a crack, which initiates in the wall of a borehole, is controlled by the manner in which the borehole interrupts a homogeneous

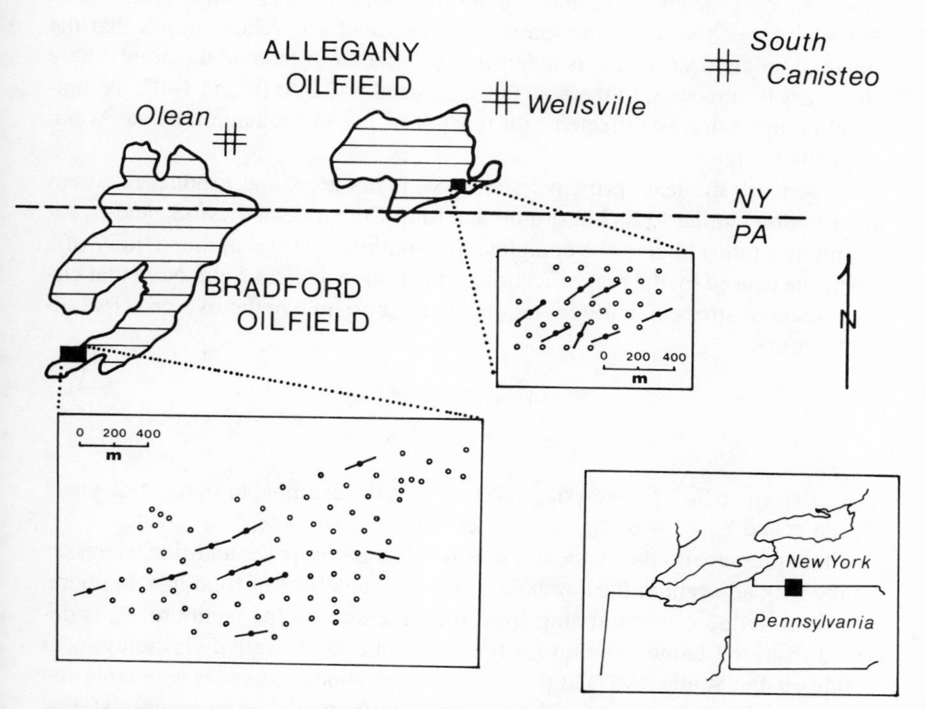

Fig. 5–1. The orientation of hydraulic fractures in the Allegany (Richburg Sand) and Bradford (Bradford Third Sand) oilfields, Pennsylvania and New York (data from Overby and Rough, 1968).

stress field in an elastic medium. In an infinite elastic plate containing a circular hole and subjected to a uniaxial far-field stress, local stresses are concentrated about the edge of the hole (Kirsch, 1898). Miles and Topping (1949) show that this theory applies to stress concentrations around deep wellbores. If rock containing a wellbore is subject to a uniform uniaxial stress at infinity, S_H, then the stresses near the borehole expressed in polar coordinates at a point θ, and r are given by

$$\sigma_r = \frac{S_H}{2}\left(1 - \frac{R^2}{r^2}\right) + \frac{S_H}{2}\left(1 + 3\frac{R^4}{r^4} - 4\frac{R^2}{r^2}\right)\cos 2\theta \qquad (5\text{–}1)$$

$$\sigma_\theta = \frac{S_H}{2}\left(1 + \frac{R^2}{r^2}\right) - \frac{S_H}{2}\left(1 + 3\frac{R^4}{r^4}\right)\cos 2\theta \qquad (5\text{–}2)$$

$$\tau_{r\theta} = \frac{S_H}{2}\left(1 - 3\frac{R^4}{r^4} + 2\frac{R^2}{r^2}\right)\sin 2\theta \qquad (5\text{–}3)$$

where R is the radius of the borehole, r is the distance from the center of the borehole, and θ is measured clockwise from the direction of the compressive stress S_H. σ_θ is the circumferential or hoop stress, whereas σ_r is the radial stress. An analysis of the redistribution of stresses about a borehole shows that the uniaxial far-field stress, S_H, is magnified by a factor of three at the point where the stress is tangential to the hole (figs. 2–3 and 5–2). At 0° and 180°, the uniaxial compression is reflected as a tension of the same magnitude as S_H but opposite in sign.

In general, the least principal stress, S_h, is nonzero, and produces a stress distribution similar to S_H but orthogonal to it. Components from S_H and S_h superimpose (add) to give the complete distribution of stress in the vicinity of a borehole caused by the magnification of the regional stress field. Note that circumferential stress magnitudes tangent to the borehole in the two principal directions are

$$\sigma_{(\theta = 90°)} = 3S_H - S_h \qquad (5\text{–}4)$$

$$\sigma_{(\theta = 0°)} = 3S_h - S_H. \qquad (5\text{–}5)$$

Note that since $S_H \geq S_h$, then $\sigma_{\theta = 90°} \geq \sigma_{\theta = 0°}$. In the example in figure 5–2 where $S_h = 1\ \sigma$ and $S_H = 1.4\ \sigma$, $\sigma_{\theta = 90°} = 3.2\ \sigma$ and $\sigma_{\theta = 0°} = 1.6\ \sigma$.

Fluid pressure on the inside of the borehole superimposes additional stresses in the rock adjacent to the borehole. To assess the effect of fluid in a wellbore, the near wellbore stress arising from fluid pressure in the wellbore, P_f, is derived from the Lamé solution for the stress in a thick-walled elastic cylinder (Hubbert and Willis, 1957). If the radius of the cylinder becomes very large and the external pressure is set equal to zero, then the radial, circumferential, and vertical stresses become

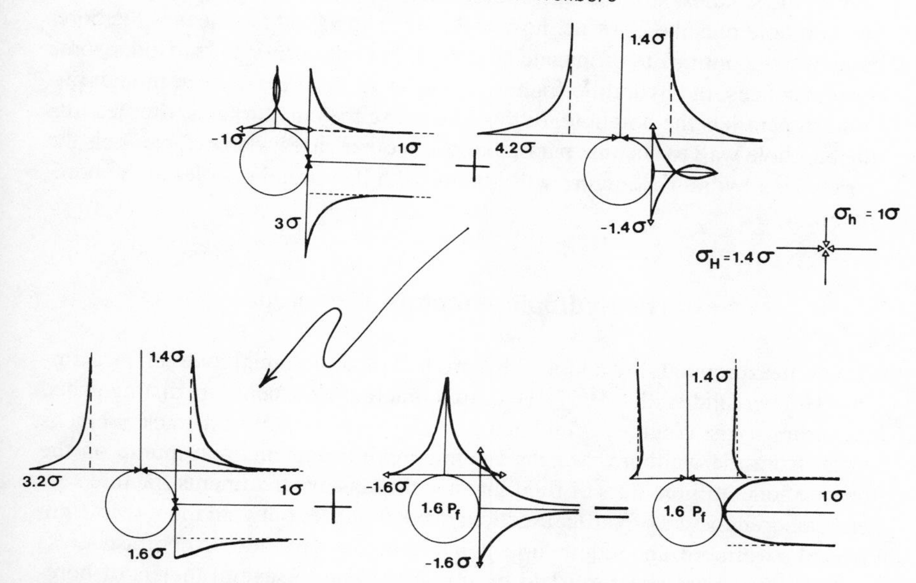

Fig. 5–2. Stress state around a wellbore subject to principal stresses S_H (σ_H) = 1.4 σ and S_h (σ_h) = σ (*top row*). These principal stresses are superimposed to give tangential stresses at the borehole wall of 3.2 σ and 1.6 σ, respectively. A fluid pressure P_f = 1.6 σ is introduced inside the wellbore which, when superimposed on the principal stresses, reduces the tangential stress in the direction of S_h (σ_h) to zero (adapted from Hubbert and Willis, 1957).

$$\sigma_r = + P_f \frac{R^2}{r^2} \tag{5–6}$$

$$\sigma_\theta = - P_f \frac{R^2}{r^2} \tag{5–7}$$

$$\sigma_z = 0. \tag{5–8}$$

In figure 5–2 a wellbore pressure P_f = 1.6 σ is superimposed on the wellbore. This fluid pressure was chosen to counteract precisely the effect of earth stress and reduce $\sigma_{\theta = 0°}$ to zero. Even with P_f = 1.6 σ, the tensile strength, T_0, of the rock would prevent hydraulic fracture propagation. Note that when P_f and $\sigma_{\theta = 0°}$ are balanced, P_f is larger than either of the far-field principal stresses. This solution assumes that water does not penetrate into the wall rock.

The rules for natural crack propagation, as outlined in chapter 2, apply to hydraulic fracture[3] propagation. Hydraulic fractures cut the borehole wall either parallel (an axial crack) or normal (a radial crack) to the borehole axis. Axial crack propagation starts once fluid pressure in a wellbore exceeds the sum of the smallest stress concentration (i.e., $\sigma_{\theta=0^\circ} = 1.6\,\sigma$ in fig. 5–2) around the borehole plus the T_0 of the host rock. Once an axial crack leaves the borehole, it will continue to propagate as a mode I crack normal to σ_3. Under some circumstances, the hydraulic fracture starts as an axial crack even though far-field σ_3 parallels the borehole. In this latter case the crack rotates after leaving the borehole wall to become parallel to σ_3. In other cases where σ_3 parallels the borehole, a hydraulic fracture will initiate with its normal parallel to the borehole.

The Hydraulic-Fracture Technique

Stress measurements are a logical by-product of commercial hydraulic fracturing (Hubbert and Willis, 1957). Hydraulic fracture treatments are distinguished according to the volume of fluid pumped into the well to drive a crack or cracks away from the wellbore. Standard commercial treatments may pump on the order of one million liters of fluid into a well. Massive treatments fracture several hundred meters of vertical section, drive fractures outward more than 1 km from the wellbore, and require up to ten million liters of fluid. In contrast, stress measurements are accomplished by fracturing one to several meters of borehole using as little as 1500 liters for minifrac measurements and as little as 40 liters for microfrac measurements.

Hydraulic fracturing involves isolating (packing off) an interval of a wellbore so that fluid, pumped into that interval under pressure, is prevented from flowing up or down the wellbore (fig. 5–3). Various devices for packing off a section of the borehole include a bridge plug, a single packer, straddle packers, or charge of cement. *Packers* are reinforced rubber bladders which are inflated by pumping water, whereas bridge plugs are stiff rubber plugs which are mechanically seated against the wellbore by moving the drill rod. Bridge plugs and packers are easily moved for repeated measurements of parameters such as permeability and stress, whereas cement is permanent and removed only by drilling.

For stress tests over intervals up to a few meters two packers are tied together in an assembly called a straddle packer. Several configurations are possible for straddle packers depending on the location of the fixed grip between the packers and the injection mandrel (fig. 5–4). In the sliding-coupled straddle-packer assembly, the top of each packer binding is screwed into the rigid mandrel and the bottom end of each packer is allowed to slide on an O-ring. This arrangement prevents slippage of the bottom packer up and down the mandrel at the contact point between the injection interval and the packers. Borehole pressurization compresses the top packer and causes slippage of the bottom end of the top packer on the mandrel. With such slippage shear stress develops at the bottom lip of the top packer where contact is made with the borehole. This shear

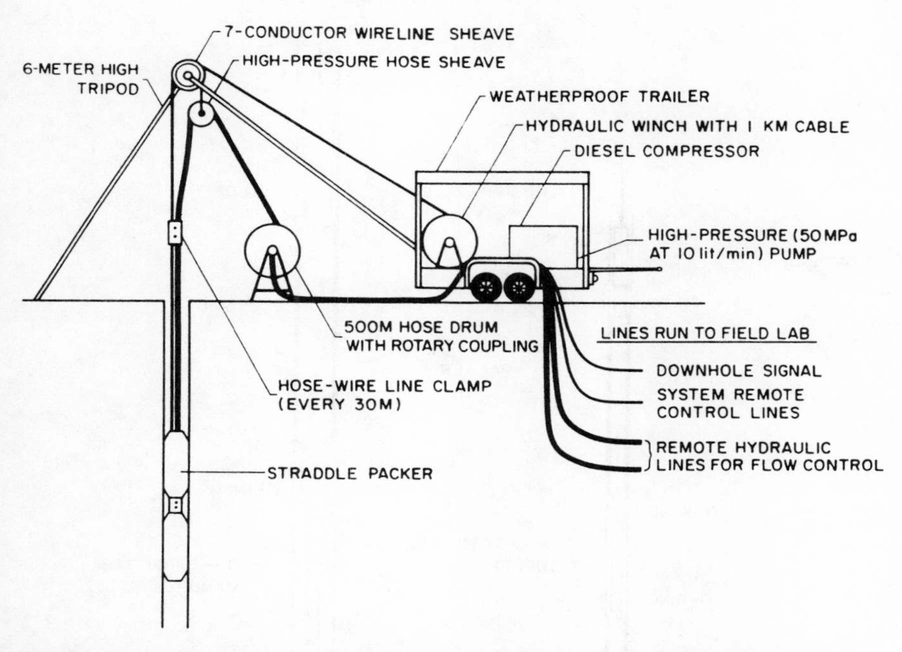

Fig. 5–3. Diagram of a wireline hydraulic fracture system designed by Fritz Rummel and Befeld Inc. in Bochum, Germany (adapted from Evans et al., 1989). A "conventional" hydraulic fracture system would consist of a drilling platform in place of the tripod, and drill rod in place of the wireline and hydraulic hose.

stress pulls on the borehole wall and places it in axial tension at the lip of the packer (Kehle, 1964). Stress measurements using this configuration include those of Evans et al. (1989a,b) and Baumgärtner and Zoback (1989). The rigidly coupled straddle-packer assembly has the top packer screwed into the mandrel assembly at the top end of the packer (Rummel et al., 1983). The bottom packer is tied to the top packer with a rigid member (rigid-packer coupler) so that the bottom packer is free to slide on the mandrel during pressurization. Connecting both packers through a rigid member prevents the tendency for slippage during pressurization and thus reduces the tendency to develop wellbore parallel tension by packer separation. Measurements using the sliding-coupled configuration are reported by Rummel et al. (1983) and Evans et al. (1988). The higher fluid pressures necessary for deeper stress measurements such as those at Cajon Pass, California (e.g., Healy and Zoback, 1988), require stronger mandrels and stiffer packers.

Some analyses of packer-induced stresses suggested that the axial tension developed immediately following packer inflation is sufficiently high to induce horizontal fractures near the end of the packer seat (Warren, 1981). This problem was further investigated using laboratory experiments in which the two

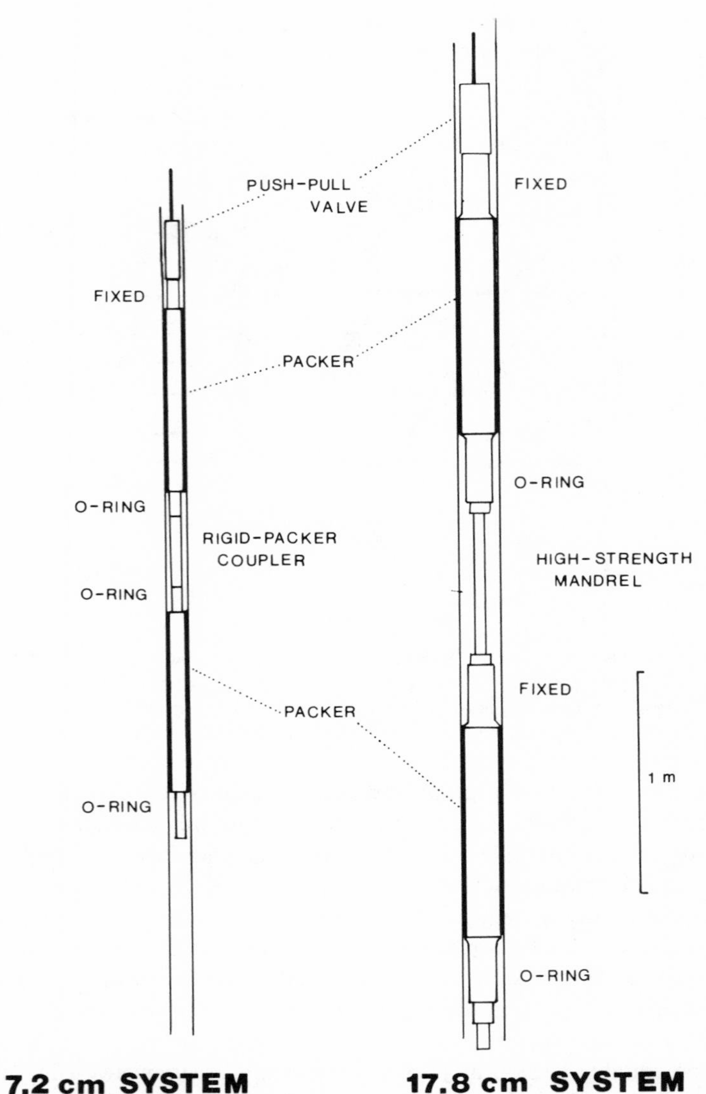

Fig. 5–4. The configuration of two types of straddle-packer grips: the rigid-coupled straddle-packer assembly (*left*); the sliding-coupled straddle-packer assembly (*right*) (adapted from Evans, 1987).

differently configured straddle-packers were pressurized within steel casing instrumented with strain gauges (Evans, 1987). Evans's (1987) results suggest that the shear tractions applied to the borehole wall during interval pressurization is small and the axial tension prevailing throughout the injection interval under such conditions is less than 0.1 MPa. Peak axial tension occurs near the packer end as was suggested by Kehle's (1964) analysis. This result is important because on occasion horizontal fractures initiate near the center of the interval during hydraulic fracture stress measurements made using the rigidly coupled system. Evidently these fractures are not the result of packer-induced stresses (Evans et al., 1988).

There are two methods for moving straddle-packer assemblies up and down boreholes. So far, the most common is to move packers or bridge plugs on the end of drill rod. This method was employed during Haimson's (1977) study of the midcontinent stress field, Zoback's et al. (1980) study along the San Andreas fault, and Warpinski's et al. (1981) study within the Piceance Basin of Colorado. Controlling packers or bridge plugs by means of drill rod offers the advantage of greater push and pull within the wellbore thereby offering some protection against the possibility that a piece of spalling rock may wedge the packers against the wall of the wellbore. The second method for positioning straddle packers within boreholes is to employ a wireline system (fig. 5–3). Examples of stress measurements using a wireline system include Rummel's et al. (1983) studies in Germany, Haimson and Lee's (1984) studies of the midcontinent of North America, and Evans's et al. (1989a) studies of the Appalachian Plateau. Wireline systems offer the advantage of speed moving within the wellbore and do not depend on the use of an oil well drilling or workover rig that can cost many thousands of dollars a day to operate. To date, wireline systems have reached depths of 1.5 km, whereas deeper measurements all employed a drill-rod system. A compromise between the wireline and drill rod uses a spool of endless tubing to drive packers down a well (Enever and Wooltorton, 1983; Bjarnason et al., 1989).

A precise record of fluid pressure in the injection interval is required during all phases of the hydraulic fracture test. Early techniques for generating pressure-time curves used pressure gauges placed in the pressure line at the surface. With this configuration, the pressure derived from the height of the water column is added to the reading of the pressure gauge to find P_f in the test interval. One ambiguity in reading pressure-time curves, particularly those generated from surface pressure gauges, comes in identifying inflection points indicating the instantaneous shut-in pressure and fracture reopening pressure. Friction between the water column and the drill rod dampen these critical inflection points on the pressure-time record to make the surface record difficult to read. To overcome this problem, pressure in the injection interval was recorded by a Kuster gauge, a mechanical gauge with a bellows driving a scribe which scratched a glass plate. The Kuster gauge record was usually read in the final analysis because pressure changes recorded within the fracture interval were much sharper. The disadvantage of the Kuster gauge is that the operator must replace the glass plate after every test, a time-consuming process that involves bringing the entire drill string to the surface. With the development of wireline stress testing using a pressure gauge in the injection interval, pressure-time data is transmitted to the surface on the wireline and recorded for viewing in real time (Rummel et al., 1983). A wireline inside a drill rod also serves for real-time pressure recording within the test interval when packers are lowered on the drill rod (e.g., Warpinski, 1984). Such arrangements do not require that the packers return to the surface after every stress test.

The Pressure-Time Curve

The hydraulic fracture test consists of a general procedure with variations depending on the special circumstances of each measurement (e.g., Haimson, 1978a; Zoback et al., 1980; Warpinski et al., 1981; Rummel et al., 1983). After a test hole is drilled for stress measurements, the first step is to locate the natural fractures as these negate a valid stress test. A borehole televiewer[4] is commonly used for this purpose. If a televiewer is unavailable, pumping tests will indicate the presence of natural fractures within the test interval as explained below. Once unfractured intervals are selected, the stress test starts with the positioning and inflating of the packers.

General aspects of the hydraulic fracture procedure may be illustrated using pressure-time curves from stress measurements in a Devonian siltstone-shale sequence within the Appalachian Basin, New York (Evans et al., 1989a). Follow pressure-time and flow rate-time curves from tests at 712.5 m and 724 m in the Wilkins well, South Canisteo, New York (fig. 5–5). After setting the straddle packers, the first pressure signal from the pressure gauge within the injection interval is P_p which at a depth of about 700 m is about 7.0 MPa. We know that P_p is hydrostatic because the density of ground water causes an increase in P_p at the rate of about 1 MPa/100m. The first pressurization of the injection interval is to test for fractures that were not detected by other means. For this test, pressure within the injection interval is increased 2 MPa above the hydrostatic pressure, then shut in by terminating pumping and closing a valve at the pump. Equilibrium is reached within a couple of minutes after a small decrease (< 0.2 MPa) in the pressure, the result of water infiltrating the siltstone. If an incipient fracture is present, the fluid pressure would continue to decrease well after the two minutes required to reach equilibrium in an unfractured interval. Assuming no fractures are detected, the test interval is then allowed to drain back to hydrostatic pressure. If the operator suspects that fractures cut the test interval, the packers are moved to a different interval and this permeability test is repeated.

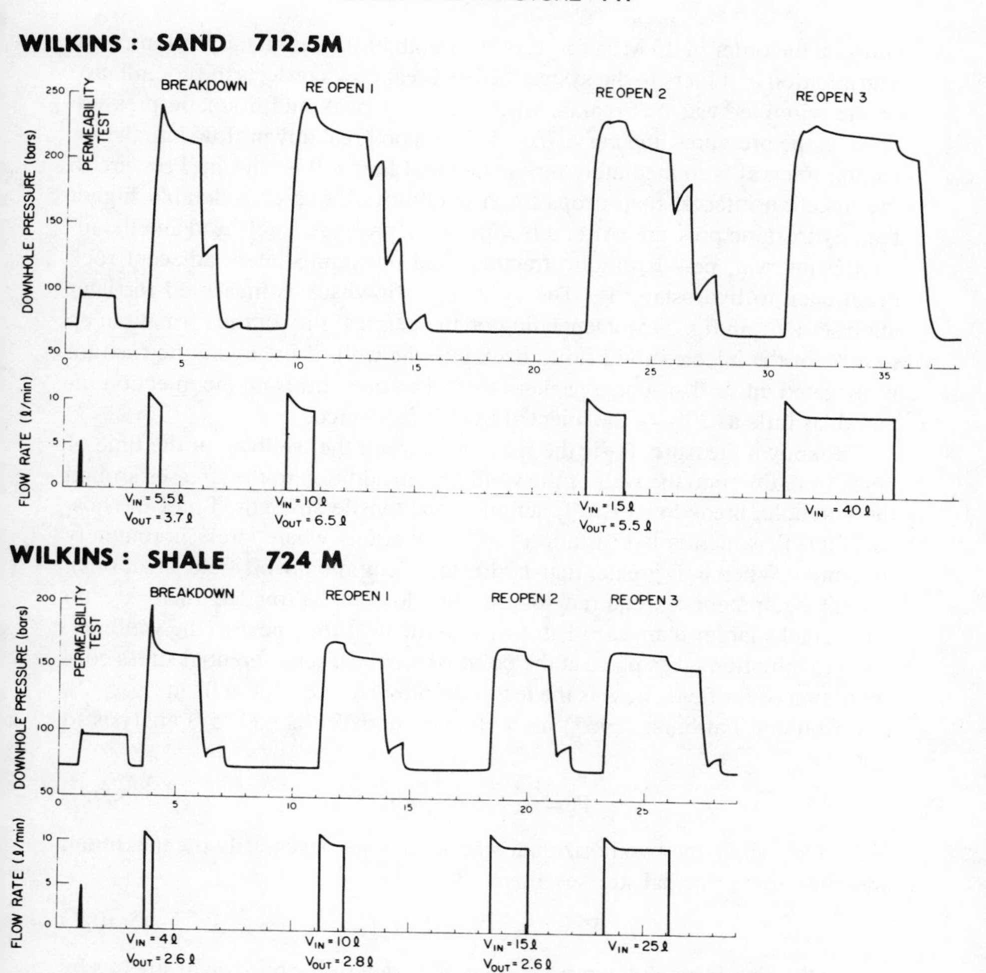

Fig. 5–5. The pressure-time and flow-rate-time records for hydraulic fracture tests in the Wilkins well at depths of 712.5 and 724 m (adapted from Evans et al., 1989).

Breakdown

After the formation is allowed to drain back to hydrostatic pore pressure for several minutes, the injection interval is pressurized very rapidly. Here the idea is to achieve breakdown (the initiation of a hydraulic fracture) as rapidly as possible so that wellbore fluids at high pressures have little time to penetrate the rock formation. If a pressure pulse penetrates the rock prior to breakdown the consequent lower effective stress will trigger a premature breakdown. At a pumping rate of 10 l/min, breakdown is achieved within 20 seconds at pressures on the order of 25 MPa for tests at a depth of 0.7 km in the Wilkins well. The addition of 4 liters to the system before breakdown reflects the compliance of the wireline system. Breakdown is the sharp peak and dramatic pressure drop on the pressure-time curve (fig. 5–5). Upon breakdown, flow into the injection interval is immediately terminated and the well is shut in. Pressure in the injection interval then drops to an equilibrium, one considerably higher than hydrostatic pressure. After a few minutes, pressure is released and the injection interval, new hydraulic fracture, and the immediately adjacent rock drain back to hydrostatic P_p. The volume of flowback is measured for later analyses (V_{out} in fig. 5–5). Depending on the permeability of the formation, up to 80% of the injected fluid flows back into the well. If the hydraulic fracture propagated upward around a packer rather than outward from the injection interval, as little as 20% of the injected fluid is recovered.

Breakdown pressure, P_b, is the fluid pressure in the wellbore at the time of crack initiation into the wall of the wellbore. In addition to the stresses around the borehole, breakdown is a function of the tensile strength, T_0, of the host rock and P_p which is hydrostatic in most situations where stress is routinely measured. When P_p is greater than hydrostatic (common in oil-field situations), special equipment is required to prevent blowouts. Provided there are no microcracks larger than, say, 1 μm, within the host rock next to the wellbore, fracture initiation takes place at the point where the circumferential stress concentration of far field stress is the least compressive (i.e., at $\theta = 0°$ in fig. 5–2). Haimson and Fairhurst (1967) used Hubbert and Willis's (1957) analysis to derive P_b

$$P_b = \sigma_{\theta = 0°} - P_p + T_0. \tag{5–9}$$

If $S_H > S_h$, where the two horizontal stresses are not necessarily the maximum and minimum principal stresses, then

$$P_b = 3S_h - S_H - P_p + T_0. \tag{5–10}$$

This is the classic breakdown equation used to determine horizontal stresses by hydraulic fracturing assuming no infiltration into a porous-permeable rock. Fluid infiltration into adjacent pore space will affect the breakdown pressure as discussed next.

The classic breakdown equations assume a smooth borehole wall. If $\sigma_1 = \sigma_3 = P_c$, the relationship between breakdown and confining pressure, P_c, is

$$P_b = kP_c + T_0 \tag{5–11}$$

where k = 2 under dry conditions. However, laboratory tests show that k < 2, which suggests that there is fluid penetration into preexisting microcracks in the wall of the borehole prior to unstable crack growth (Rummel, 1987). Assuming the presence of cracks in the borehole wall, the details of breakdown are better understood with a fracture mechanics approach to stress testing (Abou-Sayed et al., 1978; Rummel and Hansen, 1989). This approach assumes a symmetrical double crack in the borehole wall with a borehole radius, R. Here breakdown is equivalent to the critical pressure for crack propagation

$$P_b = \frac{K_{Ic}}{(h_o + h_a)\sqrt{R}} + \frac{f}{h_o + h_a} S_H + \frac{g}{h_o + h_a} S_h \qquad (5\text{–}12)$$

where K_{Ic} is the fracture toughness of the rock and f, g, h_o, and h_a are normalized stress intensity functions with respect to S_H, S_h, pressure in the borehole, and pressure gradient in the microcrack.

Fracture Reopening

Fracture reopening is the next step during a stress measurement (fig. 5–5). During the Wilkins well experiment, the pump was turned on and pumped at the rate of 10 l/min until 10 liters flowed into the injection interval. During repressurization following the first breakdown cycle, the pressure-time curve shows a linear increase in pressure until the initiation of fracture reopening where the curve becomes nonlinear. This point of nonlinearity marks the reopening pressure for the fracture which now has very little tensile strength. Now the circumferential stress acting normal to the fracture at the wellbore is counterbalanced by fluid pressure within the wellbore. Additional pumping will act to open the fracture more and, hence, drive it further out into the formation. Bredehoeft et al. (1976) suggest that S_H may also be determined using the pressure for fracture reopening, P_{ro},

$$P_{ro}(T_0 = 0) = 3S_h - S_H - P_p \qquad (5\text{–}13)$$

This equation does not depend on a knowledge of T_0 which is difficult to measure with confidence in the laboratory. Presumably the T_0 of the formation is

$$T_0 = P_b - P_{ro}. \qquad (5\text{–}14)$$

A reliable determination of S_H using P_b (eq. 5–10) depends on the condition of the borehole wall. If the wall contains irregularities or is cracked, then this wellbore damage might serve as stress risers in which case P_b is lower than would be the case for a smooth borehole wall (eq. 5–12). Such damage leads to inconsistent readings for P_b from test to test. The use of equation 5–13 to determine S_H offers two advantages: there is no need for a measure of T_0, and calculations using equation 5–13 are independent of the condition of the borehole wall. However, equation 5–13 requires an accurate measure of P_{ro} which is difficult in deep wells and when $S_H > 2S_h - P_p$ (Evans et al., 1989a).

SHUT-IN

Upon reopening, the fracture is completely supported by fluid. After 10 liters of pumping, the well is shut-in whereupon the fluid pressure drops rapidly until the fluid no longer totally supports the fracture. At the point of contact the fracture becomes very stiff relative to its compliant nature when fully open. Fluid pressure at the point of fracture contact is commonly called the instantaneous shut-in pressure or ISIP. The rapid fall in pressure during the initial shut-in phase is caused by loss of fluid through infiltration into the crack walls and by flow of fluid from the test interval into the fracture which is continuing to grow. Once asperities on the fracture start to take up the load across the fracture, the rate of fall in fluid pressure decreases markedly causing a sharp corner on the pressure-time curve (fig. 5–5). ISIP is identified based on the pressure at this sharp corner in the pressure-time curve. For minifrac measurements, the ISIP is the lowest pressure at which fluid in the hydraulic fracture carries the full load of earth stress normal to the fracture (Gronseth, 1982; Evans, 1983; Gronseth and Kry, 1983). Because the hydraulic fracture (i.e., a crack) has propagated normal to σ_3, ISIP = σ_3. Pressure decay characteristics of shut-in curves are governed by the rate of the water leak-off, the equipment compliance, and the partial crack closure (Hayashi and Sakurai, 1989).

One object of several fracture reopening tests is to sharpen the ISIP corner on the pressure-time curves. A minute or two after the ISIP is observed a valve is opened to allow the injection interval and fracture to flowback and drain to hydrostatic pressure. Each flowback lasts several minutes to assure that the formation around the hydraulic fracture is near its initial hydrostatic pressure. The hydraulic fracture is reopened a minimun of three times during stress testing largely to establish a reproducible P_{ro} and ISIP. As the hydraulic fracture is driven further into the formation during each reopening cycle, the ISIP inflection point becomes sharper. There are two explanations for this behavior. One is that sharper ISIP points reflect the increased tendency for pore space next to a hydraulic fracture to act as an accumulator. Pore fluids, rapidly draining from adjacent pore space to the hydraulic fracture just after closure, maintain fluid pressures at slightly less than ISIP for periods long enough to give a sharp corner to the pressure-time curve after ISIP. Small fractures without the benefit of the accumulator effect will rapidly drain to pressures much lower than the ISIP, and hence cause a rounded pressure-time curve. A second interpretation is that, if large amounts of fluid have been injected, a mismatch can develop between fracture walls. Such a mismatch causes a more gradual stiffening of the fracture on asperity contact, which means that fluid pressure drop is momentarily arrested to give a sharper corner on the pressure-time curve (Evans, 1983). In summary, the classic equation for breakdown pressure (eq. 5–10) contains five variables of which four—pore pressure (P_p), tensile strength (T_0), breakdown pressure (P_b), and least principal stress (σ_3 = ISIP)—

are determined directly from a pressure-time curve recorded from a hydraulic fracture test (fig. 5–5).

Interpretation of Pressure-Time Curves

STATE OF STRESS

The shape of the curves for the pressure-time history of a hydraulic fracture test is highly dependent on the ratio of S_H/S_h (i.e., the magnitude of σ_d^H). This ratio affects the magnitudes of P_b and P_{ro} relative to S_h (Hickman and Zoback, 1983). Three possibilities include

$$P_b > P_{ro} > S_h \tag{5–15}$$

$$P_b > S_h > P_{ro} \tag{5–16}$$

$$S_h > P_b > P_{ro}. \tag{5–17}$$

The pressure-time curves are distinct for all three cases as illustrated for data from two wells near the San Andreas fault, California (Hickman and Zoback, 1983). A high breakdown pressure ($P_b >>$ ISIP in Mojave 1 at 185 m) is encountered if there is a relatively small difference between the horizontal stresses, σ_d^H (fig. 5–6b). As σ_d^H increases with depth, $P_{ro} \approx$ ISIP as seen in the well, XTLR at 338 m. As S_H and P_p continue to increase relative to S_h, a situation where $P_b \approx$ ISIP is observed (i.e., 751 m in XTLR), indicating a relatively large σ_d^H. In this latter situation ISIP is not even observed in the pressure-time curves. In the XTLR well along the San Andreas fault, Zoback et al. (1980) report a transition in breakdown behavior moving from the large P_b near the surface (manifesting a low σ_d^H) to P_b of about the same magnitude as the ISIP lower in the well (manifesting a high σ_d^H).

Fracture reopening pressure is also affected by the magnitude of σ_d^H. If σ_d^H is relatively small, then P_{ro} is sharp (e.g., 185 m) (fig. 5–6b). In contrast, P_{ro} falls well below ISIP for larger σ_d^H (e.g., 751 m). Evans et al. (1989a) cast doubt on the detectability of accurate P_{ro} in cases where $S_H > 2S_h - P_p$ using current hydraulic fracture systems.

P_{ro} behavior is explained by the radial profile of normal stress across the plane of the fracture, S_N (normal to the direction of S_h). The profile is obtained by superposing the circumferential stress components arising from S_h, S_H, and P_p (eqs. 5–2, 5–7), and is

$$S_N = \left(\frac{S_H + S_h}{2} - P_p\right)\left(1 + \frac{a^2}{r^2}\right) - \left(\frac{S_H - S_h}{2}\right)\left(1 + 3\,\frac{a^4}{r^4}\right) + P_p. \tag{5–18}$$

The resulting radial profiles of normal stress along the fracture are shown in figure 5–6a for the cases described by three equations, 5–15 to 5–17, respec-

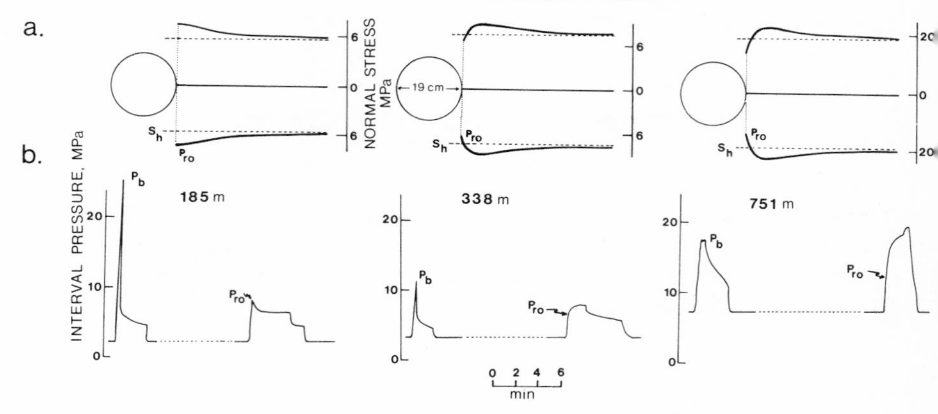

Fig. 5–6. Hydraulic fracture stress measurements near the San Andreas fault (adapted from Hickman and Zoback, 1983). (a) The calculated normal stresses (dark lines) and S_h (dashed lines) acting on hydraulic fractures propagating from 19 cm diameter boreholes. Normal stress is the tangential stress at the borehole wall. At a depth of 185 m (the well, Mojave 1) $P_b > P_{ro} > S_h$. At a depth of 338 m (the well, XTLR) $P_b > S_h > P_{ro}$. At a depth of 751 m: (the well, XTLR) $S_h > P_b > P_{ro}$ Magnitude of P_{ro} indicated by the normal stress line at the borehole. The calculations for normal stress assume that the cylindrical borehole is perpendicular to the two horizontal principal stresses. (b) Portions of pressure-time records illustrating the three different types of hydraulic fracture behavior at depths of 185 m, 338 m, and 751 m near the San Andreas fault. The fourth reopening cycle is shown for depths of 185 m and 338 m. The third reopening cycle is shown for a depth of 751 m.

tively. In all three cases the normal stress acting on a hydraulic fracture approaches S_h within a couple of borehole diameters. This behavior leads to the correlation between ISIP and S_h for a fracture which has propagated several borehole diameters from the well, in which case a large area of the fracture is subject to a normal stress exactly equal to S_h.

Accuracy in Measuring S_H

Assuming that S_h is accurately measured, two methods are available for the determination of S_H. The first uses the breakdown pressure and an inferred tensile strength of the rock (eq. 5–10). The second method uses the reopening pressure on subsequent pumping cycles and avoids the use of the tensile strength of the rock (eq. 5–13). The latter method is practical only when $S_H < 2S_h - P_p$, otherwise the P_{ro} is poorly defined as indicated above. If $S_H > 2S_h$, the elastic hoop stress across the fracture in the immediate vicinity of the wellbore is less than S_h. Although the fracture may begin to open at the wellbore when the true P_{ro} is reached, the opening does not propagate until the wellbore reaches the value of S_h which is the stress normal to the crack face beyond the immediate vicinity of the wellbore. The associated fluid loss thus becomes significant and detectable only when the pressure has reached the value of S_h well above P_{ro}. In such circumstances, the apparent P_{ro} equals ISIP and equation 5–13 yields only a lower bound to the value of S_H. This was found in all but the shallowest of the tests in the Wilkins well (Evans et al., 1989a).

Fracture Initiation

Aside from the strong effect of σ_d^H on pressure-time curves, ambiguities also arise from several factors including flow past the packers. The radial stress exerted on the borehole wall by the packers changes as the injection interval is pressurized because fluid in the injection interval loads the packer bindings and, thus, further pressurizes the packers. As the packers become further pressurized, there is the possibility of initiating a fracture under the packers before breakdown pressure is reached in the injection interval. Fracturing under the packers is often responsible for fluid bypassing the packers before breakdown pressure is achieved. Gronseth and Detournay (1979) measured pressures within the packers and injection interval and concluded that the circumferential stress around the packers always exceeded that in the injection interval. Evans (1987) concluded that this was not the case because of the finite strength of the packer elements.

Failure to recognize packer bypass causes considerable confusion during hydraulic fracture stress tests. Packer bypass appears as a slight hesitation in a linear increase of pressure with time before breakdown. Usually this is followed by a rounded peak rather than a sharp decrease in the pressure-time record. Once shut-in after packer bypass, the injection interval pressure comes to an equilibrium well above that seen when breakdown occurs. Likewise, during flowback, very little fluid is recovered, and if the well is again shut-in during flowback, there is no repressurization as is seen when a proper hydraulic fracture is still draining.

Several types of fracture initiation are possible during hydraulic fracture tests (Enever and Wooltorton, 1983; Enever, 1988). For crack initiation within the injection interval, the desired breakdown during a stress measurement is characterized by a rapid drop-off in pressure immediately upon initiation. However, crack initiation may occur under the packers. This leads to a slow drop-off in pressure after crack initiation and a range of decreasing shut-in pressures over successive pressurization cycles. Another possibility is incomplete breakdown during the first pressure cycle of a hydraulic fracture test (Hickman and Zoback, 1983). Ordinarily, there is a large pressure change between P_b and the P_{ro} of subsequent cycles. If a significant pressure change takes place between the second and third cycles, Hickman and Zoback (1983) argue that the fracture is not fully formed at the wellbore after the breakdown cycle.

In this case S_H is determined by using P_{ro} during later cycles of pressurization. This interpretation is hard to assess and Evans (1990, personal communication) suggests that a significant drop in P_{ro} between cycles has to do with progressive shear propping and fracture mismatch rather than incomplete breakdown.

If stress tests occur in a thrust fault regime where $S_v = \sigma_3$, a vertical fracture within the injection interval may rotate to the horizontal plane once it propagates away from the borehole wall. The distance a vertical fracture propagates before it rolls over to a horizontal orientation is unknown. As most hydraulic fractures are initiated axially to the wellbore, horizontal cracks are identified based on other criterion, mainly ISIP = S_v. Crack rotation is indicated by a sharp decrease in ISIP during successive tests as the fracture is driven further from the wellbore. Although each shut-in may exhibit a lower value as more fracture area is exposed to S_v, Evans et al. (1989a) show that the decline in ISIP was slight after the first fracture reopening cycle indicating that fracture rotation is early in each test. There is a lengthy discussion of the problem of fracture rollover in Evans and Engelder (1989). Under ideal circumstances the complete stress tensor can be derived from tests where the fracture rotated to horizontal (Baumgärtner and Zoback, 1989). In such a test it is necessary to separate the shut-in pressure values of the different propagation stages. If the tensile strength of bedding planes is low enough (< 0.1 MPa), horizontal initiation along these weak bedding planes may occur in regions where the σ_3 is vertical. However, Evans et al. (1989a) present evidence that axial fractures were produced at the wellbore with $S_h > S_v$ despite a significant bedding weakness.

Yet another variation in successive pressure-time curves is the gradual decrease in fracture opening pressure after the second or third pressurization cycle caused by a gradual diffusion of fluid from the wellbore into the surrounding rock (Hickman and Zoback, 1983). When pressures in the borehole are higher than the surrounding rock there is a tendency for P_p to increase in the rock immediately adjacent the wellbore. One interpretation is that this causes a decrease in effective stress at the wellbore which affects a decrease in the P_{ro} as indicated by equation 5–13. If this were indeed the case, P_{ro} measured a day or so after the testing should return to its initially high value. This was not observed when fractures were reopened more than a day later in the Wilkins well (Evans et al., 1989a)

Infiltration and Effective Stress

Fluid infiltration includes both leakage into pore space adjacent to the borehole as well as leakage from the borehole into the hydraulic fracture during repressurization (Haimson and Fairhurst, 1967; Alexander, 1983). Significant infiltration preceding breakdown reduces effective stress in the rock surrounding the borehole (eq. 5–10), and thus reduces breakdown. Infiltration along the hydraulic fracture prior to fracture reopening lowers the fracture reopening pressures, causing erroneous values for S_H unless poroelastic behavior is taken into account in interpreting the data. The rate of flow into the fracture is proportional to the fracture aperture and the pressure differences between the wellbore and the fracture. Pressurization curves become increasingly nonlinear as fluid flows into the fracture with greater efficiency. During the early repressurization cycles, the fracture is well mated and has a lower intrinsic permeability which causes a relatively small nonlinearity in the pressurization curve making P_{ro} easier to identify. During later cycles the fracture may become propped open and more permeable, which increases the nonlinear behavior of the pressure-time curves and makes P_{ro} harder to identify.

The classic breakdown equation (5–10) assumes the rock around the borehole is impermeable to the fracturing fluid even under the increased pressures just prior to breakdown. A fracture mechanics approach to breakdown (e.g., eq. 5–12) assumes fluid infiltration into cracks in the borehole wall but not beyond. In practice, fluid flows into pore space of the formation beyond the borehole wall before breakdown and the model used to interpret the data should account for this effect. Haimson and Fairhurst (1967) offered a solution to account for the poroelastic effect of increased pore pressure in the vicinity of the borehole

$$P_b = \frac{3S_h - S_H - AP_p + T_0}{2 - A} \qquad (5\text{–}19)$$

and

$$A = \alpha_b \frac{1 - 2\nu}{1 - \nu} \qquad (5\text{–}20)$$

where α_b is the poroelastic parameter defined in equation 1–41. To account for the infiltration of fracturing fluid in calculating S_H, Haimson's and Fairhurst's (1967) equation 5–19 is rewritten as

$$S_H = 3S_h - (2 - A)P_{ro} - AP_p. \qquad (5\text{–}21)$$

Using P_{ro}. Schmitt and Zoback (1989) suggest that in some cases for rocks with a low α_b, breakdown pressures are not predicted by Haimson's and Fairhurst's (1967) breakdown formula. They offer an equation using a pore pressure coefficient, β, with values less than or equal to unity. Rewriting Schmitt's and Zoback's (1989) equation using P_{ro},

$$S_H = 3S_h - (1 + \beta - A)P_{ro} - AP_p. \qquad (5\text{–}22)$$

Infiltration also affects the fracture closure pressure through the poroelastic effect (Detournay et al., 1987). A pore pressure buildup in the vicinity of the fracture causes the poroelastic rock to dilate, forcing the hydraulic fracture to close. If the hydraulic fracture is shut-in, the fracture closure pressure will exceed the far-field S_h in proportion to the poroelastic deformation of the rock.

Closure Pressure versus ISIP

Closure pressure is that pressure at which walls of the hydraulic fracture first come in contact after the well is shut-in and, hence, the pressure at the instant when crack walls are no longer completely supported by fluid. At this instant closure pressure equals S_h. For microfrac tests of the type commonly used for stress measurements the ISIP is the closure pressure (Tunbridge, 1989). However, ISIP and closure pressure do not correlate for commercial hydraulic fracture treatments. The value of ISIP for commercial hydraulic fracture treatments generally increases with injected volume due to the poroelastic deformation accompanying elevation of formation P_p through leak-off (Cleary et al., 1983; Detournay et al., 1987). For these large hydraulic fractures the ISIP marks the time at which major flow from the injection interval to the fracture stops, but it does not necessarily mark the moment when the hydraulic fracture walls come in contact (fig. 5–7a). Even at ISIP, crack propagation continues causing an increased crack volume and a further pressure drop to closure pressure. Crack propagation stops at closure pressure. Sometimes a second inflection point appears on the pressure decline curve and this is usually taken as closure pressure (Wooten and Elbel, 1988; de Bree and Walters, 1989). If formation permeability is low, the time to reach this second inflection point is long and the inflection point becomes obscure. To sharpen the second inflection point, the injection interval is often flowed back at a constant rate (Nolte, 1982; de Bree and Walters, 1989). This latter technique is sensitive to the flowback rate, but if flowback is too fast or slow, the inflection point is obscured (fig. 5–7b).

Stopping a test or opening a valve starts a waterhammer effect which is a free oscillation of pressure within the wellbore. Free oscillation is sensitive to an open hydraulic fracture because of its compliance which dampens free oscil-

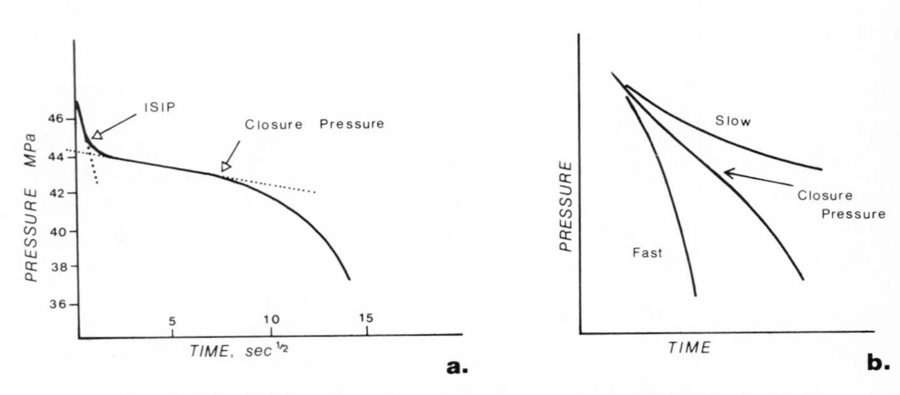

Fig. 5–7. (a) Pressure-time curve showing the difference between ISIP and crack closure pressure for a hydraulic fracture test in the Kerr-McGee Bradley No. 1 well of Smith County, Texas (data from Whitehead et al., 1989). (b) The interval pressure behavior for constant rate flowback after fracture injection (data from Nolte, 1982).

lations (Holzhausen et al., 1989). When the hydraulic fracture is closed, it is relatively stiff and has less affect on the decay of the free oscillation. Holzhausen et al. (1989) show that it is possible to measure closure pressure provided waterhammer events are properly timed during pressure fall-off immediately after the cessation of pumping.

Initiation of Horizontal Hydraulic Fractures

The debate over whether hydraulic fractures initiated as horizontal or vertical cracks started with the interpretation of data on the lifting factor (Hubbert and Willis, 1957). This debate was rekindled with the analysis of local stresses induced by pressurization of the packers (e.g., Kehle, 1964; von Schoenfeldt, 1970). Evans (1987) shows that axial tension throughout the injection interval was less than 0.1 MPa so that horizontal fractures are not caused by packer-induced stresses.

Horizontal Hydraulic Fractures

Evans et al. (1988) present an analysis of conditions favoring the propagation of horizontal hydraulic fractures. Briefly, their argument is that horizontal cracks can initiate in a vertical borehole provided that the wellbore fluid pressure, P_f^H, exceeds the vertical stress at the borehole wall, S_{vw}, by at least the tensile strength, T_0, of the rock. From this statement a failure criterion for a horizontal hydraulic fracture follows

$$P_f^H \geq S_{vw} + T_0. \tag{5–23}$$

Horizontal cracks propagate only when this failure criterion is met before vertical (i.e., axial cracks) are initiated. To induce horizontal hydraulic fractures in an isotropic rock, the far-field stresses must allow effective tension in the axial direction to develop during wellbore pressurization prior to the development of circumferential tension favoring vertical hydraulic fracturing. This means that

$$P_f^V \geq P_f^H = S_{vw} + T_0. \tag{5–24}$$

where P_f^V and P_f^H are the wellbore pressures at which vertical and horizontal fracture initiation takes place. If the wellbore pressure, P_w, is in equilibrium with the P_p in the borehole wall, then by rearranging Haimson's and Fairhurst's (1967) equation (5–19), we have an expression for the least compressive total stress, $S_{\theta\theta}$, developed at the wall of an internally pressurized vertical borehole penetrating a porous, permeable, linear-elastic medium subject to far-field total-stress components, S_H and S_h

$$S_{\theta\theta} = \{3S_h - S_H - P_p\} - \{P_w - P_p\} + \{A(P_w - P_p)\}. \tag{5–25}$$

From this equation comes the relation between the far-field principal stresses which must be satisfied if horizontal cracking is to occur

$$[3(1 -)A - 2\nu(2 - A)]S_h \geq [(1 - A) - 2\nu(2 - A)]S_H + (2 - A)S_v - AP_p + T_0. \tag{5–26}$$

This result shows that horizontal fracture initiation can occur in granite where the least horizontal stress modestly exceeds the vertical stress provided fluid pressure significantly infiltrates the wellbore wall.

Although $P_b = P_f^H$ cannot be used in equation 5–10 to determine the absolute magnitude of the horizontal stress, P_f^H is proportional to horizontal differential stress, σ_d^H. Haimson (1968) gives an expression for the horizontal fracture initiation pressure, P_f^H, for a vertical hole in a permeable rock. Following Evans and Engelder (1989), Haimson's expression is rearranged to solve for the horizontal differential stress, σ_d^H, from horizontal fracture initiation pressure

$$\sigma_d^H = S_H - S_h = \frac{1}{2\nu}\left\{S_v - (1 - A)\, P_f^H - AP_p + T_0\right\}. \tag{5–27}$$

Ljunggren et al. (1988) and Ljunggren and Amadei (1989) approach the analysis of conditions for horizontal fracture propagation using the Hoek and Brown (1980) failure criterion. This approach accounts for the effect of vertical stress which does not enter into the Hubbert and Willis (1957) analysis (eq. 5–19). For a vertical borehole subject to an internal fluid pressure, P_f, the circumferential, radial, and longitudinal stress components are equal to

$$\begin{aligned} \sigma_r &= P_f \\ \sigma_\theta &= 3S_h - S_H - P_f \\ \sigma_z &= S_v - 2\nu\,(S_H - S_h) \end{aligned} \tag{5–28}$$

The Hoek and Brown (1980) failure criterion is a specific version of equation 3–20 where

$$\sigma_1 = \sigma_3 + \sqrt{mC_0\sigma_3 + C_0^2}. \tag{5–29}$$

Bjarnason et al. (1989) point out that whenever horizontal stress is anisotropic, hydraulic fracture propagation can occur in either a vertical or horizontal plane, since from equation 5–28 either σ_z or σ_θ may be tensile. Equation 5–29 assumes four possible forms for which either σ_z or σ_θ is the minimum principal stress. For $\sigma_z < \sigma_\theta < \sigma_r$

$$P_b = S_v - 2\nu\,(S_H - S_h) + \sqrt{mC_0\,[S_v - 2\nu\,(S_H - S_h)] + C_0^2} \tag{5–30}$$

where this equation represents the Hoek and Brown failure condition for the initiation of horizontal hydraulic fractures. Although equation 5–30 cannot be solved for S_H, it can be solved for the stress difference σ_d^H to determine possible ranges of values for S_H and S_h (Bjarnason et al., 1989).

Horizontal hydraulic fractures are common in the crystalline rocks of New England. Near Moodus, Connecticut, hydraulic fractures in a 76 mm borehole drilled in gneissic rock are horizontal in 20 of 25 tests (Rundle et al., 1985). At greater depth in the Moodus area fractures started vertical and rotated to horizontal (Baumgärtner and Zoback, 1989). Hydraulic fracture stress measurements were attempted in a 76 mm diameter, 1 km deep diamond drill hole which penetrates the Conway granite near the Redstone Quarry at North Conway, New Hampshire (Evans et al., 1988). A total of twenty-one successful tests were made in fracture-free intervals to a depth of 579 m. The least ISIP during a series of repressurization cycles is plotted as a function of depth in figure 5–8. The ISIP data scatter about the overburden curve (S_v) down to a depth of 415 m below which ISIPs were found to be increasingly sublithostatic. An oriented trace of eighteen hydraulic fracture(s) at the borehole wall was obtained by using an impression packer and sixteen were single subhorizontal fractures. Equation 5–25 suggest that horizontal fracture initiation occurs whenever the S_h exceeds the vertical stress by an amount whose value is dependent upon material properties, ambient pore pressure, and maximum horizontal stress level. Equation 5–27 is used to calculate σ_d^H in the Conway granite (fig. 5–8).

Other examples of horizontal fracture initiation include Stephansson and Ångeman's (1986) report of numerous horizontal hydrofractures in cored ver-

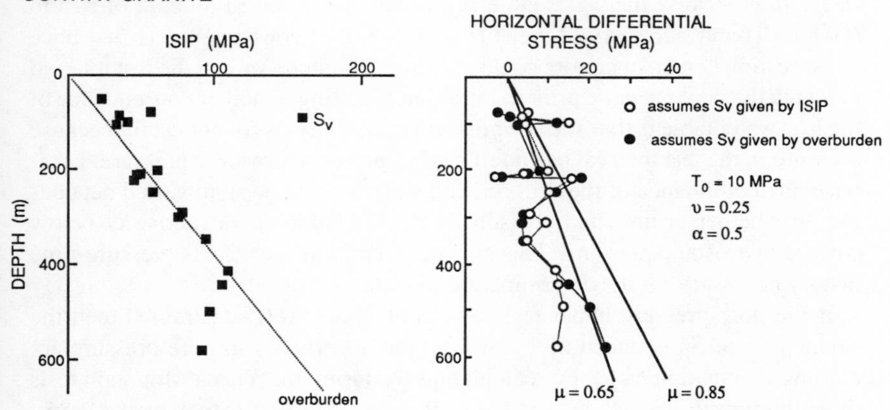

Fig. 5–8. Vertical stress versus depth in the Conway granite at the Redstone Quarry, North Conway, New Hampshire (*left*) (adapted from Evans et al., 1988). Also shown is an estimate of the overburden stress calculated assuming a density for the Conway granite of 2.65 gm/cc. Horizontal differential stress obtained from pressure at which horizontal fractures initiated (*right*) (adapted from Evans and Engelder, 1989). The two values shown at each depth correspond to different S_V-estimates obtained from computed overburden load and the observed ISIP.

tical boreholes penetrating Precambrian gneiss at Forsmark, Sweden. Overcoring measurements conducted to 500 m depth in the same borehole suggested that the S_h is everywhere greater than the S_v (Martna et al., 1983), and substantially so at depths shallower than 200 m where horizontal fracturing was the rule. Similarly, Bjarnson et al. (1986) reported horizontal fracture initiation in gneissic rock near Gideå, Sweden, where the majority of the data suggest that σ_3 is vertical. Another example is given by Warpinski et al. (1982) who reported the initiation of normal-to-axis fractures from horizontal boreholes drilled into volcanic tuff from a 420 m deep drift at the Nevada Test Site. There the vertical stress is a factor of two greater than the horizontal stress component along the borehole axis. These observations show that fracture initiation normal to the borehole axis can occur under situations where the stress normal to the borehole axis is high relative to stress parallel to the borehole axis.

The Effect of a Low Water Table

Yucca Mountain in southwestern Nevada is near the transition between the Basin and Range stress province of WNW-ESE extension and the San Andreas fault stress province of strike-slip faulting (Zoback and Zoback, 1980). Hydraulic fracture stress measurements at Yucca Mountain show S_h facing N60°W to N65°W which correlates with the direction of Basin and Range extension (Stock et al., 1985) . Stress measurements at Yucca Mountain are particularly interesting because the water table in two test holes was at a depth of 575 m (G-1 well) and 526 m (G-2 well) (fig. 3–11). The consequence is that once stress testing began, operators could not lower fluid pressure in the test interval to match the pore pressure prior to initiation of testing. Another consequence of the low water table is that surface pressure transducers were not useful because pressure in the test interval often fell below surface hydrostatic pressure. Likewise, the compliance of the drill rod and water column system varied depending on whether or not fluid pressure in the test interval was above or below surface hydrostatic pressure. This change in compliance affects pressure-time history necessary for the determination of ISIP.

If the pore pressure under test conditions is not well constrained then the calculation of S_H is uncertain by at least the uncertainty in pore pressure according to equation 5–13. For calculating S_H during the Yucca Mountain tests the minimum P_p was taken as equal to P_p prior to testing (Stock et al., 1985). This was measured off the height of the water column whose top was more than 500 m below the wellhead. Maximum P_p was equal to the height of a water column at the wellhead when the drill rod was filled prior to pressurization of the test interval. In filling the drill rod there is an uncertainty about how the additional fluid pressure in the test interval changed the effective stress in the rock surrounding the wellbore. If P_p is taken as equivalent to the surface hydrostatic head, then $S_H \approx S_h$ (Stock et al., 1985). However, televiewer images of the Yucca Mountains wells show consistent wellbore spalling which is indicative of a significant difference between S_H and S_h (Zoback et al., 1985). This difference supports the notion that P_p at the time of testing is best represented by the in situ conditions prior to testing. As the breakdown equation (5–10) shows, the smaller magnitudes of P_p equate to larger magnitudes for S_H.

Complexities of Stress as a Function of Depth in Sedimentary Basins

Early models of the upper crust predicted that stress is a smoothly varying function of depth (Hafner, 1951; Sanford, 1959). At the time boundary conditions in the upper crust were poorly understood and by default the models were simple. Over the past twenty years hydraulic fracture measurements have provided a clearer and more detailed picture of the contemporary stress state in the top 5 km of the crust. Hydraulic fracture stress measurements at various depths in boreholes commonly show considerable scatter about fitted linear regression lines (Haimson, 1978a). In some cases, these departures from linearity are quite complex as illustrated by the South Canisteo hydraulic fracture experiment in Devonian sediments of the Appalachian Plateau, New York (Evans et al., 1989a,b; Evans and Engelder, 1989) and the DOE Multiwell experiment in Mesozoic sediments of the Colorado Plateau, Rifle, Colorado (Warpinski et al., 1985; Warpinski, 1989). These and other measurements by Haimson and Lee (1980), Haimson and Rummel (1982), and Stephansson and Ångman (1986) show that both mechanical properties of rocks and their structural setting influence the stress distribution that develops in response to burial and tectonic loading, both past and present.

The Appalachian Basin

The South Canisteo experiment within the Appalachian Basin is discussed in detail to illustrate the complex nature of stress in sedimentary rocks. This complexity is manifested in three ways: topography affects the vertical stress; a burial history including a phase of high pore pressure affects horizontal stress; and mechanical properties of individual beds affects horizontal stress. These points are discussed in detail below.

The South Canisteo experiment consisted of hydraulic fracture measurements in the three wells that penetrated the distal turbidites and black marine shales of the Devonian Catskill Delta (Evans and Engelder, 1986; Evans et al., 1989a,b). To date, this experiment is unique because the bed-by-bed correlation among the wells permitted the reproduction of stress measurements in key beds. The easternmost well, the Wilkins well, lies on the floor of a prominent

NNW striking valley proximate to South Canisteo, New York. The neighboring wellhead, the Appleton, lies some 100m up the hill a distance of 1.4 km to the WSW, and the westernmost wellhead, the O'Dell, lies a further 100m higher, some 1 km due west of the Appleton. In all a total of twenty-two intervals were stress tested in the Appleton well, forty-three in the Wilkins, and ten in the O'Dell.

Stable ISIP values are plotted in figure 5–9 as a function of depth for each well with the depth axes displaced such that common stratigraphic horizons are aligned. Solid diagonal lines represent the overburden load as estimated from bulk density logs giving a mean value of 2.71 gm/cm^2. Measurement error of each ISIP datum is less than ± 0.15 MPa. With the exception of the section between the H and K sands, the ISIP data show remarkably consistent linear trends with very little scatter. Evans et al. (1989a) recognize three different stress regimes: *an upper regime* within beds of the distal turbidites above the H-silt horizon, in which the ISIP values lie on or marginally above the overburden load for each well; *a transition zone*, between the H-silt and the K-silt horizons in which ISIPs in the shales decline with depth but the ISIPs in the sand beds remain on the same trends established in the upper regime; and *a lower regime* within black and gray marine shales below the K-silt, in which ISIPs measured at the same stratigraphic horizon are identical from well to well and, with the exception of one in the Tully limestone, are significantly less than the overburden. In brief, Evans et al. (1989a) suggest that the lower stress regime arises as a consequence of the poroelastic relaxation of overpressured shales and is, therefore, a remnant stress as discussed in chapter 10.

Televiewer surveys show that in all but one case the hydraulic fractures at the wellbore are vertical. As the induced fractures are vertical, it is possible that all but one of the ISIP values are equal to S_h. This interpretation is correct for the lower regime where measurements indicate a laterally uniform horizontal stress, generally unperturbed by topographic loading. However, if the ISIP lies near the lithostat, S_h may actually exceed the lithostat in which case the vertical hydraulic fracture may turn horizontal as it propagates away from the wellbore (Warren and Smith, 1985). If so, the measured ISIP equals the vertical stress and constitutes only a lower bound for both S_H and S_h because by equation 5–10, any underestimation of S_h will result in an underestimation of S_H that is three times as great. Elastic analysis of the effect of topography on the vertical stress gradient in each of the three boreholes predicts gradients which are quantitatively similar to the trends defined by ISIPs in the upper regime (Evans and Engelder, 1989). The fracture gradients (ISIP/S_v) for the O'Dell, Appleton, and Wilkins wells in this region are 1.0, 1.07, and 1.17, respectively, which is consistent with the regional stress characterization proposed by previous workers (Haimson and Stahl, 1970; Komar and Bolyard, 1981). Using Savage's et al. (1985) model for gravitational loading of topography, Evans et al. (1989a) show that gravitational loading causes dS_v/dz under a valley (i.e., the Wilkins

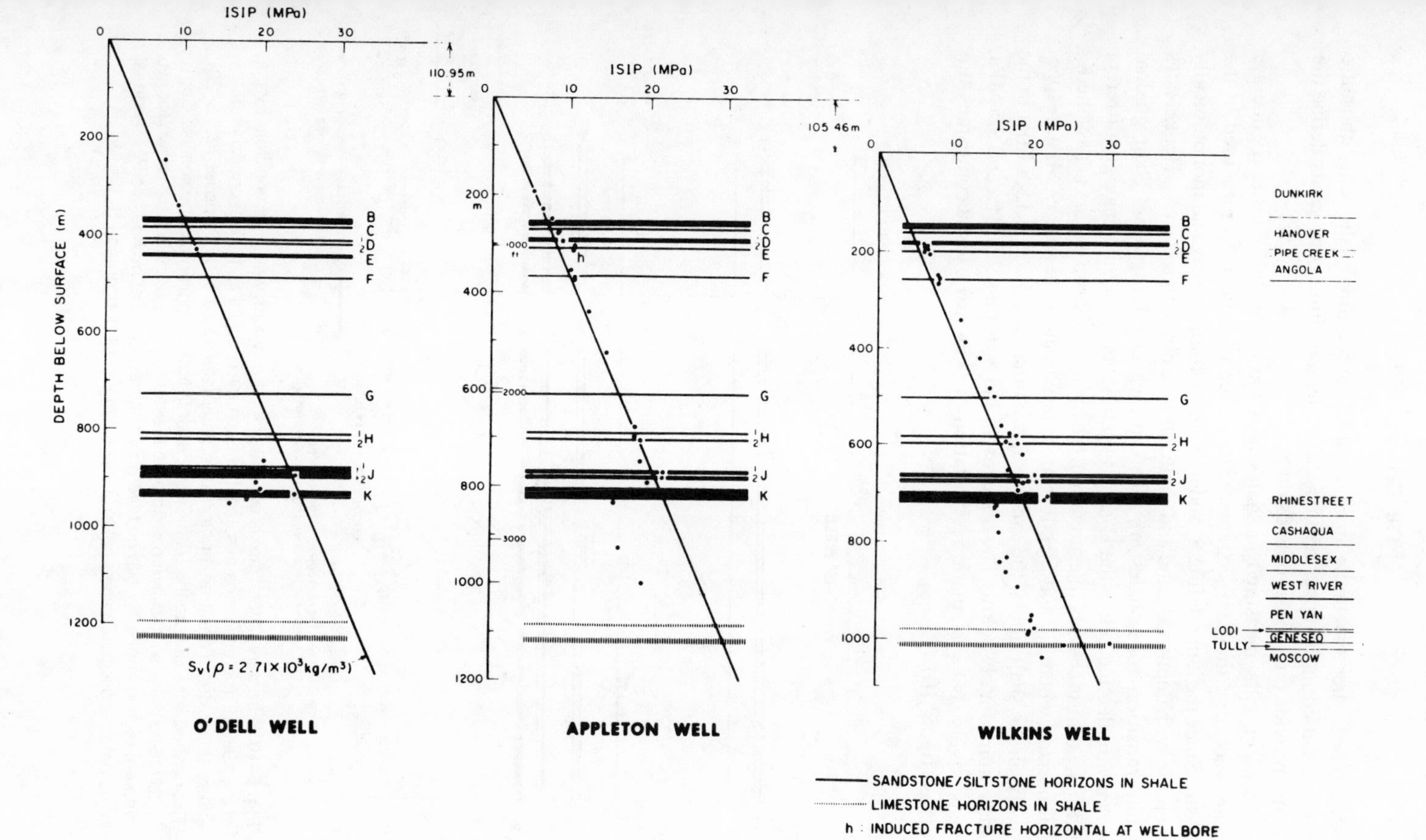

Fig. 5–9. Superposition of ISIPS from all three wells at South Canisteo on a common stratigraphic section (adapted from Evans et al., 1989a).

well) to be higher than under an adjacent hill. Accordingly, the near-lithostatic ISIPs measured in the upper regime and in the transition zone sands define the S_v and provide only lower bounds on S_h.

Although ISIPs measured in shales near the top of the transition zone are near-lithostatic, they are systematically different from ISIPs measured in the H-silt. Since the latter define S_v and as S_v must continuously and monotonically increase with depth, the sand and shale ISIPs cannot both reflect S_v. Hence, ISIPs from transition zone shales are taken to measure the true S_h. Their gradual downward decline below the H-sand horizon (becoming increasingly different from S_v) reinforces this interpretation. The principal drop occurs across the K-silt horizon where S_H and S_h fall from approximately 45 and 19 MPa, respectively, in the shale above the sand to approximately 35.5 and 15.5 MPa in the shale immediately below. The offset in S_H and S_h is thus 9.5 MPa and 3.5 MPa, respectively. No systematic reorientation of principal axes accompanies this offset (fig. 5–10).

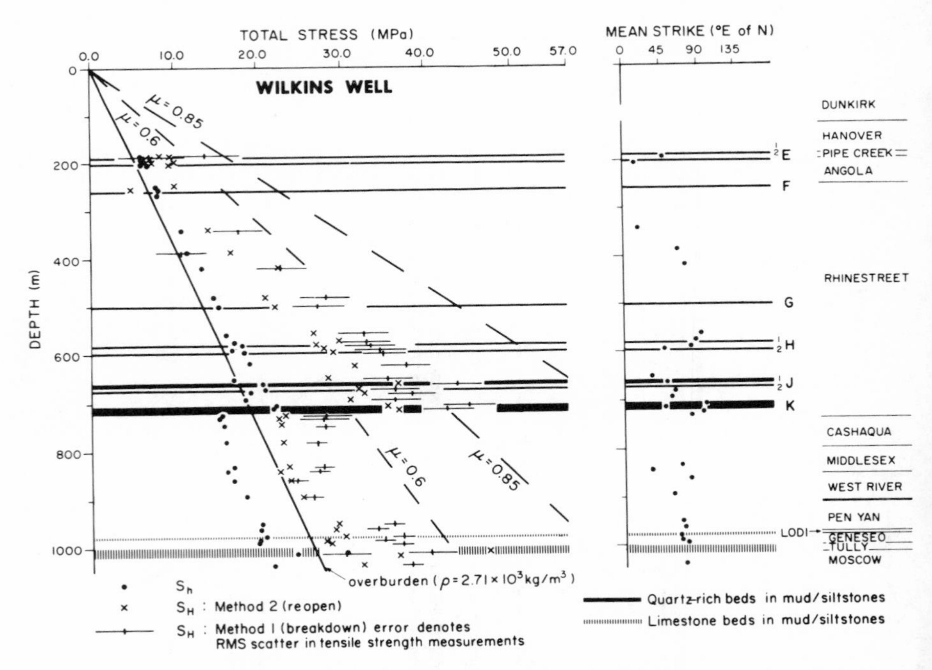

Fig. 5–10. Summary of total stress estimates for the Wilkins well (adapted from Evans et al., 1989a). The estimates assume that each ISIP is equal to the least horizontal stress at that depth and that a poroelastic parameter value of 0.75 applies (eq. 5–20). The dashed lines denote the S_H thresholds for failure assuming a cohesionless Coulomb failure criterion and hydrostatic pore pressure. The thresholds are shown for two different friction angles for slip on favorably oriented faults at hydrostatic P_p. The least principal stress is taken as equal to S_V above the K-silt horizon and S_h below.

In the transition zone S_h measured in the "shale" contrasts strongly with S_h measured in the quartz-rich siltstone beds which largely follow the ISIP trend defined in the uppermost stress regime. These stress contrasts attain a value of 5 MPa in the O'Dell well between both K and J silt horizons and the intervening shales. The Tully limestone also exhibits much higher least horizontal stresses than the surrounding shale but the precise magnitude of the lithology-correlated stress contrasts within the transition zone and about the Tully limestone is unknown because of the underestimation of S_h in the siltstones and limestone. Such stress contrasts are found in other sedimentary basins; however, in some cases such as the Piceance Basin, S_h in the shale is higher than in the adjacent sandstone beds (e.g., Warpinski, 1989). Measurements of static moduli for rocks from the siltstone and shale beds of the South Canisteo experiment show that the silt beds are systematically stiffer than the shale beds (Evans et al., 1989b). Evans et al. (1989b) conclude that the systematic contrast in stress level between siltstone/limestones and shales is due to horizontal elastic straining of beds in response to compression from the contemporary tectonic stress field. The Devonian section is responding as a layer cake in which the stiffer layers develop a greater stress as the section is uniformly shortened.

Estimates for S_H are shown for the Wilkins well in figure 5–10. High strength intervals were encountered in the shallow sections of the well with P_b as much as 30 MPa in excess of P_{ro}. This is consistent with measurements by Haimson and Stahl (1970) who report high breakdown pressures at the stratigraphic level equivalent to the B-silt horizon at South Canisteo. The profiles show that S_H generally follows the same pattern as S_h, the ratio of the two is approximately 2.5 to 1. The implied shear stress, although high, does not exceed the compressive failure envelope for Devonian shales as determined by Blanton et al. (1981). The inferred principal stresses in the upper section of the Tully limestone both lie on the super-lithostatic trends defined in the Rhinestreet sand beds, in stark contrast to the surrounding shales.

A regional study of S_h-depth variations within the Appalachian Basin is possible using commercial hydraulic fracture data (Evans, 1989). The major constraint in using commercial data, as discussed above, is that the ISIP usually overestimates S_h because the ISIP does not equal the closure pressure. In presenting a regional compilation of S_h for the Appalachian Basin, Evans (1989) normalizes his data using a least stress ratio, R_h, which is S_h/S_v. In the Appalachian Basin there are two S_h-depth trends with the boundary between these trends at the pinchout of the Silurian salt in the Appalachian Basin (fig. 5–11). Above the Silurian salt, the Devonian section generally shows an $R_h \geq 1.0$, otherwise the Devonian section shows an R_h in the range of 0.5 to 0.7.[5] An explanation of this behavior is delayed until chapter 10 where the concept of a remnant stress is introduced.

Data from the regional stress compilation are consistent with measurements at the Auburn Geothermal well, which was drilled through 1540 m of lower

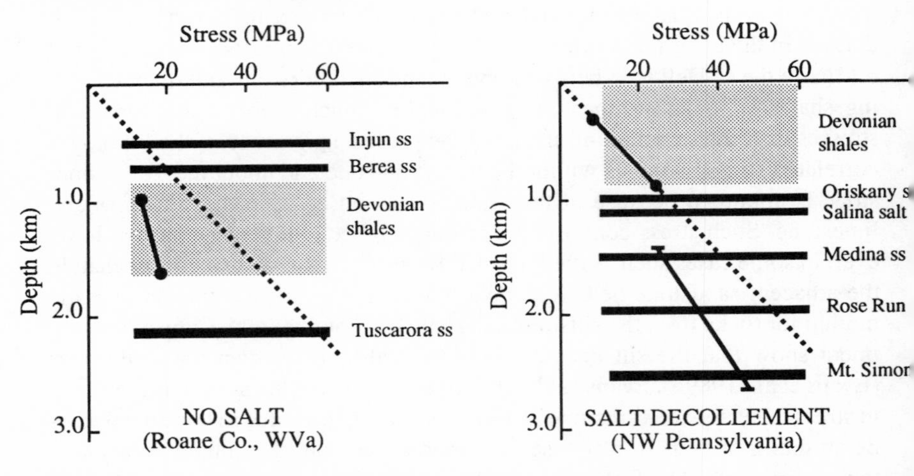

Fig. 5–11. Data on S_h from the Appalachian Basin (adapted from Evans, 1989). Formations which served as data sources are shown at approximate depths in central Ohio and western New York. Trend of S_h with depth as inferred using data from the vicinity of Roane County, West Virginia, where the Silurian salt décollement is not present (*left*). Mean trend of S_h with depth inferred from areas where the Silurian salt décollement is present (*right*).

Paleozoic rocks on the Appalachian Plateau and into 60 m of crystalline rock (Hickman et al., 1985). S_h was determined by comparing the ISIP after several pressurization cycles and S_H was determined using the P_{ro} (eq. 5–13). Fluid infiltration was minimized by using short duration pressurization cycles, high flow rates during pumping (30 l/min), and thorough flowback between pressurization cycles. Stress measurements at depths of 593 m, 747 m, 919 m, and 1482 m show the presence of a strike-slip fault regime (fig. 3–20). However, the shear stress is not large enough to initiate frictional sliding on preexisting fracture planes assuming that $\mu = 0.6$. Impression packers were used to determine the orientation of hydraulic fractures at 593 m and 919 m which strike at N83°E ± 15°.

In summary, stress data from the Appalachian Basin show that stress is neither a simple function of depth nor consistent with simple models involving gravitational and uniaxial tectonic loading as discussed in chapter 1.

The Piceance Basin

The Department of Energy's multiwell experiment (MWX) consisted of three closely spaced wells (35–65 m separation) drilled for a number of purposes including a detailed study of stress variation in highly heterogeneous rock, the

Mesa Verde Group, a sandstone and shale sequence deposited in a fluvial and deltaic environment of the Piceance Basin in western Colorado (Warpinski et al., 1985). Heterogeneities in the reservoir sandstones arise from their cigarlike shapes in contrast to flat-lying sheets of the type encountered in the South Canisteo experiment. Stress tests were performed between 1310 and 2470 m, in both marine and nonmarine rocks including shales, mudstones, siltstones, and coal (Warpinski, 1989). Hydraulic fracture tests were conducted as minifrac tests where S_h was determined using the flowback procedure of Nolte (1982). The distribution of S_h in the Mesa Verde Group contrasts dramatically with that found in the Catskill Delta during the South Canisteo experiment (fig. 5–12). In general, the sandstones and siltstones of MWX2 have much lower stresses (R_h = 0.73 to 0.82) than the marine shales (R_h = 0.87 to 0.96). At South Canisteo S_h was higher in the siltstones than in the shales.

The differences between the tectonic history of the Devonian Catskill Delta and the Cretaceous Mesa Verde Group contributed to the differences in distribution of S_h. Assuming that $\upsilon_{ss} < \upsilon_{sh}$, a uniaxial reference state (chap. 1) predicts a lower S_h within the sandstones and siltstones relative to the shales

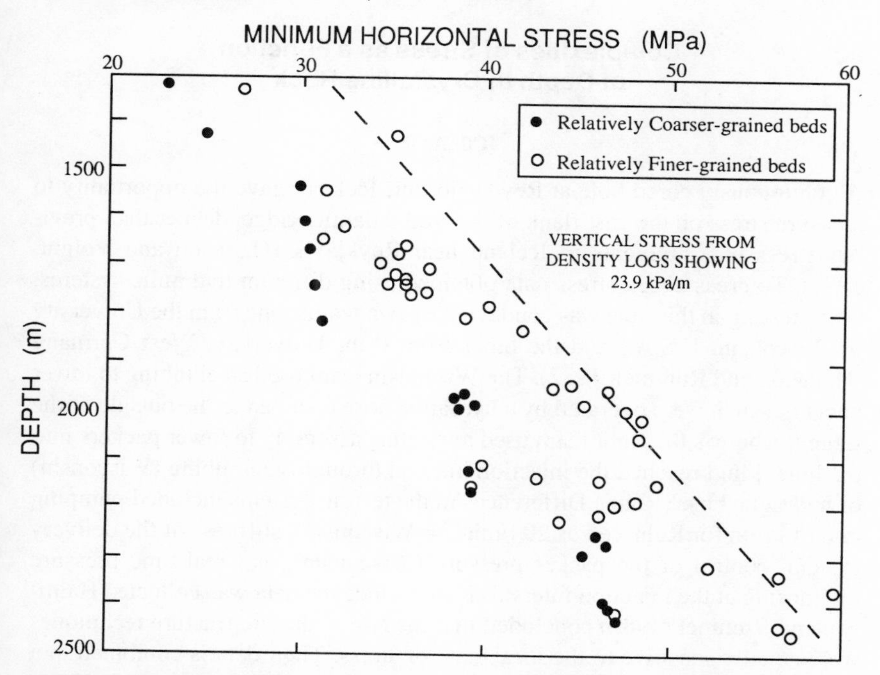

Fig. 5–12. S_h versus depth for the multiwell experiment in the MWX-2 well at Rifle, Colorado (adapted from Warpinski, 1989).

in sedimentary basins. Rather than appealing to the uniaxial strain model, Warpinski et al. (1983) suggest that the near-lithostatic S_h for shale in the Mesa Verde may have developed as a result of shale creep. Regardless of which model is favored, either is consistent with the geological history of the Piceance basin which was little affected by Laramide compressional events. In contrast, rocks of the Appalachian Plateau were compressed during the Alleghanian Orogeny, an event which was responsible for imposing a large enough horizontal stress so that all beds deformed by creeping to erase the uniaxial reference state characteristic of early basin history. Furthermore, Evans et al. (1989b) argue that the present bed-by-bed stress contrasts arise largely from elastic shortening of the section in response to the contemporary tectonic stress field superimposed on the Alleghanian remnant stress.

The details concerning the variation of stress as a function of depth are more apparent if a measurement is taken once every 20–30 m as was the case in the Wilkins well at South Canisteo and in MWX2 of the Multiwell experiment. Other data suites of comparable density include measurements in crystalline rocks in Iceland (Haimson and Rummel, 1982), Illinois (Haimson and Doe, 1983), and in North Conway, New Hampshire (Evans et al., 1988).

Complexities of Stress as a Function of Depth in Crystalline Rock

ICELAND

A continuously cored hole at Reydarfjordur, Iceland, gave the opportunity to measure stress on the east flank of the Mid Atlantic Ridge, deeper than previous stress measurements in Iceland near Reykjavik (Haimson and Voight, 1977). To cross-check stress data obtained using different hydraulic systems, stress testing in this hole was conducted by two teams: one from the University of Wisconsin, U.S.A.; and the other from Ruhr University, West Germany (Haimson and Rummel, 1982). The Wisconsin team used steel tubing to lower packers which were operated by a hydraulic hose clamped to the outside of the tubing, whereas the Ruhr team used a wireline assembly to lower packers into the hole. Fluid reached the injection interval through steel tubing (Wisconsin) or hydraulic hose (Ruhr). Differences in the testing systems included pumping rate (4 l/min for Ruhr versus 20 l/min for Wisconsin), stiffness of the delivery system, control of the packer pressure (Wisconsin), and real-time pressure monitoring at the injection interval (Ruhr). Once the data was collected Haimson and Rummel (1982) concluded that the two hydraulic fracture techniques were equally sensitive to the local state of stress. Their data is combined in a plot of stress versus depth which shows a large perturbation in the 300–500 m range (fig. 5–13). At depths of more than 500 m the stress regime is compatible

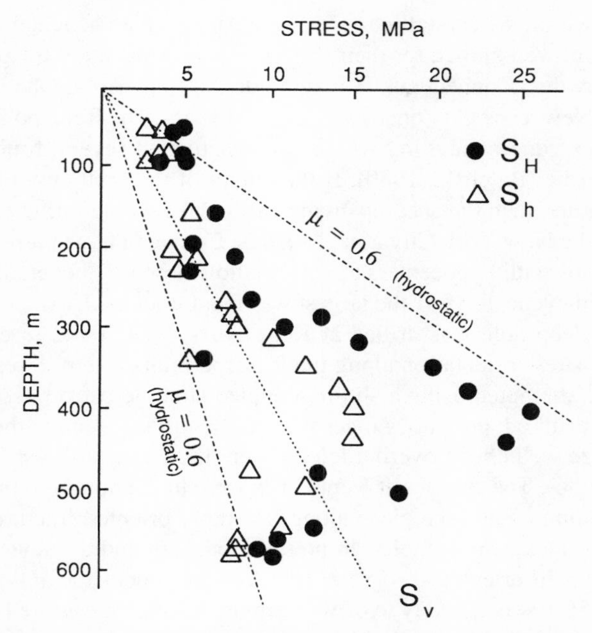

Fig. 5–13. The variation with depth of the two principal horizontal stresses at Reydarfjordur, Iceland (adapted from Haimson and Rummel, 1982).

with normal faulting. Hydraulic fracture orientations show that S_H is subparallel to local dike and fissure swarms. This correlation suggests that the deepest stresses reflect the contemporary tectonic stress field associated with sea floor spreading at a midocean ridge (Haimson and Rummel, 1982). Although its origin is unknown, the stress perturbation at the 300–500 m interval is encountered in other spots in Iceland (Haimson and Voight, 1977).

Southwest German Block, Germany

Midcontinent stress in the Southwest German Block of Europe was measured using the wireline hydraulic fracture system designed by Ruhr University, Bochum, Germany, by Rummel et al. (1983) (fig. 5–3). Stress measurements to 500 m show S_H oriented to the west of north, an orientation compatible with that found in the major portion of western Europe. Like other stress profiles, stress in the upper 200 m of the southwest German Block is in the thrust regime with a transition to strike-slip regime somewhere between 200 m and 300 m of depth. This transition from shallow thrust-fault regime to a deeper strike-slip fault is consistent with increasing S_h/S_v as the surface is approached.

Crystalline Appalachians

Two areas within the crystalline core of the Appalachian Mountains near New York City are well known for their earthquake swarms: the Ramapo fault system and Moodus, Connecticut. A magnitude 3.0 (m_b) earthquake is felt in the vicinity of New York City once every several years. The Ramapo fault system is a complex zone of faulting which was reactivated several times since the early Paleozoic (Ratcliffe, 1980). If the length of the fault system (> 50 km) were to rupture, then the accompanying earthquake would inflict considerable damage in the New York City area. Moodus, Connecticut, is known for earthquake swarms with hypocenters at very shallow depths (Ebel et al., 1982). Of five hundred events in 1981 the largest was magnitude of 2.1 (m_b).

A 1 km deep hole was drilled at Kent Cliffs, New York, to examine the question of stress orientation along the Ramapo fault system. Stress measurements were attempted using a sliding-coupled straddle-packer assembly lowered with a drill rod; pressure gauges were at the surface. Both of the horizontal stresses were well above overburden pressure indicating a thrust fault regime (Zoback, 1986). The stresses at Kent Cliffs were high enough so that incipient reverse faulting could take place along favorably oriented fractures provided that $\mu = 0.6$ along these faults. Impression packers and a televiewer survey showed that S_H is oriented at N55°E ± 10°. A major conclusion of this study was that S_H at N55°E was unlikely to drive a rupture through the entire length of the Ramapo fault system.

Prior to 1984 the state of stress in the upper crust of New England was known largely from earthquake focal mechanisms which showed no systematic orientation for the fault planes of New England earthquakes (Pulli and Toksoz, 1981). A compilation of near-surface overcoring stress measurements suggested that New England is subject to the same ENE S_H which is found in a major portion of the crust in the northeastern United States (Plumb et al., 1984a). Horizontal fracturing in the North Conway hole indicated high horizontal stresses (fig. 5–8). Two vertical fractures in the North Conway hole, taken at depths of 127.10 m and 301.77 m, strike N61°E and N63°E, respectively (Evans et al., 1988). A 1.5 km hole was drilled near Moodus to measure stress at a seismically active depth range (Baumgärtner and Zoback, 1989). Measurements show that shear stress is high enough to cause thrust faulting on optimally oriented fault planes at all depths to 1.5 km provided that $\mu = 0.6$ (fig. 5–14). S_H is oriented at N89°E ± 5° which correlates well with shallow fault-plane solutions from the Moodus area (Zoback and Moos, 1988).

Midcontinent of North America

The Illinois Deep Hole (UPH-3) was drilled in northwestern Illinois to a depth of 1607 m where 670 m of lower Paleozoic sediments cover basement (Haim-

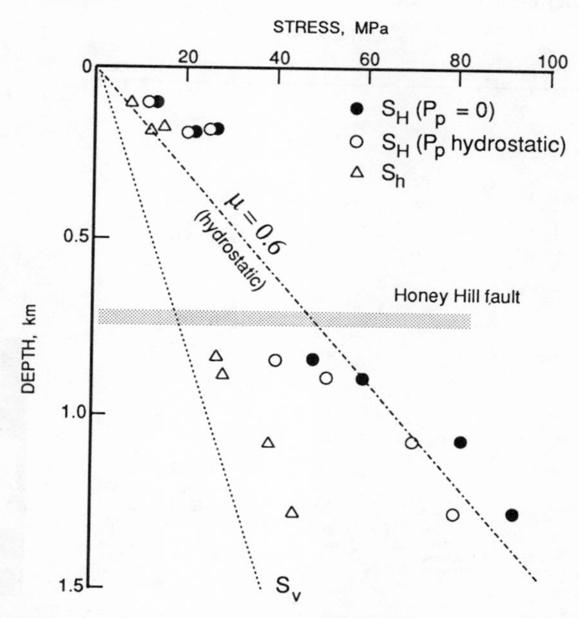

Fig. 5–14. Stress as a function of depth for the Moodus deep borehole (adapted from Baumgärtner and Zoback, 1989). The dotted line indicates the theoretical overburden stress S_v ($\rho = 2.61$ gm/cm^3). The dashed line is the theoretical strength curves for favorably oriented faults with $\mu = 0.6$.

son and Doe, 1983). Thirteen stress tests were successful in UHP-3 between a depth of 686 m and 1536 m. Another six were unsuccessful because the pressure for breakdown exceeded the safe pressure limit of the pumping system (45.6 MPa). In all but four tests S_h was larger than the overburden pressure indicating that the Illinois Deep Hole was in a thrust fault regime (fig. 5–15). The zone of four tests showing $S_h < \rho gz$ was highly fractured as indicated by core recovered from UHP-3. An impression packer survey indicates that S_H is N48°E. Haimson and Doe (1983) report that the fractures were rather thin and faint compared with induced fractures in granite at depths less than 300 m.

In a compilation of all hydraulic fracture stress measurements from within the midcontinent region, the most striking result is that the orientation of S_H is within a few degrees of ENE throughout the region (Haimson, 1974, 1978b, 1979a, 1980, 1981, 1982). All measurements less than 300 m in depth show $S_h > \rho gz$ which is representative of a thrust fault regime. Of the four wells with measurements below 300 m, three were in sedimentary rocks and all three show $S_h < \rho gz$ which is representative of a strike-slip fault regime. Later measurements by Hickman et al. (1985) at Auburn, New York, and Evans et al.

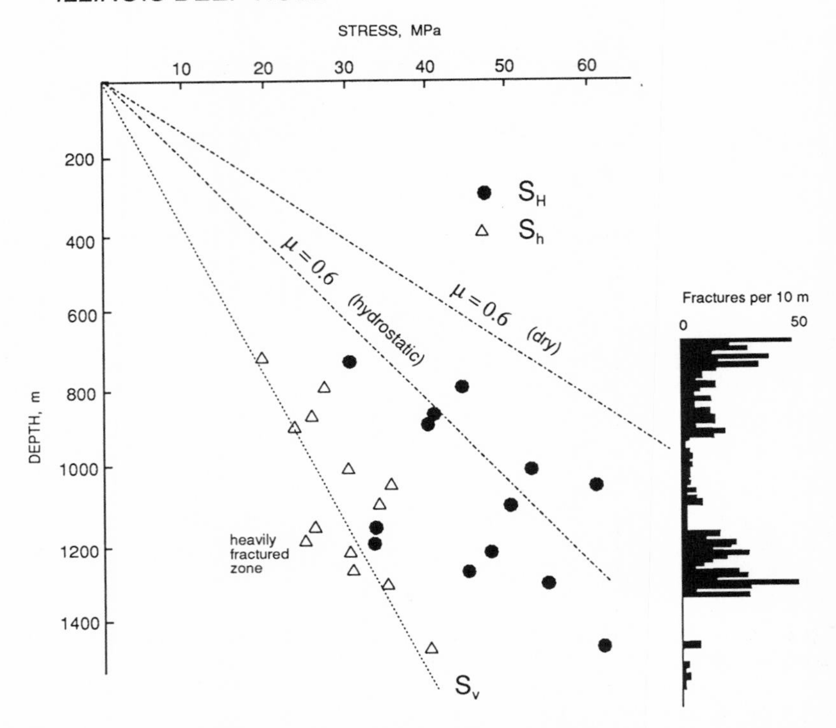

Fig. 5–15. Stress as a function of depth for the Illinois Deep Hole experiment (adapted from Haimson and Doe, 1983). Also plotted as a function of depth is the number of fractures per 10 m interval.

(1989a) also show that at depths approaching 1 km in sedimentary rocks the stress regime is strike-slip. Only the measurements in crystalline basement (UHP-3 of Illinois, the Kent Cliffs hole, and the Moodus hole) show stresses remaining in the thrust-fault regime at depths > 1 km. Even in these examples the stress in the midcontinent area is significantly lower than found in the crystalline core of the Appalachian Mountains (chap. 12).

Stress near the San Andreas Fault: A Complex Regional Pattern

The last great earthquake along the Palmdale section of the San Andreas fault north of Los Angeles, California, was in 1857. Trenching across the fault in the area of the 1857 rupture suggests that earthquake repeat time is between 150 and 300 years (Sieh, 1977). Because the Palmdale section of the San Andreas fault is approaching the lower limit for repeat time, this area is the focus of a number of hydraulic fracturing studies (e.g., Zoback et al., 1980; Zoback et al., 1987; Hickman et al., 1988; Stock and Healy, 1988). Stress measurements were attempted in several holes: one, XTLR was drilled down to 1 km while another at Cajon Pass was drilled to 3.5 km (fig. 5–16). Stress orientation and magnitude adjacent to the San Andreas fault from Palmdale to the Cajon Pass are complicated. Measurements to 1 km near XTLR indicate that the orientation of S_H is compatible with a right-lateral strike-slip fault where S_H is oriented slightly west of north, a situation expected for a fault with $\mu = 0.6$. However, measurements at Cajon Pass (Shimar et al., 1988) and Black Butte (Stock and Healy, 1988) suggest that S_H is nearly normal to the San Andreas fault (~ NE), a situation suggesting a low strength fault zone.

Variation in stress magnitude with depth and distance from the San Andreas fault is puzzling. In the vicinity of Palmdale some wells, most notably Black Butte to 650 m, show that stress is in the thrust-fault regime (Stock and Healy,

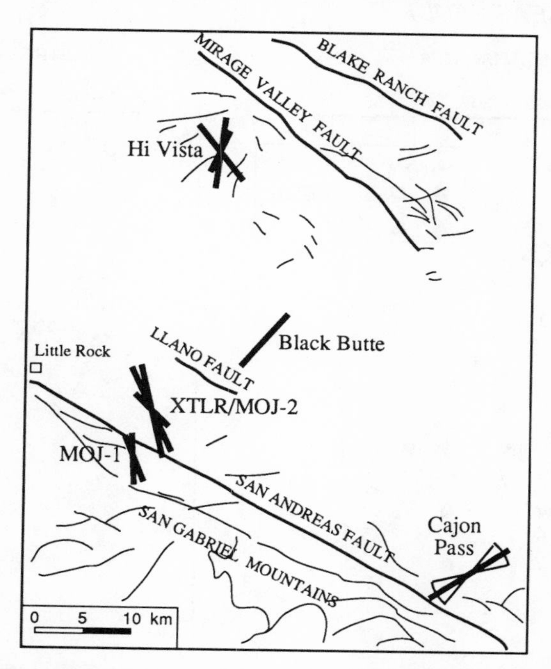

Fig. 5–16. Site locations for stress measurements in the vicinity of the San Andreas fault from Palmdale to the Cajon Pass (adapted from Hickman et al., 1988). The orientation of S_H is shown.

1988). Yet other wells, Hi Vista (Hickman et al., 1988) and XTLR (Zoback et al., 1980), show a transition from a thrust-fault regime to a strike-slip regime at depths of 250 m and 550 m, respectively (fig. 5–17). Some shallow measurements suggested that shear stress on planes parallel to the San Andreas fault increases with distance from the fault and the mean stress increases with depth in such a manner as to suggest that at seismogenic depths it exceeds several dozen MPa (Zoback et al., 1980), whereas measurements at Hi Vista suggest the opposite (Hickman et al., 1988). The implication of Zoback's et al. (1980) study was that shear stress on the San Andreas fault is near the frictional strength of the crust throughout the upper (seismogenic) portion of the crust. Later measurements at Black Butte and Cajon Pass suggest that S_H was fault normal and that the San Andreas fault is capable of moving at an extremely low shear stress (Zoback et al., 1987). However, much of the schizosphere is subject to high shear stress, leading Hickman (1990) to conclude that the high stress normal to the San Andreas fault indicates either a low-strength fault gouge or elevated pore pressures within the fault zone. Many of the S_h measurements at Cajon Pass were so much less than S_v that S_h was consistent with

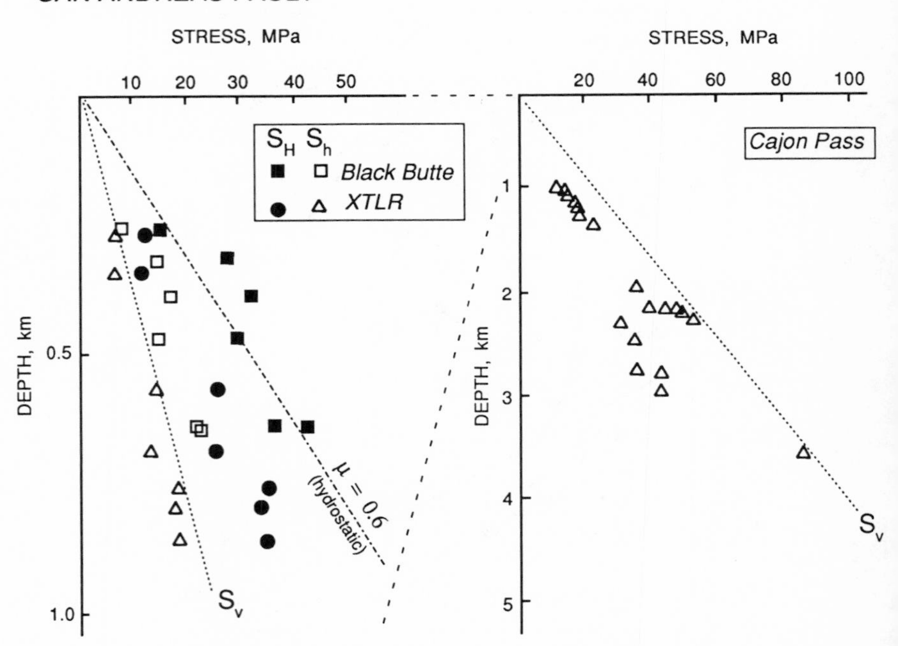

Fig. 5–17. The stress magnitudes near the San Andreas fault as measured at XTLR (Zoback et al., 1980); Black Butte (Stock and Healy, 1988); and Cajon Pass (Healy and Zoback, 1988).

the expected magnitude for slip on normal faults assuming $\mu = 0.6$ (Healy and Zoback, 1988). Like South Canisteo (Evans et al., 1989a) and Iceland (Haimson and Rummel, 1983), at Cajon Pass stress is not a linear function of depth (Healy and Zoback, 1988). In the vicinity of the San Andreas fault thrust-fault, strike-slip, and normal-fault regimes all occur within 100 km of each other so it is fair to conclude that the state of stress in the vicinity of the San Andreas fault is extremely complex.

Unresolved Issues Concerning the Hydraulic Fracture Technique

The hydraulic fracture measurement technique is complicated by a number of unresolved issues. Hydraulic-fracture stress measurements will improve with the answer to the following questions and issues.[6] (1) In some thrust-fault regimes, fractures are known to roll over after leaving the wall of the borehole in order to propagate normal to σ_3. How far does a fracture propagate before it rolls over? (2) Under conditions of high horizontal stress where both of the horizontal stress components are roughly equal and much greater than the vertical stress, it is known that horizontal fractures initiate at the wellbore. Is there a special rock fabric or packer configuration which favors the initiation of horizontal fractures? (3) Closure pressure is defined as the pressure at which cracks contact and are no longer completely supported by fluid. The least principal stress provides the force to close the fracture and, in a sense, determines how completely the fracture is closed. Are closure pressure and S_h always the same? (4) The classic breakdown formula for hydraulic fracture assumes an effective stress directly proportional to the total pore pressure. Subsequent work has shown that the poroelastic deformation within the vicinity of the borehole wall affects the breakdown and fracture reopening pressures. When should stress calculations include the poroelastic parameter (A) and the Biot parameter (α_b)? (5) For calculating the maximum principal stress either P_b or P_{ro} is required. Experience suggests that the breakdown varies depending on the conditions of the borehole wall to the point where P_b is an unreliable parameter for the calculation of maximum principal stress. Therefore, what is required for a precise measure of P_{ro}? (6) For microfractures (< 10 gal.) the fracture closure (least principal stress) is the same as the ISIP and may represent the behavior of just one sedimentary bed. Such is not the case for massive hydraulic fracture. Stresses vary on a bed-by-bed basis so that massive hydraulic fractures may provide only an average measure of in situ stress. What is the relationship among massive, mini, and micro fracture tests? (7) Stress varies significantly from bed to bed in a sedimentary rock. However, there are no general rules concerning whether shales or sandstones carry the higher stress. Fracture containment is highly dependent on higher stressed beds stopping the vertical growth of fractures. How do stresses develop in bedded sediments? (8) Wellbore storage plays a role in controlling P_{ro}. What is the relationship between wellbore storage and the opening of hydraulic fractures? (9) ISIP does not necessarily represent the least principal stress near a borehole. For example, in massive hydraulic fracturing the ISIP represents the time at which flow through the casing stops. Fracture closure, however, may occur at a much later time and lower pressure. What is the difference between instantaneous shut-in pressure and fracture closure?

Concluding Remarks

To date, hydraulic-fracture stress measurements have yielded the most complete information on stress in the upper schizosphere. In many regions, the upper crust is subject to shear stresses approaching the frictional strength of favorably oriented faults and, hence, these regions fall within the frictional-slip regime. Stress within sedimentary rocks is dependent on both lithological differences between beds as well as the tectonic history of the basin. If the stress within eastern North America reflects all cratonic areas, then the sedimentary cover is subject to lower absolute stresses than the underlying crystalline basement. A state of stress characteristic of the crack-propagation regime is more likely within sedimentary basins.

6

Borehole and Core Logging

It is possible to obtain qualitative information about the stresses in rock from observations of the fracturing of the sidewalls of horizontal and near-horizontal boreholes. For example, in boreholes drilled at great depth in the hard quartzites and conglomerates of the Witwatersrand gold mines the sidewalls tend to spall, particularly where the borehole passes through highly stressed ground.

—E. R. Leeman (1964a) after making many observations of earth stress in South African gold mines

Stress concentrations around and at the end of boreholes are large enough to favor drilling-induced cracking or fracturing in both the borehole wall and drill core. Several examples of drilling-induced brittle behavior include the development of core discs, the propagation of petal-centerline fractures through cores, and the failure of borehole walls with the subsequent breakout of wall material. Core discing and petal-centerline fractures are analyzed in core recovered from the well, whereas borehole-wall failure is analyzed using well-logging instruments. In all cases, earth stress controls the orientation and morphology of drilling-induced brittle behavior. In situ stress magnitude and orientation also controls stress-induced microcrack porosity and microcrack anisotropy in the rock behind the borehole wall. Stress-induced microcrack porosity and microcrack anisotropy affect such rock properties as sonic velocity and electrical resistivity, both of which are routinely measured by logging tools. Hence, rock properties give information on the nature of stress within rock behind the borehole wall. Well and core-logging techniques, designed to detect such cracking and fracturing in both the borehole wall and drill core, provide information on stress in the upper crust. First, this chapter deals with the analysis of the stress-induced cracking and fracturing in rocks recovered from deep boreholes. Then, it reviews the information available from the failure of borehole walls and from rock properties behind borehole walls.

Core Discing

The length of continuous core recovered from shallow levels (< 100 m) in the crust depends on either the distance between natural fractures or the length of the core barrel. Unlike core taken near the surface, core recovered from deep mines or wells (> 1 km), tends to split spontaneously into thin discs or break up into irregularly shaped fragments (Hast, 1958; Leeman, 1964a). These discs form by cracking roughly normal to the axis of the borehole and are usually concave-convex with the convex side toward the bottom of the borehole (fig. 6–1). Leeman's (1964a) stress measurements in a horizontal core hole drilled from the East Rand Mines of South Africa show that a zone of severe core discing correlates with unusually high S_v. Extreme core discing was encountered in the Kola deep hole, Russia, where at a depth of 11.5 km, core recovery was less than 30% largely because the core split about every 6 mm producing discs that were too thin to survive the drilling process (Kozlovsky, 1987). From these observations it is clear that the thickness of core discs alone is a useful criterion for indicating relative magnitude of stress normal to the borehole. Information on σ_d in the plane normal to the borehole axis comes from the disc shape; some core discs are cup shaped, whereas others are saddle shaped or potato-chip shaped (Paillet and Kim, 1987). The more irregular shape is indicative of a relatively high σ_d^H with the axis of the peaks facing the direction of S_h (Lehnhoff et al., 1982). Apparently, this relief or saddle shape of the disc becomes more pronounced as σ_d^H increases (Dyke, 1989).

Laboratory tests help to resolve questions concerning the influence of stress on core discing (e.g., Jaeger and Cook, 1963). One configuration of laboratory experiments on core discing consists of a large cylinder of rock compressed radially (i.e., the direction of loading is across the diameter of the cylinder). A core is then drilled along the axis of the test cylinder. The stress field at the base of the axial core approximates the field situation where the greatest principal stress is across the diameter of the test cylinder and the least principal stress is along the axis of the test cylinder. In the laboratory, core discing is favored only

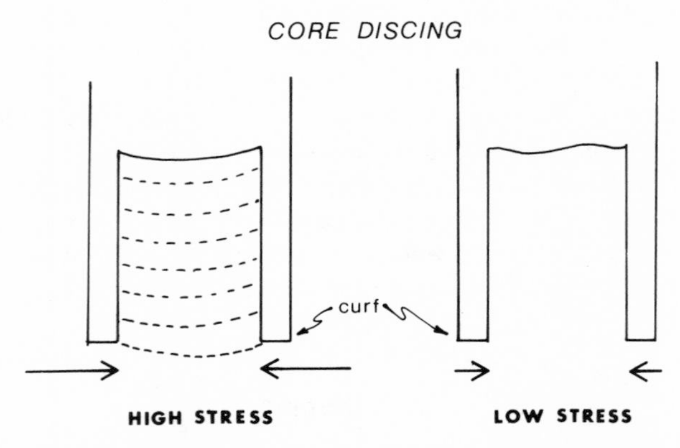

Fig. 6–1. Cross section through cores in boreholes in a high stress and low stress environment, respectively. Core discing occurs only in a high stress environment.

when stress normal to the cylinder axis (i.e., the radial stress) is greater than stress parallel to the cylinder axis (Obert and Stephenson, 1965). Hence, core discing in vertical boreholes implies that S_H is numerically greater than S_v. These laboratory experiments also show that the radial stress must be high in relation to the unconfined compressive strength of the rock. Because discing was not produced in chalk, Obert and Stephenson (1965) suggest that discing might not occur in low strength or low modulus rocks.

Discussion of the brittle failure mechanism for core discing includes reports of both shear fracture and tensile cracking (Maury et al., 1988). Hast's (1979) vision of the discing process is that failure starts at the curf of the drill bit and follows a surface of high shear stress. This shear stress is set up by the tendency of the free core above the curf of the drill bit to expand relative to that core still subject to lateral constraints. Another analysis suggests that the tensile stresses along the axis of the core permit shear to take place at a load stress lower than those necessary to produce shear in the whole system (Jaeger and Cook, 1963). Despite arguments for shear failure, the literature is converging on a consensus that failure is tensile (e.g., Kulander and Dean, 1985; Dyke, 1989). For example, Stacey (1982) argues that core discing initiates as a result of extension ahead of the borehole end. Evidence for a tension cracking mechanism includes the clean and unsheared surface of the discs. Furthermore, a stress analysis has shown that axial tension will occur under the loading conditions around a core attached to the end of a borehole (Jaeger and Cook, 1963; Dyke, 1988). Dyke (1988) shows that tensile failure initiates at the surface of the core in the direction of S_h and propagates inward and steeply downward toward the center of the core. This accounts for the saddle shape of core discs with the peak facing S_h. The strongest piece of evidence that core discing follows a tensile mechanism is that some disc surfaces have a plumose morphology which develops on propagation of mode I cracks (Kulander and Dean, 1985). Many joints in siltstones and sandstones have a well-organized surface morphology commonly called a plumose pattern (Woodworth, 1896; Hodgson, 1961; Bahat, 1979; Kulander and Dean, 1985). Key features of the plumose structure on both natural joints and core discs are the plume axis, inclusion hackles (barbs), twist hackles, and arrest lines (fig. 6–2).

Disc surfaces with a well-developed plumose morphology are useful for mapping the orientation of in situ stresses (Kulander and Dean, 1985). However, the relationship between the plume axis and in situ S_H is opposite the relationship established for natural joint surfaces. By analogy with crack propagation in glass, joint propagation takes place with the rupture front advancing parallel to σ_1 in the plane normal to σ_3 (Kulander et al., 1979). In flat-lying sedimentary rocks where $\sigma_1 = S_H$, and $S_h = \sigma_3$, vertical joints commonly propagate with the rupture front advancing most rapidly in a horizontal direction parallel to bedding. This type of rupture leaves a plume axis parallel to bedding (Bahat and Engelder, 1984). Crack propagation within drill core differs from

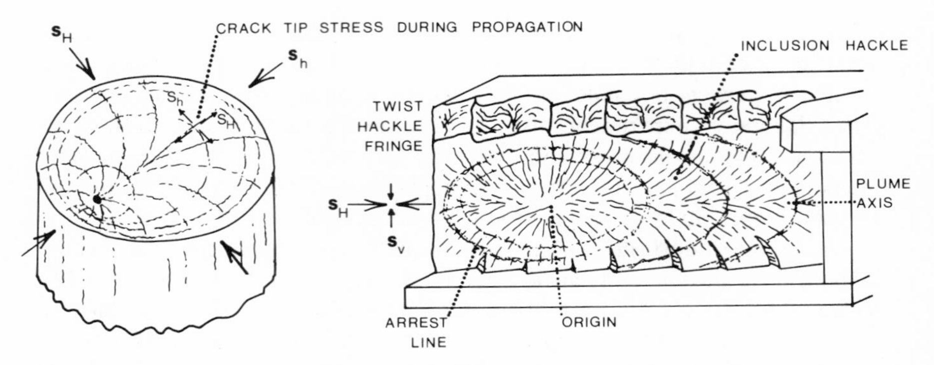

Fig. 6–2. The key elements of the plumose morphology on the surface of joints and core discs. Surface morphology of core discs shows the direction of propagation of the coring-induced joint. Note that the direction of propagation is in the direction of in situ S_h because upon stress relief this direction becomes σ_1 (adapted from Kulander and Dean, 1985).

that of a natural joint in that the plume axis on discs is controlled by the elastic strain relief during overcoring, not directly by in situ stress as is the case for a natural joint (fig. 6–2). During overcoring, stress relief of the core is such that expansion normal to the axis of the core is largest in the direction of S_H. This maximum expansion induces stress in the core to realign so that σ_2 during relief is in the direction of S_H, and σ_1 during relief is in the direction of S_h. σ_3 is subparallel to the axis of the core during discing. The plume axis, although controlled by σ_1 in the core during overcoring, is then aligned parallel with in situ S_h, a result which is not intuitively obvious but which is consistent with Dyke's (1989) analysis of the saddle shape of core discs.

Petal-Centerline Fractures

Unlike core discs some coring-induced fractures initiate at an angle to the core axis and then rotate during propagation parallel to the borehole axis (Kulander et al., 1979). Petal-centerline fractures propagate in response to core-induced principal tension which rotates downward to align normal to the axis of the core (fig. 6–3). The effect of this stress rotation is to produce a petal that "curves in a down-core direction from an initial dip angle generally between 30° and 75° at the core boundary to a vertical inclination within the core parallel to the core axis" (Kulander et al., 1979). Downward propagation follows the advance of the drill bit from an initiation point at the wall of the core. From an initial crack with a relatively steep (but not vertical) dip rotation to vertical

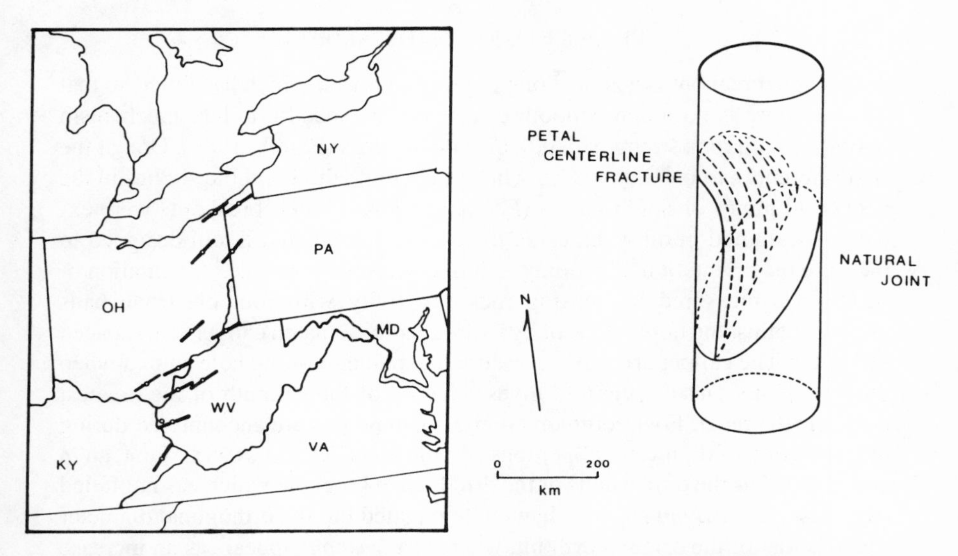

Fig. 6–3. The strike of the vertical portion of petal-centerline fractures within core taken in the Appalachian Basin (adapted from Plumb and Cox, 1987). Also shown is a core with a petal-centerline fracture abutting a natural joint in shale from the Appalachian Basin.

occurs as the crack approaches the center of the core. Petal-centerline fractures are easily indentified becasue they never cut completely through the core in a plane at an angle to the axis of the core and they often terminate within the core against a preexisting joint or bedding plane. Preexisting joints rarely curve and usually cut in a single plane through the whole core.

Information on in situ stress orientation is available through the analysis of petal-centerline fractures in oriented core. During the development of petal-centerline fractures propagation is downward into "virgin" rock so that the orientation of petal-centerline fractures is controlled exclusively by in situ stress rather than by the reorientation of stress following stress relief as is the case for core discing. The behavior of petal-centerline fractures is illustrated with core sampled in more than two dozen wells through Devonian shales in the Appalachian Basin (Cliff Minerals, 1982). That stress in virgin rock controls the orientation of the petal-centerline fractures is demonstrated by the correlation between the orientation of petal-centerline fractures and S_H for the contemporary stress field of eastern North America (figs. 6–3, 5–1). Cores from thirteen Devonian-shale wells contain petal-centerline fractures and the strike of these fractures clusters about N60°E (Plumb and Cox, 1987).

Borehole Breakouts

The Orientation of Breakouts

A borehole breakout is a zone along the wall of a well which has failed so that the well bore is no longer smooth or round. The specific failure mechanism may vary with rock strength, depth, or state of stress (Plumb, 1989). One of the first references to breakouts occurs in conjunction with borehole studies in the deep gold mines of South Africa (Leeman, 1964a). Later, breakouts were extensively studied in oil wells using the four-arm dipmeter, a tool designed to measure the orientation of bedding as it intersects the borehole.[1] Orientation of bedding is measured by sensing rock resistivity with four electrode pads pressed against the borehole wall by hydraulically actuated caliper arms spaced 90° apart. The caliper arms also measure the diameter of the hole so that when the hole is not circular the tool gives a record of the azimuth of the greatest borehole diameter. Four common borehole geometries are encountered during interpretation of dipmeter caliper logs (Plumb and Hickman, 1985). First, an *in gauge hole* has the dimensions of the drill bit in a borehole which has not failed (fig. 6–4). A *breakout* has one diameter elongated but the orthogonal diameter remains that of the original drill bit, whereas a *washout* appears as an increase

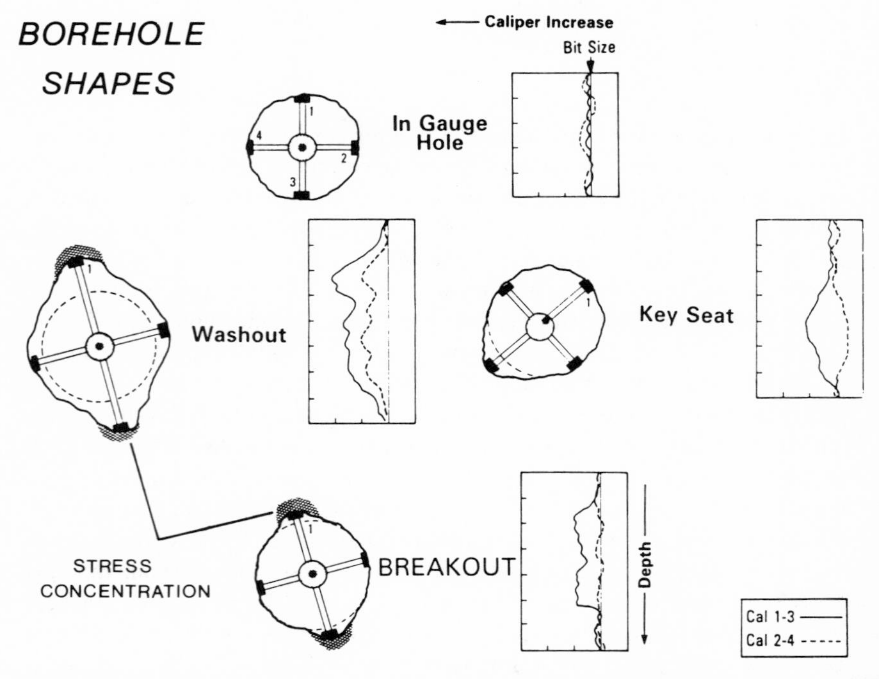

Fig. 6–4. Four common borehole geometries (adapted from Plumb and Hickman, 1985).

in all dimensions of the borehole. If the washout is limited, detection of a borehole elongation is still possible. A *key seat* occurs when the four-arm dipmeter is not centered in the borehole or the hole is pear shaped due to drill-pipe wear and one caliper reading is less than bit size. Under ordinary circumstances in a smooth-walled borehole the dipmeter continuously rotates along the borehole wall as it is pulled up the well. This rotation is produced by the twisted logging cable as it winds on the cable spool in the logging truck. Breakouts were first recognized when petroleum engineers noticed that the twisting of the tool ceased in certain intervals of the well where the pads of the tool became stuck in breakout slots (Cox, 1970).

The first regional study of breakouts was in the Upper Cretaceous Wapiabi shale of the Alberta Basin, Canada, where many wells were elongated with slots in the NW-SE direction (Cox, 1970) (fig. 6–5). Initially it was not clear if the slots were natural fractures intersecting the borehole or breakouts of the type reported by Leeman (1964a). Babcock (1978) expanded Cox's (1970) data set to show that 246 breakouts throughout the foothills of the Canadian Rockies were all generally NW-SE regardless of lithology or depth of borehole and that they had no relation to the local dip of bedding or local structures. In searching for an explanation for the borehole breakouts, Babcock (1978) correlated the slots with the pattern of regional joint sets as mapped on surface outcrops. Babcock (1978) suggested that the breakouts resulted from the drill encountering steeply dipping joints or zones of such joints. Subsequent analysis of breakouts and fractures in the Appalachian Basin did not support Babcock's (1978) hypothesis (e.g., Plumb and Hickman, 1985; Plumb and Cox, 1987).

Bell and Gough (1979) and Hottman et al. (1979) independently offered an alternative explanation for the borehole breakouts which is now generally accepted. They pointed out that, if subsurface horizontal stresses were unequal, a circular borehole would concentrate compressive stresses (e.g., fig. 5–2) which were large enough to cause shear failure at the borehole wall. Symmetric borehole elongation occurs in the direction of S_h when the circumferential stresses from the concentration of S_H in the borehole wall exceed the uniaxial compressive strength of the host rock. Hottman's et al. (1979) conclusion about borehole breakouts was based on Shell Oil Company's experience in the Gulf of Alaska where $S_H > S_h > S_v$ (fig. 2–12). Breakouts developed on portions of the wellbore where P_p is abnormally high causing relatively lower well bore strength. Assuming that wellbore failure occurs at the wellbore surface,[2] Hottman et al. (1979) model wellbore failure in a plot of octahedral shear stress, τ_{oct}, versus the effective confining pressure near the wellbore, $P_c - P_p$, where

$$\tau_{oct} = \tfrac{1}{3}\sqrt{(\sigma_\theta - \sigma_r)^2 + (\sigma_\theta - \sigma_z)^2 + (\sigma_z - \sigma_r)^2} \qquad (6\text{–}1a)$$

$$P_c - P_p = \frac{(\sigma_\theta + \sigma_r + \sigma_z)}{3} - P_p \qquad (6\text{–}1b)$$

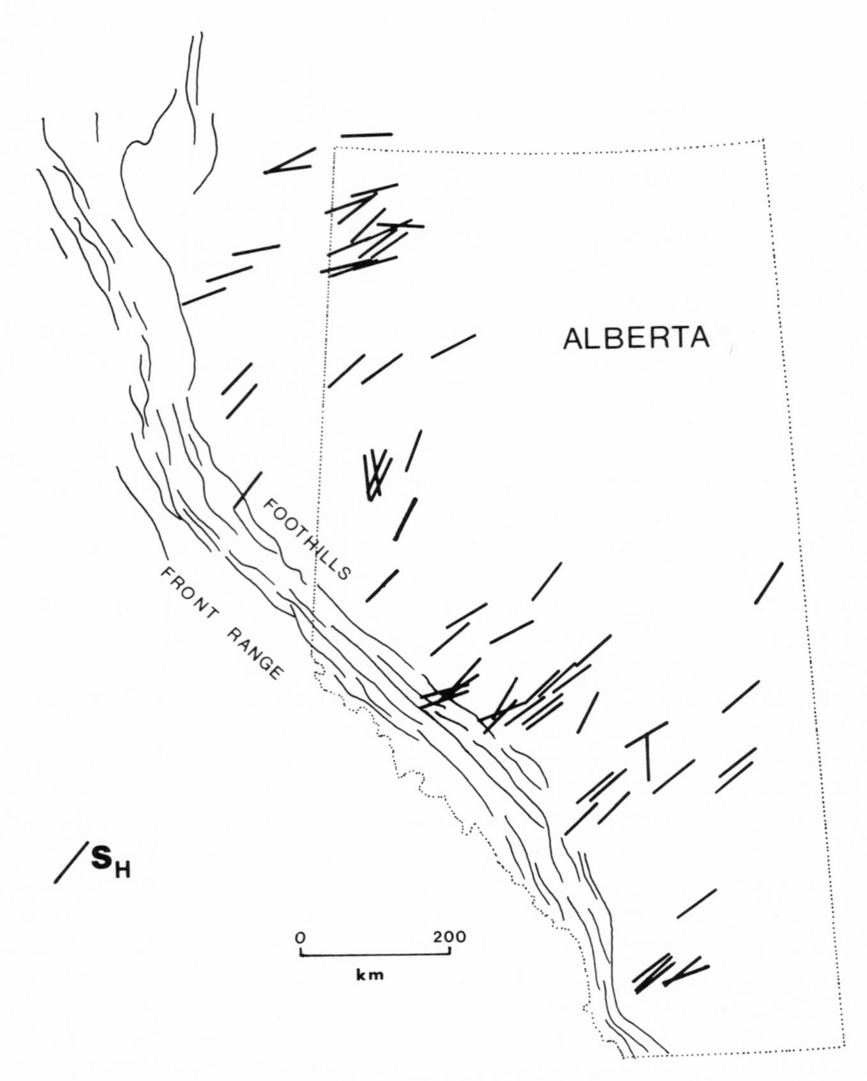

Fig. 6–5. A map of wells in the Alberta Basin showing the direction of S_H based on borehole breakouts. The data was compiled by Gough and Gough (1987) from Fordjor et al. (1983), Gough and Bell (1981), and Babcock (1978).

and σ_θ, σ_r, and σ_v are in turn functions of the three principal stresses and the mud weight in the borehole (fig. 6–6). Without changing mud weight, Shell drilled through a transition between abnormal P_p and higher abnormal P_p at about 2200 m in one well in the Gulf of Alaska. The well was stable at the lower P_p and unstable at the higher P_p as indicated by breakouts. At about

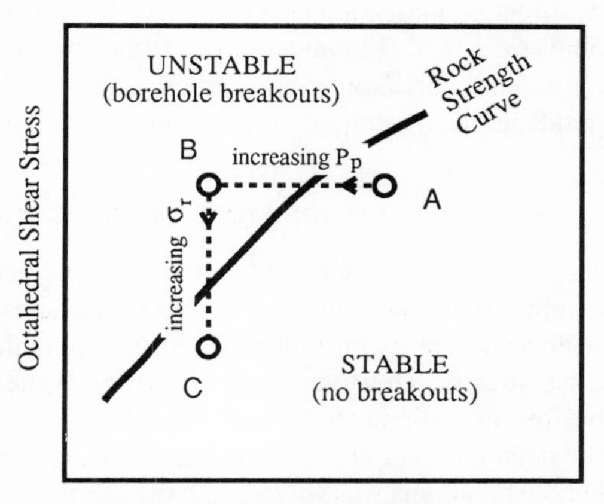

Fig. 6–6. Octahedral shear stress versus effective confining pressure for borehole breakouts in wells in the Gulf of Alaska (adapted from Hottman et al., 1979).

2230 m the mud weight was raised to counteract the lower effective stress and this again stabilized the wall of the wellbore. This sequence of events is followed on a plot of τ_{oct} versus $P_c - P_p$ where the experimental curve for rock strength is drawn (fig. 6–6). Decreasing $P_c - P_p$ caused the state of stress about the wellbore to cross the fracture strength curve for rocks of the Gulf of Alaska (A ⇒ B in figure 6–6). Increasing the mud weight adds support to the borewall of the well and, hence, stabilizes the tendency for the well to fail in shear (B ⇒ C in fig. 6–6). Supporting the borehole wall by adding mud weight increases σ_r, thereby decreasing τ_{oct} in equation 6–1. Using this failure analysis Hottman et al. (1979) conclude that S_H is between 64 and 70 MPa at a depth of 2.2 km in the Gulf of Alaska.

Bell and Gough's (1979) experience in the Alberta Basin suggests that breakouts occur in the stiffer beds such as the Cardium sandstone and the Wapiabi shale, whereas other shales tend to cave without marked elongation. The validity of the hypothesis that breakouts in the Cardium sandstone are caused by wellbore stress concentrations was supported by the correlation between the NW-SE breakouts (i.e., S_H is NE-SW) and a NE-SW S_H for the contemporary tectonic stress as documented by Sbar and Sykes (1973) in the northeastern United States. One caveat to this interpretation is that a NW-SE stress concentration may reflect a remnant stress from the NE-SW Tertiary compression of the Canadian Rockies. Gough and Bell (1981; 1982) further strengthened their argument by presenting breakout data from Colorado, east Texas, and northern Canada and comparing these breakout orientations with other data on earth stress. Fordjor et al. (1983) expanded the data set on breakouts in the Alberta Basin. In a vast area of eastern North America breakouts are oriented in the NNW direction indicating an ENE S_H which correlates with the contemporary tectonic stress field (Plumb and Cox, 1987). Inferences about stress orientation using borehole breakouts in ocean crust are consistent with stress directions inferred from local intraplate earthquakes (Newmark et al., 1984).

Information from the Shape of Breakouts

Gough and Bell's (1982) analysis of borehole breakouts predicts that the region of failure would approximate a triangle in cross section, enclosed by flat conjugate shear planes oriented at a constant angle to the azimuth of the far-field horizontal principal stresses. However, the exact shape of the breakouts is not discerned by the four-arm dipmeter because its relatively large caliper pads cannot follow the detailed contours of the breakout. The problem of mapping the shape of borehole breakouts was solved with the development of the acoustic borehole televiewer (BHTV) (e.g., Zoback et al., 1985; Plumb, 1989). A borehole televiewer is a well-logging tool that generates an image of acoustic reflectivity of the wellbore (Zemanek and Caldwell, 1969). The tool consists of a magnetically oriented rotating piezoelectric transducer which emits and receives an ultrasonic (≈ 1.4 MHz) acoustic pulse reflected from the borehole wall 600 times per revolution. The transducer rotates three times a second and is pulled up the borehole at the rate of 2.5 cm/s. *Acoustic reflectivity* of the borehole wall, a complex function of the rock acoustic impedance, is plotted as a function of azimuth and depth on an oscilloscope screen with brightness indicating the amplitude of the reflected pulse. The equivalent digital record includes the travel time and amplitude of each pulse. With a centralized BHTV, zones of low reflectance, which indicate a rough wall, show up as dark bands on the televiewer image.[3] An image of variation in reflectance is used to map features such as fractures and formation boundaries within the borehole. BHTV studies suggest that the breakouts are broad and flat depressions with less of the pointed shape than predicted by the Gough and Bell (1982) theory. However, one shortcoming of the BHTV technique is that transit-time data tend to "round" corners so that breakouts are not quite as smooth as they seem on the BHTV image (Plumb, 1990, personal communication).

Detailed observations of breakouts in several stages of development suggest that borehole failure is strongly dependent on lithology (Plumb, 1989). Failure initiates at the wellbore surface in crystalline rocks but inside the formation in sedimentary rocks. This interpretation arises from the observation that incipient breakouts in crystalline rocks are small pits in the wellbore surface, whereas breakouts first appear as well-developed fracture zones in sedimentary rocks. Using a compilation of breakout images, Plumb (1989) concludes that three types of breakouts reflect one of three different reference states of stress around the borehole:

$$\sigma_r < S_v < \sigma_\theta \quad (6\text{–}2a)$$

$$\sigma_\theta < \sigma_r < S_v \quad (6\text{–}2b)$$

$$\sigma_r < \sigma_\theta < S_v. \quad (6\text{–}2c)$$

Figure 6–7 illustrates the three types of breakouts with their associated fractures. Type 1 breakouts are associated with vertical shear and extension fractures where the largest stress is circumferential (e.g., Zheng et al., 1988). Type 2 and 3 breakouts are found at depths of several kms in sedimentary basins where S_v is the largest stress (Plumb, 1989). Such a stress state favors the development of inclined shear fractures. These three types of breakouts correspond to the three modes of wellbore failure postulated by Guenot and Santarelli (1988).

Developing a relationship between borehole breakouts and stress magnitude is difficult for a number of reasons. First, the "failed" material is rarely excavated completely during drilling so that the apparent (i.e., logged) shape is not necessarily representative of the failed region of the borehole. Second, breakouts may also get deeper with time under constant load, so again the apparent shape is not necessarily representative of the initial failed region (e.g., Plumb et al., 1988). Third, both extension and shear fracturing occur during the development of type 1 borehole breakouts (Zheng et al., 1989). Near the stress-free surface of the borehole wall, axial splitting cracks develop parallel to the borehole wall. Away from the borehole wall, where radial stress confines the rock in three dimensions, failure by shear fracture is more common. To date there is no agreement in the literature concerning which brittle failure mechanism

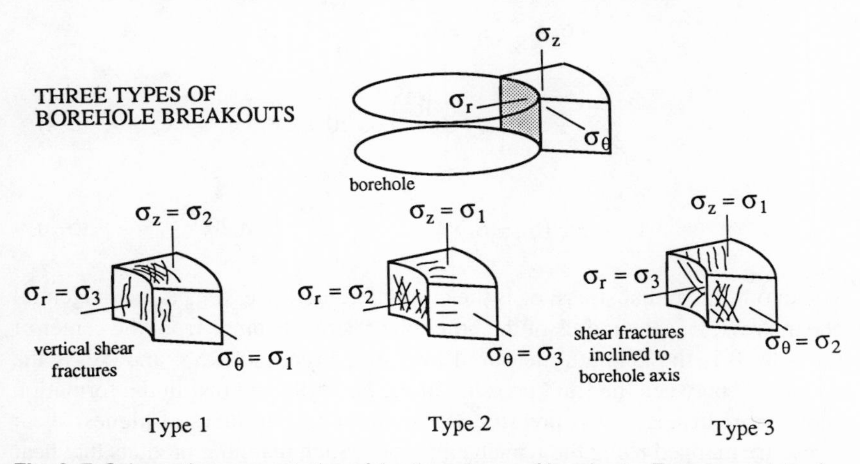

Fig. 6–7. Schematic representation of the three types of breakouts. Type 1 consists of vertical fractures parallel to the wellbore; Type 2 consists of horizontal extension fractures with shear fractures at an acute angle to the borehole; and Type 3 consists of stepped fractures developed from inclined fractures (adapted from Plumb, 1989).

plays a larger role in developing borehole breakouts. Zheng's et al. (1989) analysis suggests that type 1 breakouts grow stably by a process of axial splitting. Stability of the breakout arises because the breakout process redistributes the stresses so that these stresses are less than both the shear and tensile strengths of the rock. Zheng et al. (1989) follow Detournay and Roegiers (1986) to conclude that the shape of a stable breakout is not related uniquely to the state of stress and strength of the rock in the immediate vicinity of the borehole.

Although no unique theory between breakout shape and in situ stress exists, it is instructive to discuss one attempt to develop a relationship between breakout size and in situ stress. Zoback et al. (1985) present a theory for borehole breakouts to account for the broad and smooth shape of the breakouts observed in many boreholes. In contrast with the suggestion that axial splitting is the controlling failure mechanism, Zoback et al. (1985) assume that the breakout shape arises from shear rupture. Their analysis applies to type 1 breakouts and starts by assuming a cylindrical hole in a thick, homogeneous, isotropic elastic plate subjected to effective minimum and maximum principal stresses ($\overline{S}_h$ and $\overline{S}_H$). In terms of effective stress the Kirsch (1898) equations (eqs. 5–1 to 5–3) are:

$$\sigma_r = \frac{1}{2}(\overline{S}_H + \overline{S}_h)\left(1 - \frac{R^2}{r^2}\right)$$

$$+ \frac{1}{2}(\overline{S}_H - \overline{S}_h)\left(1 - 4\frac{R^2}{r^2} + 3\frac{R^4}{r^4}\right)\cos 2\theta + \Delta P_f \frac{R^2}{r^2} \tag{6–3}$$

$$\sigma_\theta = \frac{1}{2}(\overline{S}_H + \overline{S}_h)\left(1 + \frac{R^2}{r^2}\right) -$$

$$\frac{1}{2}(\overline{S}_H - \overline{S}_h)\left(1 + 3\frac{R^4}{r^4}\right)\cos 2\theta + \Delta P_f \frac{R^2}{r^2} \tag{6–4}$$

$$\tau_{r\theta} = -\frac{1}{2}(\overline{S}_H + \overline{S}_h)\left(1 + 2\frac{R^2}{r^2} - 3\frac{R^4}{r^4}\right)\sin 2\theta \tag{6–5}$$

where σ_r is the radial stress; σ_θ is the circumferential stress; $\tau_{r\theta}$ is the tangential shear stress; R is the radius of the borehole; r is the distance from the center of the role; θ is the azimuth measured from the direction of $\overline{S}_H$; and ΔP_f is the difference between the fluid pressure in the borehole and that in the formation (positive indicates excess pressure in the borehole). Surfaces of highest shear stress are mapped using the Kirsch equations. Such mapping predicts that near the wellbore the surfaces of high shear stress are markedly curved, thus leading to the flat depressions of the type observed with televiewers (fig. 6–8).

Breakouts develop because part of the borehole wall is subject to elastic-

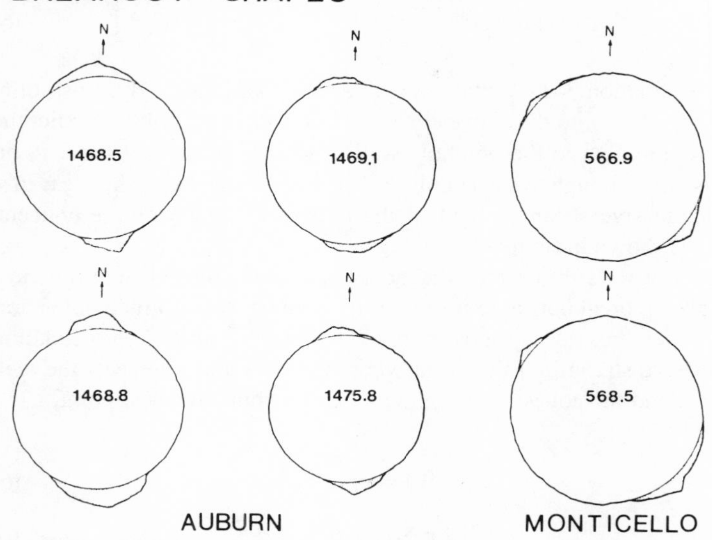

Fig. 6–8. Representative breakout shapes in wells at Auburn, New York, and Monticello, South Carolina (adapted from Zoback et al., 1985).

compressive stresses large enough to cause shear failure according to a brittle failure criterion. To predict the shape of the breakout, Zoback et al. (1985) use the extended Griffith criterion (McClintock and Walsh, 1962) in conjunction with the Kirsch equations. Although the extended Griffith criterion predicts the stress necessary for the extension of closed cracks which have a finite frictional strength in a biaxial stress field, its failure envelope is equivalent to the Coulomb criterion with a slope equal to μ^* and a τ-intercept equal to the uniaxial compressive strength of the rock (Paterson, 1978). For the breakout analysis, failure surfaces are cracks with a frictional coefficient of μ_s. Failure is treated using the Mohr's circle analysis as discussed briefly in chapter 3. To represent the in situ maximum shear stress, the radius of the Mohr's circle is $\{[(\sigma_\theta - \sigma_r)/2]^2 + \tau_{r\theta}\}^{1/2}$. Writing the failure line in cylindrical coordinates, the failure condition is given as

$$\left[\left(\frac{\sigma_\theta - \sigma_r}{2}\right)^2 + \tau_{r\theta}\right]^{\frac{1}{2}} > \frac{\mu_s}{\sqrt{1+\mu_s^2}}\left(\sigma_\theta + \frac{\sigma_\theta - \sigma_r}{2}\right) \qquad (6\text{–}6)$$

according to Zoback et al. (1985). Equations 6–3 to 6–5 are substituted into equation 6–6 to compute the size and shape of the region that is expected to fail under given in situ stresses. If the Coulomb criterion (eq. 3–2) applies, the maximum value of cohesive strength at which the material will fail is given by

$$\tau_o = (1 + \mu_s^2)^{\frac{1}{2}}\left[\left(\frac{\sigma_\theta - \sigma_r}{2}\right)^2 + \tau_{r\theta}^2\right]^{\frac{1}{2}} - \mu_s\left(\frac{\sigma_\theta + \sigma_r}{2}\right) \qquad (6–7)$$

which is equation 3–11 written in cylindrical coordinates. By substituting appropriate values into the above equations, Zoback et al. (1985) predict the size of the region next to the borehole wall in which the ratio of shear to normal stress is large enough to cause failure. The theoretical size of the areas in which the compressive shear strength of the rock is exceeded by the concentrated stresses is shown in figure 6–9.

Zoback et al. (1985) extend the theory to consider the general problem of the initial size of breakouts in terms of the τ_o, μ_s, and the magnitude of the horizontal principal stresses. By substituting equations 6–3 to 6–5 into equation 6–7 the cohesive strength at the point where the breakout intersects the wellbore $\tau_o(R, \theta_b)$, and the cohesive strength at the breakout's deepest point, $\tau_o(r_b, \pi/2)$ become

$$\tau_o(R, \theta_b) = ½ (a\,\overline{S}_H + b\overline{S}_{h)} \qquad (6–8)$$

$$\tau_o(r_b, \pi/2) = ½(c\overline{S}_H + d\,\overline{S}_{h)} \qquad (6–9)$$

where

$$a = [(1 + \mu_s^2)^{½} - \mu_s](1 - 2\cos\theta_b) \qquad (6–10)$$

$$b = [(1 + \mu_s^2)^{½} - \mu_s](1 + 2\cos\theta_b) \qquad (6–11)$$

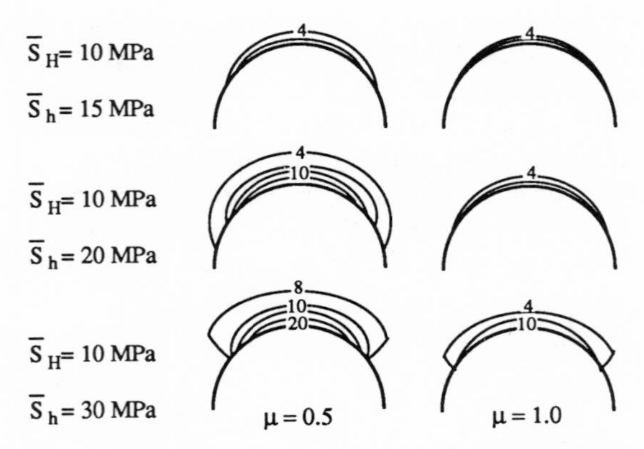

Fig. 6–9. Theoretical size of the areas in which the compressive shear strength of the rock is exceeded by stress concentrations about the borehole. For the values of the effective compressive principal stress and coefficient of friction shown, the contours in each figure define the size of the initial failure zone for a given value of τ_o and $\Delta P_f = 0$ (adapted from Zoback et al., 1985).

$$c = -\mu_s + (1+\mu_s^2)^{1/2} - (R^2/r_b^2)[(1+\mu_s^2)^{1/2} - 2\mu_s] + (3R^4/r_b^4)(1+\mu_s^2)^{1/2} \qquad (6\text{–}12)$$

$$d = -\mu - (1+\mu_s^2)^{1/2} + (3R^2/r_b^2)[(1+\mu_s^2)^{1/2} + 2\mu_s] - (3R^4/r_b^4)(1+\mu_s^2)^{1/2}. \qquad (6\text{–}13)$$

θ_b is the angle between $\overline{S}_H$ and the point where the breakout intersects the wellbore. If the breakout follows a trajectory along a given value of τ, then

$$\tau_o(R, \theta_b) = \tau_o(r_b, \pi/2). \qquad (6\text{–}14)$$

It follows that

$$\overline{S}_h = 2\tau_o \frac{a - c}{ad - bc} \qquad (6\text{–}15)$$

$$\overline{S}_H = 2\tau_o \frac{d - b}{ad - bc} \qquad (6\text{–}16)$$

$$\frac{\overline{S}_H}{\overline{S}_h} = \frac{d - b}{a - c}. \qquad (6\text{–}17)$$

From these equations a graph is constructed to estimate the stress ratio indicated by a particular set of breakouts based on the half width (ϕ_b where $\phi_b = \pi/2 - \theta_b$) and the depth of the breakout, r_b/R. Figure 6–10 graphically shows the stress ratio, $\overline{S}_H/\overline{S}_h$, which is independent of τ_o, as a function of the percent elon-

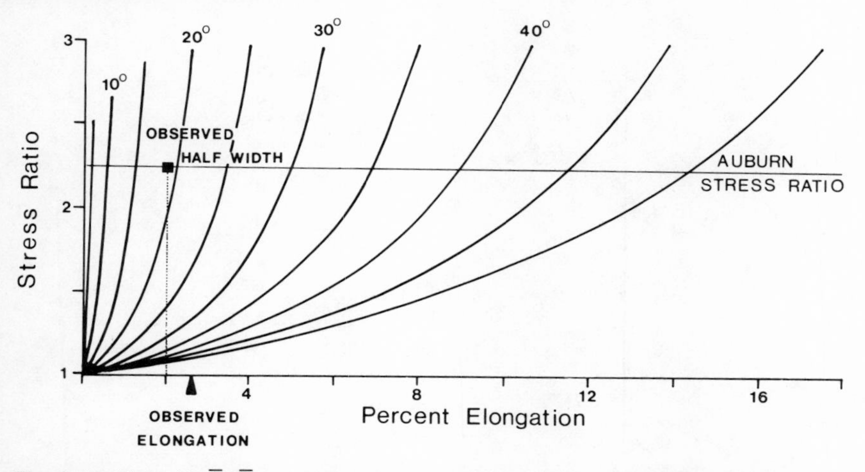

Fig. 6–10. Plotted is $\overline{S}_H/\overline{S}_h$, which in independent of τ_o, as a function of r_b/R and ϕ_b (where $\phi_b = \pi/2 - \theta_b$) for $\mu = 0.6$. Data from the Auburn, New York, tests are shown (adapted from Zoback et al., 1985).

gation, $(r_b-R)/R \times 100$, and ϕ_b for $\mu_s = 0.6$. Zoback et al. (1985) point out that extremely little spalling will occur when the two effective horizontal stresses are about equal. Although the breakouts get deeper and wider as $\overline{S}_H/\overline{S}_h$ increases, the wellbore radius increases by only about 15% when ϕ_b is as large as 50°. Although this theory of the initial formation of breakout can explain the broad, flat-sided breakouts, it cannot explain deeper breakouts such as those discussed by Zheng et al. (1989).

Zoback et al. (1985) attempted to calibrate the shape and depth of breakouts as a stress indicator by using stress measurements from the hydraulic fracture technique. Breakouts in a well at Auburn, New York (Hickman et al., 1985) are relatively deeper and have relatively smaller half widths than those in a Monticello, South Carolina, well (Zoback and Hickman, 1983) (fig. 6–8). Using the graphs in figure 6–10 we would conclude that the stress ratio $(\overline{S}_H/\overline{S}_h)$ at Auburn is larger than at Monticello. This is, indeed, the case as indicated by shallow hydraulic fracture stress measurements at Monticello where the stress ratio is very low. However, stresses at 568 m in the Monticello hole are large enough so that stress ratios at Auburn and Monticello are equivalent (fig. 3–21 vs. 6–11). Both figures 3–21 and 6–11 plot total stress, whereas figure 6–10 is a plot of the effective stress ratio, $\overline{S}_H/\overline{S}_h$.

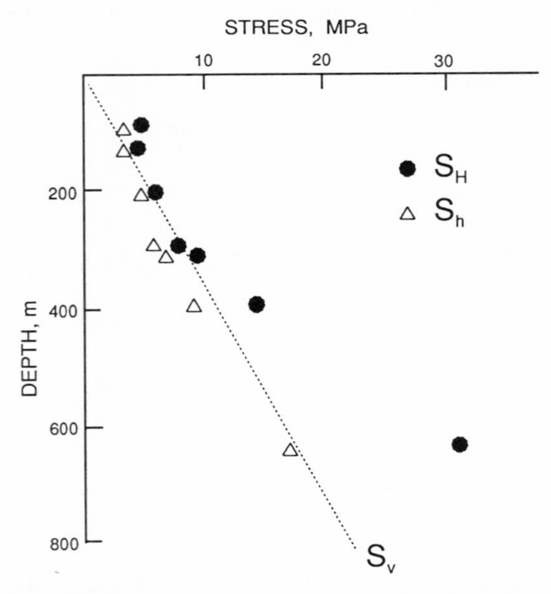

Fig. 6–11. Hydraulic fracture stress measurements as a function of depth in Monticello, South Carolina (adapted from Zoback and Hickman, 1982).

S_H may be estimated using the width of the breakout, ϕ_b (Barton et al., 1988).[4] At the borehole wall it is assumed that the circumferential stress, σ_θ (eq. 6–4), is equal to the unconfined compressive rock strength, C_o, at the maximum angle of breakout initiation, ϕ_b. By the Kirsch equations stress at the borehole wall is

$$\sigma_\theta = (\overline{\sigma}_1 + \overline{\sigma}_3) - 2\,(\overline{\sigma}_1 - \overline{\sigma}_3)\cos 2\theta - \Delta P_f. \qquad (6\text{–}18)$$

Barton et al. (1988) substitute $C_o = \sigma_\theta$ in equation 6–18, convert to principal stresses where $\overline{\sigma}_1 = S_H - P_p$, and rearrange to derive

$$S_H = \frac{C_o + \Delta P_f + 2P_p}{1 - 2\cos 2\theta} - S_h\,\frac{1 + 2\cos 2\theta}{1 - 2\cos 2\theta}. \qquad (6\text{–}19)$$

Here $\theta = 90 - \phi_b$. This analysis was applied to breakouts found in the Fenton Geothermal Site, New Mexico (Barton et al., 1988). At depths between 2.9 and 3.5 km, breakouts have a mean width, $2\phi_b$, of about 38°. Assuming that C_o is somewhere between 124 and 176 MPa, equation 6–19 predicts that S_H at a depth of about 3.3 km is between 88 and 108 MPa. Such a stress is roughly equal to the overburden and in agreement with geological indicators of stress in the region.

California

A compilation of stress indicators including borehole breakouts within the state of California show the general tendency for fault-normal crustal compression throughout much of the San Andreas fault system (fig. 6–12). These data suggested a solution to the enigmatic observation that heat flow along the San Andreas fault is low compared to heat flow from a fault slipping according to laboratory-derived friction laws. Low heat flow along the fault indicates that the fault slips at a σ_d on the order of 20 MPa (Lachenbruch and Sass, 1980). The regional borehole breakout data is compatible with low heat flow because fault-normal crustal compression means that the San Andreas fault is a nearly frictionless interface, a subject discussed in greater detail in chapter 12. Low friction causes the transpressive plate motion across the San Andreas fault to decouple into a low-stress strike-slip component and a high-stress compressive component (Mount and Suppe, 1987).

Natural Fracture Surveys

Borehole Televiewer

Previous discussion has focused on the use of the BHTV for documenting the shapes of borehole breakouts and detecting the orientation of hydraulic frac-

Fig. 6–12. Map of the state of California showing the San Andreas and other major faults and the orientation of S_H based on borehole breakouts. Data compiled by Mount and Suppe (1987) and Zoback et al. (1987).

tures. Another application of the BHTV is in open hole surveys for natural fractures (Seeburger and Zoback, 1982; Haimson and Doe, 1983; Hickman et al., 1985). Natural fracture surveys have provided some clues concerning the variation of stress magnitude in the upper crust. A graphic example is the correlation between stress magnitude and fracturing in the Illinois Deep Hole (Haimson and Doe, 1983) where a zone of lower stress at a depth of about 1200 m correlates with a section of highly fractured rock (fig. 5–14). Apparently fracture generation has a strong influence on earth stress by acting to relieve high stresses. Details concerning this phenomenon are sketchy at best.

Formation Microscanner

Another tool for fracture surveys is the formation microscanner (FMS) developed by Schlumberger (Ekstrom et al., 1987). This tool consists of several pads which contain a network of closely spaced electrodes from which an electrical image is obtained. The pads are pressed against the borehole wall where the electrodes sense the conductivity of the rock to a depth of a few centimeters into the borehole wall. Conductivity reflects such features as the porosity, permeability, and surface conduction when clays are present. Signals from the FMS are displayed as an electrical image where black represents the most conductive areas of the borehole wall and white represents the most resistive areas. The FMS has a resolution of a few millimeters which is much finer than the resolution of the BHTV largely because electrical contrasts are stronger than acoustic contrasts (Plumb and Luthi, 1986). FMS data suggest that in the Cajon Pass scientific well fractures are commonly spaced on the order of a meter at a depth greater than 1 km (Pezard et al., 1987).

Sonic and Seismic Logging

Aside from breakout analyses, other logging surveys give some indication of the variation of stress within rock. To date sonic and seismic logging appear to have the largest potential for serving as an indirect measure of earth stress.

Sonic Analysis

A 1 km profile of geophysical-log derived mechanical properties was acquired in conjunction with the South Canisteo experiment which was described in chapter 5 (Plumb et al., 1991). At South Canisteo stress measurements show a distinct discontinuity in S_h at the base of the Rhinestreet Shale with a transition to lower S_h going down through the discontinuity (fig. 5–9). Stress also varied as a function of local bedding lithology. Logs run within the Wilkins well include a density log, a gamma-ray log, a sonic log, an FMS log, a televiewer log, a resistivity log, and a temperature log. Briefly, statistical analysis of interval-averaged geophysical properties indicates that principal horizontal stress magnitudes are proportional to dynamic elastic stiffness and inversely proportional to clay content. Porosity logs show that the large drop in horizontal stress found at the base of the Rhinestreet formation in the Wilkins well occurs where shales are less compacted.

Sonic logging may detect a difference in velocities (i.e., a sonic anisotropy) as a function of the circumferential stress around the borehole. The logging tool for this type of analysis, a circumferential acoustic log, is used for determining the orientation of microcracks in fractured reservoirs. Because there is a correlation between in situ stress and sonic anisotropy (e.g., Engelder and Plumb, 1984), it is possible to assess in situ stress using circumferential acoustic tools (Mao and Sweeney, 1984).

GUIDED-WAVE POLARIZATION AND SHEAR-WAVE SPLITTING

The orientation of S_h and S_H may also control the horizontal particle motion of borehole guided waves recorded during vertical seismic profiling (VSP) experiments (Barton and Zoback, 1988). Three causes of anisotropy are intrinsic rock composition (Solomon, 1972), stress-induced in situ changes in crack dimension (Crampin, 1977), and perturbation of the ambient stress field by the existence of the borehole (Barton and Zoback, 1988). Field experiments to determine in situ stress by measuring the seismic velocity anisotropy have met with varying degrees of success (Reasenberg and Aki, 1974; Gladwin and Stacey, 1973).

The analysis of guided waves in a borehole starts with a plot of the horizontal particle motion, a hodogram (Barton and Zoback, 1988). In the analysis of guided waves it is immediately apparent that the polarization azimuths are not directed radially outward from the center of the borehole as would be expected for a Stoneley wave traveling in isotropic material (Cheng and Toksoz, 1984). Using data from Oklahoma and the Paris Basin, Barton and Zoback (1988) have shown that for a certain time interval the horizontal particle motion of a borehole guided wave is elliptical. The direction of the maximum particle velocity over the sampled wavelengths is taken as the azimuth of the guided-wave particle motion. The guided-wave particle motion is polarized in the direction of S_H, which is to say, the variation in elastic properties as a function of azimuth around the borehole appears to control the displacement of the borehole guided wave. At present it is not clear whether the particle motion is controlled by intrinsic or stress-induced anisotropy or by a combined effect (Barton and Zoback, 1988). The relationship between stress orientation and the polarization of borehole guided waves is present but neither very strong nor well defined in azimuth.

Shear waves interact with microcracks to develop orthogonally polarized fast and slow shear wave components that separate or "split" in time (Crampin, 1984). If in situ stress difference is large enough to preferentially close cracks in one orientation, then conditions may be suitable for the use of shear-wave splitting to determine the orientation of the in situ stress. At the Cajon Pass well Li et al. (1988) used a P-wave source for VSP studies with downhole geophones receiving signals from sites offset from the well between 150 and 1000 m.

S-waves generated by mode conversion of P-waves arrived at higher apparent velocities when polarized along the direction of N70°E. This behavior is consistent with open cracks and fractures which may be aligned parallel to S_H as measured by Shamir et al. (1988) at the Cajon Pass well.

Concluding Remarks

During the past ten years the analysis of borehole breakouts has added more to the worldwide inventory of data on stress within the lithosphere than any other single stress-measurement technique. This is largely due to the abundance of wells drilled for the purpose of recovering hydrocarbons and the fact that dipmeter surveys were run in many of these wells. However, the relationship between borehole breakout shape and stress magnitude is still poorly defined and, hence, will be the focus of future research.

— 7 —

Strain-Relaxation Measurements

The maximum stress was about 7500 psi, almost twice the minimum stress and in approximately the east-west horizontal direction. This stress is presumably of tectonic origin and was not anticipated in this flat-bedded limestone.

—Leonard Obert (1962) upon making strain-relaxation measurements in the Columbia Southern Chemical Co. limestone mine, Barberton, Ohio. Obert deserves credit as the first to detect the plate-scale stress field in eastern North America by a systematic stress measurement.

More than four millennia ago, tunnelers and miners were already learning to deal with the effects of earth stress. The Egyptians used rock stress to aid in the tunneling and quarrying by taking advantage of the effect of mechanically, thermally, and chemically induced stresses. For example, blocks for the pyramids were cut by driving wedges into holes. Early Egyptian kings had some galleries cut over 240 m in length into the relatively soft sandstones of the Upper Nile River. At this time the practices of tunneling and mining were extensive as indicated by ancient excavations such as the Polish salt mines of about 2500 B.C. (Bieniawski, 1984) and the drainage canals left on the east coast of Greece by the Minoans of about 2000 B.C. (Dean, 1937). One of the earliest underwater tunnels used for transportation was built by the Babylonians about 2170 B.C. (Brown, 1990). Advanced Roman tunneling techniques used fire to break the rock by thermal cracking and vinegar to chemically attack limestone and weaken it (Legget, 1939). Following the advances made by the Egyptians, Greeks, and Romans, tunneling and mining techniques did not improve greatly until about two hundred years ago when the construction of canals and railroads started in earnest at the onset of the industrial revolution.

As tunneling and mining practices developed, miners learned to prevent collapse, a version of stress relief, under the weight of overburden and lateral load. Initially, the size of open shafts and underground cavities was limited by trial and error. Room and pillar mining techniques took advantage of the strength of natural rock columns left in place to support the overburden. When the size of the cavity exceeded safe limits, the walls and ceiling were fitted with braces and trusses, a practice dating back to the Egyptians (Bieniawski, 1984). Finally, reinforced concrete and roof bolts were devised to resist the tendency for tunnels and mine shafts to close or collapse under the tremendous weight of overburden. Originally, the practice was to design tunnel linings so that reinforcing steel carried the full load of overburden largely because there was no information on the magnitude of stresses in rock surrounding the tunnel. Later, driven by the desire to save money through installing a minimum amount of reinforcing, engineers designed techniques to measure underground stresses.

Overcoring

Measurement of stress around mines and tunnels evolved around the practice of cutting core or a block from its surroundings to relieve rock stress. If rock, under compressive stress, behaves as an elastic material, then it must expand when cut loose from the ground. Because the rock changes shape when cut free, generally by expansion rather than contraction, the rock is said to undergo *strain relaxation*. The idea behind the strain-relaxation stress measurement is simply to measure the amount of expansion of a core or block when cut from the surrounding rock mass. Because the amount of expansion is directly proportional to the stress within the rock, in situ stress is calculated with a fair degree of certainty, provided the elastic modulus of the rock is known. Although strain relaxation consists, in large part, of the instantaneous recovery of elastic strains imposed by boundary tractions, many rocks continue to expand for several hours. This additional component of strain is called a time-dependent strain relaxation. Furthermore, when traction-free blocks are further cored or otherwise cut up, elastic strains are again recovered. This component of strain relaxation is driven by an internal stress called a *residual stress* (Friedman, 1972). Discussion of the time-dependent component of strain relaxation is deferred to chapter 9 and a discussion of residual stress is deferred to chapter 10. This chapter focuses on inferring in situ stress from techniques for measuring the instantaneous component of strain relaxation.

Early strain-relaxation measurements include a 1932 experiment to determine rock stress in a tunnel under Hoover Dam, Colorado River, U.S.A. (Lieurance, 1932). In this case, rock stress was relieved by drilling, with a jackhammer, a series of holes 76 cm deep on a 1.3 m diameter circle around three bench marks. A large cylinder of rock was freed by chipping out rock between the jackhammer holes (fig. 7–1). Strain relaxation was measured in three directions using the three bench marks spaced 50 cm apart. Expansional strain of the type experienced under Hoover Dam is typically on the order of -200 ± 100 $\mu\varepsilon$.[1] $\mu\varepsilon$ is a dimensionless quantity called *microstrain*, defined as a change in length, Δl, divided by initial length, l_o

$$\mu\varepsilon = \frac{\Delta l}{l_o}. \tag{7–1}$$

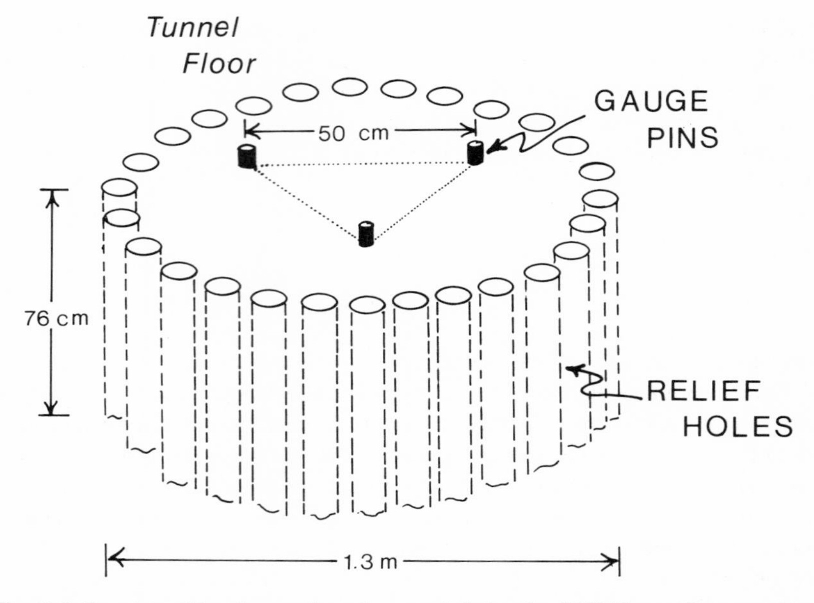

Fig. 7–1. Drawing of the stress relieved core cut during the 1932 Hoover Dam experiment to measure in situ stress. Three gauge pins allow the measurement of strain relaxation in three directions. The distance between each of the three pins defines the three components of the strain gauge rosette. From these three measurements we calculate the two-dimensional strain parallel to the tunnel floor. Assuming an elastic modulus for the rock in the tunnel, we can then calculate the stress parallel to the tunnel floor.

If an object with an initial length of 1 cm is stretched by 1 $\mu\varepsilon$, then the change in length is 1×10^{-6} cm. Measuring small changes in length upon strain relaxation of rock presented great difficulty to early twentieth-century engineers whose best ruler was a precision caliper, accurate to about ± .001 cm. As a consequence, long-strain gauges (i.e., widely separated bench marks) were necessary so that engineers could detect small changes in length. With 50 cm between bench marks, 200 $\mu\varepsilon$ is a change in separation of .01 cm, not a large distance considering the accuracy of precision calipers. Using precision calipers with a 50 cm gauge length, a change in length equivalent to 10 $\mu\varepsilon$ is nearly impossible to detect.

With the invention of small-strain gauges (≈ 1 cm in length) in the 1950s, strain-relaxation measurements became possible using small cores (≤ 15 cm in diameter) easily cut with diamond coring bits. *Overcoring* is the process by which a diamond core bit cuts a rock core containing a gauge placed inside a small pilothole[2] (≈ 4 cm in diameter) in the center of the core or a strain gauge rosette attached directly to the end of the core. The term, overcoring, applies to all stress measurements involving the cutting of cores with stress-measuring gauges attached. Both strain cells and stressmeters are inserted into a small pilothole for the purpose of measuring in situ stress. This book arbitrarily defines a *strain cell* as an instrument which measures strain relaxation during cutting a larger core centered on the pilothole. Some strain cells detect the change in shape of a solid rock core during overcoring, whereas others measure the change in shape of the pilothole during strain relaxation (Leeman, 1964a). A measurement of the expansion along a minimum of three diameters is required to assess the change in shape of the pilothole and, hence, to assess the amount and orientation of core relaxation. A calculation of the state of stress using both core-relaxation and pilothole deformation data requires the elastic properties of the host rock. Measurement of the elastic properties is accomplished by inserting the core with strain cell into a radial compression chamber and measuring the core deformation as a function of confining pressure. But, because elastic properties change on strain relaxation (Engelder, 1981), use of laboratory values for E and ν give a close but not exact estimate of in situ stress (see chap. 9).

The difference between a strain cell and a stressmeter is based on the behavior of the instrument inserted into the pilothole. In the case of the strain cell the pilothole deforms as if the instrument were not present during strain relaxation. In contrast, a *stressmeter* acts to resist deformation of the pilothole by retarding strain relaxation. Stress is transferred from the host rock to the active component of the stressmeter during overcoring. Once calibrated, the stressmeter permits the direct interpolation of absolute stress or stress changes within the rock without having information on the elastic properties of the rock. The active element in the strain cell is very compliant relative to the host rock, whereas the active element in a stressmeter is stiffer than the host rock. Strain relief is complete for the strain-cell techniques, whereas it is partial, at most, for stressmeters (Goodman, 1980). Chapter 7 deals with strain cells, whereas chapter 8 focuses on stressmeters (i.e., rigid-inclusion gauges).

The Hoover Dam experiment set in motion the evolution of strain-relaxation stress measurements, the most widely used technique for measuring in situ stress in the near surface or near the openings of mines and tunnels. Stress in the top 100 m of the crust is of interest because this is the zone accessible to the relatively inexpensive and rapid overcoring techniques. Deeper portions of the crust are sampled using overcoring techniques only near underground mines. Overcoring tests in Russia date as early as 1935 (Slobodov, 1958). Engineers in Sweden, South Africa, Great Britain, India, Australia, Canada, and the United States worked independently to develop overcoring techniques for the purpose of measuring rock stress. All had the general goal of producing a gauge which permitted rapid and inexpensive measurements of stress. Improvements throughout the 1950s by Mohr (1956), Olsen (1957), Hast (1958), and Obert et al. (1962) led to Obert's (1962) profound comment (opening epigraph to this chapter) about the magnitude of the maximum horizontal stress (S_H) in limestone of northern Ohio. At the time of his report, Obert (1962) had no global context in which to fit the high horizontal stress at Barberton, Ohio. High horizontal stresses (i.e., $S_H > S_V$) were known from Hast's (1958) work in and around the Baltic Shield, but were interpreted as a manifestation of the bending of the crust, a concept which predated plate tectonics. Obert's (1962) measurement was the first of a series of systematic measurements leading to the documentation of contemporary tectonic stress over eastern North America (Sbar and Sykes, 1973).

Strain-relaxation techniques are divided into three groups based on the location and degree of sophistication of the strain cell: direct bonding of strain gauges to surface outcrops; direct bonding of strain cells to the end of boreholes; and the insertion of strain cells into pilotholes. The first half of this chapter follows a discussion of strain-relaxation techniques, ordered by degree of sophistication of the strain cell starting with the simplest, the strain gauge bonded directly to the outcrop surface. Even with strain-relaxation measurements restricted to the upper 100 m of the crust and near mine openings, earth scientists have learned a great deal about earth stress including an understanding of thermal stresses in rock, overburden stresses, the variation of local stresses, stress concentration in and around cavities in rock, and the relationship between earth stress and natural fractures. The last half of this chapter focuses on the measurement of various components of earth stress (e.g., thermal, tectonic, and overburden stresses) using strain-relaxation techniques.

Stress Orientation from Strain-Relaxation Data

Often the orientations of the principal stresses are unknown prior to a stress measurement. However, a free surface assures that two principal stresses occur parallel to that surface. As long as the plane containing two principal stresses is known, the principal stresses in that plane may be determined with three strain gauges. So, for near-surface stress measurements, strain-relaxation data in three directions are necessary for the determination of S_H and S_h in the plane of the outcrop surface or normal to a vertical borehole. This was indeed the case for the Hoover Dam experiment where each gauge line formed an angle of $\pi/3$ with each of its neighbors (fig. 7–1).

The most common configurations for three strain gauges in a plane are the rectangular (45°) rosette and the delta (60°) rosette[3] (fig. 7–2). The rectangular rosette consists of two gauges at 90° and the third gauge at 45° from each of the 90° gauges. All three gauges are at an equal angle of 120° to each other in the delta rosette. The direction and magnitude of the principal strains are calculated from the following formulas. For the 45° or rectangular rosette

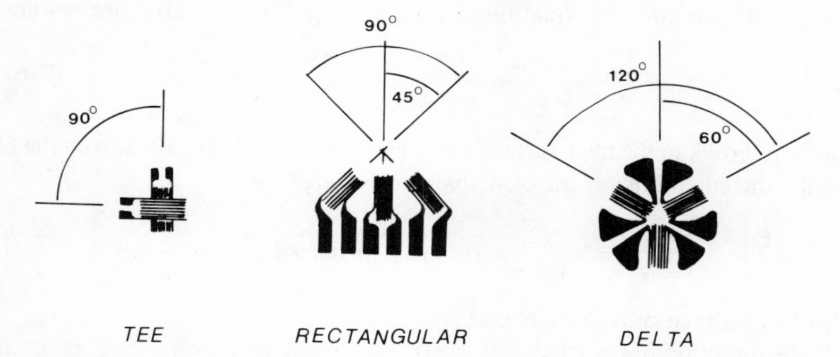

Fig. 7–2. Configurations for a two-component (Tee) and three-component (Rectangular and Delta) strain gauge rosettes.

$$\varepsilon_{1,2} = \frac{1}{2}\left[\varepsilon_A + \varepsilon_B \pm \sqrt{(\varepsilon_A - \varepsilon_B)^2 + (2\varepsilon_C - \varepsilon_A - \varepsilon_B)^2}\,\right] \tag{7–2}$$

where the angular distances, measured counter-clockwise, are $\theta_A = 0°$, $\theta_B = {}^{\pi}/_2$, $\theta_C = {}^{5\pi}/_4$. The direction of the maximum expansion, ϕ_P, is given by

$$\phi_p = \frac{1}{2}\tan^{-1}\left\{\frac{2\varepsilon_C - \varepsilon_B - \varepsilon_A}{\varepsilon_A - \varepsilon_B}\right\} \tag{7–3}$$

where ϕ_P is the angle measured in the counter-clockwise direction from the $\theta = 0°$ axis of the strain rosette. If $\varepsilon_C > {}^1/_2(\varepsilon_A + \varepsilon_B)$, then ϕ_P lies between 0° and 90°. If $\varepsilon_C < {}^1/_2(\varepsilon_A + \varepsilon_B)$, then ϕ_P lies between 90° and 180°. If $\varepsilon_C = ½(\varepsilon_A + \varepsilon_B)$ and $\varepsilon_A > \varepsilon_B$, then $\phi_P = 0°$. If $\varepsilon_C = {}^1/_2(\varepsilon_A + \varepsilon_B)$ and $\varepsilon_A < \varepsilon_B$, then $\phi_P = 90°$.

For a 60° or delta rosette ε_A is at $\theta_A = 0°$, ε_B is at $\theta_B = {}^{2\pi}/_3$, and ε_C is at $\theta_C = 4\pi/3$, measured counter-clockwise. The principal strains are given by

$$\varepsilon_{1,2} = \frac{\varepsilon_A + \varepsilon_B + \varepsilon_C}{3} \pm \frac{\sqrt{2}}{3}\sqrt{(\varepsilon_A - \varepsilon_B)^2 + (\varepsilon_B - \varepsilon_C)^2 + (\varepsilon_c - \varepsilon_A)^2}. \tag{7–4}$$

The direction of maximum expansion is given by

$$\phi_p = \frac{1}{2}\tan^{-1}\left\{\frac{\sqrt{3}\,(\varepsilon_C - \varepsilon_B)}{2\varepsilon_A - \varepsilon_B - \varepsilon_C}\right\}. \tag{7–5}$$

If $\varepsilon_C > \varepsilon_B$, then ϕ_P lies between 0° and 90°. If $\varepsilon_C < \varepsilon_B$, then ϕ_P lies between 90° and 180°. If $\varepsilon_C = \varepsilon_B$ and $\varepsilon_A > \varepsilon_B$, then $\phi_P = 0°$. If $\varepsilon_C = \varepsilon_B$ and $\varepsilon_A < \varepsilon_B$, then $\phi_P = 90°$.

Once overcoring is complete and data on principal strains are in hand, then the core is carried to the laboratory for the purpose of reloading it to determine its elastic properties, E and ν . Details are given later in this chapter. Using data from the 45° rosette, the equations for the principal compressive stresses are

$$\sigma_{1,2} = \frac{E}{2}\left[\frac{\varepsilon_A + \varepsilon_B}{1-\nu} \pm \frac{\sqrt{2}}{1-\nu}\sqrt{(\varepsilon_A - \varepsilon_C)^2 + (\varepsilon_C - \varepsilon_B)^2}\right]. \quad (7\text{–}6)$$

Rarely is a rock in the near surface in a state of tension. Using data from the 60° rosette, the equations for the principal stresses are

$$\sigma_{1,2} = \frac{E}{3}\left[\frac{\varepsilon_A + \varepsilon_B + \varepsilon_C}{1-\nu} \pm \frac{\sqrt{2}}{1-\nu}\sqrt{(\varepsilon_A - \varepsilon_B)^2 + (\varepsilon_B - \varepsilon_C)^2 + (\varepsilon_A - \varepsilon_B)^2}\right]. \quad (7\text{–}7)$$

On a free surface σ_3 is vertical and zero.

If the direction of principal stresses are known before overcoring, then only two gauges in the form of a tee (90°) rosette are necessary for assessing in situ stress (fig. 7–2). Prior knowledge of the orientation of the principal stresses is rare except in places such as in the pillars of an underground mine. Assuming the rock is isotropic, the principal stresses in two dimensions are calculated from the principal strains, ε_1 and ε_2, as measured from gauge elements aligned with the principal stress axes, σ_1 and σ_2. The principal stresses are

$$\sigma_1 = \frac{E}{1-\nu^2}(\varepsilon_1 + \nu\varepsilon_2) \quad (7\text{–}8)$$

$$\sigma_2 = \frac{E}{1-\nu^2}(\varepsilon_2 + \nu\varepsilon_1). \quad (7\text{–}9)$$

Of the various strain-relaxation techniques, direct bonding of strain gauges to the surface of outcrops is the least expensive and time consuming. Pilotholes are not necessary and very light drills run by electric motors will suffice for the purpose of overcoring. Surface-bonded strain gauges include electrical-resistance and photoelastic gauges.

Direct Bonding of Strain Gauges to Outcrop Surfaces

A discussion of surface and near-surface stress measurements must start with a caveat. Poor rock quality is a factor that often introduces spurious data into a suite of stress measurements in the near surface. The high degree of weathering and fracturing of near-surface rock contribute to the decoupling of the rock from deep crustal stresses. If there is significant decoupling, a rock fabric or residual stress might contribute a relatively large component of strain relaxation. However, there are outcrops with large pavements of unweathered and relatively unfractured rock, particularly in recently glaciated terrain or in dryer areas of the world. Such outcrops which are suitable for strain relaxation measurements are harder to find in wetter areas where vegetation covers the ground. In these latter areas quarries or underground mines often provide the only reasonable exposures for use of surface-bonded strain gauges. Surface measurements are also affected by an additional complexity, a component of thermal stress arising from either diurnal or annual temperature changes.

ELECTRICAL-RESISTANCE STRAIN GAUGES

The electrical-resistance strain gauge allows precise strain measurements over short distances (Murray and Stein, 1965). Using an electrical-resistance strain gauge, engineers can detect length changes of as little as 0.1 μm over 1 cm, a convenient gauge length[4] for overcoring. Electrical-resistance strain gauges consist of a fine wire, folded accordion-style, and bonded to a thin foil (fig. 7–2). When the thin foil and accordion-folded wire are bonded by means of epoxy to the surface of a rock, an expansion or contraction of the rock during overcoring will cause the fine wire of the strain gauge to stretch or contract and, therefore, change its cross-sectional area. Because the electrical resistance of a wire is proportional to its cross-sectional area, the amount of stretching of a strain gauge during strain relaxation is measured as a change in resistance of the accordion-folded wire. If the strain gauge is wired to a device called a *strain indicator*, a change in resistance of the strain gauge is converted by the strain indicator directly to linear strain parallel to the long axis of the gauge.

One technical problem with the use of electrical-resistance strain gauges is that they are very sensitive to moisture (Olsen, 1957; Kohlbeck and Scheidegger, 1983). The slightest corrosion to the active element will affect the resistance of the gauge and lead to erroneous conclusions about the magnitude of strain. Early strain relaxation techniques used mastic-type sealing compounds and enamel paints to protect strain gauges from moisture. In recent years manufacturers such as Micromeasurements (Raleigh, N. Carolina) and BLH (Waltham, Mass.) offer strain gauges which are coated to protect the active element from moisture even under the pressures that are encountered at drilling depths. Precautions are also necessary to protect the solder joints from moisture even though small amounts of corrosion at the solder joints will not affect the resistance of the strain gauge. Strain gauge manufacturers also warn that the epoxy between the strain gauge and rock must be thin; otherwise, the modulus of the epoxy will adversely affect coupling between the strain gauge and the rock. A thin layer of epoxy is best achieved with pressures of 0.5 MPa on the strain gauge while the epoxy is curing.

Ideal strain-relaxation measurements integrate the strain over many grains so that anomalous deformation, say at grain boundaries and other heterogeneities, do not cause erroneous strain readings. For many rocks there is a *representative measurement base* (RMB) which is the gauge length necessary to sense a statistically homogeneous strain within the rock. Such a length or RMB varies from 1 cm in chalk to as much as 22 cm in a coarse grained sandstone (Cyrul,

1983). Strain gauges 1–2 cm long are commonly used to measure rock strain during overcoring. These strain gauges are commonly shorter than the RMB for coarse-grained lithologies, in which case a single overcoring measurement may fail to sample the statistically homogeneous strain within the rock. This factor leads to scatter in the strain-relaxation data and makes several measurements necessary to collect a statistically reliable sample.

A number of stress surveys have used electrical-resistance strain gauges bonded directly to rock surfaces. An early example is Olsen's (1957) experiment using four-component rosettes bonded to rocks in the concrete-lined pressure tunnel of a hydroelectric project at Prospect Mountain, Colorado. Olsen (1957) points out that "the magnitude of existing rock stress is much higher than can be accounted for by the weight of overlying strata plus the stress concentration produced by drilling." The Prospect Mountain data show that the stress concentration around the tunnel controls the orientation of the principal stresses and that there is a general correlation between the maximum stress and the depth of overburden. Here was an early indication in the literature that rock stresses are the superposition of several components, including some that are independent of those boundary conditions arising from overburden or contemporary tectonic stresses. In this case the high stresses were believed to be "residual" as the title of Olsen's (1957) paper indicates but the exact nature of these residual stresses is unknown.

Several surveys using surface-bonded strain gauges have shown a preferred orientation for strain relaxation over regions measuring more than 10^4 km^2. Although there is little doubt about the consistent orientation, a question arises concerning which component of the earth's stress field was actually sampled. For example, the strain relaxation detected by surface-bonded strain gauges on six outcrops and fifteen samples taken to the laboratory from the Coast Range north of San Francisco show a maximum expansion at N-S ± 20° (Hoskins et al., 1972; Hoskins and Russell, 1981) (fig. 7–3). Based on the strain relaxation of core samples in the laboratory, Hoskins and Russell (1981) argued that the N-S expansion reflected a residual stress (see chap. 10). Although a N-S S_H is geologically reasonable for the active strike-slip faults in the California Coast Range north of San Francisco, Hoskins and Russell (1981) state that they did not directly measure the contemporary tectonic stress field in the six outcrop tests. Rather, they suggest that surface relaxation is controlled by a fabric of tectonically induced microcracks. Later measurements using borehole breakouts in central California showed that this core relaxation does not correlate with the orientation of S_H as determined at depth in central California (fig. 6–12).

Surface-bonded strain gauges were used to test for the behavior of a single rock type, the Potsdam sandstone, over a region of several hundred km^2 in upper New York (Engelder and Sbar, 1976; 1977). Strain relaxation tests in the Potsdam Formation, a basal Cambrian sandstone, show that maximum expan-

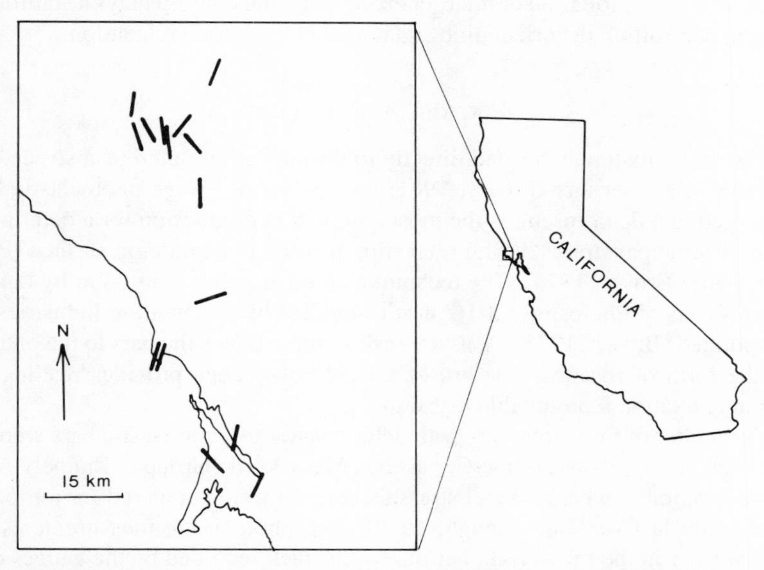

Fig. 7–3. Map of a portion of California northeast of the San Andreas fault and north of San Francisco showing the location of fifteen strain-relaxation measurements. The line through each site indicates the orientation of the maximum expansion on overcoring (adapted from Hoskins and Russell, 1981).

sion was approximately E-W on the northeast side of the Adirondack Mountains and ENE on the northwest side. Although in both cases the orientation of maximum expansion is generally consistent with S_H for the contemporary tectonic stress field for the northeastern United States, the correlation between the mechanisms controlling the strain relaxation and the contemporary tectonic stress field is poorly understood. As was the case for the northern California measurements of Hoskins et al. (1972), overcoring on outcrops of the Potsdam sandstone appears to have relieved components of a local stress field such as residual stress or that developed by local fracturing.

Other regional studies show a consistent orientation for the strain relaxation of outcrops including those in the Black Hills of South Dakota (Hoskins and Bahadur, 1971; Bahadur, 1972), in the Adirondack Mountains, New York (Plumb et al., 1984a), on the Appalachian Plateau (Engelder, 1979a), and in Vermont (Engelder et al., 1977). In the Black Hills there was a strong correlation between the orientation of maximum strain relaxation and the normal to steeply dipping foliation of crystalline rocks. Surface overcoring experiments in the Barre granite of Barre, Vermont, show a correlation between principal strains and a microcrack fabric with maximum expansion normal to microcracks (Engelder et al., 1977). From studies such as these it is clear that rock fabric (i.e., foliation or microcracks) within the outcrop plays a significant role in controlling the orientation and magnitude of strain relaxation.

PHOTOELASTIC GAUGES

Photoelastic materials bonded directly to the surface of outcrops also serve as reliable strain sensors (Emery, 1961; Pincus, 1966). Three photoelastic bars arranged in a delta rosette is the most common configuration for a determination of principal strains during relaxation parallel to an outcrop surface (Preston, 1968; Brown, 1973). The technique uses bar gauges cut 4 cm by 0.4 cm from sheets of Photostress S-16 plastic supplied by Automation Industries Incorporated (Brown, 1973). A 4HRCT resin cement bonds the bars to the outcrop in the form of rosettes. The use of a field polariscope provides readings of strain relaxation reproducible to ± 6 με.

A number of measurements with delta rosettes of photoelastic bars were attempted on sandstones of the Cretaceous Mesa Verde Group at Rangely Anticline, Colorado, where several stress techniques were compared for reproducibility (de la Cruz and Raleigh, 1972). The photoelastic measurements are noteworthy in their own right because of the drift recorded by the gauges over fourteen days after the overcoring (Brown, 1974). In these measurements drift is determined by differencing a strain gauge reading before overcoring and readings after overcoring. At one site seven rosettes were overcored with the orientation of maximum expansion grouped between N67°E and N86°E and its magnitude ranged between −191 and −871 με as measured right after overcoring. Two weeks later the maximum expansion still grouped between 69° and 88° but the range of magnitudes had decreased to between 151 and −581 με. The maximum difference between pre- and post-overcoring strain gauge readings occurred right after overcoring. Such a long-term decrease in magnitude of maximum expansion after overcoring was similar to that found in the Potsdam sandstone in upstate New York (Engelder and Sbar 1976; 1977). One interpretation of the fourteen day drift reported by Brown (1974) and Engelder and Sbar (1976; 1977) is that drilling fluid "wetted" the rock causing a short-term expansion which relaxed out during several days of drying. My hypothesis is that the immediate, large expansion is, in part, driven by the surface tension of water which exerts an opening force on partially wet microcracks. Upon drying the microcracks relax (close) and the core contracts, leaving a smaller net expansion upon overcoring.

In summary, there are several technical difficulties in using surface-bonded strain gauges including the effects of thermal mismatch, wetting and drying of the rock, reconnecting strain gauge leads, and the change in elastic properties on strain relaxation—all of which lead to an imprecise measure of strain relaxation. Photoelastic gauges require a correction for strains on the order of

30×10^{-6} / °C, due to differential thermal expansion between rock and gauge (Brown, 1973). Electrical-resistance strain gauges have a better match to the thermal expansivity of rock. The fluctuation of moisture in pores seems to cause strains that are often larger than the signal that is sought during strain relaxation. Disconnecting and reconnecting of strain gauges to strain gauge bridges (a wheatstone bridge designed to read strain changes) also leads to spurious data on strain relaxation. Changes in connection resistances can impart undetected artifacts in strain on the order of 50×10^{-6}. A large component of residual stress associated with the opening of microcracks is often found in surface outcrops (Engelder et al., 1977; Hoskins and Russell, 1981). Finally, the estimation of stress through the elastic properties of the surface rocks has been particularly troublesome because the elastic properties themselves change with strain relaxation (Engelder, 1981; Engelder and Plumb, 1984). Elastic properties are determined in the laboratory on post-relaxation samples, whereas the appropriate properties for stress calculations are those prior to strain relaxation.

Trepanning

Total stress relief is not necessary for the measurement of in situ stress at the surface of an outcrop or wall of a mine or tunnel. If a pilothole is drilled, the walls of that hole will displace inward under a compressive stress field. The amount of inward displacement is directly proportional to the in situ stress. Rather than measuring the displacement of the pilothole walls, the contraction between pins across a diameter, 2r, is measured in a technique known as *trepanning* or center hole drilling (fig. 7–4).

The following is Duvall's analysis of trepanning where the bench mark pins are at r = 12.5 cm and the center hole is reamed to a = 7 cm (Hooker et al., 1974). The compressive stress field is denoted by S_x and S_y with S_z parallel to the hole. Upon loading a rock containing a circular borehole reamed to a radius, a, the displacement (u) of a pin (at r from the center of the borehole) toward the center of the borehole, is

$$u = \frac{1}{E}\left\{\left(\frac{S_x + S_y}{2}\right)\left[(1-\nu)r + (1+\nu)\frac{a^2}{r}\right]\right\}$$

$$+ \frac{1}{E}\left\{\frac{S_x - S_y}{2}\left[(1+\nu)r - (1+\nu)\frac{a^r}{r^3} + 4\,(1+\nu^2)\frac{a^2}{r}\right]\cos 2\theta\right\} \quad (7\text{–}10)$$

provided that the boundary loads, S_x and S_y, are principal stresses and that θ is measured from the x axis. Without the borehole present (i.e., a = 0), the inward displacement of the pin upon loading is

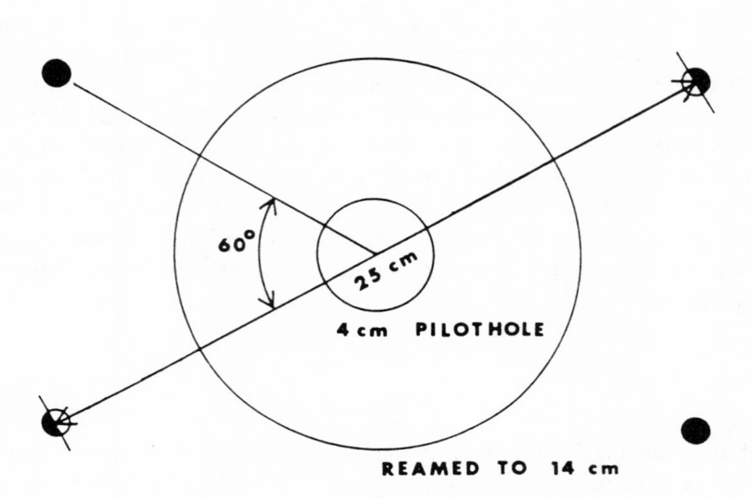

Fig. 7–4. Configuration for the center hole relief method of Duvall (adapted from Hooker et al., 1974).

$$u_o = \frac{1}{E}\left\{\left(\frac{S_x + S_y}{2}\right)(1 - \nu)r + \left(\frac{S_x - S_y}{2}\right)(1 + \nu)r \cos 2\theta\right\}. \quad (7\text{–}11)$$

Equation 7–10 minus equation 7–11 gives the displacement that results when a center hole is drilled, thus

$$\Delta u = u - u_o = \frac{1}{E}\left\{\left(\frac{S_x + S_y}{2}\right)(1 + \nu)\frac{a^2}{r}\right\}$$

$$+ \frac{1}{E}\left\{\left(\frac{S_x - S_y}{2}\right)\left[4(1 - \nu^2)\frac{a^2}{r} - (1 + \nu)\frac{a^4}{r^3}\right]\cos 2\theta\right\}. \quad (7\text{–}12)$$

If displacements, U, are measured between two pins across a diameter of the borehole, then $U = 2\Delta u$. If we let $k = a / r < 1$, then equation 7–12 becomes

$$U = \frac{d}{E}\left\{\left(\frac{S_x + S_y}{2}\right)(1 + \nu)k\right\}$$

$$+ \frac{d}{E}\left\{\left(\frac{S_x - S_y}{2}\right)[4(1 - \nu^2)k - (1 + \nu)k^3] \cos 2\theta\right\} \qquad (7\text{–}13)$$

where $d = 2a$. If $M = (1 + \nu)^k/_2$ and $N = 2(1 - \nu^2)k - {}^1/_2(1 + \nu)k^3$, then equation 7–13 becomes

$$U = {}^d/_E \{ (S_x + S_y) M + (S_x - S_y) N \cos 2\theta \}. \qquad (7\text{–}14)$$

Three measurements of the contraction between pins in three directions (i.e., U_1, U_2, U_3) at θ, $\theta + 60°$ and $\theta + 120°$ give sufficient information to solve for S_x, S_y, and θ. Thus,

$$S_{x,y} = \frac{E}{6d}\left\{\frac{U_1 + U_2 + U_3}{M}\right\}$$

$$\pm \frac{E}{6d}\left\{\frac{\sqrt{2}}{N}\sqrt{(U_1 - U_2)^2 + (U_2 - U_3)^2 + (U_3 - U_1)^2}\right\} \qquad (7\text{–}15)$$

$$\theta = \frac{1}{2}\tan^{-1}\left\{\frac{\sqrt{3}(U_2 - U_3)}{2U_1 - U_2 - U_3}\right\} \qquad (7\text{–}16)$$

where θ is measured from U.

Hooker et al. (1974) confirmed that the trepanning technique worked by testing it against several other near-surface techniques in granite quarries near St. Cloud, Minnesota. They concluded that thermal stresses were significant in near-surface rocks. However, in near-surface rocks unaffected by quarry workings or thermal stresses, S_H was oriented N 50° E with σ_d ($S_H - S_v$) about 15 MPa and σ_d^H in the horizontal plane about 5 MPa. The correlation between these data and those from hydraulic fracture stress measurements in the midcontinent of North America (chap. 5) is compelling evidence that a signal from the contemporary tectonic stress field is present in near-surface rocks.

Direct Bonding of Strain Cells to the End of Boreholes

To eliminate some of the spurious information inherent in surface outcrops, techniques were developed for placing strain gauges at the end of boreholes. These techniques were refined in South Africa during the early 1960s (Leeman, 1964a,b).

CSIR "DOORSTOPPER"

Two factors encouraged engineers to devise methods for direct bonding of strain gauges away from outcrop surfaces. The first factor included a host of technical limitations such as moisture and temperature conditions at the surface of outcrops which influence strain relaxation and, thus, complicate the interpretation of stress measurements. The second factor was the large stress concentration near the openings of mine shafts and tunnels as shown by Olsen (1957) and many others. Virgin rock stress is greatly disturbed within two to three tunnel diameters. Measurement of virgin rock stress necessitated drilling several meters into the wall or ceiling of a mine shaft to clear the zone of stress concentration around the shaft or tunnel. Hence, techniques were developed to glue strain gauges to the flattened ends of boreholes drilled through the zone of stress concentration.

An early description of working with electrical-resistance strain gauges at the end of boreholes was written by Mohr (1956). For measuring stresses in mine pillars, Mohr (1956) used two electrical-resistance strain gauges assuming that the two principal stresses were parallel and perpendicular to the pillars. Leeman (1964a,b) greatly improved Mohr's design with a device that later became well known as the CSIR "doorstopper." The name, doorstopper, was used because the shape of the device resembles a type of doorstopper commonly found in South Africa. The CSIR doorstopper or Leeman cell consists of a three-component strain gauge rosette on an Araldite shim housed in a rubber moulding with connector pins (fig. 7–5). A pneumatically operated tool was developed for installing the CSIR doorstopper at a known orientation on the end of

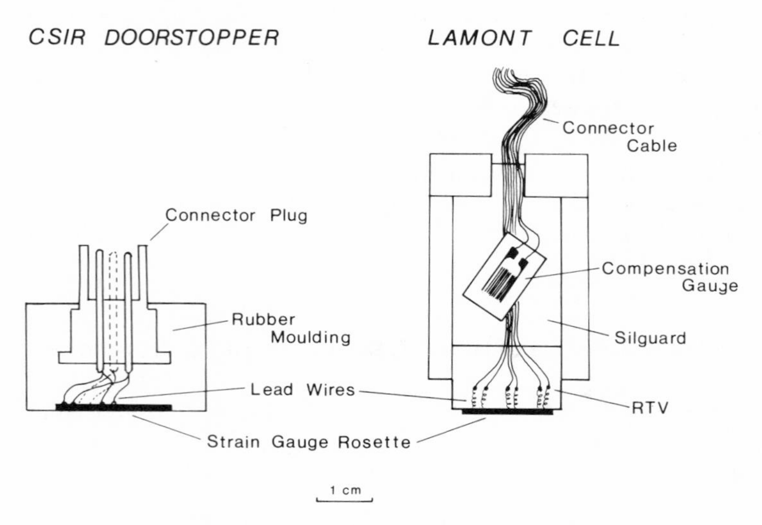

Fig. 7–5. Cross-section sketch of a CSIR doorstopper and a Lamont cell.

BX-size boreholes (6 cm in diameter). Later tools were developed for manually installing the CSIR doorstopper at the end of boreholes as much as several tens of meters long. For early models, a connector plug was necessary so that a baseline reading of the strain gauges could be established before overcoring. The connector plug was removed during overcoring but reattached for a post-overcoring strain reading. In this case, strain relaxation was the difference between strain gauge readings before and after overcoring. A later version of the CSIR doorstopper allows for continuous reading of strain relaxation during overcoring. In addition to cells using electrical-resistance strain gauges, Hawkes and Moxon (1965) developed a technique for bonding photoelastic strain gauges to the end of boreholes.

The magnitude of strain relaxation at the end of a borehole is not directly related by an elastic modulus to rock stress because far-field stress is concentrated at the end of the borehole. Before determining the actual value of in situ stress using the CSIR doorstopper, the operator must subtract the effect of the stress concentration at the end of the borehole. Laboratory experiments suggest a number of values for this concentration from a high of 1.56 (Hoskins, 1967) to a low of 1.1 (Pallister, 1969) with Bonnechere (1967) obtaining 1.25. Presently, the formula of Coates and Yu (1970) based on finite-element analysis is commonly used to correct strain relaxation data from the CSIR doorstopper. If the horizontal principal stresses are S_H and S_h and the vertical stress is S_v, then the concentrated stresses at the end of a vertical borehole are

$$S_H^* = aS_H + S_h + cS_v \qquad (7\text{–}17a)$$

$$S_h^* = aS_h + bS_H + cS_v \qquad (7\text{–}17b)$$

where

$$a = 1.366 + 0.025\,\nu + 0.502\nu^2 \qquad (7\text{–}18a)$$

$$b = -0.125 + 0.154\nu + 0.39\nu^2 \qquad (7\text{–}18b)$$

$$c = -0.52 - 1.331\nu + 0.886\nu^2 \qquad (7\text{–}18c)$$

and ν is the Poisson's ratio (Coates and Yu, 1970).

Lamont Cell

The principal drawback of the original CSIR doorstopper was that continuous strain readings were not possible during overcoring because the gauge had to be disconnected. To add this capability to doorstoppers, R. A. Plumb, then at Lamont-Doherty Geological Observatory of Columbia University, designed a version of the doorstopper called the Lamont cell (Sbar et al., 1979). The Lamont cell consists of a 38 mm diameter hard plastic cylinder filled with Silguard and a layer of Dow-Corning RTV (fig. 7–5). At the base of the RTV is a delta rosette consisting of three electrical-resistance strain gauges spaced at 120°. A thermal compensation gauge, consisting of a strain gauge bonded to a wafer of rock with the appropriate thermal expansivity, is embedded in the Silguard above the RTV. The resistance of this gauge is used to balance a resistance bridge. The thermal compensation gauge is necessary to minimize the thermal effect of drilling water which may be at a different temperature than the rock in situ.

Installation and overcoring of the Lamont cell are shown in figure 7–6. Borehole preparation is important particularly when drilled vertically down as is the case for near-surface measurements. A 7.6 cm masonry bit is used to drill the holes. These bits give larger cores than BX-coring bits using in South Africa. The disadvantage in using this bit is that cores are difficult to snap off when the hole is deeper than about 3 m. The bottom of the hole is flattened using a plug bit. This process leaves a dusting of rock flower at the bottom of the hole which must be flushed with circulating water. After flushing, the hole is drained, and any larger particles are removed by means of a sticky swab. The strain gauge rosette is epoxied to the flattened bottom of a borehole. Prior to installation, an eight-conductor cable is soldered and potted to the Lamont cell. The cable is fed through the setting tool during installation and later fed through the drill string for continuous reading of strain relaxation during over-

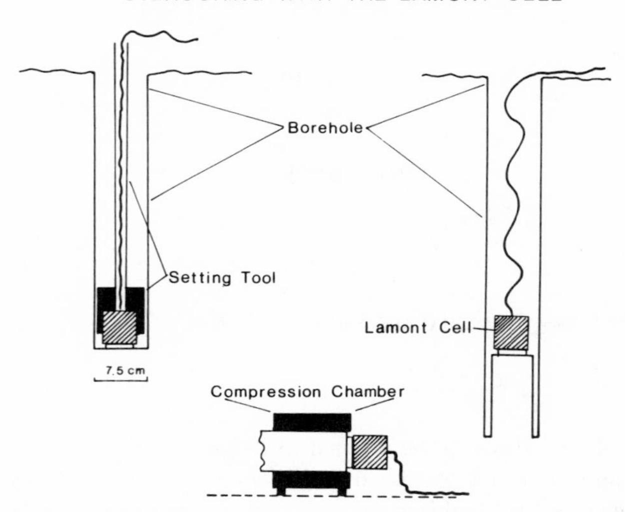

Fig. 7–6. Installation, overcoring, and testing of a Lamont cell. The Lamont cell is placed and oriented at the end of a borehole using a setting tool. After bonding, the setting tool is removed and the cell is overcored. Upon removal from the borehole, the core with cell attached is tested in a compression chamber to determine the elastic properties of the core.

coring. Drilling typically proceeds at the rate of one inch per minute. If the hole extends below the water table, it often fills rapidly with water, in which case an underwater epoxy is used for bonding. Otherwise the hole is dried and a regular strain gauge epoxy (Micromeasurements AE/10–15) is used. Before cutting the core, water is allowed to flow over the Lamont cell until strain readings stabilize. The most successful measurements are those using drilling water at the temperature of the rock (i.e., ground water) so that the drilling water does not cause thermal stresses (e.g., Engelder, 1984). The advantage of working below the water table is that drilling fluid does not shock the rock in the vicinity of the core as is likely for dry holes.

In large part, the quality of the strain-relaxation measurement is determined by examining the character of the strain-relaxation curve (Blackwood, 1978; Engelder, 1984). Examples of strain-relaxation curves of different qualities are shown in figure 7–7. The highest quality strain-relaxation curves show the relaxation divided into three parts: an instantaneous relaxation, a Poisson expansion, and a time-dependent relaxation (fig. 7–7a). Any curves that deviate from this basic shape are likely to indicate either a fractured rock near the strain gauges or poorly bonded strain gauges (fig. 7–7c). Rapid changes in the environment surrounding the Lamont cell may also lead to strain relaxation curves that do not match the shape shown in figure 7–7b.

The general shape of the strain-relaxation curve shown in figure 7–7 is followed by strain cells inserted into pilotholes as well as those bonded to the end of boreholes. Doorstopper measurements indicate that as the drill bit passes the plane of the sensor, the rock tends to expand more than expected for an instantaneous relaxation. This expansion, the Poisson expansion, is then followed by a small contraction to a value expected for instantaneous relaxation. Wong and Walsh (1985) have shown that Poisson expansion curves are useful for calculating both the Poisson's ratio of the rock and the tectonic stress to which the rock is subject.

To calculate stress from instantaneous relaxation data, it is necessary to have data on the elastic properties of samples. To obtain data on elastic properties, the core with doorstopper attached is placed into a rubber sleeve inside a hollow steel cylinder (fig. 7–6). The volume between the cylinder and rubber sleeve is filled with a hydraulic oil fed from a hydraulic pump. As the oil is pressurized, the rubber sleeve expands against the core to compress it radially. A simple data reduction scheme involves the use of an elastic property similar to linear compressibility, which is the relative decrease in length of a line when rock is subjected to unit hydrostatic pressure as defined by Nye (1957). This definition applies to a uniform three-dimensional load, whereas rock core is compressed radially but is not constrained axially. Because this is a two- rather than a three-dimensional test, the relation between fluid pressure (i.e., stress) and strain is called the pseudo-linear compressibility (PLC) (Sbar et al., 1979).

The PLC is related to the strain (ε_i) measured during radial compression and

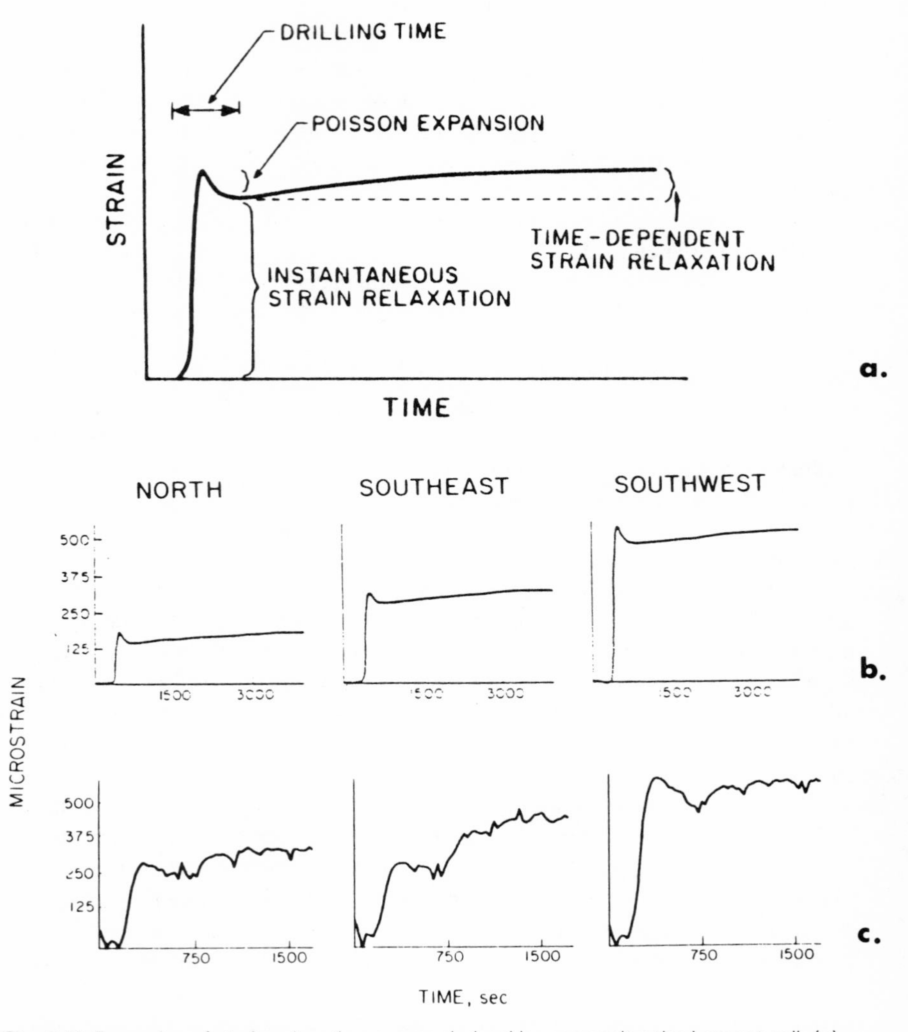

Fig. 7–7. Examples of strain-relaxation curves derived by overcoring the Lamont cell. (a) Components to the strain-relaxation curve. (b) Basic strain-relaxation curves of good quality. (c) Poor quality strain-relaxation curves where moisture is probably affecting the strain gauges. Curves for b and c come from overcoring the Algerie granite, Massachusetts, where delta rosettes on the Lamont cell give strain-relaxation components in the north, southeast, and southwest directions, respectively (adapted from Engelder, 1984). Because S_H was ENE, the southwest component shows the maximum expansion and the north component shows the minimum expansion.

to the known pressure (P_c) applied in the cylindrical test chamber, by the formula

$$PLC = \frac{\varepsilon_1}{P_c} = \frac{1 - \nu_i}{E_i} \tag{7–19}$$

where ν_i is Poisson's ratio, E_i is the Young's modulus for stress applied parallel to ε_i, when two-dimensional orthorhombic symmetry is assumed. If a core is anisotropic, strain parallel to the three gauges on the core differs at the same P_c (Sbar et al., 1979). Using the value of PLC parallel to each of the three gauges of the strain rosette, stress is calculated from the magnitude of strain relaxation in that direction

$$\sigma_{oc} = \frac{\varepsilon_{oc}}{PLC}. \tag{7–20}$$

Effectively, this is a transformation which, to a first approximation, incorporates the anisotropy of the core. From the three components of stress, the principal stresses and their orientations are calculated for a plane normal to the axis of the borehole.

A correction must be applied to the stress data to account for the concentration of stress that occurs at the end of a borehole. A simple correction comes from Leeman (1971)

$$S_i = S_i^* + 0.75\,(0.645 + \nu)S_v \tag{7–21}$$

where S_i is one component of the applied stress in the earth, S_i^* is the stress computed from strain relaxation at the end of the borehole, ν is the Poisson's ratio and σ_z is the vertical stress. Coates and Yu (1970) give a more thorough set of equations for S_i^* at the end of a borehole.

Strain Cells Inserted into Pilotholes

There are a number of techniques for measuring pilothole deformation during strain relaxation. Several are discussed briefly.

Maihak Strain Cell

The Maihak strain cell was developed to measure the change in length of one diameter of a pilothole during overcoring (Jacobi, 1958). The transducer was a wire stretched tightly across the diameter of the pilothole. An electromagnetic signal causes the wire to vibrate at a natural frequency which depends upon the length of the wire. The frequency of the wire vibration is picked up by the X plates of an oscilloscope. Changes in diameter of the pilothole effect the length of the wire and, hence, its natural frequency. Another wire of fixed length is vibrated and recorded on the Y plates of the same oscilloscope. The signal on the oscilloscope is a Lissajous figure whose configuration will be determined by the difference in frequency and phase of the two vibrating wires. Frequency changes between the wire in the borehole and the reference will give an accurate indication of the change in diameter of the pilothole. This scheme was the basis for the development of the vibrating-wire stressmeter, a rigid inclusion gauge discussed in chapter 8 (Hawkes and Bailey, 1973; Hawkes and Hooker, 1974).

Pin and Cantilever Devices

Another design used a cantilever to measure change in pilothole diameter during strain relaxation (Sibek, 1960). This cell features pins which make contact with the pilothole wall. As the pins are displaced by deformation the signal is mechanically magnified by means of a cantilever which in turn moves a contact along a resistance winding in a simple potentiometer circuit. Griswold (1963) also described a device that consisted of a set of pins placed between the pilothole wall and a transducer ring on which electrical-resistance strain gauges were mounted. These pin and cantilever systems were improved upon by the U.S. Bureau of Mines in the development of their borehole deformation gauge.

CSIR Strain Cells

One of the first CSIR (South African) strain cells measured the change in diameter of pilotholes using two LVDT transducers. Its application was in horizontal boreholes where principal stresses were assumed to be vertical and horizontal directions (Leeman, 1964a).

U.S.B.M. Borehole Deformation Gauge

Starting in the 1950s the U.S. Bureau of Mines worked on the development of a borehole deformation gauge that offered the advantage of installation without the use of epoxy (Obert et al., 1962; Hooker and Bickel, 1974; Hooker et al., 1974). The U.S.B.M. borehole deformation gauge is a cylindrical steel tool which is pushed into a 3.8 cm diameter (EX) pilothole where the friction between the gauge and the pilothole holds the gauge in place (fig. 7–8). The gauge is azimuthally oriented with setting tools which are effective for gauge emplacement to depths in excess of 30 m. Overcoring is accomplished with a 15.9 cm outer diameter by 14.3 cm inner diameter coring bit. Later the core is reloaded in a radial compression chamber to determine the physical properties of the rock which are necessary for a calculation of stress. The basic sensing elements are three pairs of cantilevers which are deflected by tungsten carbide–

U.S.B.M. BOREHOLE DEFORMATION GAUGE

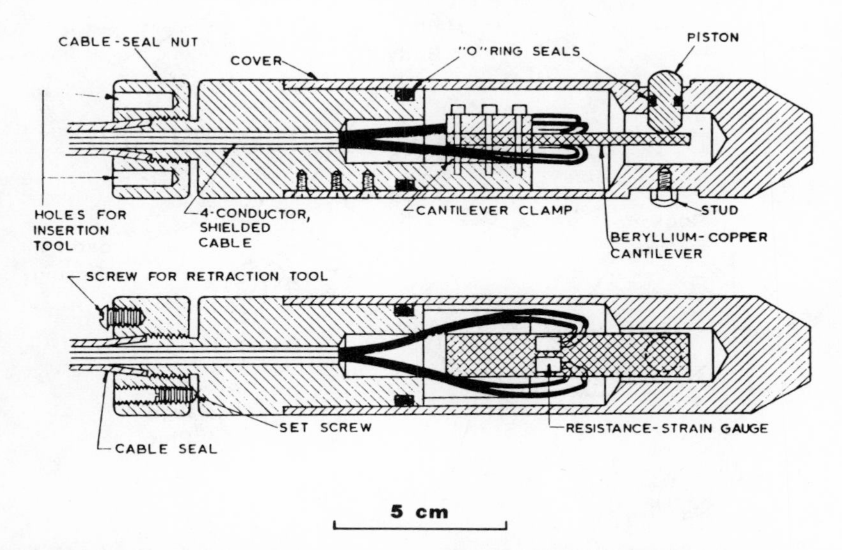

Fig. 7–8. U.S. Bureau of Mines borehole deformation gauge (adapted from Obert et al., 1962). This is an early version of the U.S.B.M borehole deformation gauge showing a single cantilever and piston. Later versions had six cantilevers for measurement of borehole deformation across three diameters. Each of the six cantilevers closely resembled the single cantilever in this device.

tipped pistons that contact the pilothole wall. The cantilever pairs are evenly spaced within the gauge at 120°. Each piston must be shimmed to give the cantilevers the proper amount of deflection so that a solid contact is maintained during deformation of the pilothole accompanying strain relaxation of the core. Each cantilever has two strain gauges mounted so that a cantilever pair is wired in a full bridge configuration to compensate for thermal drift of the strain gauges.

CSIR Triaxial Cell

The next advance was the development of a strain cell for three-dimensional stress determinations where the orientation of the stress field was unknown. The South African answer was to measure both axial as well as radial deformation of a pilothole (Leeman, 1971). A CSIR triaxial strain cell is installed in a 3.8 cm diameter pilothole at the end of a 9 cm diameter coring hole. Transducers in the triaxial strain cell are three 3-component strain gauge rosettes which are glued directly to the side of the pilothole (fig. 7–9). The three rosettes are

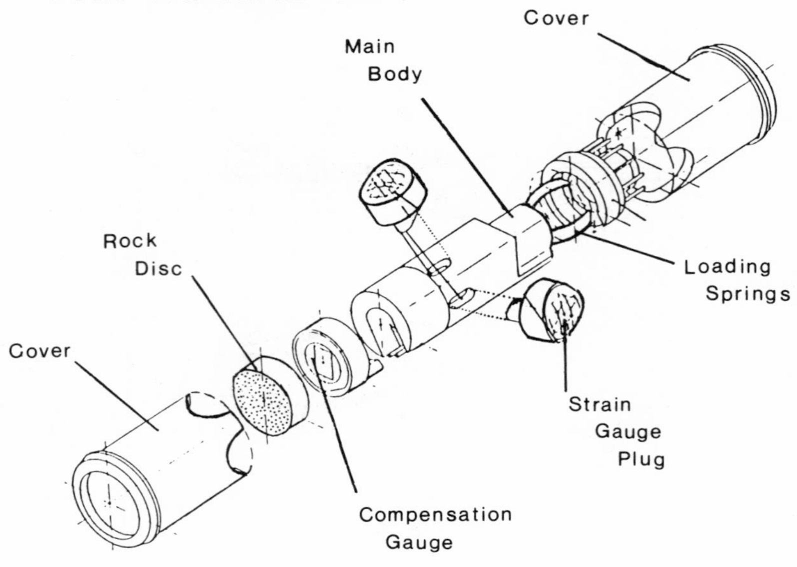

Fig. 7–9. CSIR triaxial cell.

mounted on little plugs which are forced against the wall of the pilothole by compressed air and loading springs. Thermal compensation is accomplished using a strain gauge glued to a rock disc within the triaxial cell. After overcoring, nine components of strain relaxation are used with elastic moduli to solve for the three-dimensional state of stress.

LNEC Strain Cell

The Laboratorio Nacional de Engenharia Civil in Lisbon, Portugal, has developed a soft-inclusion strain cell which consists of a hollow cylinder with three electrical-resistance strain gauge rosettes embedded within the walls of the cylinder (Rocha et al., 1974). The 45° rosettes are embedded along the equator of the cylinder at azimuthal angles of 0, $\pi/2$, and $5\pi/4$. In this configuration one gauge of the rosette is parallel to the axis of the cylinder and a second runs along the circumference of the cylinder. The 3.5 cm diameter cylinder is bonded with epoxy to the walls of a 3.7 cm diameter pilothole. Once the bonding agent has cured the LNEC gauge is overcored using a 7.5 cm diamond coring bit. During overcoring the nine strain gauges are continuously recorded. Core containing the LNEC gauge is then reloaded in the laboratory to determine the elastic properties of the rock for purposes of calculating in situ stress.

CSIRO STRAIN CELL

The Geomechanics Division of the Australian Council of Scientific and Industrial Research Organizations (CSIRO) attempted to improve on the CSIR (Leeman, 1971) triaxial strain cell (Worotnicki and Walton, 1976). The CSIRO strain cell is a hollow inclusion strain cell with strain-gauge rosettes glued to the outside of a thin-walled (3.5 cm o.d. by 3.2 cm i.d.) epoxy tube. The cell is then glued to the wall of a nominal 3.8 cm diameter pilothole with a special Araldite-based cement. The strain gauges are thus separated from the wall of the pilothole by an epoxy-filled gap, 1.5 mm in thickness. Worotnicki and Walton (1976) show that this epoxy gap does not affect the measured value of the axial strain but that the circumferential and off axial strains are slightly higher than if the strain gauges had been glued directly to the rock surface. The gauge is installed in a pilothole drilled 60 cm deeper than the large diameter hole used for overcoring (Hustrulid and Leijon, 1983). The centerline of the cell is pushed more than 20 cm down the pilothole when grouted in place. Because the epoxy grout requires a minimum of twelve hours and preferably sixteen hours to cure, data collection proceeds at the rate of one measurement a day.

LUT STRAIN CELL

The Luleå University of Technology strain cell was developed during the late 1970s in Sweden (Leijon, 1983). Its design was based on the concepts introduced by Leeman (1968) for a soft-inclusion borehole gauge to record strain on the walls of a pilothole during strain relaxation. The LUT gauge contains twelve, 5 mm long electrical-resistance strain gauges, oriented in ten different directions (fig. 7–10). The strain gauges are placed by means of pneumatically operated pistons (three pistons with four gauges each), glued directly to the pilothole wall. Once glued in place, the strain gauges offer no resistance to deformation of the core during strain relaxation which is continuously monitored during overcoring.

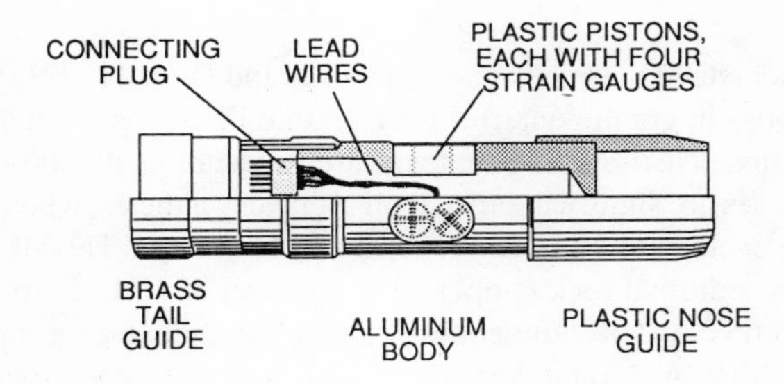

Fig. 7–10. The LUT strain cell.

CRIEPI Strain Cell

Workers at the Central Research Institute of Electric Power Industry (CRIEPI) in Japan have developed a strain cell with four radial components and an axial component (Kanagawa et al., 1981; 1986). Each of the four diametrical components is pressed against the pilothole wall by means of small hydraulic pistons. The diametrical components are then cemented in place by glue that is forced around the outside of the CRIEPI strain cell. Curing time is as much as ten days. The final configuration of cement with embedded diametrical components is a hollow-walled soft inclusion.

Thermally Induced Stresses

In rocks near the surface, the diurnal and annual temperature cycles cause large variations in thermal stress which complicate near-surface stress measurements. Near-surface rock temperature and, hence, thermal stresses are periodic and related to surface temperature by Carslaw's and Jaeger's (1959) equation

$$T_z = T_s \exp\left(-\frac{z\sqrt{\pi}}{kt}\right) \qquad (7\text{–}22)$$

where T_z is the temperature variation at depth (°C); T_s is the total temperature variation about the mean at surface (°C); z is depth (cm); k is the diffusivity constant (cm^2/sec); and t is the period (seconds). If the boundaries of a rock mass behave in uniaxial strain, the horizontal stress induced by heating or cooling is given by Hooker and Duvall (1971)

$$\Delta S_h^T = \frac{\alpha_t E(T_1 - T_0)}{1 - \upsilon} \qquad (7\text{–}23)$$

where ΔS_h^T is the horizontal stress due to temperature change (MPa) and $T_1 - T_0$ is the temperature change (°C).

Granite

To better understand thermal stresses, Hooker and Duvall (1971) installed temperature probes in granite quarries near Marble Falls, Texas, and Mount Airy, North Carolina. Near-surface temperature measurements show that diurnal thermal stresses are significant to a depth of nearly a meter, whereas the annual temperature cycle extends to depths of 8 m (Hooker and Duvall, 1971). In the Texas quarry a diurnal rock temperature variation of 12.1 °C affected the rock to a depth of 65 cm. This temperature change can induce a thermal stress of as much as 4.1 MPa. At Mount Airy, North Carolina, the annual rock temperature variation at 30 cm depth was 21.4 °C or about 10 °C from the mean temperature. During a twelve-month period a U.S.B.M. borehole deformation gauge was used to measure rock stress at a point where the rock temperature varied from 6.1 °C to 26.1 °C. The measured annual thermal stress at 30 cm was 7.9 MPa which agrees well with the thermally induced stress calculated using equation 7–23. Note that the tensile strength of some rocks is on the order of 10 MPa. If a granite is destressed at, for example, 25 °C, then thermally induced stresses during cooling less than 30°C are large enough to cause crack propagation in the near-surface.

MIOCENE SANDSTONE

During the summers of 1979 and 1980, strain-relaxation stress measurements were made in the Mojave Desert southeast of Palmdale, California, using the U.S.B.M. borehole deformation gauge to the depth of 30 m (Sbar et al., 1984). The stress-versus-depth profile for a site in the Punchbowl Sandstone (Miocene) shows substantially higher stresses between the surface and a depth of 6 m. These stresses decay exponentially to a background level on the order of 1.5 MPa (fig. 7–11). Such an exponential decay is most easily explained in terms of seasonal temperature variations at the surface, where the surface temperature T_s is given by

$$T_s = T_0 + T_a \cos(\omega t). \tag{7–24}$$

T_0 is the mean annual temperature. The amplitude, T_a, of the surface temperature variation from NOAA climatological data is 9.85 °C, $\omega = 2 \times 10^{-7}$ for a period of 1 year; and $t = 0$ corresponds to the time of maximum annual surface temperature. The temperature at depth as a function of time for a thermally isotropic half space is given by

$$T(z,t) = T_0 + T_a e^{-\mathbf{k}x} \cos(\omega t - \mathbf{k}z) \tag{7–25}$$

where $\mathbf{k} = (\omega / 2k)1/2$. The horizontally induced thermal stress for an isotropic elastic half space in a state of plane strain which is constrained against lateral expansion and free to expand vertically is given by

$$S_h^T(z,t) = \frac{\alpha_T E T_a e^{-kx} \cos(\omega t - kz)}{1 - \nu}. \tag{7–26}$$

The predicted curve shown in figure 7–11 corresponds to thermal stress of as much as 3.5 MPa superimposed on an average stress of 1.6 MPa.

Thermal stress adds a component of axially symmetric stress to the horizontal near-surface in situ stress. Near-surface thermal stress decays with depth over the first 5–6 meters below the surface. Regardless of the magnitude of the thermal stress, the contemporary tectonic stress is apparent as a biaxial component superimposed on an axially symmetric thermal stress. Note that S_H and S_h are offset the same amount by thermal stress (fig. 7–11). The effect of superposition of an axially symmetric thermal stress on top of a biaxial tectonic stress

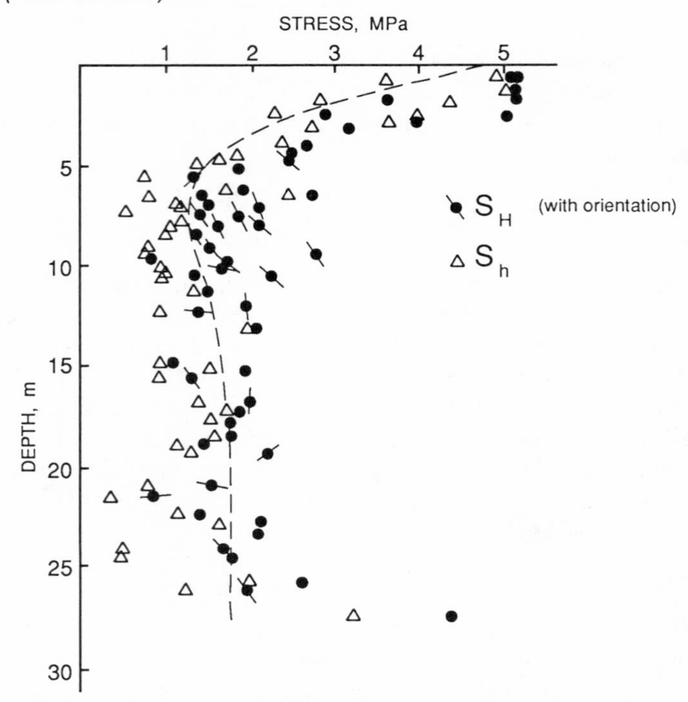

Fig. 7–11. Thermal decay curve for the Miocene Punchbowl sandstone of California (adapted from Sbar et al., 1984). Stress values were obtained by overcoring a U.S.B.M. gauge in two wells less than 1 km southwest of the San Andreas fault at Punchbowl State Park, California. Solid symbols are S_H and open symbols are S_h. Azimuth of S_H is indicated in plan view with north up and east to the right for those measurements where the stress ratio was ≥ 1.4.

is to decrease S_H/S_h as is seen in the top four meters of the Punchbowl sandstone. So, despite a significant component of thermal stress, measurements in the top four meters of the Punchbowl sandstone using the Lamont doorstopper (Sbar et al., 1979) and the U.S.B.M. borehole deformation gauge (Tullis, 1981) detected the components of horizontal stress found below the thermally disturbed zone.

The Correlation between In Situ Stress and Jointing

Near-surface strain-relaxation measurements suggest that, in jointed rock, the local stress is commonly orthogonal to the prominent joints (Preston, 1968;

Brown, 1974; Swolfs et al., 1974). S_H will often parallel the predominant joint set and the magnitude of the local stress components usually decreases along with decreasing spacing between joints. Despite several observations of this sort, the relationship among joint density, joint orientation, stress orientation, and stress magnitude is still poorly understood. Three possibilities are: (1) both joint propagation and in situ stress are controlled by the contemporary tectonic stress field; (2) the joints propagate in the near-surface where residual stress controls joint orientation with the S_H component of the residual stress remaining parallel to the joint set after propagation; or (3) the presence of the joints induces a low joint-normal modulus and, hence, causes a decrease in joint-normal compressive stress and an alignment of S_H parallel to the joint set. Examples of the relationship between in situ stress and joints follow.

Potsdam Sandstone

The Potsdam sandstone, the basal Cambrian clastic rock which mantles the Adirondack Dome, New York, contains a prominent joint set striking ENE on the northwest side of the dome. Strain-relaxation data obtained using surface-bonded electrical-resistance strain gauges show that maximum expansion parallels these joints (Engelder and Sbar, 1977). Furthermore, the magnitude of the strain relaxation correlates inversely with the extent to which intact pieces of Potsdam Sandstone have been reduced in size by jointing; the greatest strain relaxation is found in the largest blocks (fig. 7–12). The region contains many post-glacial popups with axes oriented NNW indicating an ENE S_H which parallels both the well-developed local joint set and the contemporary tectonic stress field in eastern North America (chap. 3). Although the age of the ENE joints is unknown, near-surface processes such as unloading in the presence of the contemporary tectonic stress field may contribute to their present well-developed character. The common orientation of the local joints, the post-glacial popups, and S_H of the contemporary tectonic stress field, support the argument that near-surface processes (e.g., temperature changes, weathering, and jointing) do not completely isolate surface outcrops from the deep-earth stress field (Engelder and Sbar, 1977). About 120 km to the east within the crystalline rocks of the central Adirondack Mountains, strain-relaxation data also show the same ENE S_H in the vicinity of a strongly developed ENE joint set (Plumb et al., 1984).

Mount Waldo Granite

Using the U.S.B.M. borehole deformation gauge, Lee et al. (1979) measured in situ stress down to 8 m within the Waldo granite, Maine. The purpose of their experiment was to measure stresses associated with local rockbursts and sheet fractures. A site on a hillside with widely spaced sheet fractures and fewer vertical joints was compared with a highly fractured site in a valley bottom (fig.

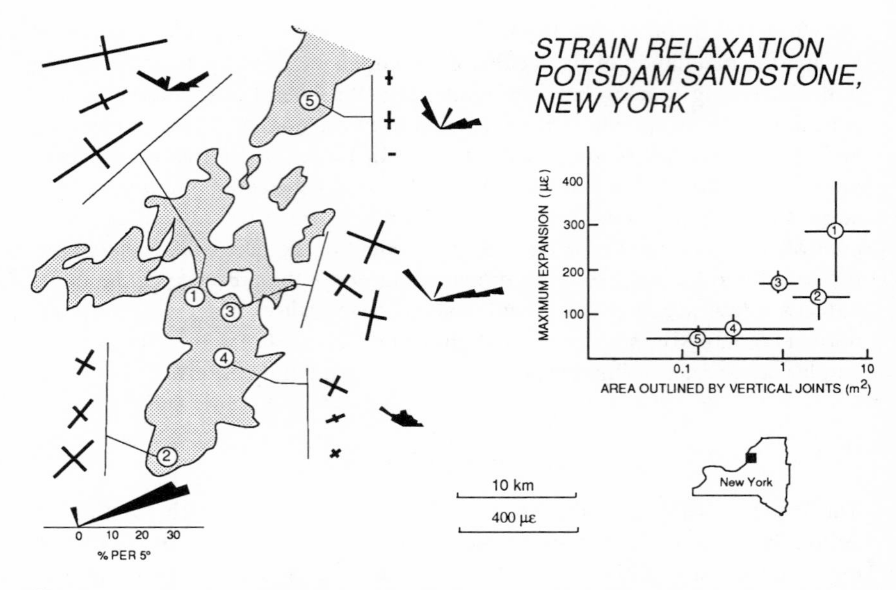

Fig. 7–12. Strain relaxation (shown as crosses representing maximum and minimum expansion) upon overcoring surface-bonded strain gauges in five outcrops of the Potsdam Sandstone, New York (adapted from Engelder and Sbar, 1977). Maximum expansion, given in units of microstrain, correlates with S_H in each outcrop. The orientation of joints within each outcrop is shown as a half rose diagram. Joint spacing was measured at each outcrop to determine the area of the outcrop surface outlined by vertical joints. This is a qualitative measure of the volume of intact pieces of rock at each outcrop.

7–13). Again an inverse correlation between joint density and stress magnitude was found with the hillside site having a lower fracture density and a higher S_H.

ELLIOT LAKE DISTRICT, CANADA

A correlation between in situ stress and jointing extends well below the surface as shown by CSIR doorstopper measurements at three sites in two mines in the Elliot Lake District, Canada (Eisbacher and Bielenstein, 1971). The depths of measurements were between 300 and 700 m where $S_H > S_h > S_v$. At all three sites S_H parallels the best-developed joint set which strikes E-W in two cases and NE-SW in a third. The age of the joints in the Elliot Lake District is uncertain. If they are Mesozoic or older, there is a fortuitous correlation between these joints and the contemporary tectonic stress field in eastern North America. If the joints are relatively young, late-stage geological processes in the

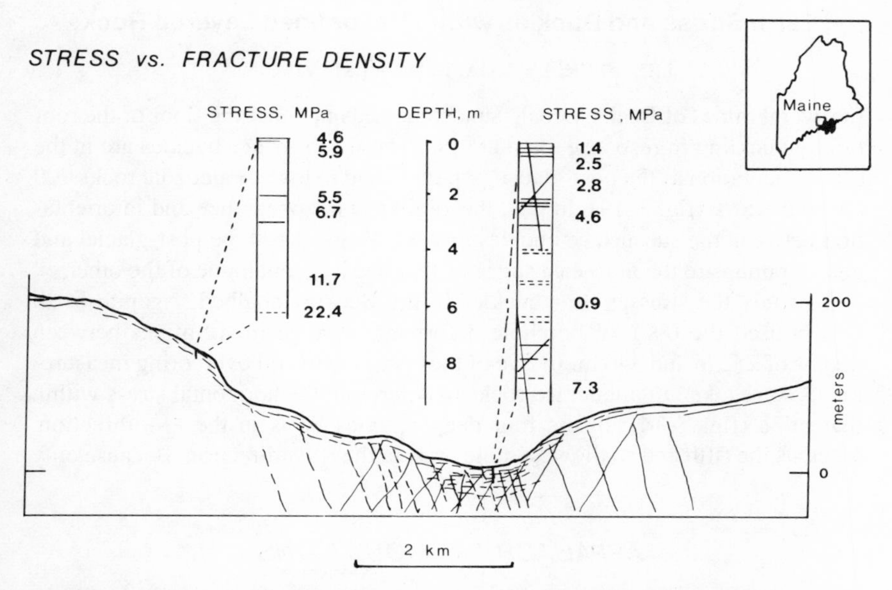

Fig. 7–13. Stress as a function of fracture spacing in the Mount Waldo granite (adapted from Lee et al., 1979). The profile shows the generalized relation of drill sites in the Mount Waldo granite to topography, vertical and sheeting joints, and horizontal stresses. Logs of cores from the two drill sites indicate fractures (solid lines) and incipient sheeting (dashed lines). Vertical scale of profile exaggerated by a factor of twelve.

presence of the contemporary tectonic stress (e.g., uplift in the case of Elliot Lake) may have contributed to the present development of these local joint sets and their correlation with S_H as is the case for neotectonic joints (e.g., Hancock and Engelder, 1989).

The Appalachian Plateau

Devonian rocks of the Appalachian Plateau, western New York, were compressed in a series of low amplitude folds during the Alleghanian Orogeny. A set of cross-fold joints mark this compression direction throughout the Appalachian Plateau (Nickelsen and Hough, 1967). In one outcrop of Machias sandstone, fourteen near-surface strain-relaxation measurements show an average S_H parallel to the cross-fold joint set within the outcrop (Engelder and Geiser, 1980). As this is an area where contemporary tectonic stress is ENE (e.g., Evans et al., 1989a), the local joint orientation and S_H in the outcrop are apparently genetically related to the Alleghanian Orogeny.

Earth Stress and Buckles within Unconfined Layered Rock

THE BECKLEY COALBED, WEST VIRGINIA

In several mines of Beckley coal,[5] sandstone beds in either the floor or the roof fail by buckling in response to an ENE S_H. These subsurface buckles are in the same orientation as the post-glacial popups found in lower Paleozoic rocks 900 km to the NNE (fig. 3–19). In fact, the similarity in appearance and in orientation between the subsurface buckles in West Virginia and the post-glacial and quarry popups to the northeast suggests that one is an analogue of the other.

To study the stresses near buckles in the Beckley coalbed, Agapito et al. (1980) used the U.S.B.M. borehole deformation gauge in six mines between depths of 252 m and 346 m. In five of these mines several overcoring measurements were taken in unbuckled rocks to determine the horizontal stress within that mine (fig. 7–14). In the four deeper mines S_H is in the ENE direction, whereas the fifth and shallowest mine S_H is in the NNW direction. Because only

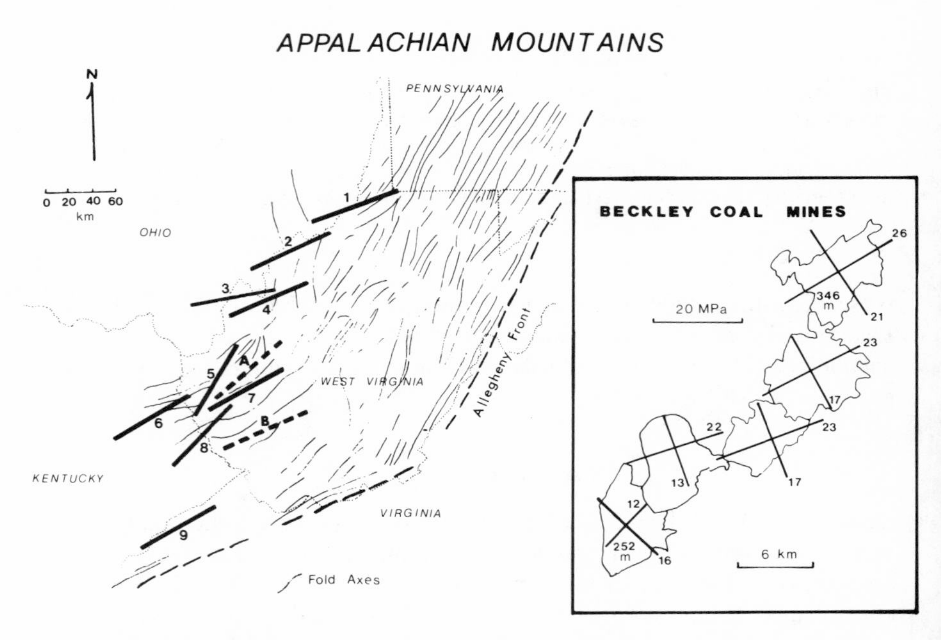

Fig. 7–14. Horizontal stress in five of the Beckley Coal Mines of West Virginia (Agapito et al., 1980). Orientation and magnitude indicated by stress crosses. Average orientation of S_H for the entire suite of Beckley Coal Mine measurements given as a dashed line on map of Appalachian Mountains at B, the location of the Beckley Mines. Orientations of coring-induced fractures from nine wells are shown as solid lines on the map of the Appalachian Mountains (Plumb and Cox, 1987). Dashed line A is the orientation of S_H determined from the hydraulic fracture measurements of Abou-Sayed et al. (1978).

two measurements were made in the fifth, compared to as many as twenty-two in another mine, that stress ellipse is considered less reliable. In these rocks σ_d ($S_H - S_v$) varies up to 18.4 MPa. Stresses within rocks adjacent to the Beckley coalbeds are of similar magnitude and orientation as those measured by Obert (1962) in northeastern Ohio fifteen years earlier. Agapito et al. (1980) suggest that the high stresses correlate with the orientation of the Allegheny Front of the Appalachian Mountains in West Virginia. However, evidence elsewhere is so strong for the contemporary tectonic stress field with S_H striking ENE in the northeastern United States that the correlation between the stress in the Beckley coalbeds and the Allegheny Front is apparently just a coincidence.

Southern Illinois Mining District

The presence of a high horizontal stress in the subsurface is further indicated by roof failures in coal mines in the southern Illinois Basin (Nelson and Bauer, 1987). These failures are buckles much like those in the Beckley mines of West Virginia. Within the Illinois Basin there is a marked tendency for shale layers in the immediate roof to buckle downward and snap along linear paths. This phenomena is most common for mines with a north-south drift where the axis of the buckle parallels the drift of the underground mine in response to an E-W S_H. In situ stress measurements throughout the southern Illinois Basin show that S_H is about E-W (Engelder, 1982) in agreement with the orientation of subsurface buckling.

In surface outcrops, buckle zones form most readily in thick, laminated shales and siltstones which have low strength, whereas limestone and other rock of high compressive strength do not develop buckles (Nelson and Bauer, 1987). These buckles resemble small thrust faults indicative of S_H oriented E-W. The tendency for shale to buckle and limestone to be stable gives a constraint on near-surface σ_d in the southern Illinois Basin. The unconfined compressive strength of claystones is between 2 and 7 MPa. Coals show an unconfined compressive strength between 11 and 26 MPa and the strength of limestone is much higher. Haimson and Doe (1983) report σ_d of up to 10 MPa in northern Illinois. Shales and siltstones in the top 100 m of the Illinois Basin apparently have the relative strength of an unconfined claystone, because σ_d from the contemporary tectonic stress field is large enough to cause thrust faulting in these weak rocks near the surface.

A Direct Measurement of Overburden (S_v)

Geologists commonly assume that S_v is equal to the weight of overburden. Sometimes this assumption is inappropriate as indicated by hydraulic fracture measurements at North Conway, New Hampshire (Evans et al., 1988) (see chap. 5). Some of the most detailed data on S_v comes from South African mines where doorstopper measurements are made at depths greater than 1 km. Leeman (1964a) describes a suite of doorstopper measurements every 1.5 m in a horizontal borehole located next to a haulage in the East Rand Proprietary Mines at a depth of 1760 m. The borehole penetrates quartzite and then hits a dike 43 m from the haulage. Within a couple of meters of the haulage the S_v is much larger than overburden as is expected for a stress concentration around an underground opening. One measurement at 1.8 m from the haulage showed S_v to be magnified to twice the calculated weight of overburden. Core discing accompanied drilling in the first 6.3 m as a consequence of the stress concentration about the haulage. At a distance of 10 to 30 m away from the haulage S_v was about 10% higher than calculated overburden. Further from the haulage a zone of fracturing was encountered where S_v dropped to less than half the expected weight of overburden. The fracturing continued into a dike where the test was terminated. For more than 10 m the fractured rock does not carry the full weight of overburden. This suite of measurements shows that fracturing also tends to unload overburden stress with the load transferred to the zone which showed a 10% excess S_v. If the quartzites in the East Rand Proprietary Mines are subject to a state of uniaxial strain then the S_h might be related to S_v through equation 1–17. In the section of the borehole where the S_v is near the weight of overburden the S_h is close to the predicted value based on a υ of 0.18 for quartzite.

Under many circumstances S_v is calculated from density measured by logging or laboratory techniques. Direct measurement of the overburden is less common. A compilation of data taken with the CSIR triaxial strain cell in deep South African gold mines gives an indication of the variation in overburden (Gay, 1975; 1977). S_v was measured at ten localities from depths between 508 and 2500 m mainly within sediments of the Witwatersrand Group (fig. 7–15). If S_v is used to calculate the integrated density of overburden from each locality, the density appears to vary from 1.56 to 3.13. Perhaps the low S_v reported by Gay (1977) reflects local conditions such as those found near the dike in the East Rand Proprietary Mines. However, an average density for the ten localities is 2.625, a number commonly derived from integrated density logs of a silica cemented sandstone (Gay, 1977). The data on overburden from the South African gold mines confirms the hypothesis that on average the overburden weight is directly related to rock density by the depth of burial.

As is the case with virtually all samples of earth stress S_v from South Africa scatters about a mean and several measurements are necessary to arrive at an average which is an accurate measure of S_v. However, one source of spurious data on a local scale is the nonideal mechanical behavior (nonlinearity and inhomogeneity) of the rocks in which measurements are made (Leijon and Stillborg, 1986). These lead to local variations in stress which occur at considerable depth in the crust of the earth as shown by Leeman (1964a). Finally, the

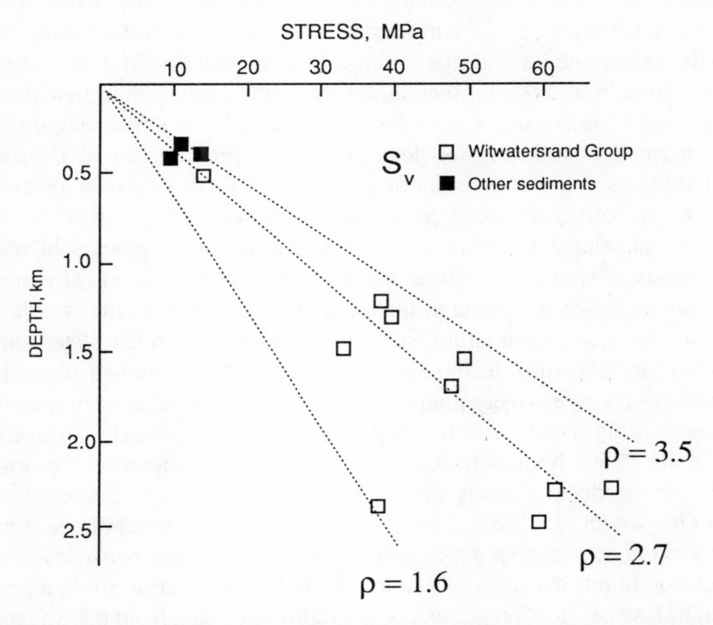

Fig. 7–15. S_v as a function of depth in South Africa measured using the CSIR triaxial cell. Data are compiled in Gay (1977). Open squares are data taken within the Witwatersrand Group sediments, whereas solid squares are data taken outside of this site.

precision of strain cell techniques (± 15%) is such that one should never have great confidence in single measurements.

Determination of Stress Magnitude Using Strain-Relaxation Techniques

Correlation of the stress magnitudes between shallow and deep measurements is problematical. Roberts (1973) argues that "objective, factual comparisons of different methods of in situ stress measurements are impossible to make, for the simple reason that it is impossible to establish any absolute standard against which the measurements may be compared." Haimson et al. (1974) compared stress data from the U.S.B.M. borehole deformation gauge and hydraulic fracture measurements and concluded that the difference between measured stresses for the two methods ranged from 9 to 30%. Using two techniques it is not clear which serves best as the "standard." Careful testing by Bonnecherre

(1967) showed that after repeated measurements using one technique in the laboratory, stress magnitudes could not be relied upon for better than 20–30% of the mean values. The near-surface data from the Rangely, Colorado, experiment (de la Cruz and Raleigh, 1972) show that the variation in stress magnitude obtained from five different techniques was much larger than that found by Bonnecherre (1967) using one technique. Although the stress magnitudes do not compare very well, it is not clear whether the problem is with the sensors, the relative position of the sensor at or below the earth's surface, or the technique for estimating the elastic parameters of the rock.

Instrument-related variation in stress magnitude also appears in regional compilations of stress data. For example, Stephansson et al. (1986) compiled stress data from Fennoscandia to show that on average stress magnitude determined by the hydraulic fracture technique was lower than that determined by several strain-relaxation techniques. In contrast, data obtained using Hast's stressmeter give a relatively high estimate of stress magnitude compared with other overcoring techniques. In comparing data from several instruments including the CSIRO hollow inclusion cell, the U.S.B.M. borehole deformation gauge, the Swedish LUT cell, the CSIR doorstopper, and a photoelastic technique, Gregory et al. (1983) conclude that the U.S.B.M. borehole deformation gauge and the doorstopper gave the best results based on consistency of data.

Rock conditions may also be a source for large variations in stress magnitude as seen in Brown's (1974) photoelastic strain-gauge data from the Rangely anticline. Brown (1974) measured the magnitude of strain relaxation right after overcoring and recorded strain at intervals over the next two weeks. His data revealed that magnitude of the strain relaxation decreased in the two weeks following overcoring. For this experiment stress magnitudes depend on which datum is selected as representative of strain relaxation.

Another source for variation in stress magnitude for strain-relaxation tests is the imprecise determination of moduli using a radial compression chamber (Hooker and Bickel, 1974). This technique for reloading the core applies an equal load in all directions about the diameter of the core. The drawback is that, while under in situ conditions, the core was not subject to an equal load in all directions. Because the reloading is not exact, the determination of proper moduli is not possible. The problem of determination of proper moduli is further complicated by the tendency of rocks to show a nonrecoverable component of strain relaxation. Nonrecoverable strain relaxation can effect a permanent change in the moduli of the rock, thus making it impossible to measure the in situ moduli of the rock (Engelder, 1981).

Concluding Remarks

Strain-relaxation techniques are useful for relatively inexpensive and rapid stress measurements. Strain-relaxation measurements have taught us a great deal about such local phenomena as stress concentrations around underground openings, near-surface thermal stresses, and the correlation between in situ stress and joints. Measurements in deep mines have given an indication that S_v is, indeed, that expected from loading of the overburden. Care is required in selecting sites for near-surface measurements where rock quality can interfere with stress measurements.

— 8 —

Stressmeters and Crack Flexure

The flatjack technique has the advantage of measuring stresses directly. . . . The results show that good quality stress trajectories (of the contemporary tectonic stress field) can be derived from surface measurements.

—Claude Froidevaux, Christian Paquin, and Marc Souriau (1980)
upon making flatjack measurements in twelve quarries
in the northern portion of France

Devices inserted into pilotholes and slots follow one of two basic schemes for measuring in situ stress. The first scheme involves devices which detect the shape change of a small volume of rock as earth stress is removed by coring or cutting. Stress is not measured directly from these shape changes but rather it is later calculated using information on the elastic properties of the rock (see chap. 7). The second scheme is to insert a device which prevents change in shape of the rock on coring or which restores the rock to its original shape. The advantage of these latter devices is that, after calibration, their data output is in terms of stress which is recorded directly without the need for information on the elastic properties of the rock. This chapter deals with stress-measurement techniques involving the direct readout of stress. Direct readout techniques fall into two general groups: those involving rigid inclusions for preventing shape change and those involving the restoration of original shape mainly by the flexure of man-made or natural cracks in rocks.

The Stressmeter

A *stressmeter* is a very rigid gauge that, when installed in a pilothole, gives a direct readout of stress in the rock and, thus, makes unnecessary the measurement of the elastic properties of the host rock (Leeman, 1964a). A direct readout of stress from the stressmeters requires that the gauge is properly calibrated in the laboratory before installation. When a stressmeter is inserted into a pilothole, the strength of the gauge prevents appreciable deformation of the pilothole during stress changes in the core around the hole. Very small changes in strain in the stressmeter are measured and expressed in terms of changes of stress in the rock.

The true stressmeter is just one of three types of gauges inserted into pilotholes for the purpose of stress measurements. Distinction among the three types is based on the ratio (E_h / E_g) of the Young's modulus of the host rock (E_h) to the Young's modulus of the gauge (E_g) (Leeman, 1964a; Riley et al., 1977). If the gauge completely fills the pilothole, usually by cementing with epoxy or grout, then it is called an *inclusion*. By this definition the U.S.B.M. borehole deformation gauge is not an inclusion. The literature makes the distinction between a "rigid" inclusion where $E_h / E_g < 1/5$ and the "soft" inclusion where $E_h / E_g > 10$. When $E_h / E_g < 1/5$, stress induced in the active element of the gauge is independent of the host-rock modulus, whereas when $E_h / E_g > 10$, deformation of the borehole is independent of the material properties of the active element of the gauge. In the former case the gauge is a stressmeter whereas in the latter case the gauge is a strain cell as described in chapter 7. The LNEC borehole stress gauge (Rocha et al., 1974) is then an example of a soft-inclusion gauge. If the ratio of E_h / E_g is between 1/5 and 10, then the active element of the inclusion and the host rock interact during strain relaxation (Riley et al., 1977). The analysis of this intermediate case is more complicated than that for either the rigid- or soft-inclusion gauges. Although not a true stressmeter, discussion of this latter type of inclusion is included in this chapter.

Stressmeters are often installed to monitor stress changes rather than complete stress relief. Following Roberts's (1977) analysis for an elastic circular inclusion cast into an elastic host rock subject to uniaxial loading, a stress change in the host rock ($\Delta\sigma$) will transmit a uniformly distributed stress ($\Delta\sigma_g$) to the inclusion. The ratio of stresses is

$$\frac{\Delta\sigma_g}{\Delta\sigma} = (1 - \nu^2)\left[\frac{1}{(1 - \nu) + \frac{E_h}{E_g}(\nu_g + 1)(1 - 2\nu_g)} + \frac{2}{\frac{E_h}{E_g}(\nu_g + 1) + (\nu + 1)(3 - 4\nu)}\right] \tag{8–1}$$

where E_h and υ are Young's modulus and Poisson's ratio, respectively, of the host material, and E_g and ν_g are Young's modulus and Poisson's ratio, respectively, of the inclusion material. The ratio, $\frac{\Delta\sigma_g}{\Delta\sigma}$ approaches a limiting value of 1.5 as long as $E_g > 5E_h$. Because $\frac{\Delta\sigma_g}{\Delta\sigma}$ is constant for high $\frac{E_g}{E_h}$, a measured change in stress in the rigid inclusion is directly proportional to 1.5 times the stress change in the host rock. If, on the other hand, $E_g < E_h$, equation 8–1 shows that the ratio $\frac{\Delta\sigma_g}{\Delta\sigma}$ becomes increasingly dependent upon the modulus of the host rock which is required to calculate in situ stress.

Unidirectional Stressmeters

The first stressmeterlike devices were installed with a wedge across one diameter of a borehole thereby measuring one component of the stress field normal to the axis of the borehole. Unidirectional gauges are not inclusions in the strict sense but are included as stressmeters because of their very high moduli. Determination of the principal stresses normal to the borehole axis required the wedging of three single-component gauges across the borehole diameter rotated $2\pi/3$ from each other.

Hast's Magnetostrictive Stressmeter

Hast's magnetostrictive stressmeter was one of the first stress-measuring devices to use an active element[1] whose E was considerably larger than the host rock (Hast 1958). The active element of the magnetostrictive stressmeter consists of a 10 mm long, nickel-alloy spool around which is wound a coil of wire (fig. 8–1). If a force is applied parallel to the axis of the nickel-alloy spool, the magnetic permeability of the spool is altered, and with it the impedance of the coil. The impedance of the coil is measured when an alternating current is passing through the coil. For a direct readout of stress, a calibration curve of stress versus impedance is constructed in the laboratory by applying known loads to the spool.

Proper installation of a stressmeter requires jamming the active element in the pilothole so that the active element exerts enough force across the diameter of the pilothole to balance the component of earth stress parallel to the active element. To accomplish this, the inside of the gauge housing has a pair of slightly tapered sides against which tapered metal strips are wedged (fig. 8–1). The housing and metal strips together form a bearing surface parallel to the

HAST'S STRESSMETER

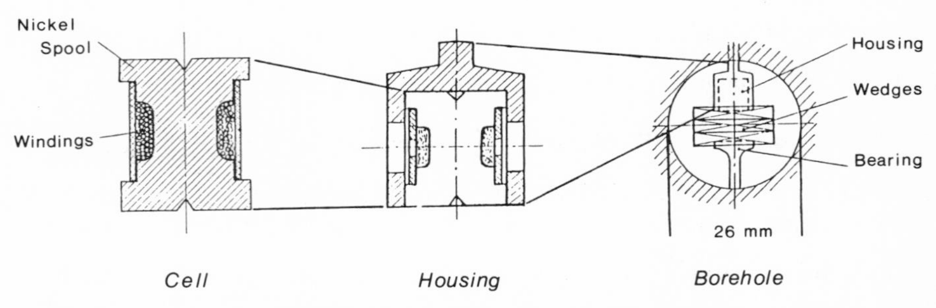

Fig. 8–1. Hast's magnetostrictive stressmeter showing the cell, housing, and installation in a borehole (adapted from Hast, 1958).

borehole wall. A stop near the bottom of a 26 mm diameter pilothole is used to jam the stressmeter and tapered strips against the borehole so that the active element is prestressed. Because the stressmeter is uniaxial, three overcoring measurements must be taken at slightly different depths in a borehole in order to determine the maximum and minimum stresses in a plane normal to the borehole. Stress measurements as close as 10 cm apart are possible provided care is taken to rotate the stressmeter by 60° from the previous measurement. An alternative technique is to stack three stressmeters for overcoring in one operation. Using his magnetostrictive stressmeter, Hast (1958) pioneered in the measurement of regional stresses by showing that horizontal stresses in the near-surface environment are large compared to S_v throughout Scandinavia.

THE VIBRATING-WIRE STRESSMETER

The vibrating-wire stressmeter consists of a stainless steel cylinder wedged tightly in a 3.8 cm diameter pilothole in much the same fashion as the magnetostrictive stressmeter (Hawkes and Bailey, 1973; Hawkes and Hooker, 1974). In this case the active element is a highly tensioned steel wire clamped diametrically across the cylinder in the direction of the bearing plates which are in contact with the pilothole. Stress changes within the rock cause minute changes in the diameter of the wedged cylinder, thereby affecting the period of the resonant frequency of the tensioned steel wire. Calibration of the vibrating-wire stressmeter amounts to generating a graph of period of the resonant frequency versus stress. The vibrating-wire stressmeter, like other stressmeters, is elastically stiffer than the surrounding rock so that little deformation occurs within the rock as external stresses change. This stressmeter is commonly used in mining applications where slow, long-term measurement of stress change is desirable. In order to determine changes in state of stress normal to the axis of a borehole, three vibrating-wire stressmeters must be wedged in close proximity with each component rotated $2\pi/3$ from the adjacent components.

The vibrating-wire stressmeter was used to record slow stress changes along the San Andreas fault (Clark, 1981). Stress changes were recorded at four sites with the vibrating-wire stressmeter installed in pilotholes less than 60 m deep (fig. 8–2). A stress transient of as much as 0.24 MPa (S_H) was recorded at the San Antonio Dam site 15 km from the epicenter of the October 19, 1979, earthquake at Lytle Creek, California (M_L 4.1). The transient was an increase in shear stress (0.18 MPa) parallel to the nodal plane of the Lytle Creek earthquake with the shear couple (right lateral) in the proper sense to have triggered the Lytle Creek earthquake. This measurement suggests that stress changes in the focal region are transmitted to shallow depths despite abundant fractures and local faults. Such stress transients have also been recorded in Utah where an anomaly on the order of 0.01 MPa preceded a large rockburst (M_L 1.0) (Swolfs and Brechtel, 1977).

VIBRATING–WIRE STRESSMETER

(a)

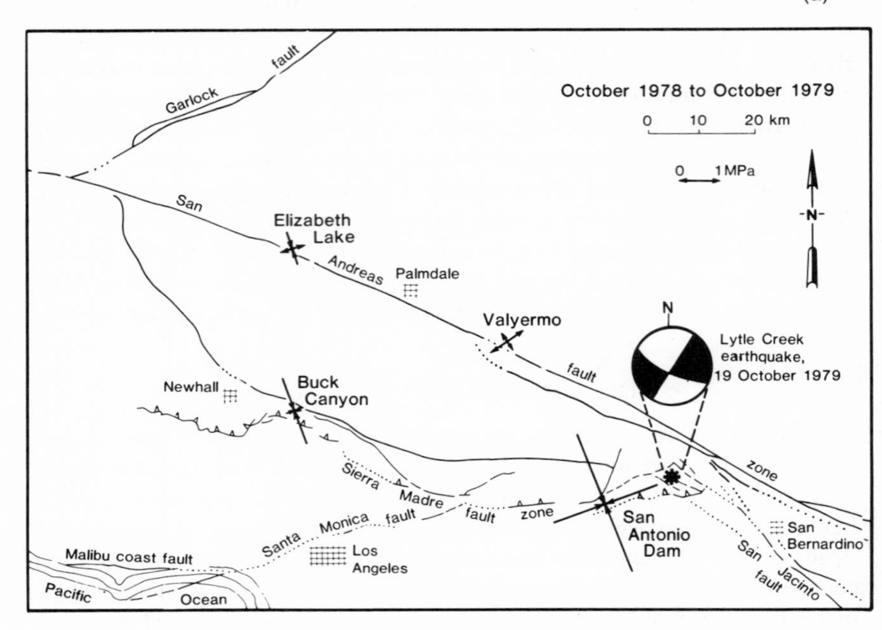

(b)

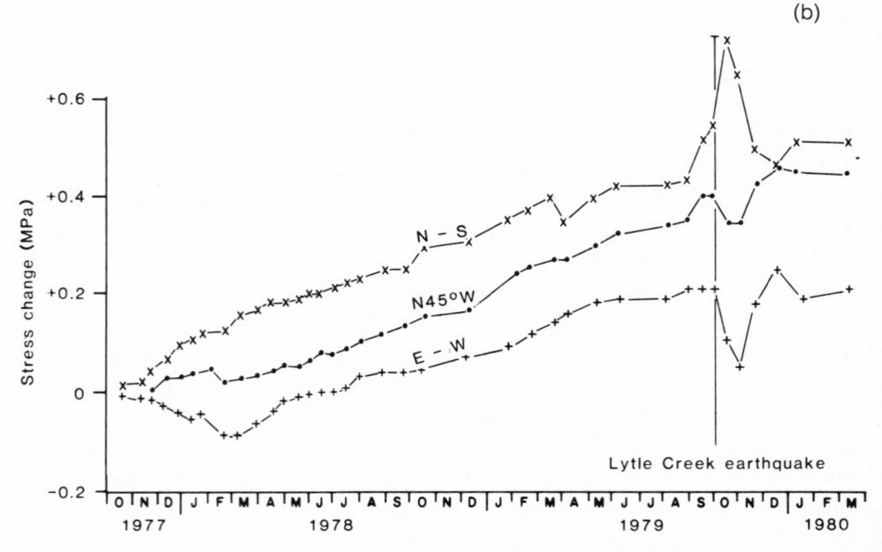

Fig. 8–2. The record of four vibrating-wire stressmeters installed in the vicinity of the San Andreas fault (adapted from Clark, 1982). (a) Orientation and magnitude of stress changes measured from 15 October 1978 to 15 October 1979 are shown as stress crosses. Also shown are the location and fault-plane solution for the Lytle Creek earthquake. (b) Preseismic and coseismic changes in stress associated with the Lytle Creek earthquake as recorded by a three-component vibrating-wire stressmeter installation at the San Antonio Dam site.

Inclusion Stressmeters

Inclusion stressmeters are designed to completely fill a pilothole in order to measure the three-dimensional state of stress in rock. However, a true rigid-inclusion stressmeter for three-dimensional measurement is very difficult to design. Although the two inclusion stressmeters described here are not rigid inclusions in the strict sense, stress is at least partially transferred to the inclusion. These are distinct from the soft-inclusion strain cells described in chapter 7.

U.S. Geological Survey Solid-Inclusion Probe

The U.S.G.S. (Nichols et al., 1968) in Denver, Colorado, has developed a solid pilothole probe. At the center of the device is a 2.5 cm alloy steel ball on which are mounted three electrical-resistance strain gauge rosettes (fig. 8–3). The steel ball and strain gauge leads are encased in an epoxy-resin cylinder, 3.8 cm in diameter by 4.1 cm long. The steel ball has a high enough modulus that this gauge is rigid relative to the rock. The steel ball is encapsulated in an epoxy with a modulus somewhat less than the surrounding rock. With a relatively soft epoxy, overcoring permits stress in rock around the probe to partially relax. The probe is cemented into a pilothole with a carborundum-filled epoxy grout which has about the same modulus as the probe epoxy. Because the rock partially relaxes during overcoring, the U.S.G.S. solid-inclusion probe does not measure stress directly and, as a consequence, the elastic properties of the rock are required to calculate in situ stress.

The U.S.G.S. probe has had limited use. It was tested during an experiment to measure stress changes in advance of excavating an underground mine in central Colorado (Lee et al., 1976). In general, the probes showed that stress

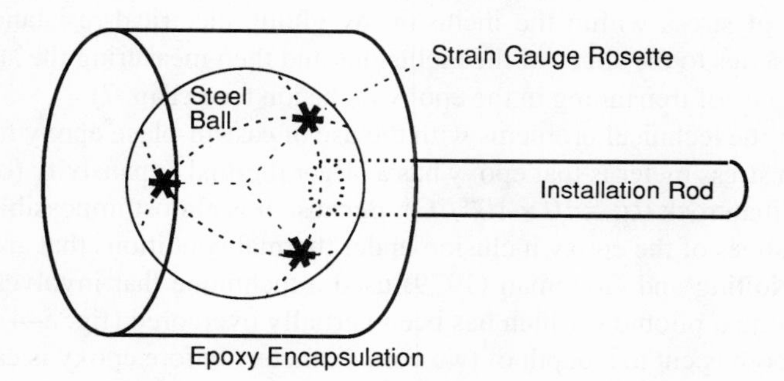

Fig. 8–3. The U.S.G.S. solid inclusion probe (adapted from Nichols et al., 1968). Three rectangular strain gauge rosettes are mounted at 90° to each other on a steel ball. The ball and strain gauge rosettes are encapsulated in a soft epoxy cylinder. For installation the probe is grouted into a borehole.

changes were larger than predicted using elastic equations and overburden loading, a result similar to that reported by Olsen (1957) in some early attempts to measure stress near an underground opening. In a direct comparison with other overcoring techniques during the Rangely, Colorado, experiment (chap. 7), the U.S.G.S. probe gave lower stress readings than indicated by four other techniques (de la Cruz and Raleigh, 1972). However, because the U.S.G.S. probe was installed at a depth of 3.4 m, deeper than the other techniques (Nichols and Farrow, 1973), the relatively high stress measured by the other techniques may have contained a large component of thermal stress (cf. fig. 7–11).

Photoelastic Inclusions

Epoxy resins serve as a convenient medium for inclusion stressmeters largely because the photoelastic properties of the resins permit assessment of stress changes without the need for electrical or other connections. Epoxy resin is inexpensive and easily cast-in-place in a small pilothole in the same manner as other strain cells and stressmeters. Such inclusions require a stable epoxy which does not creep with time. A rigid photoelastic inclusion was developed by Hawkes (1968) and Hawkes and Fellers (1969), whereas a somewhat softer photoelastic inclusion was developed by Nolting and Goodman (1979).

The technique for using photoelastic inclusions is as follows. Once cast in the pilothole, the epoxy inclusion is overcored, whereby expansion of the core places the epoxy cast in a state of tension. The core with epoxy-filled pilothole is then sliced with a diamond saw for examination under polarized light. Because clearly observable fringe patterns are necessary, a central hole is drilled in the epoxy inclusion to act as a stress riser and thereby accent the fringe pattern. An analysis of stresses within this type of soft photoelastic inclusion is given in Riley et al. (1977). Even with a stress riser, the fringe patterns are often difficult to see. For this reason Nolting and Goodman (1979) further analyzed the state of stress within the inclusion by gluing electrical-resistance strain gauge rosettes to the slices of the inclusions and then measuring the strain as a consequence of trepanning in the epoxy inclusion (see chap. 7).

One of the technical problems with the use of cast-in-place epoxy for a soft-inclusion stress meter is that epoxy has a larger thermal expansivity ($\alpha_T = 55 \times 10^{-6}/°C$) than rock ($\alpha_T \cong 10 \times 10^{-6}/°C$). Because it is almost impossible to analyze the slices of the epoxy inclusion under thermal conditions that are present in situ, Nolting and Goodman (1979) used a technique that involves casting epoxy along a pilothole which has been partially overcored (fig. 8–4). An initial overcore is cut to a depth of two core diameters before epoxy is cast in the entire length of the pilothole. The upper section of core with cast epoxy will contain only thermally induced strains and strains caused by shrinkage of the epoxy on curing. The difference between the upper prerelieved core and the

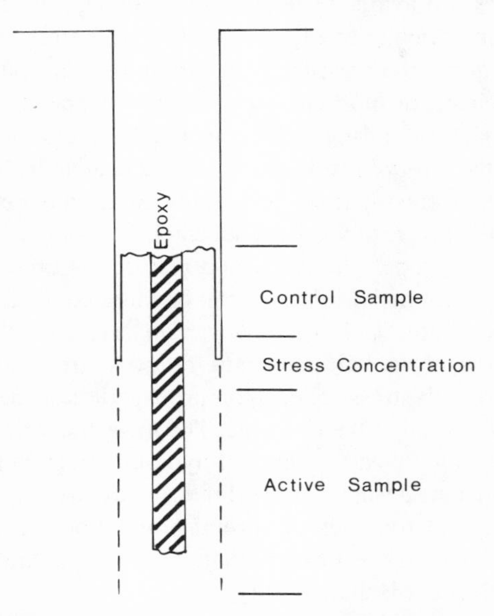

Fig. 8–4. The installation of a photoelastic stressmeter (adapted from Nolting and Goodman, 1979).

lower test core provides a good measure of in situ stress without the effect of thermal strains.

Crack Flexure Techniques

Earth stress normal to mesoscopic cracks is usually large enough to close the crack so that contact between crack walls supports the in situ stress. Fluid, introduced under high pressure, will cause the crack to open so that earth stress is supported by the fluid. This is one of the principles behind hydraulic fracture stress measurements (see chap. 5). Once the hydraulically induced cracks have extended several borehole diameters from the well, the crack-closure pressure, the instantaneous shut-in pressure for minifrac tests, is equal to the earth stress normal to the crack or cracks. The hydraulic fracture test is only one of several in situ stress techniques which involve data on pressure necessary for opening and closing of cracks in rocks. Because the hydraulic fracture usually propagates normal to σ_3, the only data available from closure pressure of hydraulic fractures is σ_3. However, man-made and natural cracks occur in orientations other than normal to the in situ σ_3, and for these, the crack closure pressure is also equal to the normal component of the earth's stress field acting on the crack. Other techniques are available for taking advantage of crack closure pressure for cracks in any orientation relative to the earth's stress field.

Techniques for measuring the pressure necessary to open cracks (crack closure pressure) and, hence, the normal stress acting on cracks, include the flatjack technique, the jack-fracturing technique, and the hydraulic test on preexisting fractures. For the *flatjack test* an artificial crack, actually a slot,[2] is cut into rock thereby relieving stress normal to the crack. Stress relief is manifested by strain associated with closure of the artificial crack. The flatjack is a bladder inserted into the crack to reopen the crack and restore the precrack state of stress. Cracks of a predetermined orientation can be initiated in borehole walls by using a borehole jack in the *jack-fracturing technique*. As is the case with the flatjack test, creation of an artificial crack relieves stress normal to the crack. Strain associated with stress relief upon propagation of these cracks is used to calculate stress normal to the borehole. The *hydraulic test on preexisting fractures* is equivalent to the latter stages of hydraulic fracture tests, particularly the fracture reopening tests. A three-dimensional stress state is determined by reopening natural fractures of several orientations. This hydraulic technique resembles the flatjack test except that cracks are natural rather than cut slots and a flatjack bladder is unnecessary.

FLATJACK STRESS MEASUREMENTS

A flatjack is a bladder usually consisting of two thin plates of steel welded face-to-face at the edges (fig. 8–5). Although most flatjacks are nearly square, other shapes are possible such as semicircles. A port is left in the upper edge of the flatjack where hydraulic fluid is injected under pressure. The principle be-

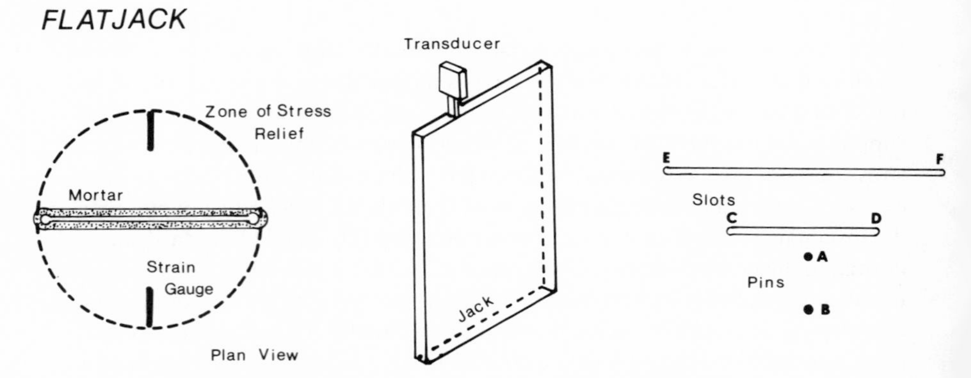

Fig. 8–5. Configuration for a flatjack test using a steel bladder. Shown are plan views of a single flatjack with strain gauges and two flatjacks, CD and EF, next to pins A and B.

hind the flatjack is simple; it reloads an artificially disturbed rock mass to restore the mass to its virgin (undisturbed) state. Disturbing the rock is most easily accomplished by cutting a slot in the rock outcrop and monitoring the strain as the walls of the slot close in on themselves under compressive normal stress. Such a slot is equivalent to a natural crack with a very large aperture. During cutting of the slot, strain is measured in order to serve as a guide for reopening the slot with a flatjack and thus restoring the rock to its virgin state.

Mayer et al. (1951) and Tincelin (1951) were first to appreciate that the flatjack pressure necessary to restore rock strain is proportional to the component of earth stress normal to the flatjack slot. Originally, strain from closure and reopening of the flatjack slot was measured by fixing pins to solid rock on either side of the planned slot and monitoring the distance between the pins as the slot was cut. Pins are fixed to the rock by cementation to the surface or in small holes on either side of the slot. The distance between the slot and the pins is not critical although the pins must be placed symmetrically with respect to the slot. The displacements between the pins is commonly measured with precision dial gauges or other devices sensitive to 1 part in 10,000. The U.S. Bureau of Mines modified the technique for monitoring the slot closure and restoration by embedding strain gauges within the rock (Panek, 1961; Panek and Stock, 1964). Placing strain gauges in the outcrop consists of cutting two small slots in a line normal to the planned position of the flatjack so that a pair of strain gauges straddle the flatjack slot. The strain gauges are then embedded with a grout.

Flatjack slots are cut by a line of drill holes or by a large diamond rock saw. Unsupported rock stress normal to the slot causes closure of the walls of the slot. This closure consists of a time-dependent as well as an instantaneous component with the time-dependent component asymptotically approaching an equilibrium value. Once the rock no longer closes in on the slot, a flatjack is inserted in the slot. In the event that the slot walls are irregular as is the case for a line of drill holes, the flatjack is supported by a grout or cement. As hydraulic fluid is injected under pressure, the flatjack walls expand against the rock. Fluid pressure is increased until the walls of the slot open to their original position as indicated when the pins or strain gauges are restored to their original readings prior to cutting a slot. The fluid pressure in the flatjack is then proportional to the original stress normal to the plane of the slot prior to cutting the slot. Like hydraulic fracturing, the flatjack method measures rock stress directly without prior laboratory calibration. The flatjack technique, then, nulls or cancels rock strain to restore an original stress state within the rock.

Because the flatjack and the slot rarely have the same dimensions there is not a one-to-one relationship between flatjack pressure, P_j, and stress normal to the slot, S_n. The cancellation pressure in the flatjack (P_j) depends on the dimensions of the slot and flatjack, the biaxial stresses, which are S_n plus the rock stress parallel to the flatjack, Q, and, to a small extent, ν. After a series of laboratory experiments, Hoskins (1966) concluded that P_j was independent of the rock's E. The formula for the displacements in the flatjack test are based on the theory of elasticity assuming an elliptical slot and plane stress (Alexander, 1960). These are:

$$W_0 = \frac{S_n c}{E}\left[(1-\nu)\sqrt{1+\frac{y^2}{c^2}} - \frac{y}{c} + \frac{1+\nu}{\sqrt{1+\frac{y^2}{c^2}}}\right] \tag{8–2}$$

$$W_1 = \frac{S_n y_0}{E}\left[(-2\nu)\sqrt{1+\frac{y^2}{c^2}} - \frac{y}{c} + \frac{1+\nu}{\sqrt{1+\frac{y^2}{c^2}}}\right] \tag{8–3}$$

$$W_2 = -W_1\frac{Q}{S_n} \tag{8–4}$$

$$W_j = \frac{S_n c_0}{E}\left[(1-\nu)\sqrt{1+\frac{y^2}{c_o^2}} - \frac{y}{c_o} + \frac{1+\nu}{\sqrt{1+\frac{y^2}{c_o^2}}}\right] \tag{8–5}$$

$$W_y = W_0 + W_1 + W_2 \tag{8–6}$$

and at cancellation pressure $W_y = W_j$ where $2W_y$ is the displacement across open slot, $2W_j$ is the displacement caused by raising jack pressure; $2W_0$ is the displacement during slot cutting due to an infinitely thin slot; $2W_1$ is the displacement due to finite slot width; $2W_2$ is the displacement due to biaxial stress; c is the half length of the slot; c_0 is the half length of the jack; y is the distance of measuring pins from the major axis of the slot; and y_0 is the half width of the slot.

Based on experiments using a 30 cm square flatjack in a 33 cm square slot, Hoskins (1966) substituted the appropriate dimensions into the above equations and assumed a $\nu = 0.2$ to arrive at the relationship between P_j and S_n

$$S_n = 0.853\ P_j + 0.056\ Q \tag{8–7}$$

where the cancellation pressure depends on the dimensions of the slot, c, and flatjack, c_0. In this equation P_j is commonly called the *cancellation pressure* because that is the pressure needed to move the measurement pins back to their original position. Under field conditions Q is determined by making orthogonal flatjack measurements. Note that equation 8–7 is not universal; it applies only to the experimental configuration used by Hoskins (1966). Dean and Beatty (1968) offer a slightly different version of equation 8–7.

A two-slot technique confirms that the flatjack method is applicable to a rock of interest (Leeman, 1964). Two pegs such as A and B in figure 8–5 are cemented into the rock. If a large horizontal slot , EF, is cut above A and B, the distance x between A and B increases by δx_1. A flatjack is inserted in the slot and the pressure increased until the distance between A and B is again x. A second slot, CD, is then cut below EF and the distance x between A and B again increases but by the amount, δx_2. A jack is inserted in CD and the pressure increased until the distance between A and B is x. The rock of interest is suitable for flatjack pressure tests if the pressure in the two jacks is identical.

Extensive testing of the flatjack was carried out by the U.S.B.M. in the Bethlehem Cornwall Corp. mine No. 4 at Cornwall, Pennsylvania (Panek and Stock, 1964). At issue were the design of the flatjack, the proper installation configuration for the flatjack plus strain gauge detectors, and the stability of the grout (mortar) and sensors. Flatjacks offer a simple device for monitoring long-term stress changes. The advantage of using a flatjack is that it is relatively rugged and can withstand blasting shocks and wet mine conditions. Seven flatjack arrays were installed in the wall of an exploratory drift at the 249 m level of the Cornwall mine. An attempt was made to measure S_v with Panek and Stock (1964) reporting values between 8.62 and 13.7 MPa. Because the theoretical S_v is about 6.7 MPa, it is clear that the flatjacks detected the stress concentration at the wall of the drift. Depending on the shape of the drift the stress concentration could be anywhere from 1.6 to 2.4 times the weight of the overburden. Even with flatjack slots on the order of 50 cm deep and strain gauge sensors embedded at depths of about 25 cm, the stress concentrations are still relatively high. Another example of the same phenomenon was reported at the Feather River hydroelectric powerplant near Oroville, California, where stress concentrations as high as four were measured within two feet of the wall of a drift (Merrill et al., 1964).

The most notable use of flatjacks in near-surface rocks was the survey of a regional stress field in limestone quarries in France (Froidevaux et al. 1980). Flatjacks were installed at twelve sites throughout France. Stress measurements were successful in eight quarries containing massive limestone, whereas sites with thin-bedded fractured limestone proved inappropriate for the flatjack technique. The orientation of S_H was in the NNW-SSE direction in eight quarries. As this is the orientation of the contemporary tectonic stress throughout northern and northwestern Europe (see chap. 12), this survey is another of many examples which demonstrate that a large component of the contemporary tectonic stress field is present in near-surface rock.

Cylindrical Jack Stress Measurements

Dean and Beatty (1968) propose a technique for measuring the biaxial stress field using a 15 cm diameter borehole rather than a slot. Pins are configured in a geometry similar to the trepanning technique with eight pins rather than six (chap. 7). Pin position is measured before the large-diameter hole is cored. A cylindrical jack is inserted in the borehole and pressurized to restore the initial pin configuration. The theory for stress calculation using this technique is given in Dean and Beatty (1968).

Jack Fracturing Technique

During the 1950s several borehole stressmeters where developed to use a change in fluid pressure within the gauge as the active element. One basic design consists of a flat steel-plate capsule mounted between two pairs of tapered wedges (fig. 8–6). This membrane-type stressmeter[3] is wedged into a pilothole so that fluid in the capsule becomes pressurized as the capsule is squeezed under the load (Creuels and Hermes, 1956). Any further change in stress around the pilothole is reflected as a change in fluid pressure inside the capsule and is measured by a pressure gauge at the surface.

More sophisticated versions of the fluid-filled stressmeter includes Potts's (1954) stressmeter and May's (1958) stressmeter. These are cylindrical devices consisting of two curved bearing plates in the form of hollow half cylinders where fluid is sealed into a slot without the need of a membrane. Between these half cylinders is a split steel bar with 0.25 mm deep grooves milled into the surface of the halves. The half cylinders are welded together so that the grooves act as a fluid-filled slit within the bar. Fluid under pressure within the slit forces the bar into the bearing plates which in turn loads the wall of the pilothole across one diameter. As stress in the rock surrounding the pilothole changes, the pressure of the fluid in the split bar changes. Fluid pressure is recorded by

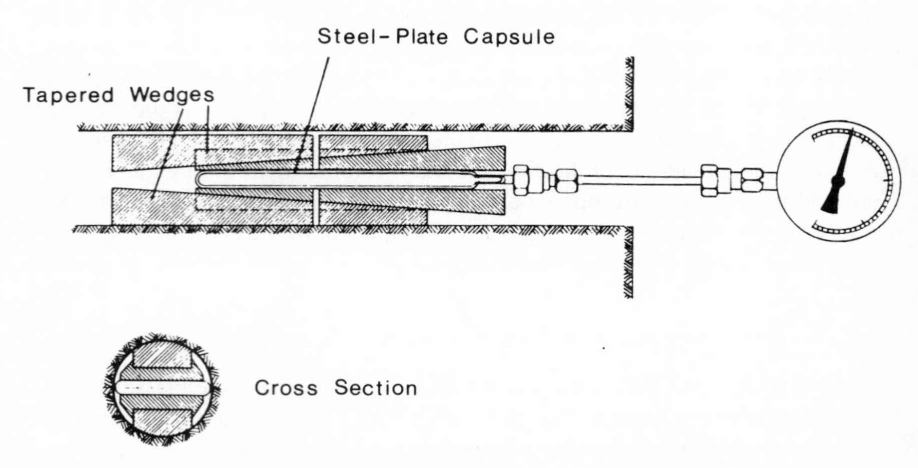

Fig. 8–6. Fluid-membrane stressmeter designed by Creuels and Hermes (1956) (adapted from Leeman, 1964).

means of a diaphragm transducer within the stressmeter, thus eliminating the necessity of running pressure tubing to a pressure gauge at the surface. These stressmeters are calibrated in the laboratory by loading a block with the stressmeter inserted. After calibration fluid pressure is read directly as a change in stress within the rock. Fluid-filled stressmeters are predecessors of a device later modified to act as a borehole jack during jack fracturing stress measurements.

For jack fracturing stress measurements, the borehole wall is cracked with a hydraulic jack consisting of two curved plates which fit the inside of the borehole in the same manner as the fluid-filled stressmeter (fig. 8–6). The idea behind this technique is that strain relaxation accompanies relief of stress next to a crack induced in a borehole wall (de la Cruz, 1977; 1978). Obviously, the magnitude of such strain relaxation is proportional to the stress normal to the induced crack. To measure strain relaxation the jack is outfitted with friction gauges which are pressed against the borehole walls at right angles to the bearing plates (fig. 8–7). Once the bearing plates are seated lightly against the borehole wall, the friction gauges are pressed against the wall by small hydraulic pistons and initial strain readings are taken. Load on the bearing plates is then increased until the wall cracks at right angles to the bearing plates. If the cracks are held open by the bearing plates, the rock in the immediate vicinity of the cracks will relax. The difference between postcrack and precrack strain readings is proportional to the stress relieved by cracking. By measuring strain on the surface of the borehole wall next to cracks driven from the borehole wall,

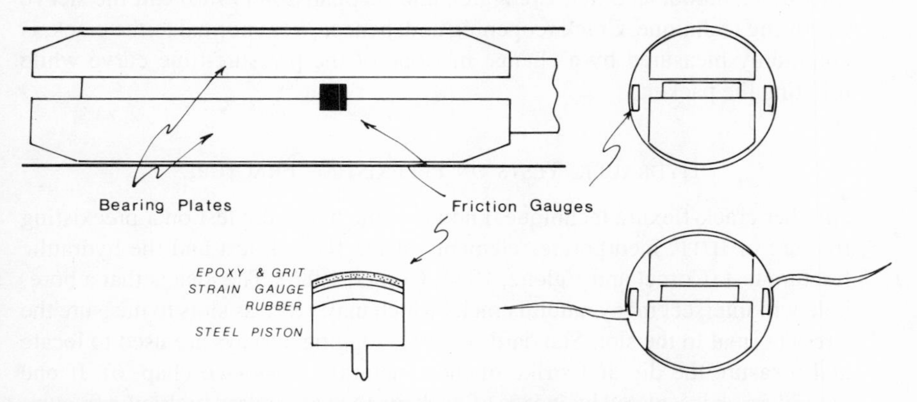

Fig. 8–7. Jack fracturing procedure of stress measurement (adapted from de la Cruz, 1977). A friction gauge consists of a strain gauge embedded between a grit-filled epoxy surface and rubber pad. The pad is attached to a steel piston which is driven against the borehole wall for the purpose of measuring strain associated with stress relief upon cracking of the borehole wall.

one can solve for the biaxial state of stress in a plane normal to that axis of the borehole, provided the experiment is repeated by driving cracks in three different directions.

An elastic plane strain analysis shows that several requirements are necessary for credible jack fracturing measurements (de la Cruz, 1977). First, the bearing plate must be at least six to eight borehole diameters long. By means of a finite element analysis, the tangential stress along the borehole wall between the crack and the center of the bearing plate 90° away has been established. de la Cruz (1977) shows that if the crack is driven into the wall as much as one borehole radius away from the borehole face, then > 95% of the tangential stress is relieved up to 8° away from the crack around the face of the borehole wall. The friction gauges must be seated against this 8° portion of the borehole wall.

The jack fracturing technique was tested on a 10-by-10 meter outcrop at the Mazzone Ranch in San Jose, California (de la Cruz, 1977). For reference, surface-bonded electrical-resistance strain gauges were used where overcoring showed a near-surface S_H oriented at N43°E with S_H = 3.6 MPa and S_h = 1.1 MPa. Jack fracturing data showed S_H oriented at N38°E with S_H = 1.3 MPa and S_h = 0.9 MPa. This is another example suggesting that thermal stresses affected surface measurements to cause a higher stress than found below the surface by another technique. Both the surface and the near-surface measurement record an orientation of S_H in agreement with the orientation of S_H recorded by deep stress measurements throughout central California (Zoback et al., 1987).

There are several variations on this technique. One experiment is to inflate a packer to crack the rock in a technique that Plumb (1983) calls the packer-fracture technique and that Ljunggren and Stephansson (1986) call the sleeve fracturing technique. Crack reopening and, hence, stress normal to the crack, is commonly measured by a change in slope of the pressure-time curve while inflating the packer.

Hydraulic Tests on Preexisting Fractures

Another crack-flexure technique, known as the hydraulic test on a preexisting fracture or HTPF, incorporates elements of the flatjack test and the hydraulic fracture test (Cornet and Valette, 1984; Cornet, 1986). The idea is that a borehole will intersect many natural cracks which may serve as slots to measure the stress normal to the slot. Standard borehole logging surveys are used to locate and measure the dip and strike of these natural cracks (see chap. 6). If one natural crack is isolated by means of packers, then a standard hydraulic fracture ISIP is used to determine the stress normal to the natural fracture. By analogy with the flatjack test, the natural fracture is a slot which is flexed open by means of a pressurized fluid injected into the slot. The difference is that flat steel-plate capsules are not used to contain the fluid for the HTPF. As a consequence, fluid slowly leaks from the natural crack. In addition, the strain of the rock accompanying formation of the natural crack (i.e., the slot) is negligible because material is not removed as is the case in cutting a flatjack slot. For the HTPF stress measurements, crack pressurization and bleed-off are necessary for determining stress normal to the crack by one of two techniques.

Techniques for measuring the stress normal to the natural crack are the ISIP test and the constant-pressure step test. The former is described in detail in chapter 5. In the latter, the fluid pressure in the crack is raised in a series of steps (Cornet, 1986). With each pressure step a slightly higher flow rate is necessary to maintain a constant pressure within the crack. When the crack opens at a pressure equal to the stress normal to the crack, a much higher flow rate is necessary to maintain constant pressure. This increase in flow rate is easily detected.

The HTPF is designed to determine the complete stress tensor, $\sigma(x_i)$, at all points, x_i, knowing the normal stress supported by the ith fracture, $\sigma_n^{(i)}$, and the normal to the ith fracture, $n^{(i)}$. Following Cornet (1986),

$$\sigma(x_i)\, n^{(i)} \cdot n^{(i)} = \sigma_n^{(i)} \qquad (8\text{–}8)$$

The complete stress tensor involves six functions of the three components of x_i at the points of measurement. If a horizontal hole is drilled from an underground mine, then the overburden stress does not change with each measurement, contrary to the case for a vertical hole. For the horizontal hole the stress, $\sigma(x_i)$, is assumed to be the same for all points x_i in which case equation 8–8 is a system of N linear equations with six unknowns where N equals the number of measurements. This system can be inverted with a least squares procedure to solve for the complete stress tensor. Analysis of tests in a vertical hole is slightly more complicated because the effect of gravity must be incorporated into the system of simultaneous equations.

For success, the HTPF technique requires a relatively large number of tests on cracks with a variety of orientations (Cornet, 1986). HTPF will not work if just one crack set persists within a borehole. Proper measurement of ISIP requires that the cracks are not interconnected otherwise the orientation of the normal to the crack responding to the fluid pressure drop becomes uncertain. Finally, except for a component from gravity the stress field cannot vary over the length of the borehole in which natural cracks are tested. Stress measurements in south central France using the HTPF technique compare well (± 2% and 20°) with stress measured using the conventional hydraulic fracture technique (Burlet et al., 1989). Data from the HTPF technique suggest that S_v was less than expected based on the rock mass density. Burlet et al. (1989) attribute this measurement to the variation of S_v as a function of topography although they do not say how topography acts to lift overburden.

Concluding Remarks

Techniques described in this chapter allow the direct readout of stress and, thus, offer a distinct advantage over those techniques which require the elastic properties of the host rock for calculation of in situ stress. The elastic properties of the host are often measured in the laboratory on rock samples which relaxed from their in situ conditions. Elastic properties of rock change on relaxation so that there will always be some uncertainty about the correlation between the laboratory and in situ elastic modulus.

9

Microcrack-Related Phenomena

> The drilled out cores were measured several times within 48 hours because deformation only ceased then. . . . It can be seen that 60 to 80 percent of the total deformation has taken place after only half an hour.
>
> —Habil. F. Mohr (1956) after monitoring the strain relaxation of cores taken from pillars in a German potash mine

When discussing granite-quarrying techniques, quarry supervisors Harry Mason (1979, personal communication) and Ivan Thunburg (1980, personal communication) use descriptive terms that indicate an implicit respect for the influence of earth stress on their operation. Both men agree that all granite, except wild rock, has a *nature*, which is the direction of preferred crack propagation in a vertical plane.[1] Based on knowledge of crack propagation there is little doubt that the nature of a granite is the direction of S_H within a granite quarry. Other terms also indicating respect for in situ stress include wild rock and pressure. *Wild rock* is, as the name might imply, a granite with no predictable orientation for crack propagation. *Pressure* is the quality of some granite to relax a great deal and, in some cases, explosively when excavated.

A microcrack fabric, present in most granites, also comes into play during quarrying operations. Quarrymen distinguish three mutually perpendicular directions in granite based on the ease of splitting a free block in these directions (Sedgwick, 1835). *Rift* is the direction of easiest splitting and, if it is vertical, is often (but not always) parallel with the in situ nature of the granite. The orientation of the rift in a free block is most important to a quarryman because that direction is used as a guide for shaping blocks of granite in the cutting mill. *Grain* is another direction of relatively easy splitting although somewhat harder to split than the rift. If the rift of a granite is horizontal, the grain usually correlates with the nature of the granite. In a free block of granite either the rift or grain constitutes the block's nature. *Hardway* is at best a nuisance to split; flame cutters or wire saws are often used to free granite blocks along this plane. In many cases, the rift and grain mark the orientation of well-developed microcrack sets. While the in situ nature of the granite might reflect the orientation of the contemporary in situ stress, Plumb et al. (1984a) hypothesize that, just like joints and dikes, rift and grain record the orientation of paleostress fields at the time of microcrack propagation. New England granites are consistent with this hypothesis as their rift and grain correlate in orientation over regions encompassing many plutons (Dale, 1923; Wise, 1964).

Removal of large blocks of granite is simplified, if quarrymen attempt to shape their blocks so that a pair of sides is parallel to the direction of the in situ nature. Thus, a new quarry, even in a familiar granite pluton, is laid out according to a simple test to determine the orientation of its nature. This test, resembling hydraulic fracturing, employs a *wedge* and *feather*. The feather is a narrow steel plate folded in the middle and then inserted in an isolated vertical drill hole, folded end first, so that the two ends of the folded plate make a slot at the open end of the drill hole (fig. 9–1a). A wedge-shaped piece of steel is then driven into the slot to force the drill hole open, thus causing a crack to propagate from the drill hole. No matter what direction the crack leaves the drill hole, it always turns parallel to the nature of the granite (fig. 9–1b). Quarrymen split the granite in the direction of the nature by drilling a line of holes parallel to the nature and spaced every 15–25 cm. The holes are linked by cracks driven from a wedge and feather inserted in each hole. This starts the process of preparing large blocks of granite for the cutting mills. Sides of the block parallel to the

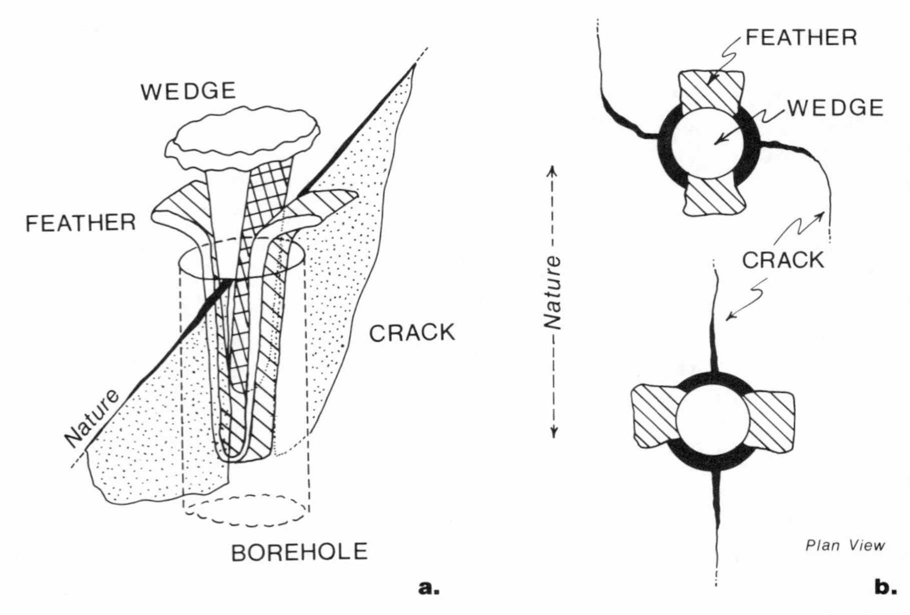

Fig. 9–1. (a) Drawing of a wedge and feather used to split granite. The crack runs from the borehole in the direction of the "nature" of the granite. (b) Plan view of a wedge and feather inserted into each of two boreholes. The configuration of the wedge and feather is such that a crack is initiated and driven from the borehole wall opposite the contact points of the feather. Once leaving the borehole wall, cracks will then turn to propagate parallel to the direction of the nature in the granite.

grain are cut in the same manner. Quarrying is impeded if undesirable structures such as *bedding*, known to the geologist as sheet fracturing parallel to the local topography, or *seams*, vertical fractures, are encountered. Both bedding and seams lead to the time-consuming process of removing smaller pieces of granite which are difficult to handle in the cutting mills and, hence, are cast aside as waste. When encountered, wild rock also slows the quarrying process operation.

The pressure of the granite relaxes on excavation sometimes causing blocks to expand and jam in place. This phenomenon is analogous to the expansion of a core on stress relief. If such expansion is coupled to microcrack propagation and opening, then it is reasonable to suppose that microcrack opening gives information on both orientation and magnitude of in situ stresses (e.g., Teufel, 1982; Engelder and Plumb, 1984). A corollary is that there is a direct relation between crack closure during reloading in the laboratory and the magnitude of in situ stress to which the rock was once subjected (e.g., Strickland and Ren, 1980; Carlson and Wang, 1986; Meglis et al., 1991). This chapter focuses on phenomena related to microcrack propagation and opening on stress relief and, then, microcrack closure on reloading.

Microcracks

Three classes of microcracks include: grain boundary cracks (located at the boundary between grains); intergranular cracks (cracks cutting more than one grain); and intragranular cracks (cracks contained within one grain) (Simmons and Richter, 1976). Microcracks propagate no further than several grain diameters (< 1 mm) and are usually less than one grain diameter in length (Tapponier and Brace, 1976; Kranz, 1983). Crack aperture is usually on the order of 1—10 μm. On average microcracks propagate as mode I cracks and, in this regard, are oriented normal to the local σ_3 just like the joints, veins, and dikes described in chapter 2. In detail the propagation path of microcracks is modified by local cleavage planes and grain boundary orientations (Martin and Durham, 1975). In the presence of pore fluids microcracks heal rapidly leaving a track of fluid inclusions to mark the plane of the previous microcrack (Smith and Evans, 1984). The rift planes of New England granites are often fluid-inclusion planes indicative of healed rather than open microcracks (Dale, 1923).

Microcrack sets propagate under one of several situations which lead to tensile stresses on the microscopic scale including: thermoelastic contraction; stress concentration at mesoscopic crack tips; stress concentration at grain-to-grain contacts during incipient failure in compression; and rapid stress relief. While a discussion of microcracking upon rapid stress relief occupies much of this chapter, I start by briefly examining the other three situations. First, thermal stresses arise from differential and incompatible thermal expansion or contraction between grains of different thermoelastic properties. Thermoelastic contraction from cooling of intrusive rocks produces large tensile stresses; columnar jointing is a manifestation of such stresses. If an anisotropic stress field is superimposed on the intrusive body during cooling, then ensuing cracks assume a preferred orientation and, if these cracks are pervasive on a microscopic scale, a rift plane develops. Thermoelastic stresses also play a major role in the development of residual stresses as discussed in chapter 10. Second, often the tip of a mesoscopic crack is surrounded by a process zone consisting of a swarm of microcracks. This local microcracking is a manifestation of a stress concentration causing large tensile stresses to develop in the vicinity of the mesoscopic crack tip. Such inelastic behavior at the mesoscopic crack tip leads to estimates of larger fracture-surface energies during crack propagation than is expected for the propagation of a single crack tip (e.g., Friedman et al., 1972).

Third, when a rock fails in compression under brittle conditions, total loss of strength is preceded by grain-scale microcracking (Scholz, 1968; Brace, 1971). During compression of an aggregate, microcracking is induced by large tensile stresses which develop despite confining pressures as high as 100 MPa. These tensile stresses arise from mechanical processes including: microcracking from stress concentrations at grain boundaries (e.g., Gallagher et al., 1974); microcracking from stress concentrations around cavities (e.g., Kranz, 1979); elastic mismatch-induced cracking (e.g., Wang and Heard, 1985); kink band and deformation lamellae associated microcracking (e.g., Carter and Kirby, 1978); and twin-induced microcracking (e.g., Olsson and Peng, 1976). The grain-grain loading geometry which induces large stress concentrations resembles the loading configuration of a Brazilian test in which a disc is loaded across its diameter by a pair of flats that make a line contact with the disc. In an aggregate intragranular microcracks will propagate between grain-grain contacts (fig. 9–2). The crack in the Brazilian test as well as most microcracks propagate normal to σ_3 and in the direction of σ_1 (Hallbauer et al., 1973; Tapponnier and Brace, 1976). Microcrack propagation leads to shear failure because the largest tensile stresses near crack tips are offset from the tip (Reches and Lockner, 1990). This offset favors the growth of en echelon microcracks in the form of an incipient fault zone (Martel et al., 1988).

The Kaiser Effect

Compression experiments have shown that extensive quantities of microcracks develop when a sample is subject to approximately 50% of its failure stress (Scholz, 1968). The propagation of a microcrack is accompanied by an acoustic emission (AE) which is a minute ultrasonic pulse. During research on AE, Kaiser (1953) recognized that, when samples are relaxed and then subject to a σ_d higher than present before relaxation, there is a significant increase in the rate of AE as the stress exceeds the previous maximum (Hardy, 1981). This

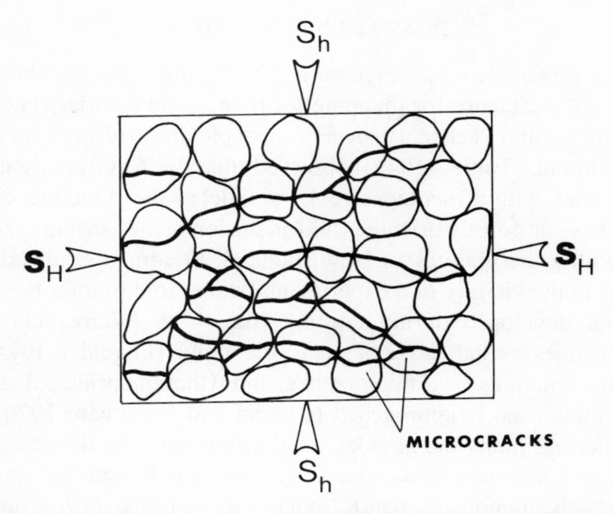

Fig. 9–2. Microcrack propagation in an aggregate of quartz grains. Intragranular microcracks propagate between grain-grain contacts and generally normal to σ_3.

behavior, called the Kaiser effect, occurs in various rocks including mudstone (Hayashi et al., 1979), dolomite (Kurita and Fujii, 1979), and granite (Hughson and Crawford, 1986). Usually the Kaiser effect is observed under controlled laboratory conditions where samples are subject to some σ_d less than the failure stress. An increased rate of AE is observed once the sample is loaded to higher σ_d, but when the sample is held at this higher stress level, the AE output decays exponentially (Hughson and Crawford, 1986).

A major issue in this chapter is the correlation between the development of microcracks and the magnitude of in situ stress. A difficulty arises in applying the Kaiser effect, as well as other microcrack techniques, as a tool for assessing in situ stress conditions because in situ stress consists of two interdependent components: mean stress, σ_m, and deviatoric stresses, $\delta\sigma_i$. The relationship between microcrack development and either σ_m or $\delta\sigma_i$ is obscure. In cutting a core, the relief of σ_3 will have a larger absolute effect on the propagation and opening of microcracks than relief of a σ_d.[2] This is so because the difference between σ_3 and ambient stress is commonly larger than σ_d. The relative effect of confining pressure versus maximum stress in setting the onset of rapid AE has been difficult to assess (Zhang, 1982). In applying the Kaiser effect to samples taken from the field, the imprecision of the technique leaves uncertainty concerning whether onset of rapid AE marks the maximum stress or just the level of confining pressure. Regardless, the Kaiser effect is a clear demonstration that microcracks propagation is stress sensitive. The onset of further microcracking is related to σ_m if not σ_d to which the rock was subjected at the moment of stress relief.

Stress-Field Orientation

Because they propagate on average normal to σ_3, microcracks, like joints and dikes, serve as a database for mapping local stress trajectories (Pecher et al., 1985; Lespinasse and Pecher, 1986). For example, in early work on the Appalachian Piedmont, Tuttle (1949) observed that uniformly oriented fluid-inclusion planes within quartzose rocks paralleled the direction of tectonic transport. These uniformly oriented fluid-inclusion planes are equivalent to the rift plane within the granites of New England. Friedman et al. (1976) used microcracks in the vicinity of a experimental drape fold to infer the stress trajectories that developed during faulting (fig. 9–3). Microcracking in and around fault zones are indicative of stress orientation (Engelder, 1974). Microscopic feather fractures (i.e., microcracks) show that the principal stresses rotate as the fault zone is approached (Conrad and Friedman, 1976). In both thrust and normal faults the angular relationship between the shear plane of fault zones and microcracks is commonly between 5° and 10° (Anders and Wiltschko, 1988, unpublished data). Anders and Wiltschko (1988, unpublished data) suggest that this angular relationship may mean that the effective confining pressure was very low during faulting, or the angle of internal friction is much larger than usual, and/or the local stress field rotated during faulting. However, information on earth stress available from microcracks goes well beyond the petrographically derived inferences about stress orientation.

MICROCRACK STRESS TRAJECTORIES

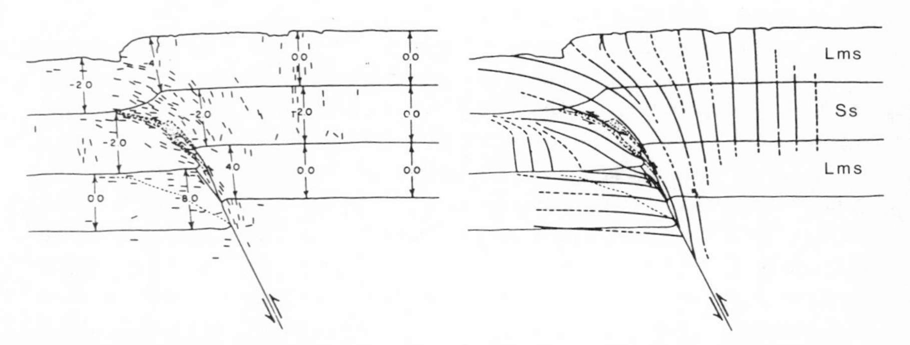

Fig. 9–3. Examples of results from laboratory studies of a faulted drape fold with a displacement of 0.18 cm along a basement fault (adapted from Friedman et al., 1976). Left diagram shows deformation features observed in thin section with the stippled area a fault-containing gouge. Microcracks are shown schematically in sandstone (Ss) and limestone (Lms) as if all are the same length. Right diagram shows stress trajectories inferred from all microcrack data. Arrows oriented perpendicular to bedding indicate percent strain calculated from bedding-thickness change with thinning counted as positive.

Time-Dependent Strain Relaxation (T-DSR)

Strain relaxation is triggered by the deformational processes activated within a rock when it is cut free of its in situ surroundings and, thereby, relieved of its boundary stresses (see chap. 7). Details concerning the mechanisms of strain relaxation are of limited interest to those making stress measurements with the assumption that elastic behavior provides a reliable indication of stress magnitude as is the case for most overcoring measurements. However, strain relaxation of rock is a complex process that goes well beyond a simple model which suggests that a rock core behaves like a single spring returning to its rest position upon the release of a restraint. The complexity of rock relaxation is reflected in a transient creep that decays for hours after a rock core is cut from its in situ environment (fig. 7–7).

Transient creep commonly follows elastic strain during loading when polycrystalline materials, including rocks, are subject to stresses that are maintained at less than 10^{-5} of their shear modulus and temperatures less than half their melting point (Weertmen and Weertmen, 1970; Kranz and Scholz, 1977). Carter and Kirby (1978) suggest that such creep is a low temperature inelasticity associated with interaction of the applied stresses and preexisting microcracks within rocks. Laboratory experiments suggest that rocks exhibit an analogous behavior when stresses are removed because specimen expansion involves both instantaneous (elastic) and time-dependent (transient creep recovery) components (Robertson, 1964). A field example of such time-dependent creep recovery upon unloading is the continuous expansion of cores for several hours after removal from deep wells (e.g., Mohr, 1956; Swolfs, 1975; Teufel, 1983). Excavation can also trigger a longer-term creep recovery as was found at the Niagara hydroelectric power project, Ontario, Canada, where walls of the excavation are known to have expanded inward over several years (Lee and Lo, 1976).

In the literature, two terms are used to indicate the presence of transient creep recovery after the excavation or coring of rock: anelastic strain recovery (ASR), and time-dependent strain relaxation (T-DSR) (e.g., Engelder, 1984; Teufel and Warpinski, 1984, Warpinski and Teufel, 1989a). Anelasticity is a time-dependent elasticity in which displacements do not take up their elastic values instantaneously after stress is applied but tend to approach them approximately exponentially in time (Jaeger and Cook, 1969). Although the distinction is subtle, this chapter chooses to use the term, T-DSR, rather than ASR because microcrack propagation, a distinctly inelastic process, is an important mechanism during the transient creep recovery of cores. A detailed discussion of the mechanisms for T-DSR comes near the end of the next section.

A useful model for visualizing the interaction between applied stresses and anelastic relaxation is that of Varnes and Lee (1972) who equate ASR with the mobilization of residual stress in rocks (see chap. 10). In the "mobilization" model, a series of interconnected springs represent closed microcracks and the elastic strain of grains plus cement in rocks. Strain relaxation becomes time-dependent because grains at a freshly cut surface relax instantaneously, whereas the interlocking of grains within the interior of the aggregate prevent instantaneous recovery. However, elastic strain in the interior of the body is upset by surface relaxation and, therefore, must eventually reequilibrate with the surface relaxation. A strain front then moves through the body giving the effect of a time-dependent behavior. Although the exact process by which the strained aggregate relaxes with time after unloading is not clear, it is clear that core expansion after drilling shows a strain-time history similar to that predicted by the "mobilization" model of Varnes and Lee (1972) and observed during transient creep in laboratory experiments such as those of Kranz and Scholz (1977).

Direct Measurement of T-DSR

The measurement of T-DSR was initially suggested by Voight (1968) for the purpose of predicting in situ stress on cores removed from deep wells. Experiments showing the creep-recovery of rock upon removal of stress were cited by Voight as the empirical justification supporting the assumption that T-DSR is proportional to the instantaneous strain relaxation of rheologically isotropic rocks and, hence, proportional to in situ stress (fig. 9–4). Robertson (1964) presented data from creep experiments by Evans (1936) where the rock expanded

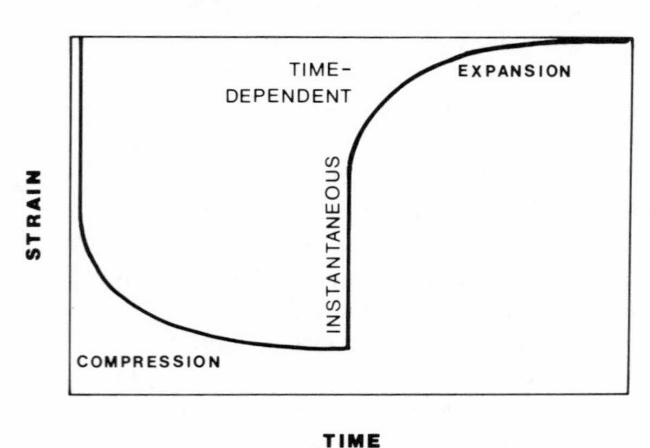

Fig. 9–4. Laboratory data on the creep recovery of rocks (adapted from Robertson, 1964). The diagram is oriented to represent loading during burial (compressional strain), and unloading during coring (extensional strain).

in a time-dependent manner following removal of the load. Such expansion is presumably the partial recovery of compressive strain accumulated during a creep experiment. Earlier Michelson (1920) observed the same type of behavior for rock upon application and release of a load. The correlation between T-DSR and other data on deep earth stress support the validity of Voight's proposal (Teufel and Warpinski, 1984; Warpinski and Teufel, 1989b).

Techniques for Measuring T-DSR

An extensive effort to measure T-DSR on deep core was by Teufel (1983, 1985) using the following techniques. For the proper interpretation of T-DSR of core from deep wells, environmentally produced strain must be minimized. Until the core barrel is brought to the surface little can be done about controlling the environment. Immediately upon recovery at the drill site, core is sealed with a polyurethane coating, which is brushed on and cured within fifteen minutes, and then with a polyurethane wrapping. Displacements associated with the time-dependent strain relaxation are measured using spring-loaded, clip-on gauges, which incorporate precision gauge heads (fig. 9–5). Three gauges, designed by Holcomb and McNamee (1984), are clipped on the core at 30° to each other in order to determine principal horizontal stresses. Another gauge is

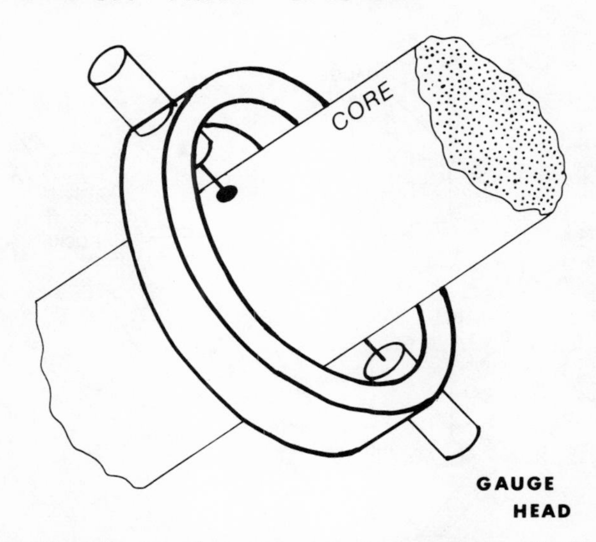

Fig. 9–5. The configuration for a ring-gauge around core as designed by Holcomb and McNamee (1984). Because the ring measures expansion across one diameter, three ring gauges must be used to recover the principal strains in a plane normal to the axis of the core.

mounted parallel to the central axis of the core. Strain resolution is ± 1 με (Teufel, 1989). Displacement of each gauge is recorded in an insulated room where temperature variation is limited to ± 1°C.

A similar technique for measuring T-DSR involves an environmental chamber designed by Engelder (1986). The environmental chamber is a thick-walled cylinder (i.e., 4″ i.d. and 20″ in length) in which the core may be sealed to maintain in situ temperature and moisture conditions (fig. 9–6). The core is suspended inside the chamber using a series of springs. Twenty DC-LVDT type D2 transducers made by RDP Electronics Ltd. are placed in the walls of the chamber. A transducer is placed in each end of the cylinder to measure axial expansion of the cores and the remaining eighteen transducers are placed with six on each of three tiers of the cylinder so that each pair of transducers face each other across a diameter of the core. On any tier core expansion is measured across three diameters 120° apart. The cylinder is configured so that each tier of transducers is rotated by 20° from adjacent tiers.

The transducers have a sensitivity of 172 mv/mm for a 6V DC input. This sensitivity translates into an output of 13 μV/με across a 76 mm diameter core. A Hewlett-Packard 3421A voltmeter/multiplexer with 5 $^{1}/_{2}$ digit resolution was used for data logging so that $^{1}/_{3}$ με is resolved without amplifying the signal.

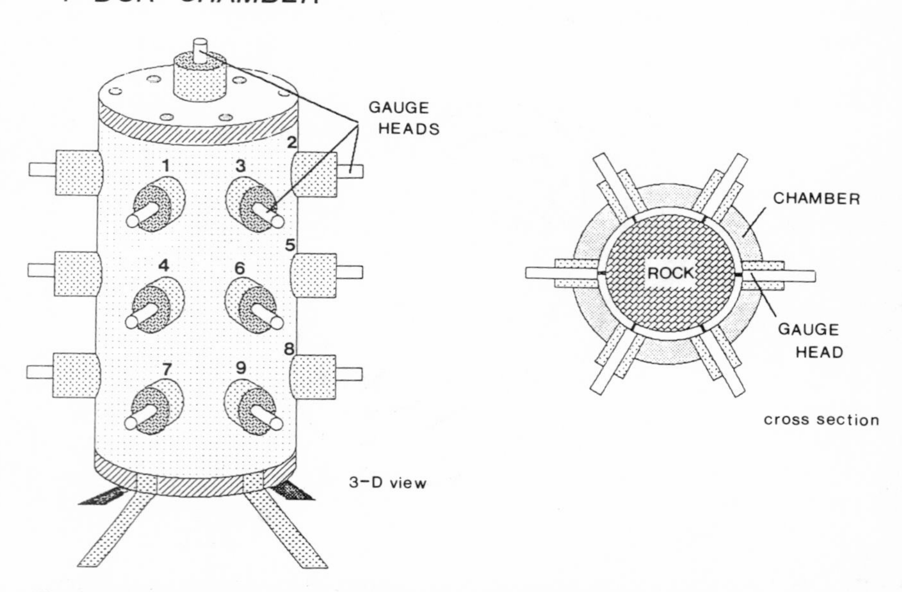

Fig. 9–6. Schematic of chamber for monitoring time-dependent strain relaxation (adapted from Engelder, 1986). The chamber has a height of 50 cm and an internal diameter of 12 cm.

Thermally compensated data are shown in figure 9–7 for a core from a depth of 511 m in crystalline rock at Kent Cliffs, New York, near the Ramapo fault system. The strain relaxation is about 60 $\mu\varepsilon$ for twenty-five hours after the core was inserted into the chamber. About one hour passed between the time the core was cut and the start of the data logging. Relaxation of the core shows a maximum expansion counterclockwise from the E-W foliation. S_H inferred from the core is about ENE which correlates with the local stress field (Zoback, 1986) and the contemporary tectonic stress field in the northeastern United States (Zoback and Zoback, 1989).

T-DSR IN SEDIMENTARY ROCKS

Blanton and Teufel (1983), Teufel (1983; 1985), and Teufel and Warpinski (1984) report on T-DSR measurements using cores recovered from deep wells in sedimentary rocks. Strain relaxation curves show that T-DSR generally takes place within forty hours after the core was cut. An example of this behavior is seen for a core cut from the Rollins Sandstone in a well near Rifle, Colorado (fig. 9–8). Because core recovery is time consuming, recording of T-DSR started seven hours after coring and so a major portion of the T-DSR was not recorded. Regardless, maximum strain relaxation took place normal to bedding where as much as 204 $\mu\varepsilon$ was recorded and in the horizontal plane the strain relaxation was between 146 $\mu\varepsilon$ and 72 $\mu\varepsilon$ depending on orientation. In four cores taken within a half meter of each other the orientation of the principal strains parallel

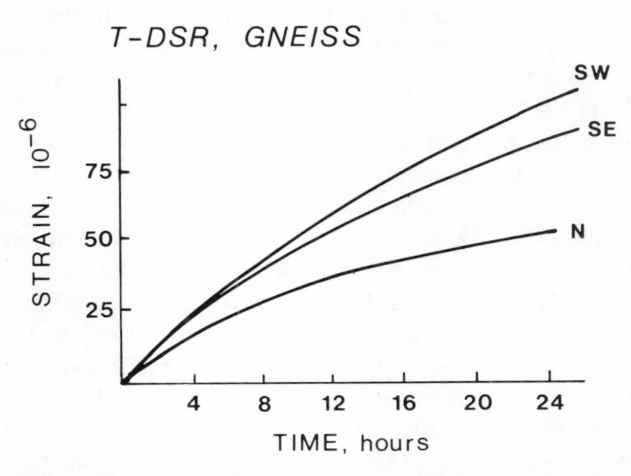

Fig. 9–7. Time-dependent strain relaxation from cores cut near the Ramapo fault, New York (data from Engelder, 1986). Plots of strain versus time for core from a depth of 1676 feet. Strain was recorded until the next core was recovered about twenty-five hours later. Data is compensated for a rock-chamber system having a thermal expansivity of 35 $\mu\varepsilon$/°F.

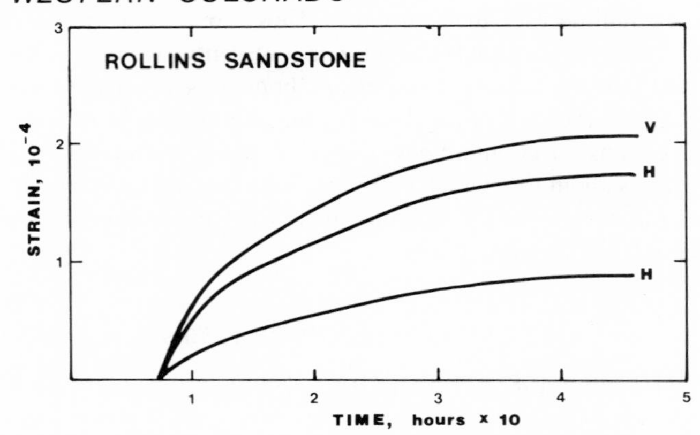

Fig. 9–8. The time-dependent relaxation of core from the Rollins Sandstone taken from an MWX well near Rifle, Colorado (data from Teufel and Warpinski, 1984). The core was 6.35 cm in diameter and came from a depth of 2304 m.

to bedding varied by ± 8°. When added to the accuracy of core orientation (± 10°), the prediction of orientation of S_H is good to ±18° for one measurement.

T-DSR is approximately exponential as shown in figure 9–8 and, thus, is compatible with a linear viscoelastic behavior (Flugge, 1967). Principal horizontal stress magnitudes are calculated from the time-dependent strains assuming that the T-DSR of sedimentary rocks is a viscoelastic response to removal of in situ stresses (Blanton, 1983; Warpinski and Teufel, 1989b). Two viscoelastic models are used to explain the exponential behavior: Blanton's (1983) direct model and Warpinski's and Teufel's (1989b) strain-history model. The direct model, based on a viscoelastic constitutive relationship devised by Shapery (1972), is easiest to apply. At the heart of the direct model is the assumption that we can calculate the strain response to a continuously varying stress history from a hereditary integral (cf., Shapery, 1972) by knowing only a standard response to one preselected stress history. Any recorded part of the T-DSR is designated as $\Delta\varepsilon$. For the calculation of principal stresses, necessary parameters are the relaxation magnitudes ($\Delta\varepsilon_H$, $\Delta\varepsilon_h$, $\Delta\varepsilon_v$) of the principal strains (ε_H, ε_h, ε_v), Poisson's ratio (ν), the poroelastic constant (α_b), pore pressure (P_p), and the overburden stress (S_v). Basic assumptions of the theory are that (1) the rock is homogeneous and linearly viscoelastic, (2) behavior is isotropic; (3) ν and α_b are not time-dependent; and (4) the in situ stresses are removed instantaneously. The direct model[3] states that if one of the principal stresses (usually the vertical stress) is known, we may calculate the other two principal stresses from the components of time-dependent strain relaxation according to the following equations:

$$S_H = (S_v - \alpha_b P_p) \frac{(1-\nu)\,\Delta\varepsilon_H + \nu(\Delta\varepsilon_h + \Delta\varepsilon_v)}{(1-\nu)\,\Delta\varepsilon_v + \nu(\Delta\varepsilon_H + \Delta\varepsilon_h)} + \alpha_b P_p \qquad (9\text{–}1)$$

$$S_h = (S_v - \alpha_b P_p) \frac{(1-\nu)\,\Delta\varepsilon_h + \nu(\Delta\varepsilon_H + \Delta\varepsilon_v)}{(1-\nu)\,\Delta\varepsilon_v + \nu(\Delta\varepsilon_H + \Delta\varepsilon_h)} + \alpha_b P_p \qquad (9\text{–}2)$$

If there is no information on the vertical stress, then the creep compliance or relaxation modulus of the rock must be measured in order to calculate the principal components of stress.

Stresses calculated using the viscoelastic theory as developed by Blanton (1983) agree well with stresses measured using other techniques such as hydraulic fracture measurements (Teufel and Warpinski, 1984). By averaging T-DSR data from four cores of the Rollins Sandstone, Teufel and Warpinski (1984) calculate: $S_H = 54.5 \pm 0.8$ MPa; $S_h = 51.9 \pm 1.1$ MPa; and $S_v = 56.4$ MPa. These estimates of in situ stress compare favorably to those derived from hydraulic fracturing and S_H = 49.5 MPa (S_h = 46.7 MPa). Orientation data also compare favorably; the hydraulic fracture was a set of narrow, en échelon cracks on one side of the well bore at a strike of N50°W to N70°W, whereas the average direction of ε_H from T-DSR tests was N63°W.

The T-DSR technique was further tested on core of the Mesa Verde Group from the MWX-3 well of the multiwell experiment near Rifle, Colorado (Teufel and Warpinski, 1984; Warpinski and Teufel, 1989a) (fig. 9–9). Hydraulic fracture tests and T-DSR measurements show the same trends for in situ stress. In a Mesa Verde sandstone-shale sequence, stresses differ between the sandstone and shale beds as is common for such bedded sequences (e.g., Gronseth, 1982; Evans and Engelder, 1986). S_h/S_v ratios indicate that the tectonic setting in the sandstones is that of normal faulting. Shale beds show a smaller σ_d and a larger S_h than sandstone beds. Variation of stress with lithology in the Cretaceous Mesa Verde Group is analogous to that encountered in the upper portion of the Devonian sediments near South Canisteo, New York (chap. 5) but in contrast to the Mesa Verde Group siltstone beds of the Devonian section carried a higher absolute stress than the shale beds (Evans and Engelder, 1986). In this latter case σ_d is again smaller in the shale beds but S_h/S_v is that consistent with thrust faulting. In both situations involving clastic sequences, horizontal stress in the finer-grained lithology (i.e., the shale) is closer to the lithostat (S_v) regardless of the stress regime.

Microcrack Propagation

The mechanisms causing T-DSR are of particular interest because of the direct correlation between T-DSR and in situ stress (Blanton, 1983; Teufel, 1982; 1983). A favorite hypothesis is that the major mechanisms for T-DSR are the inelastic processes associated with opening and propagation of microcracks

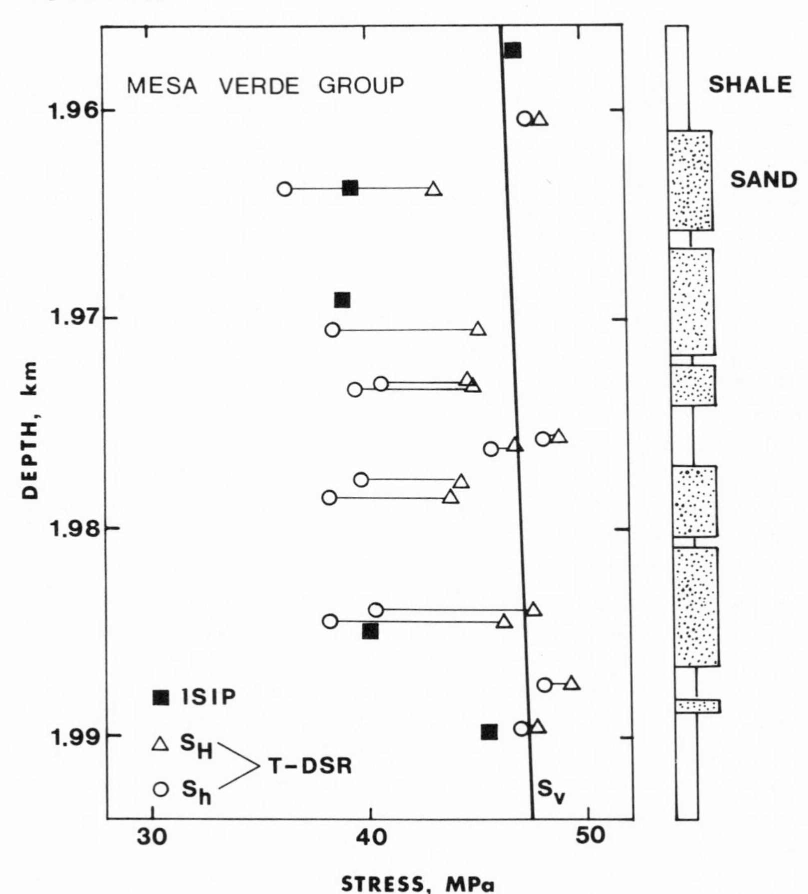

Fig. 9–9. Maximum and minimum horizontal stress magnitudes as a function of depth and lithology in sandstone/shale sequence in the Mesa Verde Group from an MWX well near Rifle, Colorado. Stress measurements include hydraulic fracture (HF) and time-dependent strain relaxation (T-DSR) (adapted from Teufel and Warpinski, 1986).

(Strickland and Ren, 1980; Ren and Rogiers, 1983; Engelder, 1984; Teufel, 1989). An obstacle in testing this hypothesis is the uncertainty in identifying the microcracks associated with strain relaxation (Kowallis et al., 1982). Early evidence supporting this hypothesis was the correlation between the orientation of microcrack sets and the principal axes of T-DSR. Another piece of circumstantial evidence comes from property changes related to the tendency for microcracks to open or propagate upon stress relief of rock aggregates (Engelder, 1984; Engelder and Plumb, 1984; Teufel, 1989). After T-DSR there is a direct correlation among the in situ stress field, the elastic anisotropy of the rock core, and the orientation of the core's microcrack population (Engelder and Plumb, 1984).

The strongest evidence supporting the microcrack propagation mechanism for T-DSR is an acoustic emission rate which decreases exponentially with time in the same fashion as the creep recovery of core (Teufel, 1989). Figure 9–10 shows the cumulative AE count and the T-DSR curves for a core recovered from a depth of 3138 m in the Central Graben region of the North Sea. Both AE activity and the T-DSR ceased when the core came to equilibrium about thirty hours after the core was cut. Like previous attempts, Teufel (1989) was unable to identify the microcracks created during T-DSR. However, he points out that only two grain-boundary microcracks per mm would have to open 0.2 μm in order for the core to relax 400 με in a given direction. Such grain-boundary microcracks are nearly impossible to identify in a petrographic microscope.

T-DSR in Crystalline Rocks

Because measurement of the instantaneous component of strain relaxation during deep coring is impossible using present technology, the relationship between instantaneous strain relaxation and T-DSR for cores from deep wells is unknown. Direct confirmation of Voight's (1968) prediction that T-DSR correlates with in situ stress is best established using conventional overcoring techniques in underground mines or in the near surface (Teufel, 1982; Engelder, 1984). A careful study of the relationship between in situ stress and T-DSR, requires measurements below the zone of weathered and highly fractured rock commonly found at the surface. Likewise, a controlled environment is necessary to eliminate noise contributed by the wetting and drying associated with

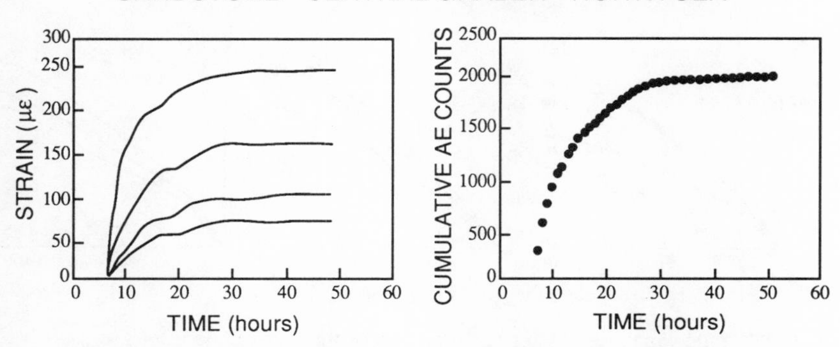

Fig. 9–10. Behavior of a sandstone core taken within the Central Graben region of the North Sea. Plots show T-DSR and cumulative acoustic emissions versus time (adapted from Teufel, 1989).

drilling and core removal. This condition is achieved with tests below the water table where the sample is left in a thermal bath of ground water at a constant temperature. Groundwater damps out the annual temperature cycle which affects near-surface rock under dry conditions. Conditions for thermal and moisture stability were met in a 20 m deep portion of the Williams Stone Company quarry in East Otis, Massachusetts (Engelder, 1984). This quarry is cut into the Algerie granite, a synmetamorphic quartz monzonite with a foliation striking N15°W and a pervasive set of healed microcracks striking N20°E. Quarrymen at East Otis report that the quarry grain (rift is horizontal) is slightly east of north and parallel to the healed microcracks.

Using the Lamont strain cell bonded to the top of cores, strain relaxation was monitored continuously for several hours after vertical cores were cut (figs. 7–7, 9–11). On average the T-DSR in the Algerie granite was about 40 $\mu\varepsilon$. Instantaneous maximum expansion, ε_H, which accounted for a major fraction of the total strain relaxation, was oriented ENE, 30° counterclockwise from the orientation of time-dependent ε_H (fig. 9–12). Based on the correlation between time-dependent ε_H and the normal to the pervasive healed microcracks, Engelder (1984) concluded that recracking of these microcracks was largely responsible for the time-dependent behavior. In addition to the healed microcracks, there is a set of open microcracks with a more uniform distribution. Within this latter set there is a cluster whose normal is in the direction of the instantaneous ε_H (fig. 9–12). This weak concentration of open microcracks is assumed to have grown during instantaneous stress relief. Such open microcracks are apparently the active element during T-DSR of sedimentary rocks which do not have a strong preexisting microcrack fabric. In the Algerie granite, open microcracks did not control T-DSR because the healed microcrack set formed a much stronger fabric.

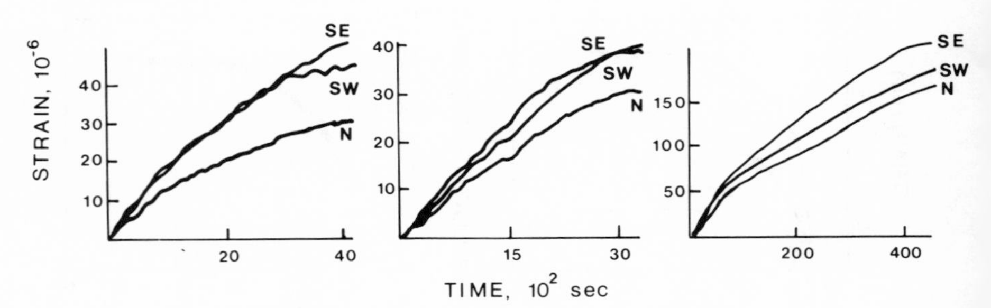

Fig. 9–11. Three examples of the strain recorded by three components of a doorstopper strain gauge attached to core of Algerie granite (adapted from Engelder, 1984). The southeast strain gauge shows the maximum expansion which correlates with a NNE grain plane in Algerie granite.

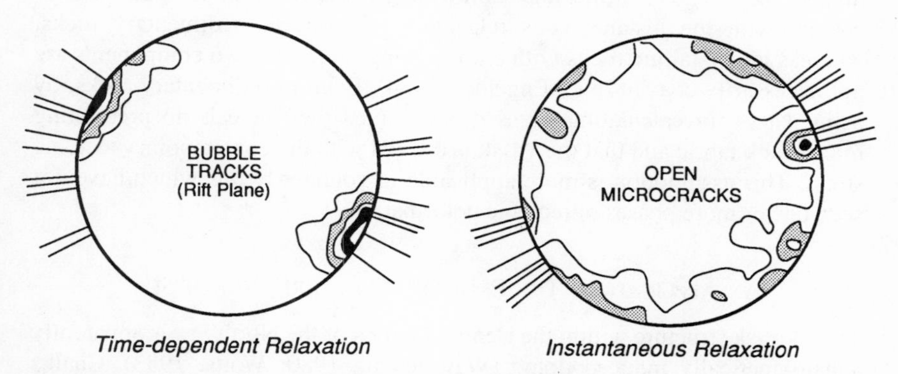

Fig. 9–12. Lower hemisphere projection of poles to healed microcracks and open microcracks in the Algerie granite in the Williams Stone Company quarry (adapted from Engelder, 1984). Contours represent 2% per 1% area. Also shown is the orientation of maximum expansion for both instantaneous and time-dependent relaxation of Algerie granite. The open microcracks are associated with the instantaneous relaxation, whereas the healed microcracks are associated with the time-dependent relaxation.

The same type of experiment was initiated at the Nevada Test Site in volcanic tuff which is axially symmetric with a compaction fabric defining bedding (Teufel, 1982). Doorstoppers (Leeman, 1964) were used to measure instantaneous strain relaxation of the welded tuff. After cutting, the cores were removed from the borehole and sealed with polyurethane wrapping to prevent environmental changes associated with core dehydration. Clip-on gauges (strain gauge type) were then mounted across the core diameter to monitor T-DSR. Temperature during the measurement was controlled to ± 1°C to minimize thermal strains. For the welded tuff at the Nevada Test Site, the instantaneous component constituted about 70% of the total strain relaxation, whereas in the Algerie granite the instantaneous component was about 90%. T-DSR in the welded tuff correlates with the instantaneous strains in both orientation and stress ratio ($\varepsilon_H/\varepsilon_h = 1.5$). Hydraulic fracturing near the overcoring site showed S_H at about N45°E which is in good agreement with the orientation of ε_H for both the instantaneous and time-dependent components. In the absence of a microcrack fabric such as found in the Algerie granite, T-DSR reflects the orientation of in situ stress in crystalline rocks. The idea is that, unless the rock contains a preexisting microcrack fabric, a sudden release of stress causes microcracks to propagate with their normals in the direction of S_H.

For measuring in situ stress, T-DSR has its limits. Experiments suggest that if a rock has a strong microcrack fabric or foliation, this anisotropy rather than the in situ stress may control T-DSR. However, measurements near the Ramapo fault suggest that a strong foliation does not always control T-DSR (fig. 9–7).

Because of the effect of a foliation, there are fundamental differences in the manner in which crystalline and sedimentary rocks relax. In general, T-DSR is coaxial with the instantaneous relaxation of isotropic sedimentary rocks, whereas in crystalline rocks with a preexisting fabric the two components are not necessarily coaxial (e.g., Engelder, 1984). Even in sedimentary rocks, by using T-DSR for calculating stress it is assumed that there is no preexisting microcrack fabric and that the T-DSR is coaxial with the contemporary tectonic stress. This assumption is more applicable in younger basins which have not seen one or more phases of tectonic deformation.

An Application: The Ekofisk Structure, North Sea

The Ekofisk structure within the Central Graben of the North Sea is apparently a halokinetically induced dome (Watts et al., 1980; Watts, 1983). Chalks within the Ekofisk dome serve as major reservoir rocks for petroleum. Once water flooding became important to the secondary recovery of oil, detailed knowledge of in situ stresses was required because of the strong influence of in situ stress on the behavior water floods. The orientation of the local stress field was mapped in the vicinity of the Ekofisk dome using T-DSR data (Teufel and Farrell, 1990). T-DSR data suggests that S_H is oriented perpendicular to the structure contours around the dome (fig. 9–13). Furthermore, ISIP data from hydraulic fracture tests suggest that S_h is lowest at the crest of the dome and highest on the north and south flanks. The current stress field orientation around the dome is consistent with analytical models for domes which show that S_H parallels the long axis of the dome near its crest and becomes radial on the flanks of the dome (Withjack and Scheiner, 1982). Indeed, a set of tectonic fractures has developed in association with the doming-induced radial stress system (Teufel and Farrell, 1990). In the Central Graben of the North Sea, the T-DSR technique gives an indication that a local stress field developed to overprint the contemporary platewide stress field seen in much of western Europe (chap. 12). This overprint shows the same type of radial symmetry as the local stress field developed in the vicinity of the Spanish Peaks, Colorado (chap. 2).

In summary, T-DSR measurements of in situ stress are most successful in quartz arenites and chalk and seem to work less well in shales where cores commonly contract during strain relaxation. The strong fabric in crystalline rocks often overwhelms any correlation between T-DSR and in situ stress.

Differential Strain Analysis

Strain relaxation of a core includes a time-dependent component which arises from the opening and growth of microcracks of a preferred orientation (e.g., Engelder, 1984; Teufel, 1989). On reloading the core in hydrostatic compres-

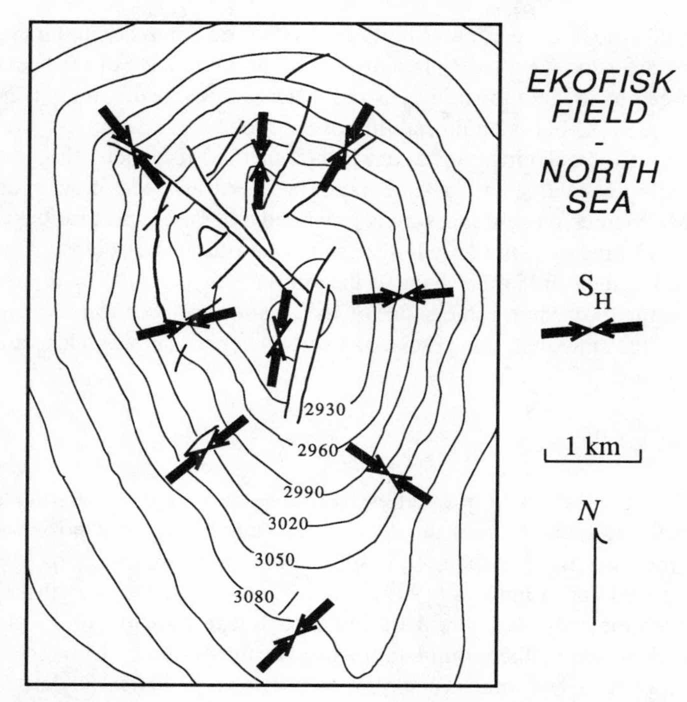

Fig. 9–13. Structure contour map for the top of the Ekofisk Formation (adapted from Teufel and Farrell, 1990). Dark arrows indicate the azimuth of S_H determined from T-DSR tests of oriented core taken from the Ekofisk field. The crest of the formation is at a depth of 2.9 km and the contour intervals are approximately 30 m.

sion, these microcracks will close and thereby contribute to a strain anisotropy within the core. If, on stress relief, T-DSR by crack opening is proportional to the magnitude of the in situ stress as indicated by Blanton's and Teufel's (1983) viscoelastic model, then on reloading the rock, microcrack closure and concomitant strain should also have a direct relationship with in situ stress conditions. Such is indicated by two techniques for assessing microcrack populations by reloading rocks in hydrostatic compression: differential strain analysis and ultrasonic measurements.

Microcracks affect the elastic properties of rocks as indicated by the early work of Adams and Williamson (1923) who observed that rocks containing microcracks are more compressible than uncracked rocks. Furthermore, the compressibility of a rock is dependent on whether its microcracks are open or closed. An open microcrack has walls which are unsupported, whereas a closed microcrack has walls in intimate contact. Rock in the immediate vicinity of an open microcrack appears to have a much higher compressibility than rock in the immediate vicinity of a closed microcrack. Therefore, during hydrostatic compression the linear compressibility of a rock (i.e., strain per unit increase in confining pressure, $\beta = d\varepsilon/dP_c$) in a direction normal to a set of microcracks is high before closure and low after closure. Microcrack closure, then, is marked by a change in the bulk compressibility of the rock.

A theory for crack closure is taken from a method for calculating the volumetric compressibility of rocks containing penny-shaped cracks (Walsh, 1965). Microcrack closure pressure is governed by crack aspect ratio so that at a lower confining pressure only longer, narrower microcracks close. Crack aspect ratio is a measure of the shape of the microcrack indicated by dividing the crack aperture (i.e., short dimension) by its long dimension. From Walsh's theory comes the statement that cracks of a certain aspect ratio, φ, close at a confining pressure[4] (P_c^c) defined by

$$P_c^c = \frac{\pi E \varphi}{4(1 - \nu^2)} \tag{9–3}$$

where E and ν are elastic properties for the crack-free rock. Morlier (1971) noted that a determination of the distribution function for φ is possible using data on rock strain (ε) as the rock is subject to increasing confining pressures (P_c). Siegfried and Simmons (1978) generalized Walsh's (1965) theory with the assumption that rock strain occurs as a linear function of P_c provided microcrack sets are either completely closed or remain open. Conversely, as a microcrack set closes, strain is a nonlinear function of P_c. The first derivative of the functional relationship between rock strain and P_c leads to the cumulative crack closure (crack strain) between atmospheric pressure and elevated P_c, whereas the second derivative of the strain-P_c function gives the rate of microcrack closure at the P_c of interest.

Differential strain analysis (DSA) is a technique for measuring microcrack closure in rocks as a function of P_c (Simmons et al., 1974). The DSA technique is based on the simultaneous measurement of strain in a crack-free material and in the rock of interest when both are subjected to the same change in P_c. Strain is measured using electric-resistance strain gauges attached to cubes of jacketed rock and crack-free material, usually fused silica. The crack-free material must have a compressibility similar to the intrinsic compressibility of the rock. If the intrinsic compressibilities are similar, then the major difference in compressibility between the crack-free sample and the rock arises because microcracks and pores within the rock are closing as a function of P_c. The term, differential strain, arises from the fact that the data of interest is the difference in strain between the reference cube of crack-free material and the rock sample over a certain change in P_c. Strain measurements on both the reference cube and the rock of interest are subject to errors caused by instrument drift and changes in temperature or lead resistance. By differencing strain from the reference cube and rock sample, these errors are nearly eliminated. The most useful information from DSA is the P_c at which various microcrack sets close and the orientation of those microcrack sets. Both closure P_c and orientation data relate directly to in situ stress.

THEORY

Strain of the rock (i.e., the cracked solid), ε_{ij}, is the sum of the strain in the intrinsic rock (i.e., the uncracked solid), ε^{i}_{ij}, and the strain due to the presence of cracks, η_{ij}. The incremental strain, $\delta\eta_{ij}$, due to the presence of microcracks in a rock is

$$\delta\eta_{ij} = \delta\varepsilon_{ij} - \varepsilon^{i}_{ij} \qquad (9\text{–}4)$$

Following Siegfried and Simmons (1978) the pressure derivative of this equation for rock strain is

$$\frac{d\varepsilon_{ij}}{dP_c} = \frac{d\varepsilon^{i}_{ij}}{dP_c} + \frac{d\eta_{ij}}{dP_c}. \qquad (9\text{–}5)$$

Since the compressibility of the uncracked rock is

$$\beta^{i}_{ij} = -\frac{d\varepsilon^{i}_{ij}}{dP_c} \qquad (9\text{–}6)$$

then the compressibility of the microcracked rock is

$$\beta_{ij} = -\frac{d\varepsilon_{ij}}{dP_c} = \beta^{i}_{ij} - \frac{d\eta_{ij}}{dP_c}. \qquad (9\text{–}7)$$

Siegfried and Simmons (1978) recognized three useful parameters arising from data on the compression of rock: the microcrack spectra, $\xi_{ij}(P^c_c)$; the strain associated with only those microcracks left open after the application of a certain hydrostatic pressure, $\eta_{ij}(P_c)$; and the cumulative microcrack strain, ζ_{ij}. A microcrack spectrum [e.g., $\xi_{11}(P^c_c)$] is defined as the microcrack strain in one direction due to microcracks which close completely over an infinitesimal pressure increment. The idea is that the hydrostatic compression marked by the major peak of the microcrack spectra in a particular orientation correlates with the stress normal to that crack prior to relief by coring. The microcrack spectra are a distribution function $\xi_{ij}(P^c_c)dP^c_c$. Using the distribution function for the microcrack spectra, the strain due to open microcracks is defined as

$$\eta_{ij}(P_c) = \int_{P_c}^{\infty}\left(1 - \frac{P_c}{P^c_c}\right)\xi_{ij}(P^c_c)\, dP^c_c \qquad (9\text{–}8)$$

where P^c_c is the closure pressure for various microcracks. This expression sums up the contributions due to all cracks that remain open at pressure P_c. Equation

9–8 is not an expression for cumulative microcrack strain, ζ_{ij}. Differentiating the microcrack strain function with respect to pressure gives

$$\frac{d\eta_{ij}}{dP_c} = -\int_{P_c}^{\infty} \left(\frac{1}{P^c{}_c}\right) \xi_{ij}(P^c{}_c) dP^c{}_c \, . \tag{9–9}$$

Substituting for compressibility in equation 9–7

$$\beta_{ij} = \beta^i{}_{ij} + \int_{P_c}^{\infty} \left(\frac{1}{P^c{}_c}\right) \xi_{ij}(P^c{}_c) dP^c_c \, . \tag{9–10}$$

Again differentiating with respect to pressure

$$\frac{d\beta_{ij}}{dP_c} = -\left(\frac{1}{P_c}\right) \xi_{ij}(P_c) \tag{9–11}$$

$$\xi_{ij}(P_c) = -P_c \left(\frac{d\beta_{ij}}{dP_c}\right) \tag{9–12}$$

or in terms of strain

$$\xi_{ij}(P_c) = -P_c \left(\frac{d^2\varepsilon_{ij}}{dP^2{}_c}\right) . \tag{9–13}$$

This indicates that a microcrack spectrum is the second derivative of rock strain with respect to pressure.

The other useful parameter is the cumulative microcrack strain (ζ_{ij}) at zero pressure due to the microcracks closing completely at some $P_c < P^c_c$. By definition of $\zeta_{ij}(P^c_c)$

$$\zeta_{ij}(P_c) = -\int_0^{P_c} \xi_{ij}(P^c_c) dP^c_c \tag{9–14}$$

and so

$$\zeta_{ij}(P_c) = -\int_0^{P_c} P^c_c \left(\frac{d^2\varepsilon_{ij}}{dP^2{}_c}\right) dP^c_c \, . \tag{9–15}$$

Integration by parts gives

$$\zeta_{ij}(P_c) = P_c \frac{d\varepsilon_{ij}}{dP_c} - \varepsilon_{ij}(P_c) \, . \tag{9–16}$$

This relationship indicates that ζ_{ij} has a simple geometric interpretation; it is the zero-pressure intercept of the tangent to the differential strain curve at pressure P_c (Siegfried and Simmons, 1978) Volumetric parameters such as the total crack porosity, ζ_v, are scalar invariants of the corresponding tensor

$$\zeta_v = \zeta_{11} + \zeta_{22} + \zeta_{33}. \tag{9–17}$$

Because ζ_v is a scalar invariant, total crack porosity is calculated from any three mutually orthogonal strain gauges on the faces of a rock cube reloaded in the laboratory.

Examples of DSA Measurements

Samples of crystalline rock from the Illinois Deep Hole,[5] UPH-3, illustrate the nature of the information available from DSA tests, particularly the correlation between microcrack strain and in situ stress (Kowallis and Wang, 1983; Carlson and Wang, 1986). Typical curves for differential strain (η_{ij}) versus confining pressure show a negative linear strain at lower confining pressures (fig. 9–14). This reflects a higher rate of strain due to microcrack closure in the rock compared with the reference cube. At low pressures the curves are often nonlinear because microcracks are continuously closing with increasing pressure. Once a majority of the microcracks have closed the differential strain curves become linear with either a positive or negative slope depending on the relative values of the intrinsic compressibilities of the rock and the control sample. The cumulative crack strain $\zeta(P_c)$ for each strain gauge on the sample is the zero-pressure intercept of a tangent line drawn from the differential strain curve at a specific pressure (fig. 9–14). At a pressure of 200 MPa $\zeta(P_c)$ in the vertical direction of the core from UPH-3 is about 700 $\mu\varepsilon$, whereas both horizontal $\zeta(P_c)$

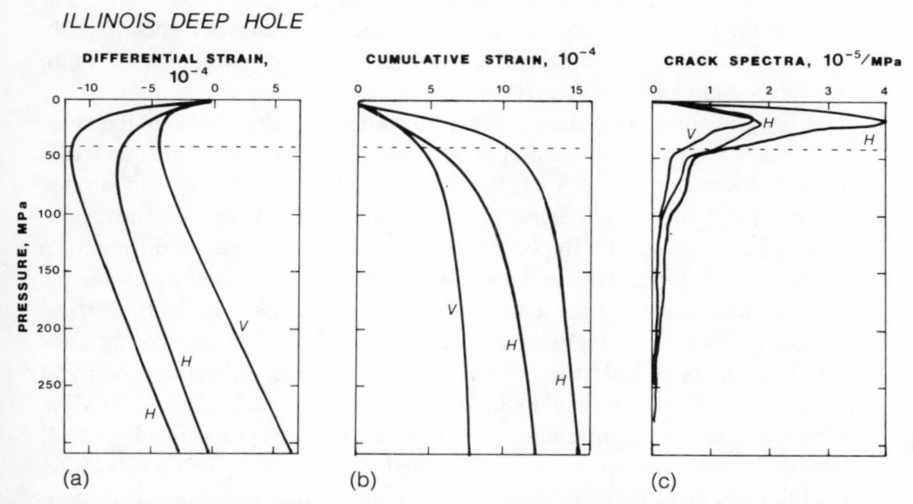

Fig. 9–14. (a) Differential strain versus confining pressure for a sample (1601.8) from the Illinois Deep Hole, (UPH-3). (b) Cumulative crack strain (ζ_{ij}) versus confining pressure for sample 1601.8 from the Illinois Deep Hole (UPH-3). (c) Crack spectra (ξ_{ij}) for sample 1601.8. Each spectrum represents the amount of crack strain in a single direction due to crack closure between pressure P and P + dP (all data after Carlson and Wang, 1986). Dashed line indicates S_V assuming a density of 2.7 g/cm^3.

are larger than 10^3 $\mu\varepsilon$ (fig. 9–14). At P_c between 100 and 120 MPa, the differential strain curve becomes nearly linear and the $\xi(P_c)$ increases very slowly. The samples continue to show a small crack strain associated with increasing pressure even after the last sharp knee in the η_{ij} below 100 MPa. The crack spectra, $\xi_{ij}(P_c^c)$, obtained by differentiating $\zeta(P_c)$ with respect to P_c, indicate the relative number of cracks which close completely over an infinitesimal pressure increment. For a sample taken from a depth of 1601 m in the UPH-3 well the crack spectra peak is at $P_c \approx 20$ to 25 MPa which is equivalent to the weight of the overburden minus the in situ pore pressure (fig. 9–14). The explanation for this correlation is that fluid-filled microcracks were present before coring and opened on stress release and, hence, close when reloaded to conditions matching the effective confining pressure (Carlson and Wang, 1986). This interpretation does not apply if microcracks propagated on stress release; such microcracks should respond by closing on reloading to $P_c \approx 43$ MPa. Assuming that the former explanation is generally valid the peak in the crack spectra marks the effective lithostatic stress to which the sample was subjected (i.e., depth from which the sample was taken). However, the data from UPH-3 are not sensitive to relative magnitudes of the principal stresses.

Microcrack closure continues up to $P_c \approx 120$ MPa in samples from UPH-3 suggesting that some microcracks are present in situ which are not directly related to the present lithostatic stress (fig. 9–14). If 120 MPa represents a total lithostatic stress, then cracks closing at this pressure might have formed at depths on the order of 4 km. Because microcracks are known to heal rapidly (e.g., Smith and Evans, 1984), the data suggest that any microcracks forming deeper than 4 km have apparently long since healed.

DSA tests on cores taken during overcoring stress measurements in the near-surface show a rather complex correlation between in situ stress and crack closure pressures (Plumb et al., 1984b). In situ stress was measured in surface outcrops of the Adirondack dome where hornblende gneiss has a visible foliation striking about N60°E. The behavior of the hornblende gneiss differs from UPH-3 samples as indicated by a distinct difference between closure pressure for horizontal and vertical microcracks in the Adirondack core with vertical microcracks closing at higher pressures. Strain relaxation overcoring tests showed that S_H is 20 MPa and normal to the foliation in the outcrop. Vertical microcracks, which are normal to S_H, have a closure pressure of about 30 MPa. To interpret the 50% difference between the closure pressure of vertical microcracks and outcrop stress, as indicated by overcoring strain relaxation tests, Plumb et al. (1984b) suggest that while microcrack opening was responsible for rock strain on stress relief, the microcrack set was already present in the outcrop as an integral part of the foliation. DSA tests show that a maximum crack porosity comes from microcracks parallel to the outcrop surface with a closure pressure of about 20 MPa (fig. 9–15), indicating that horizontal microcracks in the hornblende gneiss were present prior to the strain relaxation.

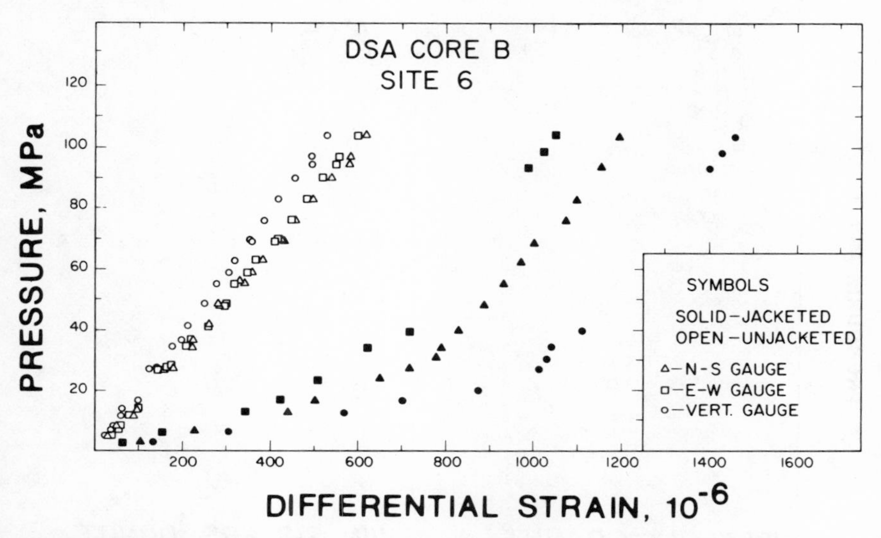

Fig. 9–15. Pressure-strain curves for the jacketed (solid symbols) and unjacketed (open symbols) DSA runs on a core of hornblende gneiss from the Adirondack Mountains (adapted from Plumb et al., 1984b). The north-south gauge is normal to an east-west foliation.

In contrast, samples from the Illinois Deep Hole show crack closure pressure within a few percent of the effective lithostatic stress.

The intrinsic compressibility of a rock is measured using an unjacketed sample. In principle, all microcracks fill with confining fluid so that only the intact minerals between microcracks are compressed when the sample is subject to hydrostatic compression. Intrinsic compressibility tests show that, although a bulk sample of Adirondack hornblende gneiss is most compliant in the vertical direction because of microcracks, the foliation causes an anisotropy in the intrinsic behavior of the samples with the vertical direction the stiffest (fig. 9–15).

Six components of strain from DSA tests allow the calculation of the orientation of microcrack sets as is illustrated for Westerly granite and Twin Sisters dunite (Siegfried and Simmons, 1978). DSA tests show that there is significant crack porosity in all three principal directions of Westerly granite, whereas virtually all microcracks in the Twin Sisters dunite are in one orientation (fig. 9–16). The closure pressure for the major microcrack set in the Westerly granite is on the order of 20 MPa. A minor suite of microcracks closes continuously to about 180 MPa, a considerably higher pressure than found for closure of a minor suite of microcracks in the crystalline rock from UPH-3 (fig. 9–14). The

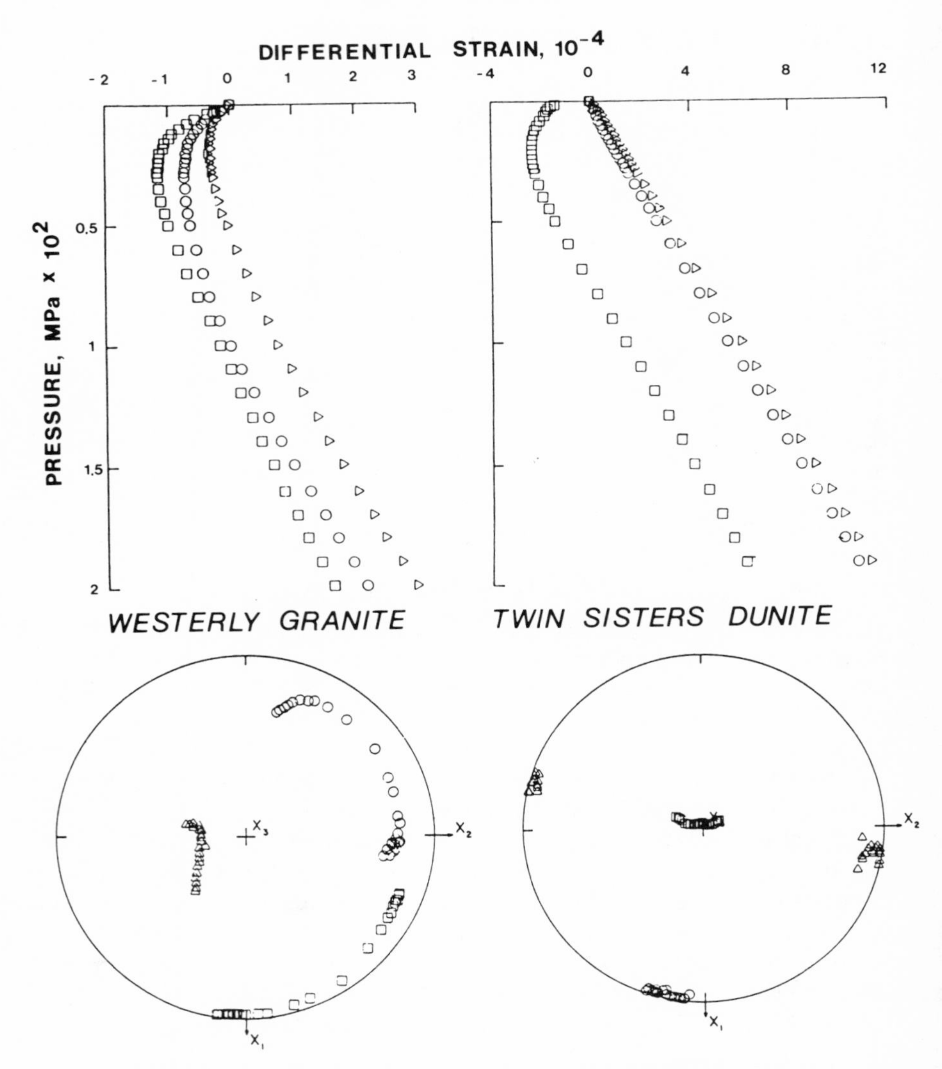

Fig. 9–16. Principal values of the differential strain tensor $\eta_{ij}(P)$ for Westerly granite and Twin Sisters dunite versus confining pressure (adapted from Siegfried and Simmons, 1978). Orientation of the principal axes of $\eta_{ij}(P)$ are shown in lower hemisphere projection.

orientation of the principal differential strains for Westerly granite are plotted on an equal-area projection with the rift plane normal to x_1 (fig. 9–16). Below 50 MPa the principal incremental crack strains cluster together near the x_1, x_2, and x_3 directions.[6] At higher pressures the orientations of the greatest and intermediate incremental principal strains rotate roughly counterclockwise around x_3. This higher pressure incremental strain is due mainly to mineral deformation or deformation of pores closing at pressures up to 180 MPa, and the principal incremental strain axes rotate to reflect this high-pressure anisotropy which does not correlate with the rift plane. Knowing both the magnitude and orientation of the cumulative crack strain, Siegfried and Simmons (1978) confirm that the microcrack porosity, $\zeta_{11}(P_c)$, the largest cumulative crack strain, is in the direction normal to the rift plane of Westerly granite. Although the incremental microcrack strain rotates at high pressures, the orientation of cumulative crack strain remains relatively constant. Simmons and Siegfried (1978) suggest that most of the microcrack porosity in the sample of Westerly granite was produced by the erosional unloading.

Many rocks show a crack spectrum with one peak (Feves et al., 1977). Exceptions to this rule include the Twin Sisters dunite which has a bimodal closure pressure with a strong peak at very low pressures and a weak peak at 20 MPa (fig. 9–17). Likewise, samples taken from a shocked granodiorite from the site of the Piledriver nuclear test, Nevada, show crack spectra with two distinct peaks, one of which is assumed to be associated with the shock of a nuclear blast (Simmons et al., 1975).

VOLUMETRIC CRACK POROSITY AND MEAN STRESS

Volumetric crack porosity (ζ_v) is measured by summing the cumulative crack strain in three orthogonal directions as measured by strain gauges on two or more faces of cubic samples. Knowledge of the orientation of the principal crack strains is not necessary for calculated ζ_v. Using DSA, Carlson and Wang

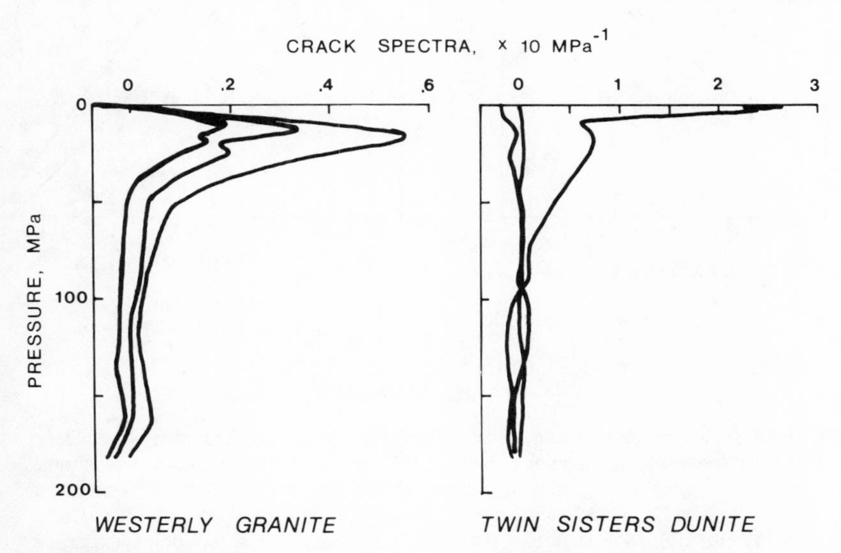

Fig. 9–17. Principal values of $\xi_{ij}(P_c)$ for Westerly granite and Twin Sisters dunite versus confining pressure (adapted from Siegfried and Simmons, 1978).

(1986) were able to show a positive correlation between the mean stress and crack porosity in core samples recovered from the Illinois borehole UPH-3 to depths of 1600 m. Microcrack porosity varied from as low as 1.5×10^{-4} at a mean stress of 20 MPa to as high as 24×10^{-4} at a mean stress of 45 MPa (fig. 9–18). Likewise, for core of crystalline rocks from the Kent Cliffs, New York, and Moodus, Connecticut, scientific holes, Meglis et al. (1991) observed a strong correlation between mean in situ stress and microcrack porosity. Interestingly, the zero porosity intercept varies from well to well. For reasons that are not presently clear, such a correlation was not observed for core from the Devonian shales of the Appalachian Plateau (Meglis et al., 1990).

DSA tests on core removed from the Illinois deep borehole UPH 3 showed that many open microcracks had closure pressures between 15 and 22.5 MPa. Furthermore, microcracks occur in greater concentrations in cores from greater depths. Because the closure pressures were equivalent to effective overburden pressure, and because there seemed to be a direct correlation with crack con-

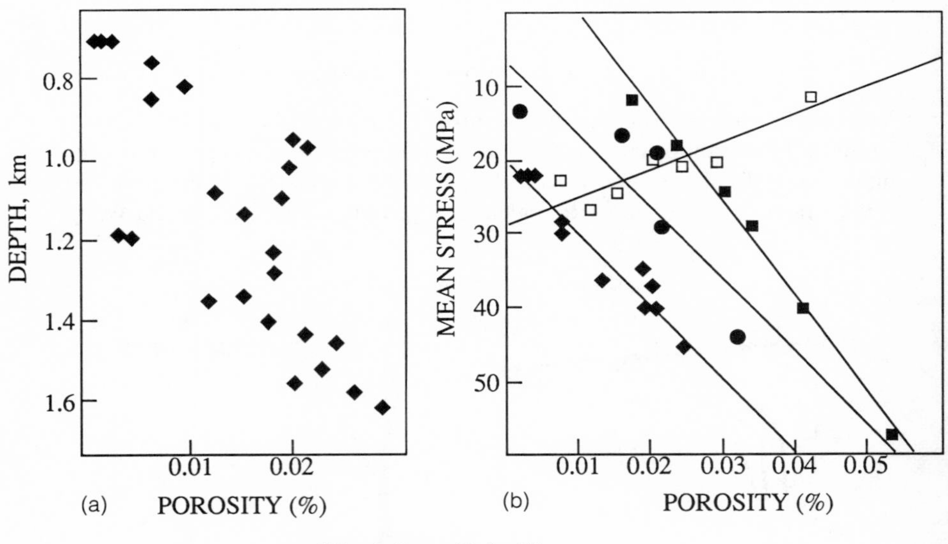

Fig. 9–18. (a) Microcrack porosity versus depth as found in core from the Illinois Deep Hole (UPH-3) (data from Carlson and Wang, 1986). (b) Microcrack porosity versus mean in situ stress in UPH-3; the Moodus scientific hole, Connecticut; the Kent Cliffs scientific hole, New York; and EGSP hole NYS-1 from Alleghany County New York (data from Carlson and Wang, 1986; Meglis et al., 1990, 1991). The latter plot shows only microcrack measurements were made within 11 m of corresponding stress measurements on UPH-3.

centration and depth of burial, Kowallis and Wang (1983) concluded that microcracks within the core propagated upon stress relief.

The positive correlation between mean stress and microcrack porosity for core samples taken at depth suggests that microcrack porosity of outcrop samples might also relate to the mean stress at some point during burial. For comparison, the microcrack porosity of the hornblende gneiss from the the Adirondack dome was 18×10^{-4}, Westerly granite was 6.5×10^{-4}, and Twin Sisters dunite was 3.7×10^{-4}. All three were taken from surface outcrops following maximum stress relief due to erosion. The hornblende gneiss shows the same microcrack porosity as the Moodus deep well samples projected to an expected mean surface stress of 10 MPa at the surface. The porosity of surface samples of Westerly granite and Twin Sisters dunite fall between the porosity of samples from Moodus and Kent Cliffs when projected to a mean surface stress of 10 MPa. This technique is also difficult to calibrate because the effect of lithology on microcrack porosity is yet unknown. For example, the present mean stress on the Adirondack hornblende gneiss is on the order of 10 MPa, yet it has a microcrack porosity as large as the sample from UPH-3 subject to a mean stress of 45 MPa. The behavior of crystalline rocks suggest that, in contrast to the sudden release of stress by coring, the slow removal of stress by erosion allows time for the healing of microcrack porosity (Meglis, 1990, personal communication).

In Situ Stress Ratios

A comparison of laboratory and field measurements has confirmed that DSA of core recovered from deep wells provides information about stress ratios as well as mean stress. An analysis of stress ratios assumes that microcrack strain on recompression is largest in a direction normal to the maximum in situ stress (Strickland and Ren, 1980; Ren and Rogiers, 1983; Thiercelin et al., 1986). The difference between crack densities in orthogonal orientations reflects the ratio of normal stresses in those two orientations.

The crack density distribution is determined by measuring the total strain tensor $[\varepsilon_{ij}(P_c)]$ of the rock as a function of P_c (Thiercelin et al., 1986). From this, the zero pressure strain tensor or cumulative crack density distribution $[\zeta_{ij}(P)]$ may be calculated according to equation 9–16. The strain tensor due to microcracks closing between two pressures, P_c and $P_c + \Delta P_c$, is measured as

$$\xi_{ij}\,(P_c, P_c + \Delta P_c) = \zeta_{ij}(P_c, P_c + \Delta P_c) - \zeta_{ij}(P_c). \qquad (9\text{–}18)$$

If $\Delta P_c = dP_c$, then ξ_{ij} is the crack spectra at the pressure P_c. To determine the ratio of stresses, Ren and Rogiers (1983) have proposed that the principal effective stress ratios are proportional to the principal strains resulting from crack closure. By virtue of Poisson's effect, this stress ratio is equal to

$$\frac{\bar{\sigma}_i}{\bar{\sigma}_j} = \frac{(1-\nu)\xi_i + \nu(\xi_j + \xi_k)}{(1-\nu)\xi_j + \nu(\xi_i + \xi_k)} \tag{9–20}$$

where ξ_i are the principal crack spectra.

Direct Calibration of DSA

In DSA tests on core of fine-grained sandstones from depths 1150–1800 m in Texas and 1500–2440 m in Colorado, Ren and Rogiers (1983) observed a good correlation between orientation of stress predicted by both hydraulic fracture and DSA tests. To further assess the DSA technique, Dey and Brown (1986) measured stress using core recovered from a Fenton Hill, New Mexico, geothermal project well. Working with core at Fenton Hill offers the advantage that higher temperatures in situ favor rapid microcrack healing. This condition assures that strain on recompression reflects only those microcracks formed on stress release. In general, the DSA tests agree with hydraulic fracture data, although principal stress estimates vary ± 20 MPa. In the Fenton Hill core, the stress estimates become less precise as the maximum compression rotates away from vertical. Dey and Brown (1986) find that the orientation estimates are accurate to within ± 15° when compared with data from earthquake fault-plane solutions.

Ultrasonic Techniques

Use of ultrasonic techniques for inferences about in situ stress stems from Birch's (1960) findings that the travel time of ultrasonic waves within rock decreases on reloading in hydrostatic compression, and at very high pressures ultrasonic wave velocities approach values expected for the intrinsic constituents of the rock. At the time of Birch's studies, it was known that the travel time of ultrasonic waves was very sensitive to relatively subtle variations in the elastic properties of rock. Later studies showed that microcrack closure was largely responsible for controlling the travel time of ultrasonic waves (e.g., Nur and Simmons, 1969). Early ultrasonic studies that pertain to stress relief of rock include those of Simmons and Nur (1968) using granites from the 3.0 km deep Wind River well in Wyoming and from the 3.8 km deep Matoy well in southeastern Oklahoma. Simmons and Nur (1968) observed near-intrinsic velocities at a P_c corresponding to the depth from which the core was retrieved. Thus, they concluded that microcracks were either absent or saturated under in situ conditions and that microcracks formed on stress relief, but they made no correlation between microcrack density and magnitude of in situ stress.

Scientists eventually correlated ultrasonic property changes (i.e., growth of microcracks) and stress relief. Buen (1977) reports that the sonic velocity of several Norwegian granites decreased with time after coring. Holzhausen and Johnson (1979) point out that continuing deformation after coring suggested gradual relaxation of stresses, accompanied by further reduction of strain energy in the rock. Wang and Simmons (1978) compared ultrasonic and DSA data from core material taken at the 5.3 km depth of a deep well drilled into the Michigan Basin and concluded that the core contains open microcracks which close at burial pressures. Wang and Simmons (1978) explicitly stated that open cracks, observed mainly in chlorite, were not present in the rock in situ but rather were produced as a result of the sudden stress relief on removing a sample to the surface.

Just like microcrack porosity, the ultrasonic travel time within core varies markedly as a function of the depth from which the core is recovered. Empire State Electric Energy Research Corporation (ESEERCO) drilled a 1 km deep well at Kent Cliffs in southeastern New York to measure in situ stress and, hence, assess seismic hazards associated with the Ramapo fault system (Zoback, 1986). Seven core samples of highly foliated crystalline rock were recovered from depths between 184 and 1008 m for later laboratory analysis. Meglis et al. (1991) measured the ultrasonic velocities in samples of the core which were reloaded in hydrostatic compression to 20.7 MPa. The cores showed a velocity anisotropy controlled mainly by preferred crack orientation parallel to the plane of foliation. The general trend of this suite of samples was that the shallower samples showed higher average ambient velocities than did deeper samples (fig. 9–19). However, reloading the deep core to 20.7 MPa was not sufficient to recover the velocities observed for the shallow samples. As microcrack porosity was the main factor controlling the velocities (the lithology being relatively constant in all samples), the Kent Cliffs ultrasonic data suggest that the microcrack porosity in the core samples increased with initial sample depth as was verified by the direct measurement of microcrack porosity (fig. 9–19). The same conclusion was reached by Carlson and Wang (1986) in their study of cores from the Illinois Deep Hole, UPH-3. The laboratory samples showed lower V_p and V_s than were observed in situ from sonic logs (Moos, 1986). In addition, a general decrease in average ambient velocity with sample depth was not seen in the in situ measurements. From these observations and calculations, Meglis et al. (1991) conclude that the bulk of the microcracks which affect ultrasonic velocity in the core were not present in situ, but formed on removal of in situ stresses.

Not all microcracks in core form during stress relief. Microcavities in basaltic core from Iceland were examined both ultrasonically and under the microscope (Kowallis et al., 1982). In contrast to studies of other crystalline rock, the pores or microcracks of this Icelandic basalt did not close when loaded to pressures equivalent to burial. In this case strain relaxation was not responsible for the microcrack porosity detected ultrasonically. However, this study did suggest that there were many low aspect ratio cracks that were difficult to detect.

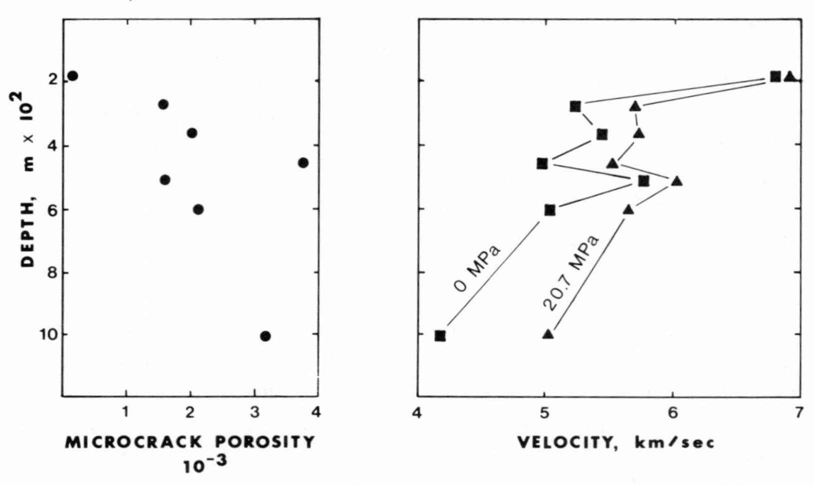

Fig. 9–19. Sample porosities plotted as a function of initial sample depth for core recovered from the Kent Cliffs deep hole near the Ramapo fault zone, New York (data from Meglis et al., 1991). Average compressional wave velocities at 0 and 20.7 MPa confining pressure on core taken from the Kent Cliffs deep hole. Velocities are plotted as a function of initial sample depth.

The inability to observe low aspect ratio microcracks has caused discrepancies between the observed and calculated crack spectra.

Anisotropy in Ultrasonic Travel Time

Understanding the effect that stress relief has on changes in ultrasonic properties of rock is easiest using concurrent measurements of in situ stress and ultrasonic properties in near-surface rock. Much of the initial work on the effect of stress relief on ultrasonic properties of near-surface rock was by Swolfs and coworkers (Johnson et al., 1974; Swolfs and Handin, 1976; Swolfs, 1976; Swolfs et al., 1981). Their experiments on sandstones consisted of stress relieving blocks greater than a meter on a side; the ultrasonic travel time within the block increased upon relaxation. Swolfs (1976) reports the same effect on examining cores of Barre granite.

Early experiments show that if open microcracks have a preferred orientation, then travel time of P-waves in the host rock varies with P-waves traveling faster parallel to the open microcrack set than normal to the set (e.g., Nur and Simmons, 1969). The host rock is said to have a velocity anisotropy which is here called a V_p anisotropy (i.e., the difference in V_p between the fast and slow directions). Three general types of V_p anisotropy are evident in situ (Engelder and Plumb, 1984): (1) outcrops of limestone and siltstone show little or no variation in V_p with azimuth (± 0.1 km/sec); (2) outcrops of granites show a well-defined anisotropy; and (3) other outcrops show nonsystematic variation in velocity with azimuth. The latter behavior is attributed to local joints in the outcrop or inhomogeneities in stress and microcrack density. The V_p anisotropy within stress-relieved cores also varies with lithology (Engelder and Plumb, 1984). The limestone shows no anisotropy and little variation in absolute velocity among cores (± 0.1 km/sec). Many of the other lithologies have a marked anisotropy and a larger variation in velocity as measured in the same direction on several cores (± 0.5 km/sec). Some of the best information on the nature of the development of microcracks upon stress relief come from experiments which measure the change in V_p anisotropy as a rock is cut free of its boundary tractions. Several examples are discussed below.

The first example of V_p anisotropy is one for which the anisotropy does not change on stress relief. The effect of stress relief on P-wave velocity (V_p) is shown for the Milford granite at the Barretto Quarry in Milford, New Hampshire (Engelder and Plumb, 1984). The ultrasonic data are displayed using plots of P-wave velocities as a function of azimuth for both in situ conditions and the relaxed core (fig. 9–20). S_H at the Barretto quarry has an azimuth of about 110° (the azimuth of maximum expansion) with a stress difference of about 23 MPa (S_H = 27 MPa and S_h = 4 MPa). In situ V_p varied from a maximum of 5.6 km/sec along the azimuth of 110° to a minimum of 4.9 km/sec at right angles to the maximum. After stress relief, V_p of the Milford granite varied from a maximum of 4.3 km/sec along the azimuth of 110° to a minimum of

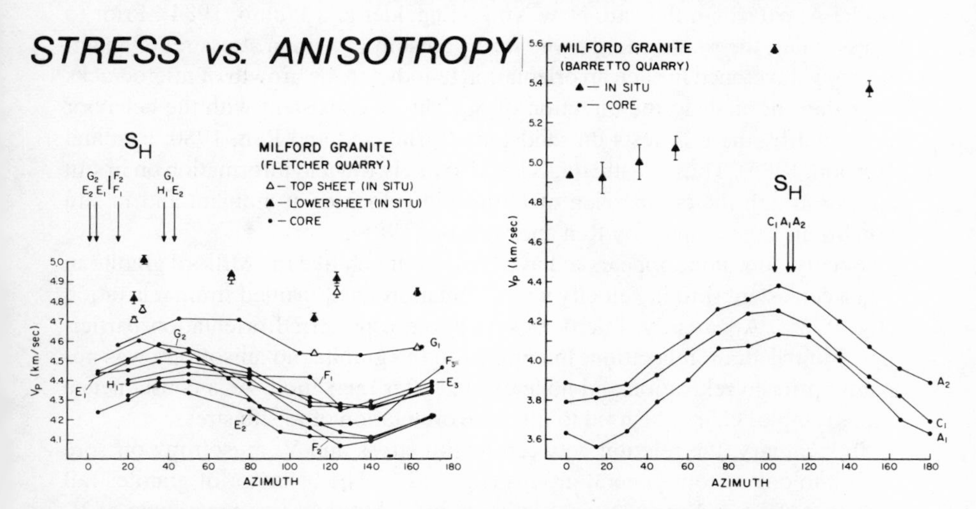

Fig. 9–20. Plot of compressional wave velocity (V_p) versus azimuth for the Milford granite from two different quarries. Both in situ and core velocities are shown as well as azimuth of maximum expansion on overcoring (adapted from Engelder and Plumb, 1984).

3.7 km/sec. In the case of the Milford granite, the magnitude and orientation of the V_p anisotropy remained constant despite a large decrease in absolute V_p. Opening of microcracks upon stress relief had a dramatic effect on the V_p of the Milford granite as predicted by laboratory experiments of Nur and Simmons (1969).

A comparison of two measurements in the same granite (the Milford of southern New Hampshire) but at different quarries shows that the magnitude of ΔV_p upon strain relaxation correlates with the magnitude of the in situ stress (fig. 9–20). At the Barretto granite quarry, data were taken from a deep quarry floor 50 m below land surface, whereas data at the Flecther quarry were taken less than 5 m below surface level in a zone of sheet fractures. The natural processes of sheet-fracturing near the surface at the Flecther quarry relieves some of the stress within the granite as is indicated by the magnitude of the in situ stress. This latter outcrop of Milford granite exhibits a smaller ΔV_p on strain relaxation compared with the more highly stressed granite in the floor of the Barretto quarry.

A second example of the nature of V_p anisotropy is seen for the Barre granite in Vermont. The Barre granite, like the Milford, has a strong anisotropy under in situ conditions. The characteristic that distinguishes the Barre granite is that upon stress relief the anisotropy becomes much stronger with the maximum ΔV_p in the direction of S_H as indicated by overcoring. Such behavior is characteristic of microcracks growing with their normal facing S_H, the orientation assumed for the growth of microcracks during T-DSR.

A third example is seen for the behavior of the Devonian Machias sandstone of the Appalachian Plateau, New York (Engelder and Plumb, 1984). Prior to stress relief, the sandstone showed no V_p anisotropy. Upon stress relief an anisotropy developed in such an orientation to indicate the growth of microcracks with their normals to the direction of S_H. This is consistent with the behavior predicted by the DSA tests on sandstone (Strickland and Ren, 1980; Ren and Hudson, 1985). Thus, clastic rocks are also likely to yield information on in situ stress through the comparison of ultrasonic velocities at ambient and in situ conditions as attempted by Ren and Hudson (1985).

Finally, limestone appears to have behaved much like the Milford granite in that a consistent shift in velocity with orientation accompanied strain relaxation (fig. 9–21). Apparently, microcracks without a preferred orientation participated in the strain relaxation. In contrast to the granite, no anisotropy was apparent prior to relaxation and none existed after relaxation. As a consequence these samples cannot be used to infer the orientation of in situ stress.

In summary, the relation between in situ stress and V_p anisotropy on core falls into one of four general groups (fig. 9–21). The behavior of granites fall into two classes: those showing decrease in V_p but the same magnitude of V_p anisotropy as was found in situ (Milford granite); and those showing a larger V_p anisotropy than was found in situ (Barre granite). The other two groups include

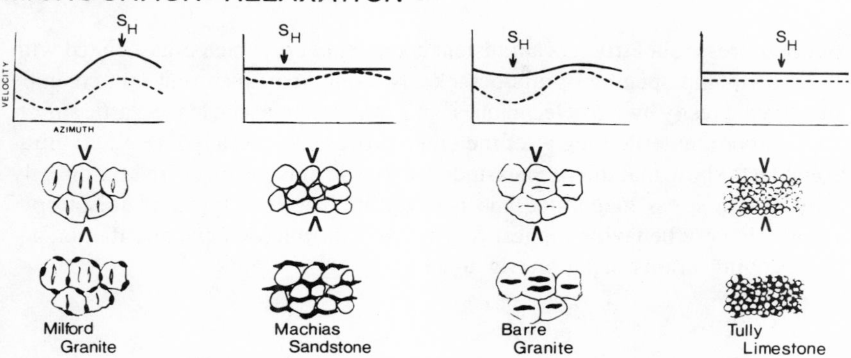

Fig. 9–21. A schematic diagram showing four different types of behavior for the change in ultrasonic velocity with respect to in situ stress (adapted from Engelder and Plumb, 1984). The top row consists of plots of velocity versus azimuth where the solid line is the in situ velocity and the dashed line is the velocity measured on a core in the laboratory. The orientation of the maximum compressive stress is shown by the arrow above each set of velocity versus azimuth curves. The middle row shows four rocks under load where the maximum compressive stress is indicated by arrows. Preexisting microcrack fabrics are shown where appropriate. The bottom row shows the growth or opening of microcracks (indicated in black) upon removal of the in situ stress by overcoring.

siltstones showing a minimum V_p parallel to S_H when there was no anisotropy in situ, and limestones with no relation between anisotropy and S_H because the rock lacks a homogeneous anisotropy. These experiments show that stress relief of all rocks causes a decrease in V_p and thus that the microcrack growth or amount of microcrack opening is a direct record of the magnitude of stress relief.

Testing models for microcrack behavior on stress relief (e.g., figure 9–21) is difficult because, despite extensive thin section observations, the identification of the microcracks that actually formed during the stress relief is difficult. This difficulty is best illustrated for Milford granite where the in situ and core anisotropy are of the same magnitude. On relaxation, there is a consistent shift in the velocity curves (the velocity decrease is independent of azimuth). The anisotropy both in situ and in the cores correlates with a set of transgranular cracks that are readily apparent in thin section. However, the consistent shift in the velocity curves implies that microcracks opened on relaxation but that no particular orientation of microcracks was favored. In contrast with the Barre granite, a well-aligned set of microcracks seemed not to participate in the relaxation. If other microcracks are responsible for relaxation, they are not readily apparent when the granite is viewed in thin section.

Concluding Remarks

Sudden stress relief triggers a transient creep recovery which is associated with the growth and opening of microcracks. Not only may such transient creep be measured directly by T-DSR techniques but it is also manifested in the elastic anisotropy of core material long after the creep process has decayed. DSA and ultrasonic tests show that microcrack-induced elastic anisotropy correlates directly with in situ stress magnitude and orientation. The four types of anisotropy-creep recovery behavior suggest four models for microcracks distribution accompanying strain relaxation (fig. 9–21).

10

Residual and Remnant Stresses

Clearly, "in situ stress" and "residual stress" should not be regarded as synonymous terms: the latter is merely one component of the former.
—Barry Voight's (1966) review of the various components of earth stress

Components of the earth's stress field may arise from small-scale residual stresses as well as large-scale boundary tractions (e.g., frictional resistance at strike-slip boundaries of lithospheric plates) and body forces (e.g., sinking of a dense lithosphere in a subduction zone). Of these components the significance of residual stress is by far the most difficult to assess. *Residual stress* is manifest by elastic strain within elements of an isolated body after all boundary tractions are removed. The elastic distortion of any element in the body is held in place by surrounding elements which may also be elastically distorted. All of the distorted elements are locked together in such a manner that the body as a whole is in a state of static equilibrium and does not spontaneously fly apart to restore each elastic element to its unstrained state.

Examples of the presence of residual stress in rock are found in several engineering applications, particularly in quarrying and tunneling. Probably the most famous rock containing a large component of residual stress is the Carrara marble, for centuries a favorite Italian building stone. In selecting pieces of Carrara marble for statues, Michelangelo carefully avoided blocks from quarries well known for residual stress for fear that the marble would spontaneously crack or explode after months of sculpting. Despite Michelangelo's diligence, statues such as his *David* contain cracks which John Logan (1990, personal communication) attributes to the relaxation of residual stress.

Some of the first thoughts about residual stress in a geological context come from the Swiss geologist, Albert Heim (1878), who was reacting to his observations of the inside of an early trans-Alpine tunnel. Heim noted that walls in the tunnel would collapse under the tremendous stress in the surrounding rocks. From these observations, Heim formulated his ideas on residual stress by assuming that S_H and S_h were similar in magnitude to S_v. Although Heim correctly attributed S_v to the weight of the overburden, he thought that the present horizontal stresses in rocks of the trans-Alpine tunnels were left from the "gigantic forces" responsible for the Alpine Mountains. Hence, Heim reasoned that the present horizontal stresses were ancient or "residual" and not related to modern processes. Heim's notion of residual stress is more like remnant stress, a concept discussed in detail later in this chapter. Briefly, one example of *remnant stress* is the component of the present stress field which arises as a consequence of erosion under uniaxial-strain conditions. By such a definition, the increase in the ratio, S_h/S_v, in the near surface (see chap. 3) is a manifestation of the development of remnant stress rather than a reflection of modern stress within the deep lithosphere. Remnant stress arises as a consequence of processes not directly related to modern tectonics. This usage of the term remnant stress is similar to that employed by Ranalli and Chandler (1975).

Residual Stress

Whether or not a body contains residual stress is based on arbitrary definitions which are not applied uniformly in the literature. The arbitrary nature of these definitions is illustrated using examples of two-element bodies: the prestressed beam of concrete and the bimetal strip used as a temperature indicator (fig. 10–1). Through the core of a prestressed-concrete beam is a steel rod which holds the concrete in compression.[1] Two techniques are used to place the concrete in compression: a "pre-tension" and a "post-tension" technique (M. Friedman, 1990, personal communication). A "pre-tension" beam is manufactured by first placing a steel rod in tension and then pouring concrete around it. The concrete is allowed to set and then the tension in the rod is relaxed. Because the rod is cemented in place, relaxation of the rod throws the concrete

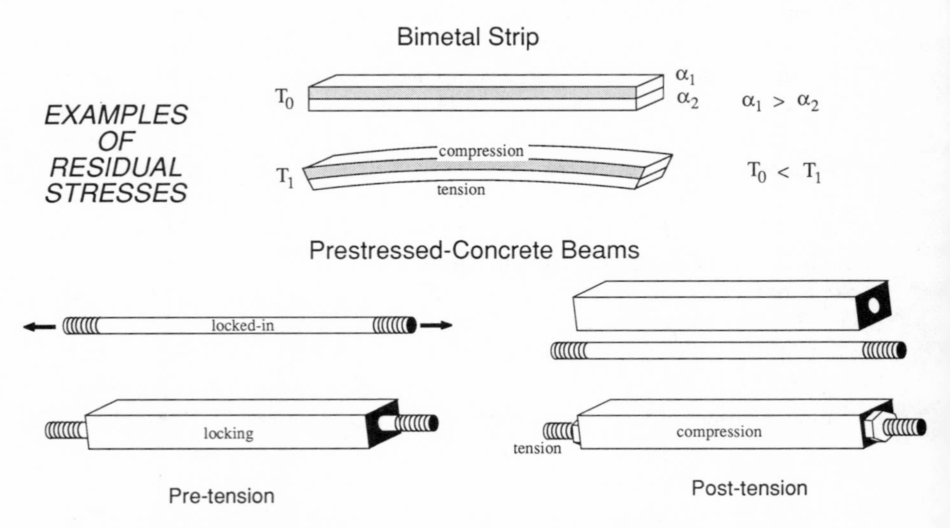

Fig. 10–1. Bimetal strip and prestressed-concrete beams. The two concrete beams are known as the "pre-tension" and the "post-tension" beams. These are three examples of types of residual stress in nongeological bodies.

into compression. This two-element body is free of boundary tractions and, yet, each element contains elastic strains which would relax should one element be cut free of the other. Following Friedman's (1972a) definition the tensile stress in the steel rod is locked-in by the concrete beam which contains locking stresses as a consequence of being compressed by the steel rod. Although the locked-in stress came first, both the locking and locked-in stresses are residual stresses.

A "post-tension" beam is manufactured by pouring concrete about a strain-free rod or plastic tube in which a steel rod is placed after the concrete has cured. Once the concrete has cured, nuts are threaded to both ends of the rod and tightened against the concrete. Tightening the nuts places the steel rod in tension and the concrete beam in compression. Although this is also a traction-free body with stressed elements, the concept of locking and locked-in stress does not apply because the stresses are imparted simultaneously to both elements. In both examples of prestressed concrete, tensile forces in the steel rod equal compressive force in the concrete beam. However, the absolute magnitude of the stress in the steel rod and beam differs depending on the cross-sectional area of each element. If a small piece of the prestressed concrete beam is cut free it would spontaneously expand to its initial or undeformed shape. Friedman (1972a) would argue that the "post-tension" beam does not illustrate residual stress because an earlier loading event, the prestressing of the steel rod, is not locked in place. The literature is more generous with its definition by including analogues of the "post-tension" beam as a type of residual stress; thermally induced stress is one such analogue (e.g., Savage, 1978).

The bimetal strip consists of two sheets of metal with different thermal expansivities, welded in contact, face-to-face. As the temperature of the strip increases one metal sheet expands faster than the other. Because of the weld, the differential expansion causes the bimetal strip to bend. The sheet on the outer radius expands faster but is constrained in compression by the sheet on the inner radius. The same constraint forces the sheet on the inner radius into tension. Again, the concept of locking and locked-in stresses does not apply. The bimetal strip contains residual stress induced by thermal changes, whereas the concrete beam contains residual stress induced by mechanical changes. In contrast to the preceding examples which have just two elements, the complexity of residual stress in rocks is magnified by a large number of elements.

In a rock containing residual stress, equilibrium is achieved by counterbalancing elements which are in tension with elements that are in compression. The *equilibrium volume* is the smallest unit of a rock within which all forces balance (Varnes and Lee, 1972). Other terms for the equilibrium volume include locking domain (Varnes and Lee, 1972) and self-equilibrating unit (Russell and Hoskins, 1973). Mathematically, the equilibrium volume is the smallest rectangular parallelepiped for which each of the Cartesian components of residual stress: (1) averages to zero over its interfaces; and (2) exerts no net moments (Tullis, 1977, after Bishop and Hill, 1951). In the example of the bimetal strip and the prestressed concrete-beam residual stresses balance one another over a body of mesoscopic dimensions. In rock, an equilibrium volume may vary in scale from the microscopic to the size of granite plutons. Residual stress may vary in magnitude and orientation from place to place within the rock, particularly if the rock consists of many elements. Still, an integration of all forces within the equilibrium volume will equal zero indicating that the forces are balanced. However, integration of forces within selected parts of the equilibrium volume may yield a significant net force. It is this net force in a small part of the prestressed-concrete beam which causes it to relax when cut free. Identifying the size of the equilibrium volume is difficult; without knowledge of the equilibrium volume, the interpretation of residual stress data is subjective.

Portions of the equilibrium volume may contain forces consistent in magnitude and orientation. These portions are called *residual-stress domains*. This definition for residual-stress domains in rock is scale dependent. For example, each element of the bimetal strip and prestressed concrete beam is a residual-stress domain because the same magnitude of strain relaxation would occur regardless of the position of the piece cut from the initial element. If the integration of forces within adjacent portions[2] of a rock is consistent in magnitude and orientation, then each portion of the rock may belong to a single residual-stress domain at a macroscopic scale. While the rock may contain macroscopic residual stress domains, residual stress within grains may differ from that in adjacent cement and so the rock also contains the domains of consistent stress orientation which are very small. Hence, we must conclude that a rock may contain residual-stress domains on both microscopic and mesoscopic scale. Regardless of the scale of the domains, they are always smaller than the associated equilibrium volume.

The terms residual stress and residual strain are used interchangeably. Many, including Friedman (1972a), prefer to use residual strain largely because strains are actually measured and stresses are only inferred depending on the appropriate scaling factor (usually an elastic modulus). Residual stress is used in this chapter largely because lithospheric stress is the subject of this book.

The Grain-Cement Model

Residual stress in rocks is perhaps best illustrated by a grain-cement model which closely resembles a sandstone (Friedman, 1972 a,b). This model involves the generation of residual stress by both mechanical and thermal processes accompanying burial, diagenesis, and uplift (e.g., Gallegher et al., 1974). First, grains are loaded (i.e., the aggregate is buried) without pore fluid or cement, in which case stress is transmitted between grains through grain-

grain contacts (fig. 10–2). Such stress arises from tractions due to overburden loads and tectonic deformation. Once buried, cement is introduced to the aggregate which in this hypothetical model remains free of pore fluid. Prior to and during the introduction of cement the entire load is carried by the grains. Even after cementation, all boundary tractions are still carried by the grains unless further deformation takes place. The second step in generating residual stress by the grain-cement model is the removal of the boundary tractions from the cemented aggregate, a process equivalent to the erosion of overburden rocks. Upon removal of the overburden, the grains will start to relax against the constraining cement. Such a constraining action allows only a partial unloading of the stress within the grains and transfers some of the load to the cement. On complete removal of overburden, the aggregate equilibrates without surface tractions, yet each element (i.e., grain and cement) within the rock remains stressed. Because cement prevents stresses in the grains from completely relaxing, the grains contain a locked-in stress[3] (Friedman, 1972a). Once the boundary loads are removed, part of the grain-on-grain stress is transferred to the cement which then contains "locking" stresses. The rock now contains residual stresses with grain-scale domains. Cutting a small portion of the rock free (i.e., removing a grain) causes the rearrangement of stress within the rest of the grains and cement.

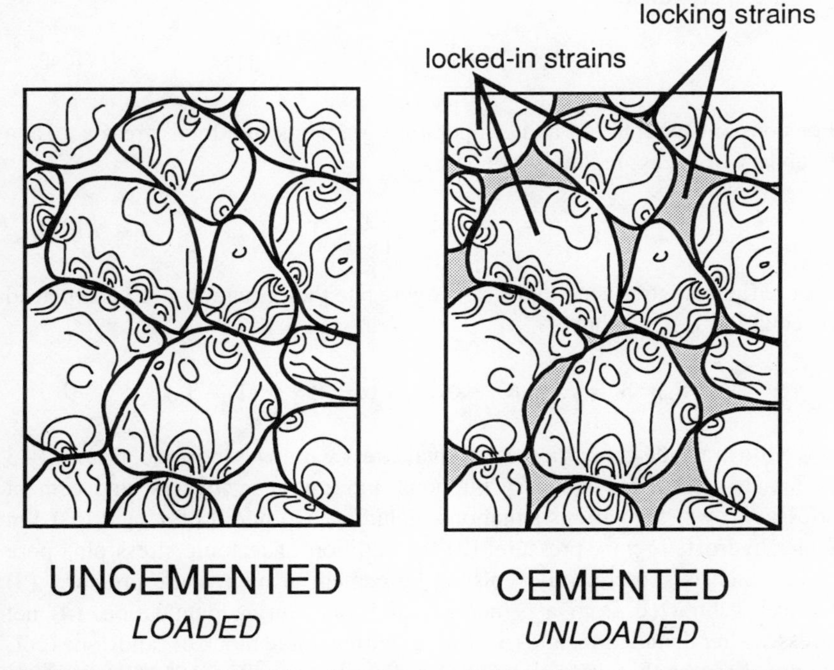

Fig. 10–2. Friedman's (1972b) grain-cement model for residual stress in a sandstone.

Voight and St. Pierre (1974) presented a quantitative version of the grain-cement model, where a cement is again introduced only after a sand aggregate has reached its full depth of burial. This highly idealized model does not consider the effect of burial diagenesis on such depth-dependent properties as density and elasticity. Despite this drawback, it is a useful model for illustrating the grain-scale genesis of residual stress. The following is a variation of the Voight–St. Pierre analysis in which the effect of pore fluid pressure on stress within grains and cement is considered. Although the photoelastic model of figure 10–2 shows sharp stress gradients near grain-grain contacts, the following calculation assumes a homogeneous state of stress within grains and cement. The Voight–St. Pierre analysis starts with a calculation of boundary tractions on a cube of sand buried at a depth of 1 km. If uniaxial strain conditions are assumed for a sand aggregate during burial, then by combining equations 1–17 and 3–49

$$dS_H = dS_h = \frac{\nu}{1-\nu} dS_v + \frac{\alpha_T E dT}{1-\nu} \tag{10–1}$$

where dS_i are incremental total stresses, α_T is the linear coefficient of thermal expansion, and dT is the incremental temperature change. If the aggregate is buried to a depth, z, where $S_v = \rho g z$, then the lateral stresses on the sand aggregate are approximately

$$S_H = S_h = \int_0^{S_v} \frac{\nu}{1-\nu} dS_v + \int_{T_0}^{T_1} \frac{\alpha_T E}{1-\nu} dT. \tag{10–2}$$

For constant overburden and temperature gradients, the total stress across a boundary of the aggregate is

$$S_H = S_h = \frac{\nu}{1-\nu} S_v + \frac{\alpha_T E}{1-\nu} (T_1 - T_0)\,. \tag{10–3}$$

If pore fluid is introduced to the sand aggregate then equation 1–17 is replaced by equation 1–43 so that

$$S_H = S_h = \frac{\nu}{1-\nu} (S_v - \alpha_b P_p) + \alpha_b P_p \frac{\alpha_T E}{1-\nu} (T_1 - T_0)\,. \tag{10–4}$$

To illustrate the generation of residual stress within a sandstone, figure 10–3 is a matrix showing boundary tractions, intragranular stresses, and cement stresses (rows) for various situations including: (1) burial of a sand to 1 km under hydrostatic pore pressure; (2) the addition of tectonic stress plus pore fluid which is subsequently replaced by cement at hydrostatic pressure; (3) stresses subtracted from a cemented sandstone during denudation; (4) net stresses after denudation; and (5) stresses within a free block of sandstone (columns). Assume that during burial of a saturated sand E = 10^3 MPa, υ = .21,

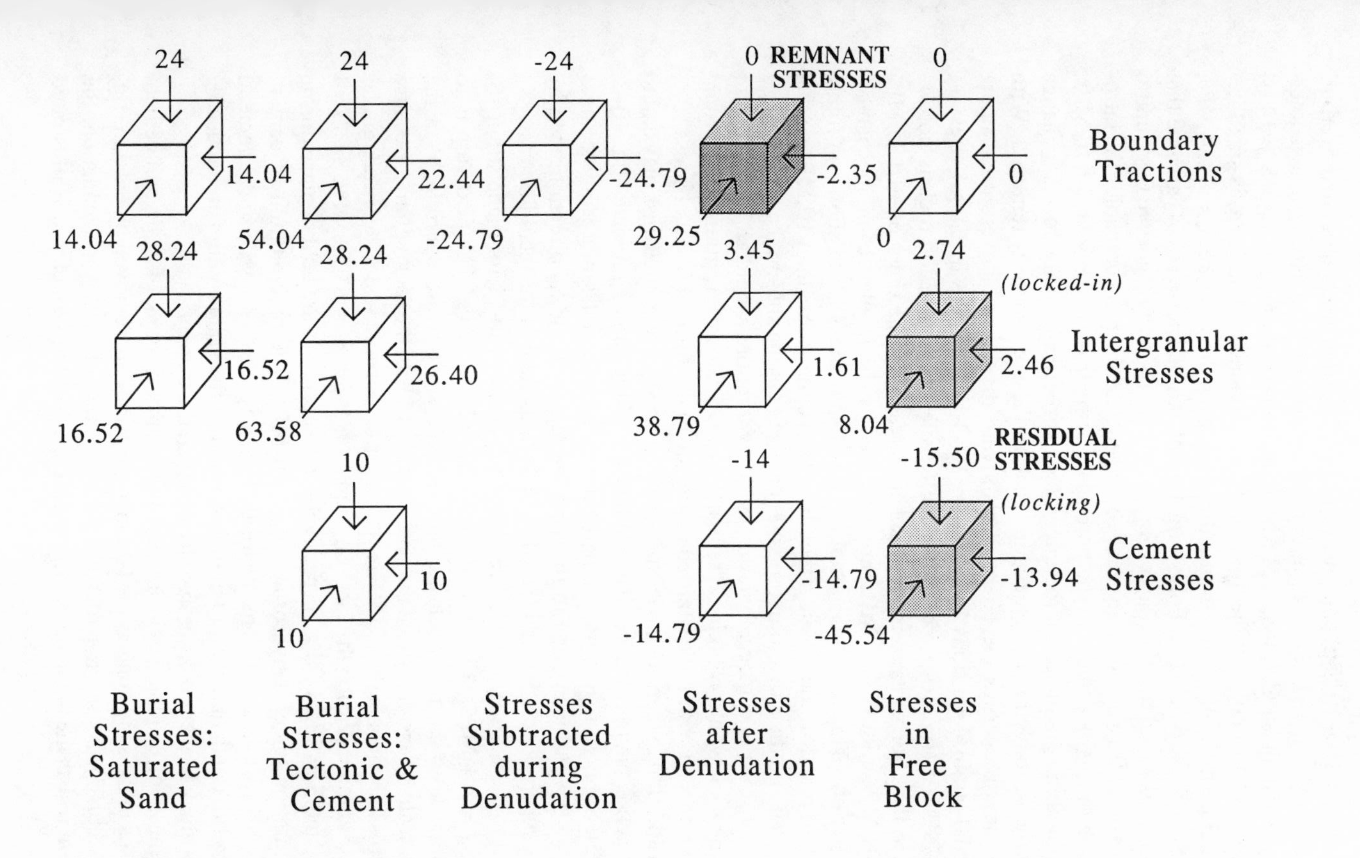

Fig. 10–3. Voight and St. Pierre's (1974) analysis of the grain-cement model. See the text for a detailed explanation.

$\alpha_b = 1$, $\alpha_T = 10 \times 10^{-6}$/°C, and porosity, $\phi = 15\%$, so that the integrated overburden density of the sand is 2.4 gm/cm^3. At a depth of 1 km assuming a geothermal gradient of 25°C/km, $S_v = 24$ MPa and from equation 10–4, $S_H = S_h = 14.04$ MPa (fig. 10–3, col. 1). The thermoelastic component, 0.32 MPa, is small because E is small. At 1 km intragranular stresses in the sand are $S_v = 28.24$ MPa, and $S_H = S_h = 16.52$ MPa. Following Voight and St. Pierre (1974) assume that E increases to 4.5×10^3 MPa, α_T goes to 10.8×10^{-6}/°C, α_b goes to 0.85, and υ decreases to 0.13 as a consequence of diagenesis which accompanies the introduction of quartz cement in the pore space at hydrostatic pressure of 10 MPa. For simplicity, assume that the integrated overburden density does not change with the introduction of cement and that all the pore space is removed. With these new properties the effect of removing the 1 km of overburden under uniaxial-strain conditions by equation 10–3 is a change in lateral stress of –24.79 MPa (fig. 10–3, col. 3). If the tensile strength of the rock is not exceeded then horizontal remnant (tensile) stresses on the order of –10.75 MPa (14.04 MPa minus 24.79 MPa) are generated by thermoelastic and mechanical contraction.[4] This particular example of remnant stress is not shown in figure 10–3. In practice, the tensile strength of sandstone is less than 10 MPa so that this model predicts that joint propagation occurs in the subsurface. If a block of this sandstone is cut free, then the boundary tractions would all return to zero and the rock would contain no remnant stresses. However, grain-stress would remain locked within the lithified aggregate in the form of residual stress.

In addition to thermoelastic stress, a component of tectonic stress may add to the tractions across the vertical boundaries of the sand aggregate (fig. 10–3, col. 2). Suppose that before cementation at a depth of 1 km, a tectonic stress ($'S_H = 40$ MPa) is applied to the sand aggregate in the S_H direction, then by stress superposition $S_H = 54.04$ MPa. The component of the tectonic stress, $'S_h$, added to the S_h direction is calculated using equation 1–21 for plane strain, $'S_h = \nu' S_H = 8.4$ MPa, so that total stress, $S_h = 22.44$ MPa. The next step is to examine how the thermoelastic and tectonic stresses are distributed on the grain scale by assuming that the pore space is 15% of the rock volume. Because the cross-sectional area is 15% less on the microscopic scale, the intragranular stresses are proportionally larger by equation 1–39: $S_H = 63.58$ MPa, $S_h = 26.4$ MPa, and $S_v = 28.24$ MPa, respectively. These stresses are held in place during cementation under a hydrostatic $P_p = 10$ MPa. The precipitation of cement under hydrostatic conditions means that $S_H = S_h = S_v = 10$ MPa in the cement. Under full depth of burial, 1 km in this example, the grains within the sandstone carry much higher compressive stresses than the cement.

Genesis of residual stress is tied to the rearrangement of grain and cement stresses in the sandstone as the boundary tractions are removed by erosion (fig. 10–3, col. 3). As the macroscopic stress system changes on denudation, the cement resists the tendency for the stresses in the grains to relax as is illustrated in figure 10–2. Removal of the overburden reduces the vertical boundary traction to zero and decreases the horizontal tractions by 24.79 MPa so that remnant stresses on the sandstone cube in a flat outcrop are $S_H = 29.25$ MPa, $S_h = -2.35$ MPa, and $S_v = 0$ MPa (fig. 10–3, col. 4). Even though the tectonic event causing a stress difference in the horizontal plane is ancient history, the record of that event is present to this day as indicated by a remnant stress whose principal components show a difference. This remnant stress is also more highly compressive than the remnant stress left after denudation of a sandstone not subjected to a large tectonic compression. At this stage stress in both the grains and cement is reduced at the same rate as the boundary tractions are removed so that $S_H = 38.79$ MPa, $S_h = 1.61$ MPa, and $S_v = 3.45$ MPa in the grains and $S_H = -14.79$ MPa, $S_h = -14.79$ MPa, and $S_v = -14$ MPa in the cement (fig. 10–3, col. 4).

Residual stresses in the sandstone must balance so that static equilibrium is maintained once surface tractions are removed. This means that the integral of residual stresses within the cement and grains must vanish. Although the balance of forces within the grains and cement is complex, Voight and St. Pierre (1974) simplify the balance with the approximation that is modified in this book to

$$\sigma_g^i A_g + \sigma_c^i A_c + \Delta\sigma_g A_g + \Delta\sigma_c A_c = 0 \qquad (10\text{–}5)$$

where σ_g^i is the intragranular stress just after cementation, σ_c^i is the cement stress just after cementation, $\Delta\sigma_g$ and $\Delta\sigma_c$ are stress changes in the grains and cement induced by relaxation of the boundary stresses, and A_g and A_c are the proportion of area of grains and cement to the total area of an average cross section. Because stress on the grains and cement is reduced at the same rate as boundary tractions are removed, $\Delta\sigma_g = \Delta\sigma_c$ provided grain and cement materials have the same elastic properties, as assumed here, so

$$\Delta\sigma_g = \Delta\sigma_c = \frac{\sigma_g^i A_g + \sigma_c^i A_c}{A_g + A_c}. \qquad (10\text{–}6)$$

From equation 10–6 the change in stress in the S_H direction is −55.54 MPa. Now the average residual intragranular stress, which is locked-in in the direction of the ancient tectonic stress, is 63.58 − 55.54 = 8.04 MPa for the rock free of boundary tractions (fig. 10–3, col. 5). Following the same procedure, other components of the locked-in stress are $S_v = 2.74$ MPa and $S_h = 2.46$ MPa. Once the block is free of boundary tractions, the locking stress in the cement becomes as large as −45.54 MPa. Tensile stresses of this magnitude should assure that microcracks develop within the cement. Significant residual tensile stresses are also present in the other directions in the cement. A quick check will show that the absolute value of the residual cement stresses and the intragranular stresses are equal when multiplied by A_c and A_c, respectively; so, each component of the compressive intergranular stress is balanced by a tensile cement stress as required by the definition of residual stress.

Inclusion Models

Inclusion models deal with the development of thermally induced residual stresses caused by mismatches in the thermal properties of adjacent grains of a multimineralic rock. Thermoelastic mismatches work in the same manner as the bimetal strip (fig. 10–1). Initial conditions for the inclusion model assume that all grains in a multimineralic intrusive rock fit together without exerting tractions across grain boundaries. This may happen during the cooling of an igneous intrusion just after the rock has completely crystallized. Any grains that crystallized above the solidus maintain thermomechanical equilibrium with surrounding melt by contracting while cooling. Once all melt has crystallized, the grain boundaries weld so that thermal contraction is no longer accommodated by the surrounding melt. Further thermal contraction is constrained by neighboring grains in such a manner that grain-scale residual stresses develop in the interlocking grains.

One model which illustrates thermally generated residual stress is the bisphere, a two-element elastic sphere with an inner core of a different material (fig. 10–4). Savage (1978) calculates stress in the bisphere due to temperature changes by using the elastic approach of Sokolnikoff (1956) in spherical coordinates r, ϕ, and θ. Of particular interest to Savage (1978) is the residual stress developed in a spherical granite "intrusion" in an infinite matrix of country rock. The two components of this bisphere model are assumed to cool from a

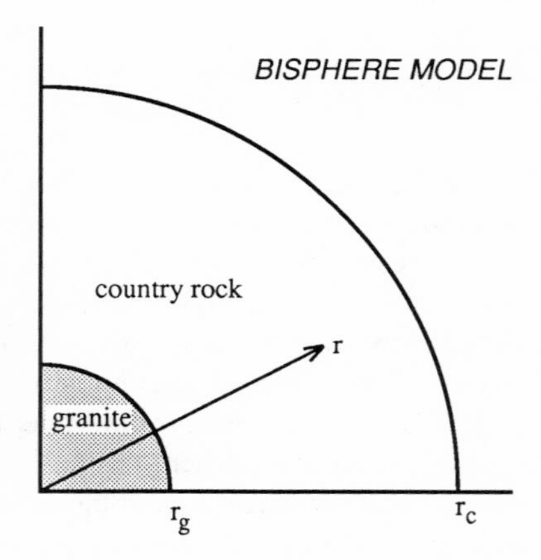

Fig. 10–4. The bisphere model for residual stress (adapted from Savage, 1978).

temperature at which both were unstressed. Stress in the granite inclusion (g) is uniform

$$\sigma_r^g = \sigma_\theta^g = \frac{12\kappa_g\zeta_c(\alpha^c{}_T - \alpha^g{}_T)}{3\kappa_g + 4\zeta_c}\Delta T, \tag{10–7}$$

whereas stress in the country rock (c) is a function of the radius of the granite pluton, r_g, and the distance into the country rock from the center of the granite pluton, r,

$$\sigma_r^c = \frac{12\kappa_g\zeta_c(\alpha^c{}_T - \alpha^g{}_T)}{2\kappa_g + 4\zeta_c}\left[\frac{r_g}{r}\right]^3 \Delta T \tag{10–8}$$

$$\sigma_\theta^c = \frac{6\kappa_g\zeta_c(\alpha^g{}_T - \alpha^c{}_T)}{3\kappa_g + 4\zeta_c}\left[\frac{r_g}{r}\right]^3 \Delta T. \tag{10–9}$$

κ and ζ are elastic constants and α_T is the coefficient of thermal expansion. Stresses within the pluton are constant, whereas in the country rock stress drops in proportion to $1/r^3$. Savage (1978) calculated that a thermoelastic tensile stress of 23 MPa develops in the granite inclusion for a 300°C temperature drop. Under this condition the granite might still be at a depth of more than 1 km where the compressive stress from overburden loading would superimpose on this tensile stress to suppress cracking. Tensile cracking is likely if the granite is completely eroded and uplifted to the surface. Despite Friedman's (1972a) definition, Savage (1978) classifies these thermoelastic stresses as residual.

In general, residual stresses can occur in bodies which yield plastically during heating and react elastically during cooling. This behavior is analogous to the "post-tension" concrete beam (fig. 10–1). To analyze this behavior Holzhausen and Johnson (1979b) carry the inclusion model for residual stress a step further by considering the effect of plastic deformation within the inclusion. Their analysis starts with a homogeneous inclusion of circular cross section embedded in another homogeneous material. The two-dimensional stress-strain relations for an elastic body subjected to temperature change, ΔT, and to plastic deformation, ε^p, are:

$$\varepsilon_{xx} = \frac{\sigma_{xx}}{\omega} - \frac{\sigma_{yy}}{\chi} + \alpha_T\Delta T + \varepsilon^p{}_{xx} \tag{10–10}$$

$$\varepsilon_{yy} = \frac{\sigma_{yy}}{\omega} - \frac{\sigma_{xx}}{\chi} + \alpha_T\Delta T + \varepsilon^p{}_{yy} \tag{10–11}$$

$$\varepsilon_{xy} = \frac{\sigma_{xy}}{\zeta} + \varepsilon^p{}_{xy} \tag{10–12}$$

where ζ is the shear modulus, α_t is the coefficient of thermal expansion, and:

$$\omega = \frac{E}{1-\nu^2} \quad \text{and} \quad \chi = \frac{E}{\upsilon(1+\nu)} \tag{10–13}$$

for plane strain, and

$$\omega = E \quad \text{and} \quad \chi = \frac{E}{\nu} \tag{10–14}$$

for plane stress. These equations are valid for radial symmetry if x is replaced with r, and y is replaced with θ.

The elastic-plastic inclusion has a circular cross section and is bonded to a hole in an infinite elastic matrix, where the inclusion and matrix are composed of different materials. The inclusion fits into the hole in the matrix in a stress-free state (fig. 10–4). To develop a residual stress the inclusion is strained plastically, where $\varepsilon^p_{rr} = \varepsilon^p_{\phi\phi}$ so that it no longer fits into the hole. In nature this might occur if the bisphere was heated from the inside out as would be the case for buried nuclear waste which serves as a radioactive heat source. For the inclusion to fit into the hole after plastic deformation, the inclusion, the matrix around the hole, or both, must deform. Once the deformed inclusion is fit into the hole, the inclusion then loads the matrix giving rise to residual stresses in both the inclusion and matrix. According to Holzhausen and Johnson (1979b), if the inclusion has a radius, a, and the matrix is not subjected to tractions on its external boundaries, then the residual stresses in the inclusion are homogeneous and equal to

$$\sigma^i_{rr} = \sigma^i_{\phi\phi} = \frac{-[(\alpha^i_T - \alpha^m_T)\,\Delta T + \varepsilon^p_{\phi\phi}]}{\dfrac{\chi^i - \omega^i}{\chi^i\omega^i} + \dfrac{R\chi^m + \omega^m}{\chi^m\omega^m}} \tag{10–15}$$

in which superscript (i) refers to the inclusion, superscript (m) refers to the matrix, R is unity, and $\varepsilon^p_{\phi\phi}$ is the plastic strain in the inclusion. In this analysis positive values are tensile. This equation has the same form as that presented by Savage (1978) in equation 10–7. The residual stresses in the matrix are

$$\sigma^m_{rr} = -\sigma^m_{\phi\phi} = \sigma^i_{rr}\left(\frac{a}{r}\right)^2 \tag{10–16}$$

where r is the radial distance from the center of the inclusion. If the inclusion and matrix have the same thermal and elastic properties, the residual stresses in the inclusion and matrix are

$$\sigma^i_{rr} = \sigma^i_{\phi\phi} = -\varepsilon^p_{\phi\phi}\frac{\omega}{2} \tag{10–17}$$

$$\sigma_{rr}^{m} = -\sigma_{\phi\phi}^{m} = -\varepsilon_{\phi\phi}^{p} \frac{\omega}{2} \left(\frac{a}{r}\right)^{2}. \qquad (10\text{–}18)$$

Holzhausen and Johnson (1979b) point out that these equations indicate two general features of thermoelastic residual stress: (1) there is no single state of residual stress in a body; and (2) the residual stress is constant within the inclusion and varies inversely with the square of the radial distance in the matrix (i.e., Savage's 1978 conclusion). The sign of the residual stress in the inclusion depends upon the relative thermal properties of the inclusion and matrix and upon the signs of the plastic strains and the temperature change. Tensile stresses are induced in the inclusion if the plastic strain of the inclusion reduces the volume of the inclusion, or if the inclusion contracts at a faster rate than the matrix. Residual stress in the inclusion depends upon the relative sizes of the inclusion and the body containing the inclusion. However, feature #2 for residual stress is due to the specific boundary conditions: a finite sphere and infinite radius of matrix.

Multilayered Bodies

Layered sedimentary rocks have the potential for developing residual stresses if the layers have different elastic properties. Holzhausen and Johnson (1979b) illustrate the process of inducing residual stresses in a layered body of sandstone and shale by assuming that alternating the beds of sandstone have different elastic properties than the shale beds (fig. 10–5). Initially the boundaries of horizontal beds are frictionless while each bed is loaded to the same initial external stress (S_H and S_v) under plane-strain conditions. While under load the beds are cemented to each other to make a multilayered body. Equation 10–10 is rewritten to account for strain arising from the superposition of both external (S_H and S_h) and internal (σ_H and σ_v) stresses

$$\varepsilon_{H}^{sh} = \frac{\sigma_{H}^{sh} - S_H}{\omega_{sh}} - \frac{\sigma_{v}^{sh} - S_v}{\chi_{sh}} \qquad (10\text{–}19a)$$

$$\varepsilon_{H}^{ss} = \frac{\sigma_{H}^{ss} - S_H}{\omega_{ss}} - \frac{\sigma_{v}^{ss} - S_v}{\chi_{ss}}. \qquad (10\text{–}19b)$$

Under initial conditions at the time the beds were cemented together the internal stresses and external stresses were equal and so by equation 10–19 the block was in a state of zero strain. Holzhausen and Johnson point out that when the sedimentary rock is freed from all boundary tractions there is nothing to constrain expansion or contraction normal to the layers so that $\sigma_{v}^{ss} = \sigma_{v}^{sh} = 0$. This leaves residual stresses parallel to bedding which arise because horizontal

RESIDUAL STRESS (layered model)

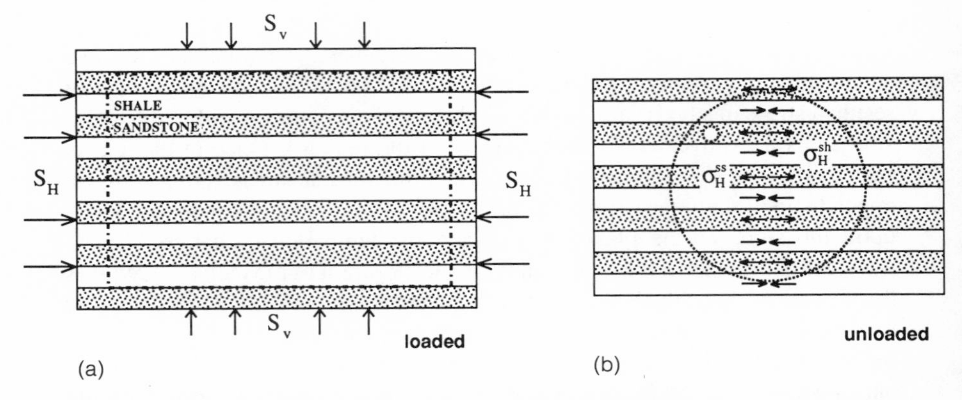

Fig. 10–5. (a) Infinite elastic multilayer composed of stiff sandstone (ss) and soft shale (sh) layers. Contacts between layers are initially frictionless, but layers are bonded together while under the homogeneous initial external stresses S_H and S_v. (b) A block is cut from the multilayer along dashed line in (a) so that it is now free of external stresses. This block represents an equilibrium volume. Residual stresses in the stiff layers are tensile, whereas residual stresses in the soft layers are compressive, assuming that S_H was more compressive than S_v originally. Each layer acts as a single residual stress domain. Dashed circles represent position of overcores described in the text (adapted from Holzhausen and Johnson, 1979).

strain in the sandstone and shale layers are equal after removal of all boundary tractions. By equating 10–19a and 10–19b

$$\sigma_H^{ss} = \frac{\omega_{ss}}{\omega_{sh}}\sigma_v^{sh} + S_v\omega_{ss}\left(\frac{1}{\chi_{ss}} - \frac{1}{\chi_{sh}}\right) - S_H\left(1 + \frac{\omega_{ss}}{\omega_{sh}}\right). \qquad (10\text{–}20)$$

If the sandstone beds gave a thickness, t_{ss}, and the shale beds have a thickness, t_{sh}, then the equilibrium condition for residual stress requires that there is no net force across the vertical plane and

$$\sigma_H^{ss}t_{ss} + \sigma_v^{sh}t_{sh} = 0. \qquad (10\text{–}21)$$

Substituting equation 10–21 into equation 10–20, it is a simple matter to compute the residual stress in the sandstone and shales beds

$$\sigma_H^{ss} = \frac{S_v\omega_{ss}\left(\frac{1}{\chi_{ss}} - \frac{1}{\chi_{sh}}\right) - S_H\left(1 + \frac{\omega_{ss}}{\omega_{sh}}\right)}{1 + \frac{t_{ss}\omega_{ss}}{t_{sh}\omega_{sh}}}. \qquad (10\text{–}22)$$

From equation 10–22, it is apparent that the magnitudes of the residual stresses depend on the external stresses, the ratio of the elastic properties of the beds, and the thicknesses of the beds. If the sandstone beds are stiffer than the shale beds and if $S_H > S_v$, then the residual stress in the sandstone layer is tensile. This is consistent with the fracture distribution of bedded sediments which indicates that joint development in sandstone layers is more common (e.g., Engelder, 1985).

Several characteristics of overcoring into a multilayered body with residual stress are apparent. First, each layer acts as a residual stress domain so that if the diameter of an overcore is smaller than one layer, then residual stress in the core would be completely relieved (Holzhausen and Johnson, 1979b). The sign of the strain relaxation would depend on which layer was drilled. In contrast, if several layers were overcored then the interior of the core would partially relax and residual stresses would remain locked in place. Residual stress accumulated in this manner is highly anisotropic with large strains parallel to bedding. This is one mechanism for the development of the apparent orientation of residual stress (Hyett et al., 1986).

Large-Scale Residual Stresses

In some models, the scale of the residual-stress domain is very large (e.g., Voight, 1974). Imagine a large thrust sheet pushed cratonward above a décollement of very low effective stress (fig. 10–6). Compressive stress within the thrust sheet is derived from tractions developed within the hinterland of the mountain belt as a result of plate-tectonic processes. These tectonic stresses

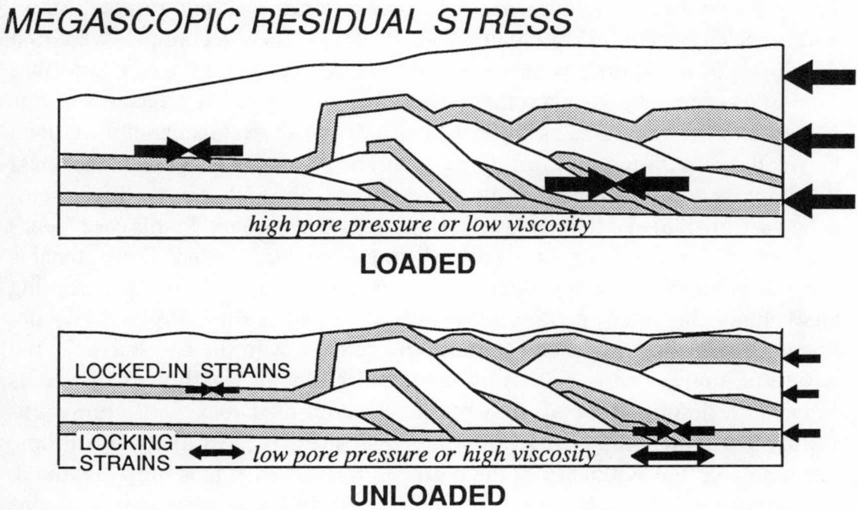

Fig. 10–6. A thrust sheet with compressive stresses being locked-in upon increase of effective stress along the basal detachment (adapted from Voight, 1974).

may arise from a continent-continent collision, tractions along a subduction zone, or gravitational forces due to uplift in the core of the mountain belt. Immediately following tectonism (i.e., removal of the boundary tractions at the back end of the thrust sheet) the compressed thrust sheet will tend to relax by back thrusting toward the core of the mountain range. However, suppose the high fluid pressures bled off from the décollement just prior to the relaxation of tectonism within the core of the mountain belt. Then higher friction along the basal detachment prevents backsliding and acts to "lock-in" compressive stresses within the thrust sheet. To balance the locked-in stresses, rock below the thrust sheet is stretched and, hence, serves as the host for "locking" stresses. This model consists of a pair of elements or residual stress domains (i.e., the thrust sheet and basement) on the scale of an entire mountain range.

According to Bielenstein and Barron (1971) a key drawback to the proposal that residual stresses exist on a regional scale is that large volumes of rock must exist in tension. Because the crust is heavily fractured on the regional scale, such a model requires that these discontinuities transmit tensile stress, an unlikely situation. Martin et al. (1990) point out that models for residual stress at the scale of bedded-sedimentary rocks (i.e., the mesoscopic scale) have the same drawback and, therefore, residual stress may exist only on the microscopic or intergranular scale.

Detection by Means of X-Ray Diffraction

X-ray diffraction techniques are also useful for measuring residual stress within rock aggregates (Friedman, 1967a,b; 1972). These techniques were first developed in metallurgy where it was discovered that, upon plastic deformation, grains in many metals remain elastically distorted. This elastic strain is locked into the aggregate as a consequence of the shape mismatch developed between grains during plastic deformation. The mismatch gives rise to an elastic component of strain held in the grains by neighboring grains, which themselves are not allowed to relax by their immediate neighbors. In this case "locking" and "locked-in" stresses are indistinguishable. Welded grain contacts prevent portions of the aggregate within an equilibrium volume from coming apart under the force of these elastically distorted grains. A plastically deformed aggregate composed of grains and cement as in the case of rocks behaves the same way as plastically deformed grains of roughly equal size as found in a deformed metal. In a plastically deformed rock with grains uniformly distributed in all orientations, residual stress is most likely anisotropic depending on the symmetry of the plastic deformation. If anisotropy is found, the magnitude of the elastic distortion of a certain lattice plane within grains varies with the orientation of the grain.

The d-spacing or distance between lattice planes {hkl} in a crystal is well known and thus can serve as a precise gauge length. Any elastic distortion of the crystal will cause a shift in the d-spacing of lattice planes so that elastic strain is $\frac{\Delta d}{d}$. Because the angle of X-ray diffraction is directly proportional to the d-spacing between lattice planes by Bragg's Law, a distorted lattice is detected by a shift in the diffraction angle. These shifts are large enough to be detected by conventional X-ray techniques. For residual stress measurements by the X-ray technique, the most useful lattice planes satisfy two requirements (Friedman, 1967a,b; 1972). First, the lattice plane must have a relatively small spacing and, hence, a relatively large diffraction angle (e.g., d-spacing changes of 10 parts in 10^{-14} cm are detectable at large diffraction angles (M. Friedman, 1990, personal communication). At larger diffraction angles, changes in lattice distances cause larger changes in diffraction angle. Hence, small changes in lattice distance are more easily detected at large diffraction angles. Second, the lattice plane must give a relatively intense diffraction peak even though it has a large diffraction angle. These two requirements are satisfied by the $\{32\bar{5}4\}$ lattice plane in quartz which, because it is a common mineral in many rocks, is convenient for residual stress measurements by the X-ray technique.

The best rock aggregate for an X-ray analysis of residual stress is fine-grained with the orientation of crystals uniformly distributed. Quartzite, silica-cemented sandstone, and other quartz-rich rocks such as granites are common subjects for residual stress analysis by the X-ray diffraction technique (Friedman 1967a; 1972a). The X-ray technique measures the elastic distortion in grains of a quartz-bearing rock by comparing the d-spacing of $\{32\bar{5}4\}$ in quartz (d_{OBS}) with that of the strain-free material (d_u), i.e., $\varepsilon = (d_{OBS} - d_u)/d_u$. Shifts in the d-spacing for crystals in a particular orientation depend on the strength of the anisotropy in residual stress. Orientations and magnitudes of the three principal strain axes are calculated from strain components measured in each of five directions inclined at 45° to the surface of the sample. The amount of lattice distortion for crystals in various orientations is measured by mounting a 5 cm diameter disk of rock in a goniometer, a device for rotating the disk to various orientations relative to the X-ray beam. Commonly, the disk is tilted at 45° to the X-ray beam so that X-rays are diffracted by lattice planes whose normal is 45° to the surface. In each position of the disk, only a small fraction of the crystals are oriented favorably for diffraction into the collimated detector. A large number of grains in the correct orientation for diffraction are required for proper enhancement of the intensity of the diffracted X-rays. Accordingly, fine-grained rocks are used so that X-rays are diffracted by many grains at one time. The d-spacings are determined by analysis of fixed-count data obtained by step scanning the diffraction profile.

X-ray data do not detect equally both the "locked-in" and "locking" stresses in rock aggregates (i.e., the grain stress and cement stresses of figure 10–2). The intensity of the diffraction is dependent upon the volume of material satisfying the Bragg condition. Thus in a sandstone, the grains (locked-in stresses) occupy a larger volume of the rock than the cement (locking stresses). Hence, the X-rays "see" disproportionately the "locked-in" stresses.

Residual stresses have a profound effect on the mechanical properties of rock as indicated by two mechanical tests. First, Friedman and Bur (1974) measured an ultrasonic velocity anisotropy in several rocks and concluded that there was a geometric and probably genetic correlation between ultrasonic data and residual stress in these rocks. The most compressive residual stresses correlated with the elastically stiff orientation of the sample. Second, experiments on the fracture of quartz-rich rocks show that residual stresses are, indeed, a component of the total stress field as suggested earlier by Voight (1966). Using X-ray diffractometry, Friedman and Logan (1970) measured residual stresses in Tennessee sandstone, Uinta Mountain quartzite, and Tensleep sandstone. Assuming that residual stress acts as a component of the total stress to which test specimens are subject during compression tests, Friedman and Logan (1970) predicted the orientation of either shear fractures or joints based on the orientation of the residual stress in the specimen. Each prediction was then tested by experimentally fracturing the rocks either by point loading disks to induce tensile cracks, or by polyaxial compression of cylinders to induce shear fractures. All test samples were cut with their axes normal to bedding. With this test geometry the components of residual stress parallel to bedding controlled fracture orientation. Indeed, both tensile cracks and shear fractures propagated normal to the most tensile residual stress in the bedding plane. By analogy some have suggested that joint propagation during unloading by erosion and removal of overburden is controlled by the presence of a residual stress (e.g., Preston, 1968; Engelder and Geiser, 1980).

Detection by Means of Double Overcoring

Any elastic strain in a rock, that is free of surface tractions, is attributed to residual stress. The presence of residual stress in rocks is indicated by the strain relaxation which commonly accompanies the cutting of small cores from larger cores and overcoring in a detached block of rock which is free of all boundary tractions. *Double overcoring* is a technique used to detect residual stress in rocks, whereby a smaller core is cut from the axial region of a larger core. The idea is that the initial overcore frees the rock sample from boundary loads so that any strain relaxation accompanying the cutting of the smaller core from the larger core is driven by internal (i.e., residual) stresses. Experiments on blocks of granite show that cutting up to four concentric cores releases some residual stress with each cut (Nichols, 1975; Martin et al., 1990). The implication of this experiment is that residual stresses are completely relieved only after the granite is sliced down to individual grains (i.e., elements). In general, rocks free of boundary tractions will relax if sliced, cut, overcored, or otherwise subdivided. Limestone constitutes the most notable exception to this rule (Engelder, 1979a).

Interpretation of overcoring data is difficult because of the complex distribution of residual stresses within a rock with both microscopic and macroscopic domains. Holzhausen and Johnson (1979b) point out that the absolute magnitude of residual stresses is measured only if overcoring occurs within a single residual stress domain (fig. 10–5). This is equivalent to overcoring a strain gauge on a small portion of the concrete in the prestressed-concrete beam. Overcoring stress relief is only partial if the strain gauge sits over several microscopic domains. In this case, stress relief causes changes in residual stress within domains but it is difficult to relate those stress changes to the absolute magnitude of residual stress. Tullis (1977) concluded that the amount of strain relieved by the second overcore depends upon both the distance from the new cut and the microscopic domains of the residual stress which in a rock are the diameter of constituent grains. Cutting a new surface through a grain causes a grain-scale stress concentration which according to St. Venant's principal (cf. Love, 1934) falls off rapidly away from the concentration. The effect of St. Venant's principal on the rearrangement of residual stresses in the second overcore may be shown using a Fourier analysis of a two-dimensional elastic half-space with stresses arising from surface tractions developed from the residual stresses in the individual grains (Tullis, 1977). In this analysis, the wavelength of sinusoidally varying tractions is on the order of grain diameters. When the domains of the residual stress in a rock consist of grains that are much smaller than the second overcore, a stress change of 100 MPa at the surface is reflected by a 0.1 MPa stress change at three grain diameters into the newly cut core. The interlocking of grains acts to prevent stress relaxation even a few grain diameters into the interior of the rock core. No stress changes occur at the center of the core. Tullis (1977) points out that many double overcoring measurements are not compatible with this expected elastic behavior because strain relaxation occurs more than five grain diameters from the edge of the second core.

Another approach to understanding residual stress measurements is to consider the size of an overcore compared with the initial equilibrium volume, assuming an outcrop is isolated from regional stresses possibly by jointing. Three situations arise: (1) the size of the overcore is large relative to the equilibrium volume; (2) the overcore is the same size as the initial equilibrium volume; and (3) the overcore is small relative to the equilibrium volume (Tullis, 1977). For the first situation overcoring will cause no change in the strain at any distance away from the edge of the overcore as long as the behavior is elastic. This is the same type of behavior as modeled by Tullis (1977) using St. Venant's principle for the behavior of microscopic domains. If the size of the overcore is approximately that of the equilibrium volume, then average stress change integrated over the core should be zero. In the third situation there are two possibilities: The initial core was placed inside one domain; or the initial core was smaller than the equilibrium volume but placed over several domains (fig. 10–5). In the former case the initial overcore relieves all the residual stress so that no strain relaxation takes place during a double overcore. In the latter case, there remains a residual stress which may be further relaxed by a double overcore following the initial overcore. Residual stress is not completely relieved until individual domains are isolated. From these observations one is forced to the conclusion that, if stress is relieved during double overcoring, the second core sat inside the equilibrium volume of the initial core volume. Often a rock cut from an outcrop becomes a new equilibrium volume in which the average of stresses sum to zero. To cut a piece from this equilibrium volume by double overcoring would require the rearrangement of stresses inside the grains and cement of the free piece as well as the initial sample. The smaller core would in turn become a new equilibrium volume.

Stress is a property at a point and there is a high potential for scatter in inhomogeneous rocks (Hudson and Cooling, 1988). Because of this scatter, the regional stress field is reliably measured at much larger scales than most overcoring measurements. Martin et al. (1990) suggest that the stress measurements in rock mass volumes as large as 10^5 m^2 are required before the regional stress field is reliably measured. A stress measurement at the scale of an equilibrium volume should cause the relaxation of stress arising only from the regional stress field. However, a stress measurement within an equilibrium volume relieves an additional component associated with residual stress within the aggregate. Because residual stresses vary from point to point within an equilibrium volume, a suite of stress data from small-scale overcoring measurements should scatter about the magnitude of the regional stress field (fig. 10–7). Indeed, Martin et al. (1990) show that overcoring measurements scatter about a mean which presumably is the regional stress and that the amount of scatter is inversely proportional to the size of overcore. 60 cm overcores may approach the size of the equilibrium volume. In a Canadian granite pluton at a depth of 240 m Lang et al. (1986) use information on the scatter about a mean to infer that the magnitude of residual stress is 3.3% of the magnitude of the regional stress field. Presently, the relationship between grain-scale residual stress and the larger-scale equilibrium volume is not understood.

An Example of the Difference between X-Ray and Overcoring Data

Residual stresses as measured by the X-ray technique do not necessarily correlate with those measured by overcoring. There are a number of reasons for lack of correlation as listed in table 10–1 (Friedman, 1972a). In brief, the most significant difference between the techniques is that upon strain relief strain

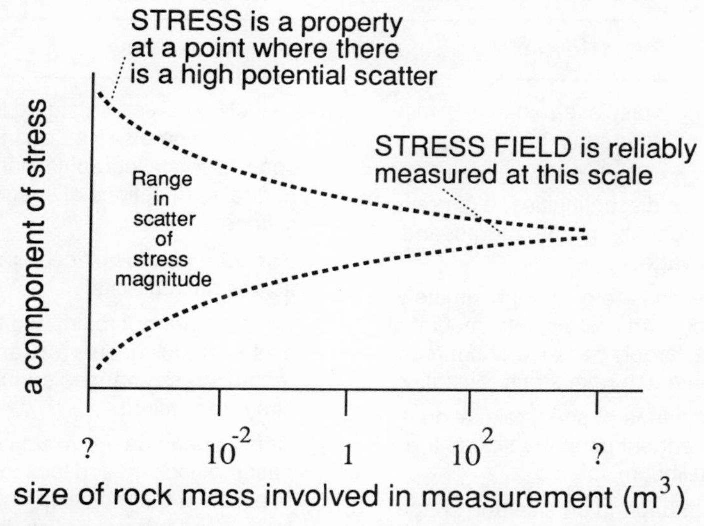

Fig. 10–7. Illustration of the effect of scale on the scatter of in situ stress data (adapted from Hudson and Cooling, 1988).

gauges detect displacements along and across discontinuities such as microcracks in the rock, whereas the X-ray technique only records elastic strain within grains. This difference is illustrated for measurements from the Barre granite of Vermont.

The Barre granite of Vermont is a Late Devonian pluton that was differentiated at depth and emplaced in part by stoping and in part by shouldering aside its wall rock. Foliation of the country rock defines a regional grain associated with the Acadian Orogeny and striking about N20°E, whereas the dominant fabric in the granite is a rift plane of vertical microcracks striking about N30°E. These microcracks have long since healed and are now fluid-filled bubble tracks.

X-ray analyses show that the quartz grains within the Barre granite are elastically stretched such that the average tensile residual stresses are oriented N50°W or subnormal to the rift of the granite (Friedman, 1972a). Friedman (1972a) estimates that the elastic stiffness of quartz normal to $\{32\bar{5}4\}$ is 8.9×10^4 MPa. Using this figure and the average elongation from four samples of Barre granite as measured by the X-ray diffraction technique (Friedman, 1972a, Fig. 5), the average tensile residual stress in Barre granite is 13.5 MPa normal to the rift plane. This tensile stress approaches the average tensile strength for crystalline rocks (Paterson, 1978).

Overcoring surface-bonded strain gauges initiates a strain relaxation in the Barre granite with a maximum expansion (S_H) about N55°W (Engelder et al.,

Table 10–1. Differences in Measurements of Residual Strains by Strain-Relaxation and X-Ray Techniques (Friedman, 1972a)

Strain Relaxation	X-ray
1. Detects displacements in all minerals under strain gauge.	1. Detects mean strain stored in specific minerals (e.g., quartz), in specific crystallographic planes normal to directions of specific stiffness.
2. Detects displacements along and across discontinuities, e.g. grain boundaries, microfractures, and cleavages.	2. Cannot record displacements across microcracks.
3. Stresses are only approximately calculated using elastic moduli of rock largely because of nonrecoverable nature of strain relaxation.	3. Stresses are approximated from elastic moduli of quartz. Cannot record stress changes during strain relaxation.
4. Magnitude of strain relaxation is dependent upon the size of the equilibrium volume.	4. Data depend on the exact combination of locking and locked-in stresses in the rock elements suitably oriented to satisfy the Bragg condition for diffraction. Data biased by the most voluminous component.
5. Data depend on size and shape of block and on geometry of new free surfaces (Nichols, 1971).	5. Once the polished specimen is made, the recorded strains remain unchanged for further change in specimen shape and size except in the immediate vicinity (5 mm) of induced free surfaces.

1977) which is the direction of tensile residual stress (S_h) as measured by the X-ray technique (Friedman, 1972a). Double overcores either expanded normal to the rift or contracted parallel to the rift plane thereby indicating that an equilibrium volume the size of the initial 15.2 cm core is established upon initial overcoring. The S_H component of the residual stress is normal to the rift plane in Barre granite as measured by strain relaxation (fig. 10–8). Concurrent measurements by Swolfs (1976) corroborate this result.

Clearly, overcoring and X-ray measurements were sampling two different phenomena within the Barre granite, both of which are given the general label, residual stress. X-ray measurements detected an elastic residual stress which Tullis (1977) suggests cannot be relieved by double overcoring. Engelder et al. (1977) suggest that the opening of microcracks in the Barre granite is the major

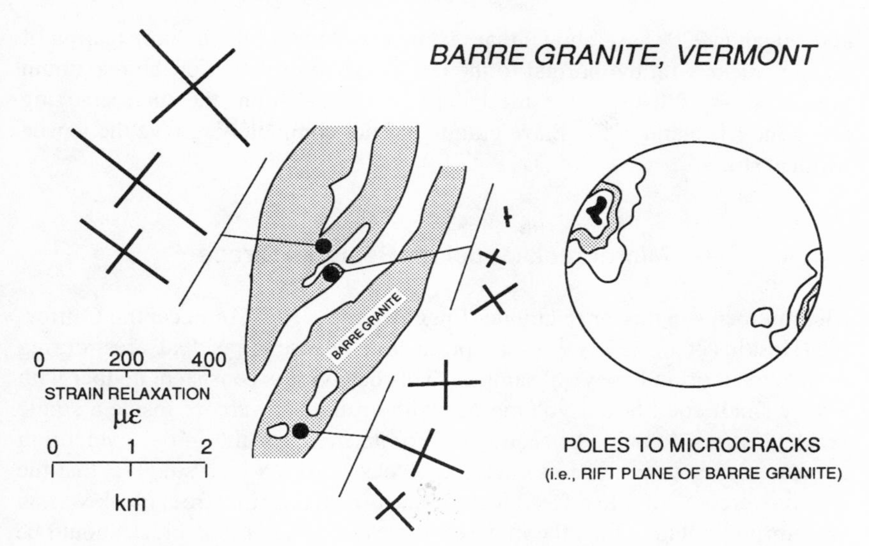

Fig. 10–8. Geology and in situ strain in the vicinity of the Barre granite quarries, Vermont (adapted from Engelder et al., 1977). Outcrops of Barre granite are indicated by stipples. A Paleozoic metasediment surrounds the granite. Sample sites are indicated by large dots. The magnitude and orientation of the strain relieved by the initial overcore are indicated by crosses. Solid lines represent expansion. A scale for the magnitude of the relieved strain is given in units of microstrain. Orientation of 100 poles to microcracks within quartz grains of Barre granite as plotted in equal area, lower hemisphere projection. Contours are at 15%, 10%, 5%, and 1% per 1% area.

mechanism for strain relaxation within the Barre granite as measured by overcoring. This conclusion is supported by the correlation between the density of the microcracks and the magnitude of strain relaxation from site to site within the Barre granite plutons. The samples with the highest microcrack density show the largest expansion normal to rift upon overcoring. One interpretation of the difference between the X-ray and overcoring data is that the X-ray data is measuring the locking stress which would have the opposite sign from that experienced by cores upon relaxation. Another possibility is that overcoring activated a microcrack-controlled, time-dependent strain relaxation as described in chapter 9.

The genetic relationship between the regional tectonic grain and the subparallel rift plane of the Barre granite is poorly understood. One hypothesis is that the rift of the Barre granite develops as a consequence of the relaxation of the regional compression associated with the Acadian Orogeny. However, the stress field in which the microcracks propagated has the same orientation as the present grain-scale residual stress in the Barre granite. Therefore, it is not clear whether the microcracks owe their orientation to the present elastic residual stress or to a regional tectonic stress at the time of cooling and uplift. Friedman and Logan (1970) have shown that residual stress controls the propagation of tensile cracks with the normal to the cracks in the direction of the maximum tensile stress. Whatever the mechanism of propagation, the microcracking along the rift plane in the Barre granite did not completely relieve the tensile residual stress.

Microcrack Model for Residual Stress

Measurements in the Barre granite (Engelder et al., 1977) or in central California (Hoskins et al., 1972) show a reproducible release of residual stress during double overcoring of several samples. This behavior is consistent neither with a very small equilibrium volume nor with an initial overcore inside a single residual stress domain. In central California many of the initial overcoring measurements were made in fractured blocks of rock which suggests that the initial overcore cut into a free block. If each fractured (i.e., free) block was an equilibrium volume, then the integrated stress throughout the block should be zero. If the free block is an equilibrium volume, the average of several overcoring measurements within a single free block should also equal zero, and the orientation of strain relaxation measurement should vary from sample to sample and block to block. Contrary to this prediction, several overcoring measurements in each of these fractured blocks show the same orientation for strain relaxation (Hoskins and Daniells, 1971). From these data Tullis (1977) concluded that such rock did not behave elastically, that perhaps microcracking was responsible for relaxing the residual stress within double overcores, and that the preferred orientation of the microcracks from block to block was responsible for controlling the orientation of strain relaxation within each block.

On a regional scale where double overcoring data are reproducible, Hoskins and Russell (1981) argue that residual stress is controlled by an array of microcracks within the rock. Hoskins and Russell (1981) suggest that, if a set of microcracks propagates in response to a regional tectonic stress field, the microcracks cause the reorganization of residual stress so that it has the same orientation and magnitude from point to point. This reorganization occurs only if microcracks relieve the component of residual stress normal to the microcracks while leaving a large component parallel to the microcracks. Stress relief which accompanies the microcracking event then defines the direction of the minimum residual stress within the rock body which is, of course, normal to the set of microcracks.

To check the behavior of the microcrack model for residual stress Hoskins and Russell (1981) use a model in which the scale of the microcracking (i.e., the characteristic size of residual stress domains) is very small in comparison with the scale of the strain gauge as is the case for most overcoring experiments. In a numerical experiment a block is cut from a rock mass after it was subject to a tectonic or thermal process that induced an array of oriented microcracks. Microcracking relieved 80% of the initial strain in the direction normal to the strike of the microcracks. Cutting a portion of this block by overcoring relieves more strain parallel to the microcracks than in the direction normal to the plane of the microcracks. Hoskins and Russell (1981) go on to cite several field examples where stress relief on double overcoring is maximum in the direction parallel to the set of microcracks. Examples include measurements in the Black Hills of South Dakota (Hoskins and Bahadur, 1971) and in northern California (Hoskins et al., 1972) (fig. 7–3).

Oddly, the overcoring measurements in the Barre granite (fig. 10–8) are not compatible with the Hoskins-Russell microcrack model for residual stress where maximum expansion is parallel to the microcracks. However, the stress ellipsoid for the granite, as measured using the X-ray technique, is compatible with the residual stresses predicted by the Hoskins-Russell microcrack model. Double overcoring of the Barre granite is consistent with crack-related behavior controlled by time-dependent relaxation as described in chapter 9 (Nichols and Savage, 1976).

Residual Stress in Rocks of the Appalachian Plateau

The Appalachian Mountains are ideal for developing some understanding of the correlation between residual stress and an ancient tectonic event largely because compression (the orientation of S_H) during the Alleghanian Orogeny is at 60°–90° to the orientation of S_H from the contemporary tectonic stress field. The parallelism of S_H associated with both the contemporary tectonic stress field and the compression direction of a major tectonic event in other mountain ranges such as the Alps of Europe and the Canadian Rocky Mountains, presents an ambiguity which is not present in the Appalachian Mountains. Documentation of residual stress within rocks of the Appalachians comes from data collected using overcoring and X-ray techniques (Engelder, 1979a; Engelder and Geiser, 1980; Engelder and Sbar, 1984).

Grimsby Sandstone and Onondaga Limestone

At the time residual stress was first measured in rocks of the Appalachian Plateau, the presence of a uniform contemporary tectonic stress field in the northeastern United States was noted but not well documented (e.g., Sbar and Sykes, 1973). Although early near-surface overcoring measurements by Engelder and Sbar (1976; 1977) were consistent with the notion that the northeastern United States was subject to a contemporary tectonic stress with S_H oriented ENE-WSW, some doubt lingered about whether a significant component of this stress field existed in the near surface. Surface overcoring measurements were attempted in western New York to measure earth stress associated with some shallow earthquakes triggered during mining of salt by fluid injection. The overcoring data suggested that S_H in the near surface was oriented just west of north in both the Onondaga Limestone and the Grimsby Sandstone (Engelder, 1979a). This was not the orientation of the contemporary tectonic stress field indicated by fault-plane solutions from nearby earthquakes (Fletcher and Sykes, 1977). Double overcoring measurements in the sandstone showed the same stress orientation as the initial overcore (fig. 10–9). Onondaga Limestone showed no strain relaxation on double overcoring, a characteristic common to limestones.

The double overcoring measurements in the Grimsby Sandstone strongly suggested that the major stress component detected in the near surface was residual. The presence of residual stress in the Grimsby sandstone was further confirmed by X-ray analysis which showed that quartz grains were elastically distorted such that the maximum elastic shortening was west of north and the maximum elongation was south of west (M. Friedman, 1979, personal communication) (fig. 10–10). Unlike the Barre granite, elastic distortion of quartz grains in the Grimsby Sandstone is compatible with the overcoring data both in orientation and sign. Furthermore, ultrasonic tests on the sandstone show an elastic anisotropy with the stiff direction of the samples just west of north. Ap-

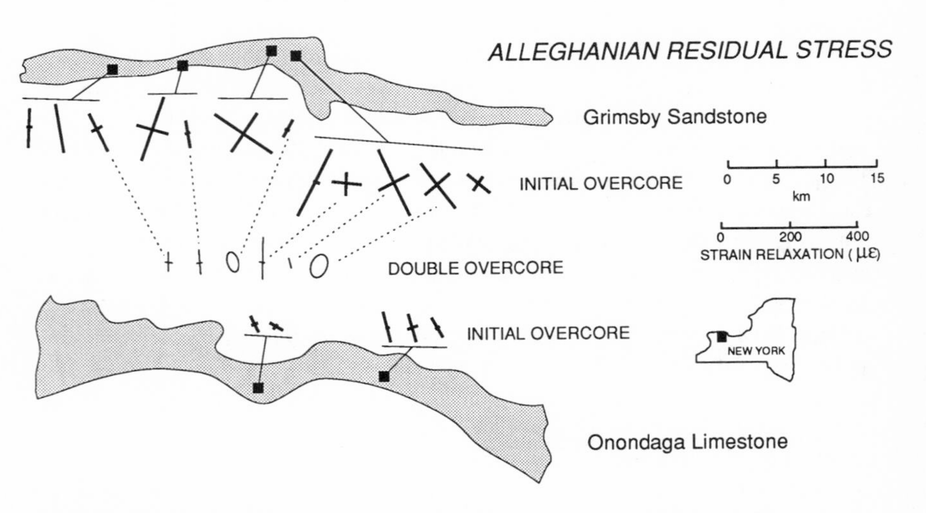

Fig. 10–9. Geology and in situ strain in the vicinity of Brockport and Batavia, New York (adapted from Engelder, 1979a). Six overcoring sites are denoted by squares. The magnitude and orientation of the strain relieved by the initial overcore are indicated by dark lines. The magnitude and orientation of the strain relieved by a double overcore are indicated by the light lines. Solid lines represent expansion and ellipses represent contractions. The long axis of the ellipse is oriented in the direction of maximum expansion (minimum contraction) where contraction took place along both axes. The magnitude of relieved strain is represented by a scale in microstrain.

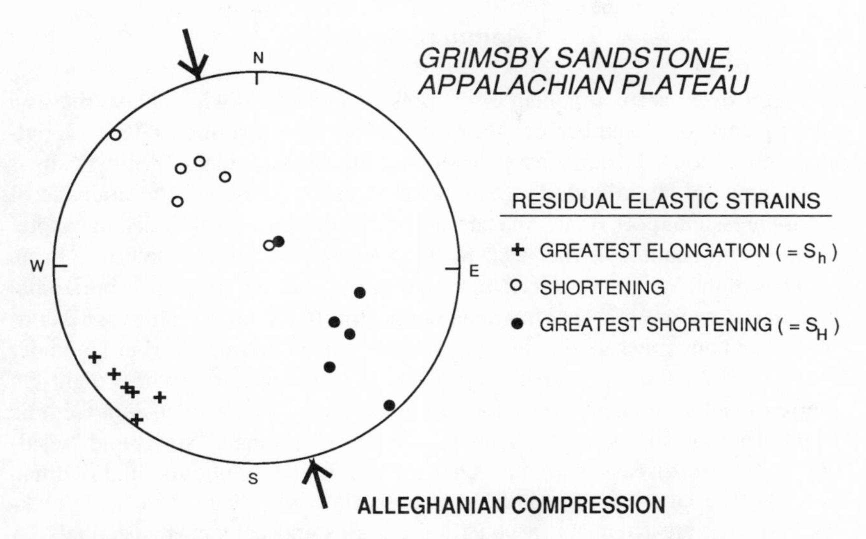

Fig. 10–10. Lower hemisphere projection of the state of residual elastic strain in Grimsby sandstone (adapted from Engelder, 1979a). Six sets of principal strain axes were determined from elastic distortions in quartz grains using the technique of Friedman (1967) (M. Friedman, personal communication, 1979). The symbols, cross, circle, and dot denote the greatest, intermediate, and least principal elongations, respectively. Average strain magnitudes are given.

parently, an excess stiffness developed parallel to a compressive residual stress. Using the elastic properties of the Grimsby sandstone measured by static compression, I infer that the components of residual stress are $S_H = 6.5$ MPa and $S_h = 3.5$ MPa.

The component of S_H from residual stress in rocks of western New York is normal to fold axes of the Appalachian Mountains to the south.[5] This observation stimulated a search for finite strain markers indicative of an Alleghanian-aged compression of the Appalachian Plateau. Prior to this search, finite strain markers were known only at the inner edge of the Appalachian Plateau 150 km to the south (Nickelsen, 1966). Within a year, Engelder and Engelder (1977) discovered finite strain markers showing a NNW layer-parallel shortening throughout the Appalachian Plateau of western New York; this strain is a manifestation of the Alleghanian Orogeny. Other tectonic structures also share a common origin with residual stress in the rocks of the Appalachian Plateau; these include joint sets and the mechanical twinning of calcite in either cement or matrix of rocks (Engelder, 1979b; Engelder and Geiser, 1980). The presence of these tectonic features reinforced the initial interpretation that the surface overcoring had detected a significant grain-scale stress locked within rock as a residual of the Alleghanian Orogeny.

Remnant Stress

Remnant stress is a component of the present stress field which may arise as a consequence of a number of situations including unloading under uniaxial-strain conditions. Two processes leading to the development of remnant stress under uniaxial strain include the removal of overburden and the drainage of highly pressured pore fluid. An example of the former was discussed in chapter 3 under the heading, "The Near-Surface Horizontal-Stress Paradox." In an analysis of the Voight–St. Pierre model, remnant stresses are seen as horizontal boundary tractions present after denudation (fig. 10–3, col. 4). This is an example of remnant stress arising as a consequence of removing overburden under uniaxial-strain conditions. Although tectonic stress appears in the Voight–St. Pierre model, a remnant stress does not necessarily reflect an ancient tectonic event. There is an important distinction between "remnant" stress and "residual" stress as discussed in this chapter and found implicitly in Friedman (1972a). The latter is defined on the grain scale where "locked-in" and "locking" stresses are balanced in constituent grains and cement, respectively. A body may contain residual stresses without the aid of boundary tractions. Remnant stresses are imprinted as a component of the present stress field and arise as a consequence of large-scale boundary tractions, necessary, for example, for maintaining uniaxial strain conditions during unloading. However, there is no evidence that remnant stresses were at one time balanced in an equilibrium volume even on a scale as large as several hundred km^3 as envisioned by Voight (1974) for residual stresses locked into thrust sheets.

The nature of a remnant stress developed by a pore-pressure change is perhaps best illustrated using the stress profile to a depth of 1 km through the Upper Devonian sediments of the Appalachian Plateau where there is a major stress discontinuity at the base of the Rhinestreet Formation (fig. 5–9). See chapter 5 for a detailed discussion of hydraulic fracture stress measurements[6] in the Wilkins well near South Canisteo, New York (fig. 5–10). At the base of the Rhinestreet Formation there is an abrupt decrease in both S_H and S_h (Evans et al., 1989a). Geologic evidence suggests that during the Alleghanian Orogeny, the lower portion of the Devonian Catskill Delta (the section below the Rhinestreet Formation) was overpressured, whereas the overlying section was normally pressured (Engelder and Oertel, 1985; Evans et al., 1989b). The abrupt decrease in the present components of the horizontal stress field correlates with the inferred transition from normal to abnormally high fluid pressure.

The decrease in horizontal stress is consistent with the effect of bleeding, under uniaxial-strain conditions, the high pore pressure to the hydrostatic levels found today (Evans et al., 1989b). Drainage of the highly pressurized pore fluid to normal levels results in a tendency for the rock to contract elastically in proportion to the difference between the bulk and intrinsic compressibility of the rock (Nur and Byerlee, 1971). However, if the rock is constrained to behave according to the uniaxial-strain model, it contracts only in the vertical direction. The response in the horizontal plane is to develop a tensile stress component which reduces the preexisting total horizontal stress according to equation 1–42. The magnitude of this poroelastic component, S_h^p, is given by

$$S_h^p = \alpha_b \frac{1 - 2\nu_d}{1 - \nu_d} dP_p \tag{10–23}$$

where υ_d is Poisson's ratio measured under drained conditions and dP_p is the change in pore pressure.

An analysis of the development of a remnant stress by pore-pressure drainage starts with the assumption that the horizontal components of the stress field increased linearly with depth throughout the Catskill Delta during the Alleghanian Orogeny (Evans et al., 1989b). This linear increase can be seen in the upper portion of the Wilkins well (fig. 5–10). Assuming a value for ν_d of 0.25 for rocks of the Catskill Delta, the reduction in horizontal total stress affected by poroelastic contraction was at least 43% of the pore pressure reduction. Pore pressure is constrained to have been less than the overburden because no horizontal joints are present on the Appalachian Plateau. Thus the maximum possible drop in P_p is the difference between the overburden and hydrostatic pressure. Stratigraphic considerations and conodont color indices (Epstein et al., 1977) suggest that 0.5–1 km of section was eroded from the South Canisteo area of the Appalachian Plateau which implies that the maximum depth of burial of the base of the Rhinestreet shale was between 1.4 and 1.9 km. For an overburden density of 2.65 g/cm^3, the corresponding maximum drop in pore pressure at the Rhinestreet base was 22.7–30.8 MPa. If such an extreme drop occurred after cementation, the concomitant reduction in horizontal stress was at least 43%; that is, 9.7–13.2 MPa. The observed offsets in S_H and S_h at the base of the Rhinestreet are 9.5 and 3.5 MPa, respectively, for the Wilkins well.

If the reduction in horizontal stress at the base of the Rhinestreet was indeed caused by the dissipation of a paleo-overpressure alone, then the resultant field should have been axially symmetric; that is, the offset in S_H should equal the offset in S_h. The stress data suggest that this was not the case (fig. 5–9). However, during the Alleghanian Orogeny the NNW compression was larger than the ENE compression. When this stress state (i.e., S_H = NNW) is superposed with a dominant, axially symmetric poroelastic component, this will leave a smaller net offset in the NNW oriented principal stress component than in the ENE oriented component below the base of the Rhinestreet shale. A smaller offset in the NNW component (i.e., the ISIP data) is seen at the stress discontinuity in the Wilkins Well. This smaller offset reflects a component of tectonic stress remnant from the Alleghanian Orogeny.

In summary, a marked discontinuity in horizontal in situ stress at the base of the Rhinestreet shale corresponds to the top of a section which is inferred, on the basis of local and regional finite strain data, to have once hosted abnormally high pore pressures (Evans et al., 1989b). Dissipation of the abnormal pressure following lithification is shown to be qualitatively if not quantitatively consistent with the observed stress discontinuity. A poroelastic model predicts that the resultant perturbation to the stress field is a consequence of drainage of paleo-pore pressure and, hence, is a manifestation of a "tensile" remnant stress developed by pore fluid drainage under uniaxial-strain conditions.

Another example of a remnant stress appears in Evans's (1989) compilation of commercial hydraulic fracture data for the Appalachian Basin. A map of S_h within the Devonian shales of the Appalachian Basin reveals two distinct trends depending on whether or not the Silurian salt is found below the shale (fig. 5–11). Such a compilation of S_h in the Appalachian Basin leads one to conclude that where the shales were subjected to an Alleghanian layer-parallel shortening over the Silurian salt décollement of the Appalachian Plateau, S_h is considerably higher than in Devonian shales unaffected by major Alleghanian strain. Evans (1989) suggests that the present differences in S_h within rocks, which otherwise appear identical, reflects the presence of a "compressive" remnant stress from the Alleghanian shortening of the shale above the Silurian salt. This interpretation is consistent with the uniform pattern of relatively low S_h in the lower Paleozoic rocks of the Appalachian Basin regardless of whether the Silurian salt is present. Basically, if the deeper rocks of the Appalachian Plateau were affected by the Alleghanian Orogeny, shortening was uniform throughout the basin and considerably less than seen by Devonian shales above the salt décollement.

The Ambiguity between Remnant and Residual Stress

Near-surface stress data are often difficult to interpret, particularly when trying to distinguish between the presence of a mesoscopic-scale residual stress domain and a remnant stress. This is so despite the presence of either residual or remnant stress as a significant proportion of the stress measured in a typical strain relaxation measurement (e.g., Hyett et al., 1986; Hudson and Cooling, 1988). One example of the subtle nature of the difference between residual and remnant stress comes from data taken at a pavement outcrop of the Tully Limestone in a riverbed near Ludlowville, New York (Engelder and Geiser, 1984). This riverbed outcrop forms the lip of a waterfall 20 m due south of the measurement point so that the entire 6 m thickness of Tully Limestone is exposed as a vertical free face with no surface tractions. Like the clastic rocks of the Catskill Delta, the Tully Limestone is also strongly affected by a northward layer-parallel shortening of the Appalachian Plateau during the Alleghanian Orogeny.

Of the ten stress measurements in the Tully Limestone using the "Lamont-cell" doorstopper technique, S_H of nine were within a 15° cluster about N04°E. This is a remarkably tight cluster for any type of in situ stress measurement. The average direction of maximum expansion during strain relaxation of the Tully Limestone is within a couple of degrees of the compression direction of the Alleghanian Orogeny as indicated by the teeth of tectonic stylolites and the strike of nearby Alleghanian-age cross-fold joints (fig. 10–11).

This data set illustrates some difficulties in identifying the origin of stress within the Tully Limestone. A preliminary interpretation suggests that these data are indicative of either a remnant or residual stress for two reasons. First, S_H correlates with the Alleghanian orogeny and not the ENE S_H of the contemporary tectonic stress field of the Appalachian Basin. Second, S_H is normal to the free face of the waterfall which should have unloaded any contemporary tectonic stress in that direction. In attempting to distinguish between residual and remnant stress, some possibilities are ruled out. First, because the measure-

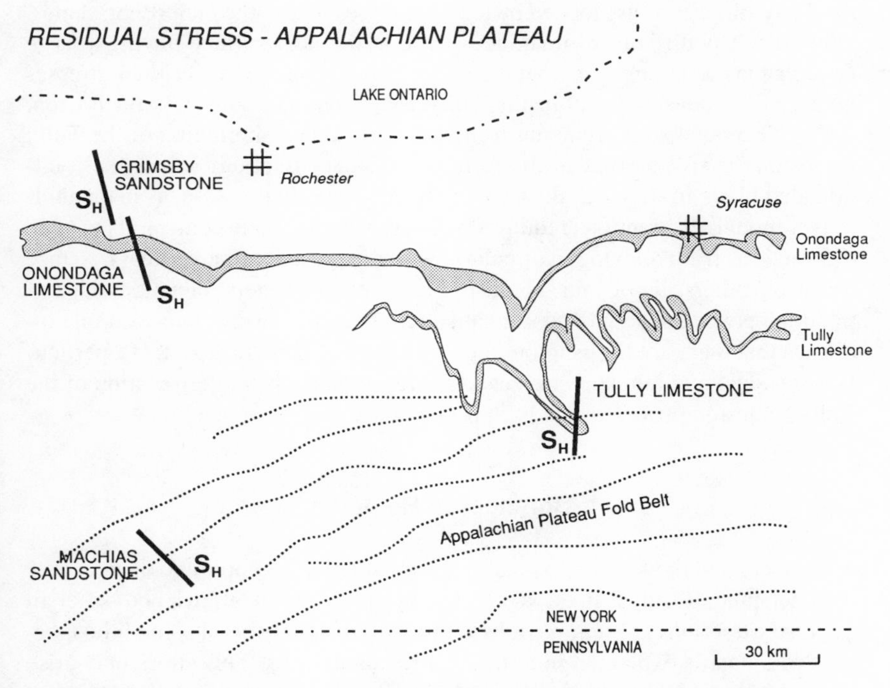

Fig. 10–11. The location of strain-relaxation experiments in the Tully Limestone, Grimsby Sandstone, Machias Sandstone, and Onondaga Limestone (adapted from Engelder and Geiser, 1984). Dotted lines are anticlinal axes on the Appalachian Plateau.

ments were made in a joint bounded block where the spacing of the orthogonal joint sets exceeded 5 m, a genetic relationship between jointing and residual stress seems unlikely. Second, an expansion normal to tectonic stylolites is also ruled out because care was taken not to bond the doorstoppers on top of stylolitic seams which are spaced rather than pervasive. Third, in thin section, no noticeable preferred orientation of grains or microcracks appears, so that a fabric-controlled relaxation similar to the Barre granite is unlikely. Fourth, intragranular deformation of the matrix accompanied by stress solution and recrystallization could have locked in a residual stress by a process much like the plastic deformation of metals. Although this is a mechanism for locking residual stress on the grain scale, it would lead to microscopic equilibrium volumes which are not detected by overcoring.

Based on the definition of residual and remnant stress at least two interpretations are possible for stress in the Tully Limestone. First, the cluster in orientation and magnitude suggests that all measurements were made within one residual stress domain which is consistent with the Holzhausen-Johnson multilayer model (fig. 10–6). If this is the case, the compressive stresses within the Tully Limestone are locked by tensile stresses within the underlying shales. However, it is difficult to imagine a shale in the near surface having a large enough tensile strength to maintain the large compressive residual stresses which were measured within the Tully Limestone (Bielenstein and Barron, 1971). Second, a stress remnant from the Alleghanian shortening of the Tully Limestone is still present in the form of boundary tractions across the joint-bounded block from which the measurements come. In this case, as in the multilayer model, the tractions require some sort of larger scale constraints to keep the tractions from relaxing, particularly with the presence of a vertical free face nearby. At the scale of joint-bounded block several meters on a side, the distinction between residual and remnant stress becomes fuzzy. This example indicates that there are limits to our understanding of lithospheric stress, particularly residual and remnant stresses. As a consequence, an interpretation of the Tully Limestone stresses data is, at best, vague.

Concluding Remarks

The conceptual models for residual stress are relatively simple and, hence, easy to understand. Residual stress is manifest by elastic strain left in a body after all boundary tractions are removed. In addition to residual stress, some lithospheric stress data is explained in terms of a remnant stress. This stress may arise through changes in stress accompanying removal of overburden or draining of high pore pressures. Remnant stress is present as surface tractions which can be removed upon excavation of a block of rock. The distinction between a macroscopic residual stress domain and a remnant stress is vague other than the stipulation that a locked-in residual stress domain requires a specific locking domain. Because rock stress data rarely fit the simple situation envisioned by models for residual and remnant stresses, the interpretation of stress data leaves a feeling that something has been overlooked. So far, our models have not become sophisticated enough to fully understand residual and remnant stresses.

— 11 —

Earthquakes

The results (of calculated stress release for some well-known surface faults) lead us to conclude that the maximum stress released by strike-slip faults in the upper part of the crust lies within the range 1 to 10 MPa.
—Mike Chinnery (1964) on modeling the stress released during strike-slip faulting

During the time when Anderson (1905; 1942) formulated his mechanical concepts for the origin of faulting, earthquake seismologists realized earthquake waves contained information on the orientation of the faults hosting earthquakes. To infer the orientation of an earthquake fault seismologists developed the *fault-plane solution* which is a mapping of the radiation pattern of a particular wave type (i.e., P or S waves) on a sphere about the *hypocenter* of an earthquake (i.e., the position of the initial rupture of an earthquake). A fault-plane solution is often presented as a stereographic projection showing the orientation of earthquake nodal planes (i.e., the earthquake-fault plane and its auxiliary which is orthogonal to the fault). Construction of the fault-plane solution is based on the principle that motion on a fault controls the pattern of seismic wave radiation, particularly the first motion of the P wave arriving, at distant seismometers.

Until 1949, fault-plane solutions were produced by one person or a small group reading seismograms from stations throughout the world. Collecting these seismograms was a slow process prior to the invention of quick and inexpensive copying machines. The advantage of collecting seismograms was that one reader could apply a uniform code of rules while reading the seismograms. Even under these conditions, picking the polarity of the first P-wave motion is a subjective business. In 1949 Dominion Observatory of Ottawa, Canada, started to construct fault-plane solutions at a much higher rate (Hodgson, 1957). Their technique was to depend on data supplied by seismologists who answered questionnaires. No longer was a uniform code applied to picking the polarity of the first arrival, because the seismologists at Dominion Observatory had to rely on someone else's judgement. For this reason, some fault-plane solutions were less reliable. However, the rate of producing fault-planes from earthquakes throughout the world increased significantly. By 1956 there were as many as seventy-five solutions available through the Dominion Observatory. At the same time three other schools (Dutch, Japanese, and Russian) were producing fault-plane solutions using a variety of techniques.

With the mass production of fault-plane solutions seismologists started comparing the orientation of earthquake faulting with the three fault regimes: normal, thrust, and strike-slip. Based on earthquake fault-plane solutions seismologists recognized that large regions of the earth were characterized by a particular type of faulting. These compilations gave a first glimpse of modern global tectonics and, in particular, the orientation of forces responsible for modern orogenesis. In 1957 there were several theories to explain orogenesis including both expanding and contracting earth models. The surprise accompanying the compilation of the first seventy-five fault-plane solutions was that sixty-seven of them were associated with earthquakes along strike-slip faults (Hodgson, 1957). Neither expanding nor contracting earth models were supported.

The Dutch school for fault-plane solutions under A. R. Ritsema had focused its attention on the southeast Asia area where the Dutch had a colony in Indonesia. From fault-plane solutions, Ritsema (1957) also recognized the three modes of faulting described by Anderson (1942). Upon grouping earthquakes from the Indonesian Arc according to depth, Ritsema (1957) found that earthquakes in the range of 0 to 60 km of depth were dominantly strike-slip events as was found by the Dominion Observatory for other parts of the earth. Although thrust faulting along the Indonesian Arc was compatible with a homogeneous contraction of the earth, the large number of strike-slip earthquakes was not consistent with such a theory. Taking 265 fault-plane solutions distributed worldwide, Scheidegger (1959) concluded that the ratio of strike-slip to dip-slip earthquakes was 2.4. Observations such as these set the stage for a general theory of plate tectonics which was developed during the 1960s (Isacks et al., 1968).

Fault-Plane Solutions

An early graphical technique for construction of fault-plane solutions (i.e., the method of extended position) was developed by P. Byerly and his students starting with a study of a 1925 Montana earthquake (Stauder, 1962). Byerly's work was stimulated because seismologists recognized that fault motion accompanied large earthquakes such as the 1906 San Francisco event. Using thoughts developed by Reid (1910), Byerly (1926) reasoned that the motion or "fling" along a fault is distributed into four quadrants in which the motion is alternately a compression adjacent to a rarefaction. The fling in the vicinity of the earthquake is recognized as the direction of the first motion of a seismic wave as it arrives at a distant seismic recording station. Ground motion at local seismic stations surrounding the 1927 Tango earthquake in Honshu, Japan (Honda, 1957) is shown in figure 11–1. Compression is indicated by first ground motion up and away from the earthquake *epicenter* (i.e., the point on the earth's surface vertically above the hypocenter), whereas dilation is indicated by first ground motion down and toward the epicenter. Byerly's model for earthquake motion has two *nodal planes* separating the four quadrants: one nodal plane is the fault plane and the other is an auxiliary plane having no geological significance. On a map of the Tango earthquake, nodal lines are easily drawn as those lines separating stations showing dilation from those showing compression (fig. 11–1). Although the Tango earthquake was located neatly inside a local seismic network, most earthquakes are not so conveniently situated. Byerly reasoned that if he could identify the direction of the ground

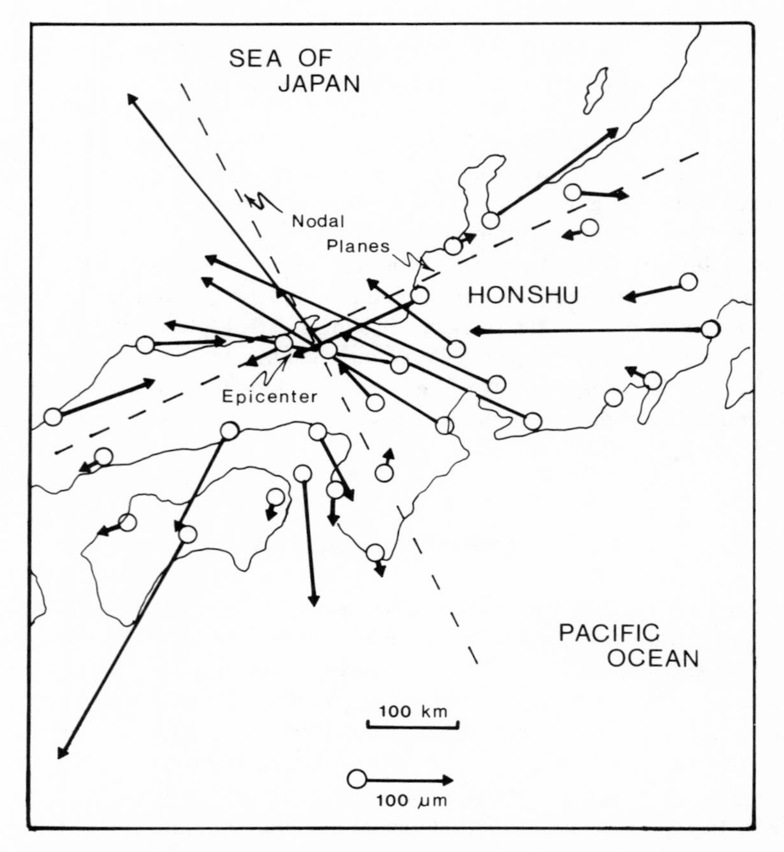

Fig. 11–1. Ground motion on the order of a few dozen to several hundred μm (vectors) caused by the arrival of the initial P-wave from the 1927 Tango earthquake in Honshu, Japan (data taken from Honda, 1957). Dashed lines show the nodal planes.

motion around the earthquake source based on the first motion of P-wave motion at distant seismic stations, he could graphically construct the orientation of the nodal planes.

In constructing a fault-plane solution based on worldwide data, one must deal with seismic ray paths which are curved (fig. 11–2a). Because seismic ray paths through the earth are not straight, the first motion indicated by a seismic wave, if plotted as traveling in a straight line from the epicenter (E) to the recording station, might not correctly represent the ground motion in the vicinity of the earthquake. Byerly recognized that the proper way of plotting first-motion data for each station (S) is to assume that the seismic ray path was straight but tangent to the curved ray path as it leaves the hypocenter. The tangent angle or angle of emergence (i_h) is easily determined since

$$\sin i_h = \frac{a_o dT}{d\Delta} \qquad (11\text{–}1)$$

where Δ is the distance in degrees from the recording station to the epicenter, T is the travel time of the P wave, and a_o is a constant. The i_h-Δ relationship is based on the velocity distribution within the earth (Bullen, 1953). On the spherical surface of the earth, the process of locating a seismic station based on a straight path from the angle of emergence places most seismic recording stations at their extended position (S′) which is often several hundred km or more from their geographic position (S) (fig. 11–2). Byerly plotted the extended position, S′, on an equatorial plane through the center of the earth whose pole touches the epicenter (E) and whose axis, EP, is normal to the plane. The ex-

EXTENDED POSITION OF A SEISMIC STATION

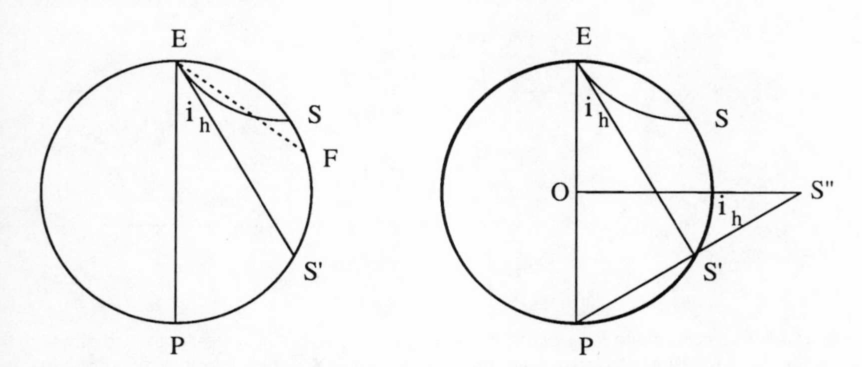

Fig. 11–2. The extended position of a seismic station, S′. Dashed line is the fault plane. E is the epicenter and S is the seismic recording station. S″ is the projection of the extended position onto the equatorial plane. In this example the epicenter and hypocenter are so close that they are indistinguishable.

tended position, S″, is located on the equatorial plane at a distance from the axis, EP, equal to the cotangent of the incident angle of a ray at the focus. The distance, OS″ = cot i_h, is defined as the extended distance of the station on the equatorial plane.

The Method of Stereographic Projection

The most convenient fault-plane solution uses a stereographic projection of the ray paths as they emerge from the hypocenter of the earthquake (Ritsema, 1959). This method assumes that the *focal sphere*, a sphere of unit radius, surrounds the hypocenter. A fault-plane solution is constructed from a stereographic projection which shows where the ray paths to distant stations cut the focal sphere. A seismic wave reaches a distant station S along an azimuth ϕ from the hypocenter. With knowledge of the Δ of the station, i_h is calculated from the the travel-time vs. Δ table for the velocity distribution within the earth. The earthquake hypocenter is located at the center of the stereographic projection representing the focal sphere. ϕ and i_h give the bearing and plunge of S′ for a particular station relative to the center of the stereographic projection (i.e., the hypocenter). S′ for a number of stations recording the same earthquake are plotted in either lower or upper hemispheric projection. Hopefully, enough data are available so that four first-motion quadrants of the earthquake are clearly defined. Compressional first arrivals of P waves are plotted as solid circles and dilatational first arrivals are plotted as open circles (fig. 11–3). A procedure for drawing the two nodal planes is given in Kasahara (1981). Given two nodal planes, geological or other evidence such as distribution of aftershocks is necessary for proper identification of the actual fault plane.

Fault-plane solutions for one earthquake (i.e., the main shock) are con-

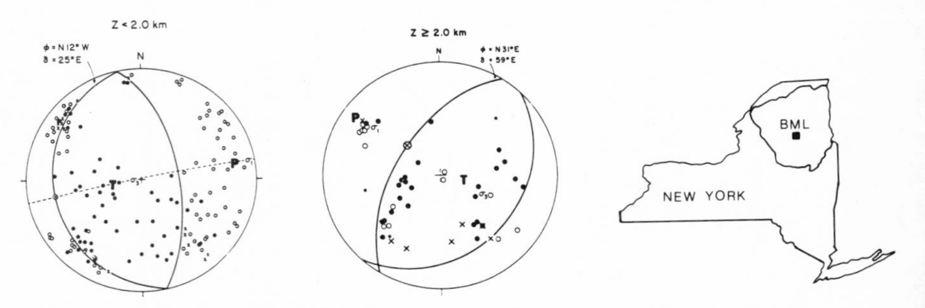

Fig. 11–3. Two composite fault-plane solutions from microearthquakes in the vicinity of the May 23, 1971 Blue Mountain Lake Earthquake, New York (BML). Left is a composite fault-plane solution for microearthquakes at depths of less than 2 km, whereas the right fault-plane solution shows earthquakes from the same swarm but at depths greater than 2 km (adapted from Sbar et al., 1972).

structed from either local network data or from worldwide seismic stations. In these constructions the major source of error is the incorrect reading of polarity of the first motion (Kasahara, 1981). If this happens, compressional and dilitational first motions are incorrectly plotted. The source of this error may arise because the polarity of the station was incorrectly recorded or because the seismogram had a low signal-to-noise ratio. Such errors make it more difficult to locate nodal planes with confidence. One characteristic of the nodal planes is that the P-wave amplitude in the vicinity of the nodal plane is quite small compared with other stations. Hence, a small ratio of the P-wave amplitude relative to the S-wave amplitude is indicative of a station located near a nodal plane.

When the main shock is so small that detection does not occur outside a limited region, *composite fault-plane solutions* are constructed by superimposing data from aftershocks (Sbar et al., 1972). Aftershocks, which may include mainly microearthquakes, are often recorded by portable networks carried near the epicenter of the main shock after that earthquake is located using data from a few distant stations. Superposition requires that the operator locate each aftershock in order to calculate ϕ and i_h for each portable recording station. A composite plot of ray paths cutting the focal sphere is made by moving the center of the stereonet to the hypocenter of each aftershock. Because ray paths for microearthquake studies usually follow an upward path, an upper hemisphere projection of ϕ and i_h is more convenient. For very shallow earthquakes (< 2 km) the angle of emergence varies by less than 5° depending on the crustal structure model. In this case straight-ray approximations are convenient for construction of the fault-plane solution. Calculating the appropriate angle of emergence becomes more critical for larger and deeper earthquakes in areas with a more complex crustal structure.

One assumption in constructing a composite fault-plane solution is that all aftershocks have the same earthquake focal mechanism. This assumption is reasonable if aftershocks occur along the same fault as the primary earthquake. In practice, aftershocks do not necessarily occur along the fault plane responsible for the primary earthquake as is usually indicated by the distribution of hypocenters. Hence, composite fault-plane solutions rarely show a perfect separation of compressional and dilatational arrivals (fig. 11–3). Some aftershocks may occur on faults of a much different orientation from the main shock. A swarm of aftershocks from the 1971 Blue Mountain Lake, New York, earthquakes show hypocentral depths between 0.5 km and 3.5 km. Yet, Sbar et al. (1972) distinguished between shallow (0.5–2 km) and deeper (2–3.5 km) earthquakes to construct two thrust-type composite fault-plane solutions with compression directions separated by about 40°. A complicated tectonic process is required to explain these two fault-plane solutions, provided they are the correct interpretation of the data.

SURFACE-WAVE TECHNIQUES

Alternative techniques for fault-plane solutions were proposed by Brune (1961) using surface waves and by Gilbert and MacDonald (1960) using the amplitude of free oscillations. These techniques are largely based on the comparison of theoretical and observed amplitude and phase-spectra data (Tsai and Aki, 1970). Theoretical amplitude and phase-spectra data are generated for a particular focal mechanism which the operator believes is similar to the real focal mechanism. On the basis of a comparison, the theoretical data are then perturbed in order to minimize phase errors. This process is repeated until a best fit is found. Herrmann (1979) refined this technique using a multiple-filter analysis to estimate spectral amplitudes. The spectral amplitudes are corrected for anelastic attenuation and then a systematic search technique is applied to fault-plane parameters of strike, dip, and slip to find the set of focal mechanisms which have the best fit between observed and theoretical spectral amplitudes (Herrmann, 1979). To evaluate the result, the preferred fault-plane solution is checked against that generated using P-wave first-motion data, provided the P-wave data are of fair to good quality.

The interpretation of the seismotectonics of a region is critically dependent on the location of nodal planes in a fault-plane solution. The picking of nodal planes may depend upon the technique for generating the fault-plane solutions as well as the earthquakes used for the solution. An example of disparate results arises in studies of earthquakes in the vicinity of the Clarendon-Linden Fault of western New York (fig. 11–4). This area is known for moderately strong earthquakes with the largest, a 1929 event, assigned a Modified Mercalli intensity of VIII (Fletcher and Sykes, 1977). A composite fault-plane solution based on P-wave first motions for earthquakes triggered by salt mining near Dale, New

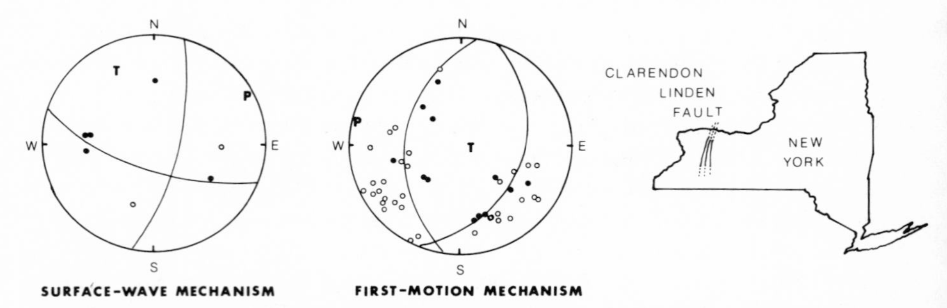

Fig. 11–4. Left is a fault-plane solution for the January 1, 1966 earthquake near the Clarendon-Linden fault in western New York using surface-wave data (adapted from Herrmann, 1979). Right is a composite upper-hemisphere plot of first motions from microearthquakes occurring near the Clarendon-Linden fault (adapted from Fletcher and Sykes, 1977).

York, shows a thrust-type solution with nodal planes striking about 15° east of north (Fletcher and Sykes, 1977). The hypocenter of these microearthquakes is at a depth between 0.5 km and 1.1 km. Using a surface-wave technique, Herrmann (1979) determined a strike-slip-type solution for a $m_b = 4.4$ (January 1966) earthquake in the vicinity of the Clarendon-Linden Fault (fig. 11–4). Its hypocenter was at 2 km. One nodal plane had the same orientation as that predicted by the composite fault-plane solution determined by Fletcher and Sykes (1977). However, the location of the second nodal plane in the surface-wave solution affects an interpretation of the type of faulting along the Clarendon-Linden Fault.

Stress Orientation Based on Fault-Plane Solutions

Because earthquakes were a manifestation of the release of stresses, seismologists inferred that the preferred orientation of earthquake faults over a broad region might indicate the presence of systematic stresses within that region. From these intuitive notions grew the analysis of regional stress fields based on the determination of earthquake fault-plane solutions. Seismologists approached the problem in the same way as Anderson (1942) who pointed out that conjugate faults breaking intact rock are geometrically related to the orientation of the principal stress causing the faulting. In reality, fault-plane solutions are a representation of slip on a fault and the pressure (P) and tension (T) axes represent the axes of maximum shortening and maximum extension. The pressure or P axis is located in the middle of the dilitational quadrant, whereas the tension or T axis is located in the middle of the compressional quadrant. The B axis is located at the intersection of the fault and auxiliary planes. Stress orientation is inferred from this information by assuming that σ_1, σ_2, and σ_3 fall near the P, B, and T axes, respectively. If the P and T axes are near the orientation of the principal stresses, then there is a large difference in the orientation of earth stress depending on which of the two fault-plane solutions from Blue Mountain Lake, New York (fig. 11–3) and the Clarendon-Linden Fault, New York (fig. 11–4) one chooses to represent the regional stress field. In both cases one fault-plane solution shows that the strike of the P-axis is about N60°E, whereas the other fault-plane solution shows the strike of the P-axis is about N75°W.

The physical process accompanying earthquake rupture was not clear when seismologists started mapping P and T axes from fault-plane solutions. An early assumption was that earthquakes reflected the rupture of virgin rock in which case seismologists were justified in mapping the P axis as σ_1. Contrary to this assumption, evidence for earthquakes on preexisting faults came from both the field and the laboratory. In the field, earthquakes which ruptured at the surface clearly followed preexisting faults. Experiments indicated that stress drops accompanying fracture of virgin rock are 100 MPa or larger, whereas seismic source theory predicts that earthquake stress drops are generally less than 10 MPa (e.g., Chinnery, 1964; Wyss and Brune, 1968). From data on rock fracture McKenzie (1969a) argued that the stress drops associated with earthquakes are too small to have been produced by fracture of virgin crust. McKenzie (1969a) pointed out that the Andersonian model for a consistent stress orientation relative to faults does not apply to earthquakes which occurred along preexisting faults. Hence, there is no physical basis for the a priori orientation of these faults at 45° or any other angle to principal stresses. A consequence of the association between earthquakes and preexisting faults is that fault-plane solutions are not precise devices for inferring stress orientations in the crust. In fact, McKenzie (1969a) showed that the only restriction for mapping stress orientation based on fault-plane solutions is that σ_1 must lie in the quadrant containing rarefactions (i.e., the P axis), whereas σ_3 must lie in the quadrant containing compressions (i.e., the T axis). Shear stress along a fault must exceed some yield stress to activate slip along the fault. In an area of preexisting faults those faults oriented so that they have the highest resolved shear stress will slip. Nevertheless, fault-plane solutions are an important ingredient in our understanding of regional stress fields.

The application of fault-plane solutions to understanding intraplate stress patterns was first attempted by Sbar and Sykes (1973) who compiled data on the orientation of stress in the northeastern United States. At the time only three fault-plane solutions were available from earthquakes recorded in the eastern United States. In mapping the regional stress pattern for the eastern United States, Sbar and Sykes (1973) located the compression (P) axis in the center of the dilatational quadrants and at 45° from the nodal plane. Despite the fact that the P axis of a fault-plane solution should not necessarily correlate with the orientation of σ_1, regional compilations show that the average orientation for P axes determined from a number of earthquakes gives a good indication of the maximum compressive stress orientation throughout a region (Sbar and Sykes, 1973; Zoback and Zoback, 1980; Sbar, 1982).

To devise an averaging technique which was more quantitative for determination of stress orientation from fault-plane solutions, Gephart and Forsyth (1984) borrowed from the structural geology literature (chap. 3) where stress tensors were calculated from field measurements of fault orientation, the slip vector, and the sense of slip (e.g., Carey and Brunier, 1974; Angelier, 1979; Etchecopar et al., 1981). Their analyses of fault planes is based on the constraint that fault slip occurs in the direction of maximum resolved shear stress (Bott, 1959). This constraint narrows the possible stresses responsible for slip on a particular fault. If slip has occurred on a family of faults of different orientations, there is a relatively small set of stress tensors which could have activated slip on all of the faults involved. The idea behind fault-slip analyses is to find that family of stresses which is common to the activation of slip along most if not all faults in a region.

Gephart and Forsyth's (1984) objective is to determine a best-fitting regional stress tensor from observed earthquake fault-plane solutions. Briefly, stress tensor is characterized by six independent parameters: σ_1, σ_2, σ_3, and three Euler's angles which give the orientation of the three principal stresses relative to geographic axes. This stress tensor reduces to two parts: a hydrostatic compression and a deviatoric stress. The analysis of fault-plane solutions yields information on neither hydrostatic compression nor the absolute magnitude of stress. This is not a limitation because only the deviatoric stress contributes to the resolved shear stress on a fault plane. Hence, a unit vector in the direction of the resolved shear stress depends on four parameters: the three principal stress directions and the scalar quantity, R, which is the relative ratio of principal stresses (eq. 3–43). Gephart and Forsyth (1984) use an inversion method based on a grid search for all possible tensors which satisfy the fault-plane solutions over a region. The use of a grid search allows for an error analysis and the establishment of confidence limits for the preferred regional stresses. The analysis employs an objective means for identifying which of the two possible nodal planes is more likely to have slipped.

New England, U.S.A., is characterized by earthquake fault-plane solutions which show a large variability in the orientation of P-axes (Pulli and Toksoz, 1981). Using their grid-search technique to maximize shear stress on fault planes from ten fault-plane solutions in New England (fig. 11–5), Gephart and Forsyth (1985) show that there are many sets of stresses consistent with all ten events, but that the orientation of σ_1 was generally consistent with the ENE orientation for S_H established with much of the northeastern United States. The large variation in orientation of P-axes is attributed to the broad range in orientation of preexisting fault zones throughout the crust of New England. In principle, averaging techniques based on fault-plane solutions give the same results as found by averaging other stress data. For example, a compilation of overcoring measurements for New England (chap. 12) shows that S_H is N50°E with a wide variability in orientation (Plumb et al., 1984a).

Initial compilations of fault-plane solutions in intraplate areas such as the eastern United States progressed slowly because the frequency of earthquakes is very low. Only with the installation of local microseismic networks did the inventory of fault-plane solutions increase. Later compilations of focal mechanisms from instrumentally recorded microearthquakes in the northeastern United States confirmed Sbar and Sykes's (1973) initial conclusion that S_H was oriented about ENE throughout a large region (Seborowski et al., 1982; Quittmeyer et al., 1985). However, the fault-plane data in New England appear to show more variability in the orientation of S_H than is found closer to the craton.

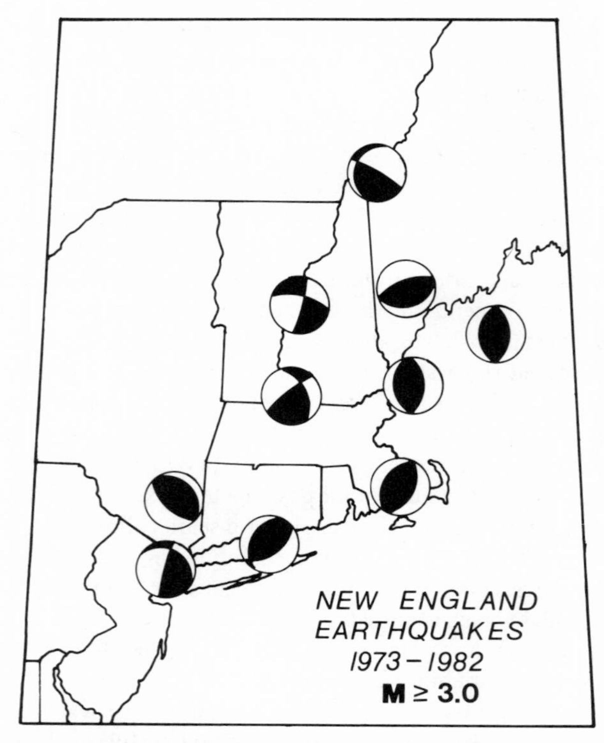

Fig. 11–5. Ten fault-plane solutions for recent earthquakes in New England. Dark quadrants are compressional first motions (i.e., T axes). Data compiled by Gephart and Forsyth (1985) from Herrmann (1979), Pomeroy et al. (1976), Yang and Aggarwal (1981); and J. Pulli (unpub.).

Platewide Stress Maps Based on Fault-Plane Solutions

Earthquake fault-plane solutions are the most convenient devices for inferring stress orientation in the oceanic portions of lithospheric plates where other data on stress orientation are rare. In a recent compilation of stress data on a global basis, Zoback et al. (1989) report that 50% of all stress orientation data come from fault-plane solutions.

Indo-Australian Plate

The orientation of stress in the Indian Ocean portion of the Indo-Australian plate is known largely from earthquake fault-plane solutions (fig. 11–6). Characteristics of the Indo-Australian plate which make it unique include: a high level of intraplate seismicity (Bergman and Solomon, 1980); strong ($M_s > 7$)

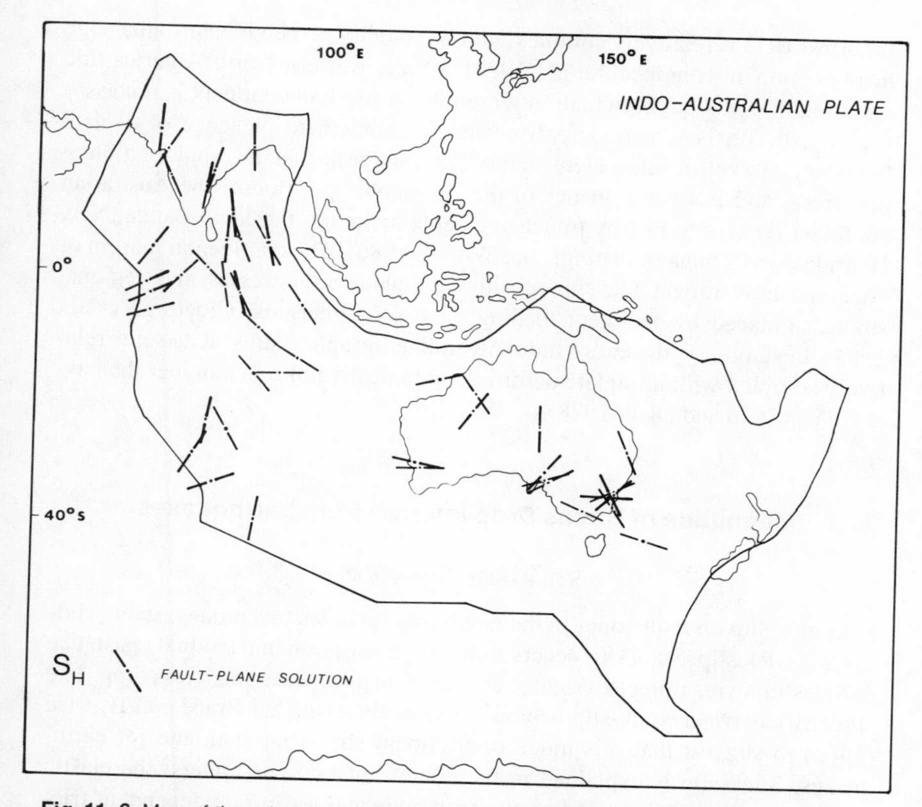

Fig. 11–6. Map of the orientation of S_H based on intraplate focal mechanisms within the Indo-Australian Plate. Data compiled by Cloetingh and Wortel (1986).

oceanic intraplate earthquakes which are not associated with plate margins (Bergman, 1986); and compressional folding of oceanic lithosphere (Weissel et al., 1980; Stein et al., 1989). These characteristics are consistent with a plate-wide stress field having anomalously high stresses. Modeling shows that the highest compressive stresses in the Indo-Australian plate develop parallel to the Ninetyeast Ridge (Cloetingh and Wortel, 1986). This is the area of the Indo-Australian plate where σ_d on the order of 500 MPa are necessary to account for the folding of the oceanic lithosphere (McAdoo and Sandwell, 1985). S_H in the Indian subcontinent is oriented roughly north-south (Chandra, 1977). Fault-plane solutions adjacent to the Central and Southeast India Ridges show S_H normal to the spreading center (Wiens and Stein, 1984). Off the Java trench modeling suggests that S_h is oriented normal to the trench axis.

Fault-plane solutions also constitute a major portion of the stress orientation data set from the continental crust of Australia. S_H varies from east-west in southwest Australia to north-south in central Australia. In southeast Australia the stress field is relatively complex with S_H roughly perpendicular to the edge of the continent (Lambeck et al., 1984). Unlike western North America, the variation in stress field orientation within the Australian continent is not associated with Tertiary tectonic activity or the immediate presence of a plate boundary. However, finite-element modeling suggests that the rapid rotation of the stress field is a consequence of the geographic position of the Australian continent relative to nearby trench segments including the Java, Banda, New Hebrides, and Tonga (Cloetingh and Wortel, 1986). The continental margin of Australia may reflect the superposition of platewide stresses and regional stresses induced by sediment loading on passive margins (Cloetingh et al., 1983). In summary, the entire Indo-Australian intraplate stress data set is relatively complex with intraplate deformation indicative of much higher than average σ_d (Stein and Okal, 1978).

Magnitude of Stress Drop Inferred from Earthquakes

Stick-Slip Behavior

Frictional slip on fault zones in the laboratory includes two modes: stable sliding and stick slip. Stick slip occurs if there is a variation in frictional resistance during sliding resulting in a sudden acceleration in slip rate and stress drop. The similarity between stick-slip behavior and earthquakes led Brace and Byerlee (1966) to suggest that this mode of frictional slip is an analogue for earthquakes. Stick slip is explained in the context of two parameters: the elastic rigidity of the earth surrounding the fault zone and a slip-rate dependent friction of the fault zone. First, to understand the effect of rigidity, consider that the earth surrounding a fault zone is elastic; a decrease in shear stress causes a significant relaxation of shear strain. The earth next to a fault zone behaves as a spring loading the fault zone in shear. Such shear tractions are counterbalanced by frictional resistance along the fault zone. Assume that when slip rate on a fault increases, the frictional resistance of the fault decreases (points A to B in fig. 11–7). With a large decrease in frictional resistance, the fault zone is said to unload because it can no longer support shear tractions exerted by the surrounding earth. After this point (B in fig, 11–7) the total earth-fault zone system is unstable, because forces between the fault zone (heavy curve) and the surrounding earth (light curve) are no longer in balance. The surrounding earth will follow suit and unload by slipping along the fault zone during the recovery of elastic shear strain. As long as the shear forces exerted by the surrounding earth are larger than the tractions developed by the reduced frictional resistance, the earth will accelerate as it slides along the fault zone (points B to C in fig. 11–7). Eventually shear strain within the earth is relieved to the point that shear force drops below frictional resistance, at which point (after point C) the

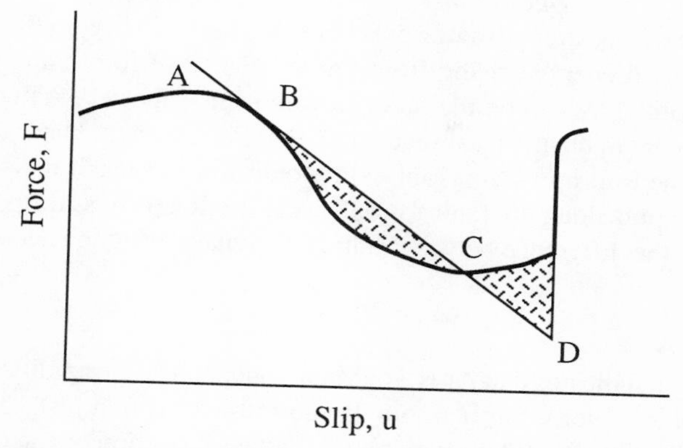

Fig. 11–7. A force-fault slip diagram illustrating the stick-slip instability which is believed to explain earthquake behavior. The earth surrounding a fault zone is represented by a spring with stiffness K whose slope is a constant, $^{dF}/_{du}$. Frictional resistance of the fault zone is represented by a solid curve which shows that friction varies with slip. If the frictional resistance, F, falls off faster than K (point B), then an instability develops because of the force imbalance between the surrounding earth and the fault zone. Slip accelerates as long as the frictional force is less than force allowed by unloading of the spring, K (points B to C). Slip is arrested when the area between the curves from B to C is equal to the area between the curves from C to D.

earth starts to decelerate. Slip along the fault zone is arrested (point D) once the excess work generated during deceleration of the earth (slip on the fault zone) equals work generated during the acceleration phase.

Velocity weakening, a behavior in which frictional strength is inversely proportional to slip velocity, is the key mechanism leading to a very rapid unloading of a fault zone. Velocity dependence is one of the rate effects governing the frictional properties of rocks as summarized by Scholz (1990). In brief, higher slip rates lead to an instantaneous increase in friction (the direct effect) followed by a gradual decrease in friction (the evolving effect). If the evolving effect is larger than the direct effect, then a stick-slip instability develops due to an imbalance in forces as explained above (Tse and Rice, 1986). To this point in the discussion of velocity dependence, the frictional strength in question is a dynamic friction (μ_d). If fault surfaces are in contact without slipping a static friction (μ_s) develops. In general, stick-slip events are marked by a drop in friction from static to a dynamic value at constant σ_n where both μ_s and μ_d follow a rate-dependent behavior. Rabinowicz (1958) observed that μ_s increases as a function of the time that surfaces are held in contact, whereas μ_d is dependent on velocity of the sliding surfaces. There is a critical slip distance in order for μ_d to change from one value to another if the slip rate changes.

As earthquake seismologists measure earth stress in terms of a stress drop during an earthquake rupture, it is important to understand the relationship between stress drop and friction. In a simple stick-slip cycle the static friction is defined as the peak strength at the onset of slip. Once slip becomes unstable the excess potential energy arising from the imbalance of forces along the fault zone is absorbed by slip on the fault. Bowden and Tabor (1964) were among the first to point out that, on average, this behavior causes the shear stress along the fault zone to drop twice as far as it would have to reach the steady-state dynamic friction along the fault zone. Thus, a frictional stress drop, $\Delta\tau$, is proportional to the difference between static and dynamic friction according to

$$\Delta\tau = 2\sigma_n(\mu_s - \mu_d). \tag{11–2}$$

This equation indicates that $\Delta\tau$ is a constant independent of rigidity of the earth (Hoskins et al., 1968). Furthermore, because there is often just a few percent difference between μ_s and μ_d, with typical values of friction on the order of 0.6, $\Delta\tau$ is at most a modest fraction of the absolute τ on the fault.

The frictional stress drop during stick-slip is associated with a time-dependent behavior for μ_s (Dieterich, 1972) with μ_s increasing in proportion to the log of the time that the sliding surfaces are held in static contact. Scholz et al. (1972) confirmed Dieterich's results by observing that μ_s appeared to increase with decreasing rate of loading of the fault zone (an order of magnitude change in slip rate produces about a 2% change in μ_s). Stress drop is proportional to the time in stationary contact which suggests that the surfaces become more strongly locked with time and require larger shear stresses to rupture through the contacts to reinitiate slip. This behavior accounts for the relatively larger stress drops associated with intraplate earthquakes along faults with very low slip rates. Hence, time-dependent μ_s is of interest.

A time-dependent μ_s is attributed to the indentation of asperities on the sliding surface where the area of indent increases as a function of time (Scholz and Engelder, 1976). This result is in accord with the same behavior for a wide variety of nonmetallic materials (Westbrook and Jorgensen, 1968). Such behavior is termed microindentation creep and appears to result from the hydrolytic reactions which produce stress-corrosion cracking in silicates and other nonmetals. When two surfaces are placed in frictional contact under normal force, **N**, the asperities of the harder materials will gradually penetrate the softer surface through indentation creep. The real area of contact, A, will thus increase with time as

$$A = (1 + a \log t)\frac{N}{P_1} \tag{11–3}$$

where a is a constant, t is time in contact, and P_1 is the penetration hardness measured at unit time. Frictional sliding will then occur when

$$\mathbf{F} = (1 + a \log t)\frac{S}{P_1}\mathbf{N} \tag{11–4}$$

where S is the shear strength of the softer material and **F**, the shear force. Therefore, the static coefficient of friction, μ_s, will be time-dependent according to

$$\mu_s = (1 + a \log t)\bar{\mu}_s \qquad (11\text{–}5)$$

where $\bar{\mu}_s$ is the friction coefficient at unit time of contact. The velocity dependence of μ_d has a similar form assuming that memory of the previous frictional state fades over a critical slip distance, Λ,

$$\mu_d = \mu_0 - a \log\left(\frac{\Lambda}{t}\right). \qquad (11\text{–}6)$$

Using data from Dieterich (1978) and Scholz and Engelder (1976), Scholz (1990) makes a comparison of the μ_s-time and μ_d-velocity relations by assuming that μ_s after holding at time, t, is equivalent to μ_d at a steady velocity, $^{\Lambda}/_{t}$, for time, t (fig. 11–8).

Laboratory experiments indicate that stick-slip should not occur throughout the crust of the earth. The prediction is that stable sliding should occur at the shallowest levels and that the depth of the transition from stable sliding to stick-slip depends on several parameters including lithology, normal stress, temperature, and the presence of fault gouge (Brace and Byerlee, 1970; Byerlee and Brace, 1968). In a biaxial shear experiment using Westerly granite the onset of stick-slip is at a σ_n of 0.5 to 1.5 MPa (Scholz et al., 1972). In triaxial experiments on Westerly granite having rough gouge-covered sliding surfaces the transition to stick-slip occurs at much higher σ_n (100 MPa). Stress drops also show the same dependency on experimental conditions where $\Delta\tau$ was

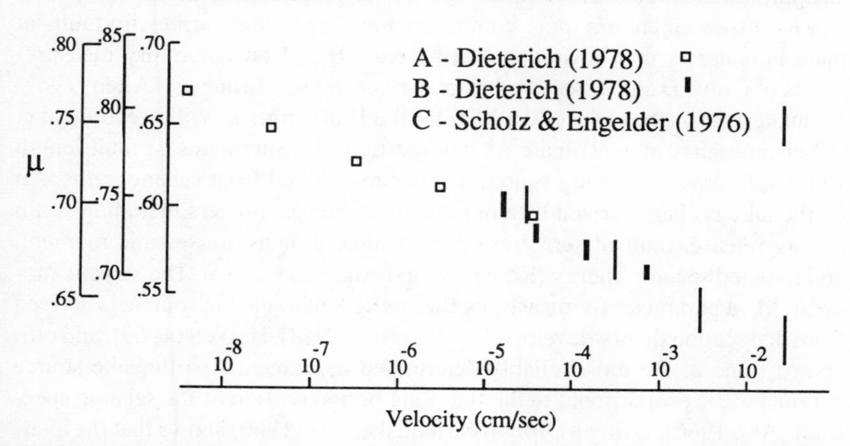

Fig. 11–8. A comparison of the μ_s-time and μ_d-velocity relations. (A) μ_s-t, quartz sandstone, σ_n = 18.7 MPa; (B) μ_d-v, Westerly granite, 1.96 MPa; (C) μ_d-v, Westerly granite, 20 MPa. Critical slip distance, Λ = 5 μm. Friction scales are offset because base friction is not reproducible (adapted from Scholz, 1990).

about 1 MPa at 70 MPa σ_n (Scholz et al., 1972) versus 300 MPa at 400 MPa σ_n (Byerlee and Brace, 1968). Data from biaxial tests are in better agreement with conditions during natural earthquakes. Higher temperatures stabilize sliding so that the transition to stick-slip occurs at higher confining pressures (Stesky et al., 1974).

Source Mechanisms

Information on absolute stress magnitude in rocks in the vicinity of an earthquake is not available in the seismic record. However, dislocation theory does allow the calculation of change in shear stress accompanying slip along a fault zone during an earthquake. This change in shear stress is called stress drop ($\Delta\tau = \tau_i - \tau_f$) where τ_i is the shear stress on the fault plane before the earthquake and τ_f is the shear stress after the rupture. Knopoff's (1958) formula for $\Delta\tau$ applies to an infinitely long vertical fault with strike-slip displacement

$$\Delta\tau = \frac{1}{2}\zeta\frac{D}{r} \tag{11–7}$$

where D is the average displacement along the faulted area, ζ is the shear modulus of the country rock, and r is a characteristic length associated with the narrowest dimension of the fault (Madariaga, 1977). If the fault is circular then r is the radius of the fault. This formula is conveniently applied where earthquake faulting has broken the ground surface because field measurements give an approximate fault dimension and average displacement. The size of the fault and its displacement are more commonly measured along strike-slip faults at plate boundaries than in midcontinent areas. Based on laboratory measurements of ζ plus D and r measured from surface breaks, Brune and Allen (1967) calculate a $\Delta\tau$ on the order of 0.1 MPa for the 1966 Imperial Valley earthquake.

Seismologists may estimate $\Delta\tau$ without field measurements of fault length and displacement by using source parameters derived from seismograms. An earthquake is characterized by a number of source parameters including strain energy release, fault offset, stress drop, source dimension, seismic moment, and radiated seismic energy (Kanamori and Anderson, 1975). The seismic moment, M_o, a parameter for measuring the strength of a seismic source, is derived from dislocation theory developed by Steketee (1958), Haskell (1963), and others. M_o, one of the most reliably determined instrumental earthquake source parameters, is proportional to the flat, long period portion of the seismic spectrum (Aki, 1966). Using M_o for an earthquake, Aki (1966) shows that the average displacement, D, is

$$D = \frac{M_o}{\zeta A}, \tag{11–8}$$

provided the area of the fault surface (A) is known from the distribution of aftershocks. To find $\Delta\tau$ a source dimension, r, determined from the area of the aftershock sequence, and D from equation 11–8 are substituted directly into equation 11–7. If the source dimensions are determined from the seismic spectrum (Brune, 1970) then stress drop is approximately

$$\Delta\tau = k\frac{M_o}{r^3} \tag{11–9}$$

where k is a dimensionless form factor that depends on the geometry of the fault and the direction of slip. Formulas for k are given in Madariaga (1977). From equation 11–9 Wyss and Molnar (1972) estimate that stress drops for the 1967 Denver earthquakes range between 0.3 and 2.2 MPa. Chinnery (1964) pointed out that stress drops for most large earthquakes do not exceed 10 MPa. The January 1975 Brawley earthquake swarm is characterized by a wide range in estimated $\Delta\tau$ from 0.1 to 63.6 MPa, over a magnitude range from $M_L = 1.0$ to 4.3 (Hartzell and Brune, 1977).

Several other source parameters relate in one way or another to the shear stress (τ) along a fault zone. The total work (energy) during a rupture is

$$E = \frac{1}{2}(\tau_i + \tau_f)DA = \tau_a DA \tag{11–10}$$

where τ_a is the average stress along the fault (Wyss and Molnar, 1972). The energy during earthquake rupture is partitioned at least three ways: some is dissipated as seismic waves (E_s); some is used to create new surface area during the grinding of fault zone materials (E_c); and some appears as frictional heat generated along the fault zone (E_f). Seismic efficiency (η) is the fraction of total work which goes into seismic radiation

$$E_s = \eta E. \tag{11–11}$$

Combining the equations for total work (11–10), seismic moment (11–8), and seismic efficiency (11–11),

$$\tau_a\eta = \frac{E_s\zeta}{M_o} \tag{11–12}$$

which is a formula used by Aki (1966) to calculate apparent stress ($\tau_{app} = \tau_a\eta$). Apparent stress is distinguished from stress drop (Wyss and Molnar, 1972). Following Scholz (1990) stress drop is related to seismic energy by

$$\Delta\tau = \frac{2E\zeta}{M_o}. \tag{11–13}$$

The dissipation of energy along a fault depends on the nature of frictional resistance during faulting (Sibson, 1977b). If the mean kinetic shearing resistance during slip stays constant, then much of the energy along the fault is dissipated as frictional heat, E_f. If the shearing resistance decreases with slip, then more energy is dissipated as seismic energy rather than frictional heat.

Seismologists have long debated the relationship between $\Delta\tau$ and absolute τ on fault zones. Equation 11–2 suggests that the former is normally a small fraction of the latter. Some of the most reliable direct measurements of the two come from South African gold mines where mine tremors are indistinguishable from natural earthquakes (McGarr et al., 1979). Analysis of P and S waves from those mine tremors suggest that all stress drops are between 0.5 and 5.0 MPa (Spottiswood and McGarr, 1975). McGarr et al. (1975) calculated that shear stress in the vicinity of the mine tremors was on the order of 70 MPa, an order of magnitude greater than the stress drop. This analysis supports the conclusion that earthquakes act to relieve only a fraction of the total shear stress along the faults.

Using observational data, Kanamori and Anderson (1975) show that stress drop follows the equation

$$\log M_o = \frac{3}{2}\log A + \log\left(\frac{16\Delta\tau}{38.98}\right)A^{\frac{3}{2}} \qquad (11\text{–}14)$$

in a plot of seismic moment, M_o, versus area of the fault zone, A (fig. 11- 9). Such a compilation suggests that most earthquakes have a stress drop between 1 and 10 MPa and that stress drop does not increase with magnitude of the earthquake (Aki, 1972). A compilation of smaller earthquakes showed that stress drop was on average a few MPa and that this average is independent of source strength across twelve orders of magnitude of seismic moment (Hanks, 1977). The physical consequence of the stress drop for earthquakes being independent of earthquake size is that slip correlates linearly with length of earthquake rupture (Scholz, 1982). The correlation between slip and rupture length was first suggested in a compilation by Bonilla and Buchanan (1970).

Earthquake stress drop is not the same throughout the lithosphere. In a study of shallow-focus earthquakes in the Tonga-Kermadec Arc, the arc that defines the boundary between the Pacific and Indo-Australian Plates, Molnar and Wyss (1972) show that earthquakes with larger stress drops did not occur along the main underthrusting zone. Rather, the largest stress drops were associated with earthquakes located within one plate of the lithosphere.

Interplate earthquakes have distinctly lower stress drops than intraplate earthquakes (3 MPa versus 10 MPa according to Kanamori and Anderson, 1975) (fig. 11–9). From a plot of log seismic moment versus log rupture length Scholz et al. (1986) conclude that stress drops for intraplate earthquakes are systematically about five times greater than interplate earthquakes (fig. 11–10). The source of the difference between stress drops for large interplate earthquakes and smaller intraplate earthquakes may arise from such mechanisms as time-dependent frictional slip as discussed above (Kanamori and Allen, 1986). Fault zones hosting interplate earthquakes have an average slip rate greater

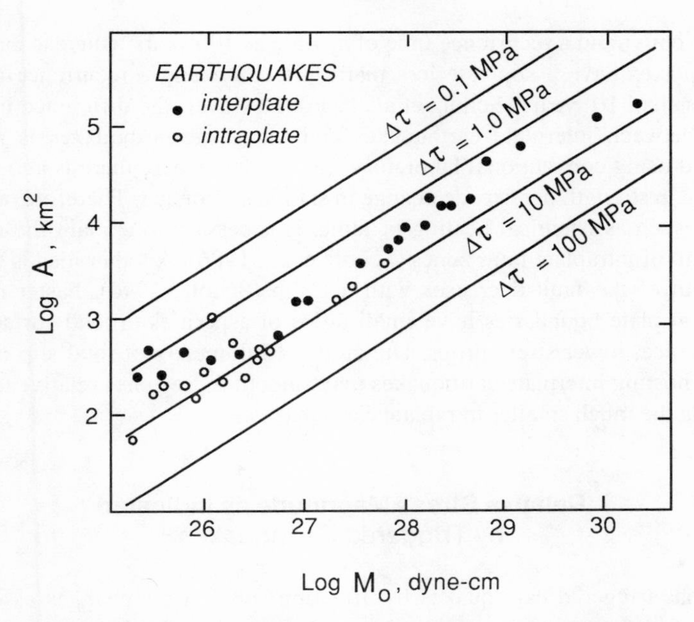

Fig. 11–9. Relation between the fault surface area (A) and the seismic moment (M_0) for large earthquakes. The straight lines give the relations for circular cracks with constant stress drop (adapted from Kanamori and Anderson, 1975).

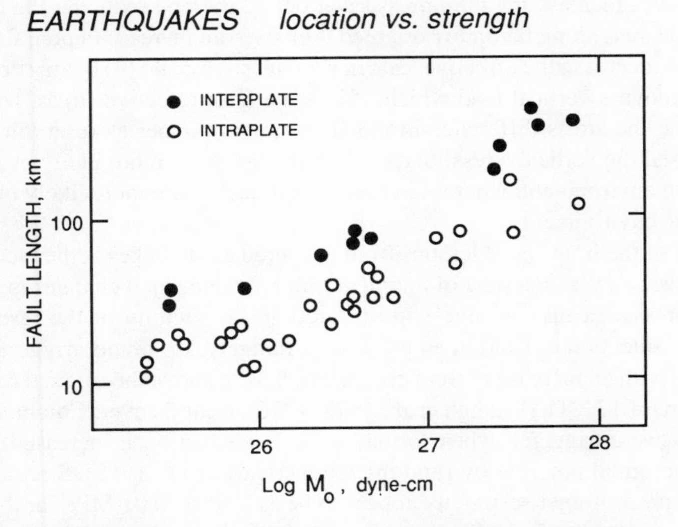

Fig. 11–10. Log fault length versus log moment for large interplate and intraplate earthquakes (adapted from Scholz et al., 1986).

than 1 cm/yr and a recurrence time of as little as 100 years, whereas intraplate earthquakes have a slip rate less than 0.01 cm/yr and a recurrence time of greater than 10^4 years (Scholz et al, 1986). However, the difference in stress drop between interplate earthquakes and intraplate earthquakes is not explained using conventional laboratory friction data where there is about a 5% change in strength per decade change in stationary contact. Therefore, a mechanism such as chemical healing of faults is necessary to explain the relative strength of intraplate fault zones (Scholz et al., 1986). Another idea is that the strength of the fault decreases with net slip (Scholz, 1990). Faster moving faults at plate boundaries have small ratios of asperities to total surface area and, hence, lower stress drops. Hundreds of kilometers of total slip on fault zones hosting interplate earthquakes may smooth those zones relative to faults hosting the much smaller intraplate earthquakes.

Relative Stress Magnitude as Indicated by Triggered Earthquakes

Man has triggered earthquakes by affecting small stress changes associated with loading or unloading at the surface of the earth or fluid injection. Induced seismicity may be the consequence of at least four mechanisms, three of which are associated with reservoirs and the fourth associated with quarrying (fig. 11–11) (Simpson, 1976): (1) the weight of the impounded water adds to S_v which, if it is σ_1, causes an increase in the stress difference and, hence, shear stress along favorably oriented faults; (2) increased hydrostatic head in the reservoir will increase the pore pressure at depth, thereby reducing the effective normal stress along favorably oriented faults; (3) an increase in pore fluid pressure can decrease the effective least horizontal stress; and (4) quarry operations can remove a vertical load which, if it is the least effective stress, will act to increase the stress difference in much the same manner as reservoir loading increased the vertical stress. In case 1 earthquakes are more likely in a normal faulting environment, whereas in case 4 earthquakes are more likely in a thrust faulting environment.

One of the impressive lessons from triggered earthquakes is the fact that the earth seems to be in a state of incipient failure where small changes in stress or pore pressure cause seismic slip on faults in the vicinity of the reservoir or quarry. Injection of fluid in an oil field at Rangely, Colorado, triggered earthquakes with an increase in fluid pressure of 9 MPa above the original formation pressure of 17 MPa (Raleigh et al., 1972, 1976). Near Denver, Colorado, earthquakes were triggered when formation pressure had been increased 12 MPa above original pressure by fluid injection (Healy et al., 1968). Stress changes necessary to trigger seismicity appear to be as little as 0.01 MPa for the Nurek Reservoir in the Tadzhikistan, Central Asia (Simpson and Negmatullaev,

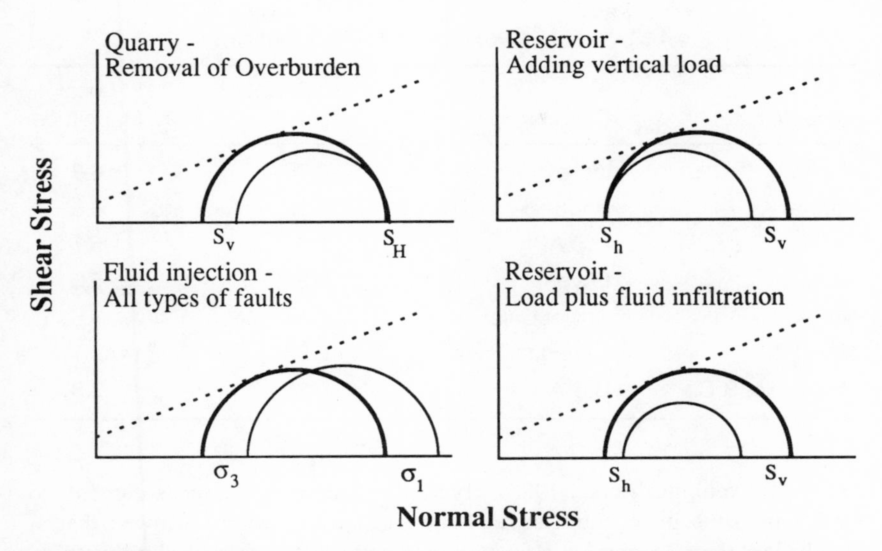

Fig. 11–11. Mechanisms for induced seismicity based on changes in position or diameter of Mohr circles relative to a Coulomb-Navier failure (frictional slip) criterion (dashed line). Lighter Mohr circles show initial stress states, whereas darker Mohr circle shows stress state after change in load or pore pressure (adapted from Simpson, 1984).

1981). Some of the more famous cases of reservoir triggered seismicity are listed in table 11–1 (compiled by Simpson, 1986). All triggered events associated with reservoirs are affected by stress (or pore pressure) changes of 2.3 MPa or less. The point is that induced stress changes are small in comparison to the magnitude of ambient stress at the depth of faulting. Interestingly, these stress changes are on the same order as the stress drops associated with earthquakes. From this it seems that significant portions of the crust are at less than the equivalent of one stress drop (< 10 MPa) away from failure and modest crustal stress changes are all that is required to trigger an earthquake (Gough and Gough, 1970).

Monticello Reservoir

Although the models for induced seismicity are straightforward, the relationship between triggered earthquakes and in situ stress is enigmatic. The Monticello, South Carolina, Reservoir was associated with a magnitude 2.8 earthquake within days of the time it reached its impoundment level of 35 m. The spatial association of shallow earthquakes under the reservoir and temporal association with the earthquake swarms with impounding of the reservoir suggested that this was a classical case of triggered seismicity (Talwani and Rastogi, 1979) with the time delay after filling related to local hydraulic diffu-

Table 11–1. Reservoir-Induced Seismicity

Reservoir	Country	Stress Change (MPa)	Local Magnitude
Koyna	India	0.75	6.5
Kariba	Rhodesia	1.25	5.8
Oroville, Calif.	U.S.A.	2.3	5.7
Aswan	Egypt	0.9	5.3
Nurek	Soviet Union	1.1	4.6
Manic-3	Canada	0.85	4.1
Monticello, S.C.	U.S.A.	0.35	2.8

sivity (Talwani and Acree, 1985). Hydraulic fracture stress measurements to 0.96 km in the hypocentral area of the earthquake swarms showed that at depths less than 200 m slip is favored on thrust faults, whereas slip by strike-slip faulting is favored at depths greater than 200 m (Zoback and Hickman, 1982). Furthermore, the measurements suggest that failure is most likely in the upper 200 m of the crust where σ_d exceeds the frictional strength of the local rocks. Below 200 m stress measurements suggest that the crust is stable; but earthquakes indicate that it is not. Furthermore, earthquake fault-plane solutions are not compatible with the in situ stress measurements as the earthquakes all appear to have thrust-faulting mechanisms from close to the surface to below 1 km despite an indication that the stress state should favor strike-slip faulting below 200 m (Fletcher, 1982). Slip throughout the top km suggests that rock friction is much less than inferred from conventional laboratory values.

Concluding Remarks

Earthquake fault-plane solutions account for the largest fraction of all data on stress orientation in the schizosphere (Zoback et al., 1989) and in many places the only data from the oceanic portions of lithospheric plates. The upper portion of the schizosphere is in a critical state where small stress changes (as little as 1 MPa) can induce earthquakes (Simpson et al., 1988). The schizosphere relaxes through earthquakes with stress drops on the order of 3 to 10 MPa but these stress drops are only a relatively small fraction of shear stress along fault zones.

12

Data Compilations

The maximum compressive principal stress trends easterly to northeasterly and is nearly horizontal for a region extending from west of the Appalachian Mountain system to central North America.

—Marc Sbar and Lynn Sykes (1973) after making the first regional compilation of data on earth stress in the North American lithospheric plate

Compilations of stress data vary in detail depending on the scale and purpose of the compilation. On the largest scale, a global compilation, data may come from several types of stress measurements and an individual datum may, itself, be some average of many stress measurements. Compilations of data from individual wells usually consist of one type of stress measurement. Continent-wide and regional compilations may include large quantities of orientation data, whereas compilations from individual wells may focus more specifically on the nature of the variation of stress magnitude with depth in the lithosphere. Global compilations of stress data are helpful in understanding of such phenomena as the driving mechanisms for plate tectonics and earthquakes in the lithosphere. A major issue in global compilations is the degree to which stress is consistently oriented within lithospheric plates, for this consistency (or lack thereof) is important in defining boundary conditions that set up the contemporary tectonic stress field.

One application of regional compilations is their use for earthquake prediction. In brief, if tectonic stress has a preferred orientation over a large region then it is possible to identify major structures oriented favorably for stress release by earthquake slip. Although earthquake stress drop does not vary appreciably from earthquake to earthquake, earthquake magnitude is likely to vary as a function of the fault dimensions of favorable orientation. Earthquake frequency is more likely a function of the rate at which stress accumulates in the lithosphere, with stress recovery and release most rapid at plate boundaries.

One of the inherent problems with data compilations concerns what to do with data sets containing seemingly spurious data. This is illustrated by examining stress orientation data from the Rangely, Colorado, earthquake prediction experiments (Raleigh et al., 1972). In an attempt to demonstrate that modest increases in fluid pressure along a fault zone will trigger earthquakes, the U.S.G.S. conducted a number of water-flood experiments in an oil field at

Rangely anticline, Colorado (Raleigh et al., 1972). Concurrently, an attempt was made to understand the tectonic stress responsible for these induced earthquakes. Information on stress came from hydraulic fractures, borehole breakouts, earthquake fault-plane solutions, and five near-surface techniques (fig. 12–1). The near-surface techniques were compared in sandstones of the Mesa Verde Group at the nose of Rangely anticline (site 1 of de la Cruz and Raleigh, 1972). These techniques included electrical-resistance strain gauges (Handin, 1971) and photoelastic bars bonded to an outcrop surface (Brown, 1974) plus a solid-inclusion borehole probe (Nichols and Farrow, 1973), the South Africa "doorstopper," and U.S.B.M. borehole deformation gauge placed in pilotholes less than 5 m deep (de la Cruz and Raleigh, 1972). The orientation data for all techniques correlates within 30°, showing a maximum stress in the ENE direction. This leaves little doubt that the signal carried by the outcrop is an ENE S_H and it correlates with S_H determined by a single hydraulic fracture stress measurement at the depth of 1.8 km in the Rangely anticline (Raleigh et al., 1972; Haimson, 1973). However, de la Cruz and Raleigh (1972) were hard pressed to explain the significance of an ENE S_H which was neither parallel nor perpendicular to the local fold axis of Rangely anticline nor consistent with the orientation of S_H (N70°W) determined by hydraulic fracturing in the Piceance Basin (Bredehoeft et al., 1976). The answer is probably that both hydraulic fracture test and data from site 1 were spurious in the sense that local boundary loads

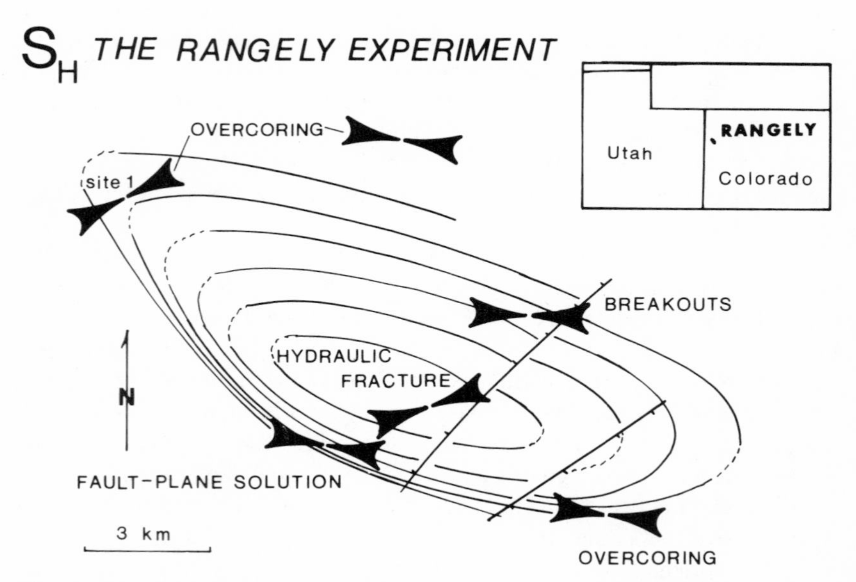

Fig. 12–1. Structural contour map of the Rangely Anticline, Colorado, showing orientations of S_H. The map shows contours (50 m) drawn to the top of the Weber Sandstone and illustrates faults which offset this surface (adapted from Bell and Gough, 1983).

induced stresses which were not compatible with regional trends. The hydraulic fracture test shows that local stress conditions are not unique to the near surface.

In contrast to site 1, surface overcoring measurements at two other points around the Rangely anticline, a fault-plane solution, and borehole breakouts all suggest that S_H is WNW and in agreement with the regional stress field (Bell and Gough, 1983). Taken at face value, near-surface measurements from two out of three outcrops at Rangely detected a signal compatible with the regional stress field in northwestern Colorado. Deeper techniques also scored two out of three at Rangely. For both shallow and deep techniques confidence in the measurement is gained through additional data points. To make some judgement concerning which data sets are useful for understanding tectonic processes, quality control criteria are necessary.

Quality Control

Like any scientific measurement, the quality of in situ stress measurements depends on both accuracy and precision. Because stress in the lithosphere may have several components, accuracy of a particular technique depends on whether or not the component of interest is sampled. Precision of stress measurements is indicated by the reproducibility of a series of tests in close proximity. Obviously, stress measurements can be relatively precise without being accurate.

To best understand the driving mechanisms of plate tectonics, a reliable knowledge of platewide stress fields is necessary. Such knowledge is gained through platewide and global compilations based on a variety of stress measurements including earthquake fault-plane solutions, borehole breakouts, hydraulic fracturing, overcoring, recent fault slip, neotectonic joints, and volcanic dike alignments (e.g., Zoback and Zoback, 1989; Zoback et al., 1989). Not all data in such compilations are of equal accuracy and so the data must be culled using a quality control criteria. In judging the accuracy of stress data for assessing platewide stress fields, the location of the measurement is often regarded as the paramount factor because all components of earth stress are not equally apparent at all locations. If platewide stress fields are of most interest, high quality measurements are those unaffected by stress concentrations, thermal stresses, residual stresses, and local fracturing. To map the orientation of platewide stress fields through data compilation, Zoback and Zoback (1989) constructed a quality control ranking system in which deeper measurements are given higher quality (table 12–1). Their assumption is that deeper measurements are more accurate, which is to say they are more likely to sample the platewide stress fields. Because of this assumption, overcoring data, for exam-

Table 12–1. Quality Ratings for the Measure of Deep Lithospheric Stress (Zoback and Zoback, 1989)

	A	B	C	D
Focal Mechanism	Average P-axis or formal inversion of four or more single-event solutions in close geographic proximity (at least one event $M \geq 4.0$, other events $M \geq 3.0$)	Well-constrained single-event solution ($M \geq 4.5$) or average of two well-constrained single-event solutions ($M \geq 3.5$) determined from first motions and other methods (e.g., moment tensor wave-form modeling, or inversion)	Single-event solution (constrained by first motions only, often based on author's quality assignment) ($M \geq 2.5$) Average of several well-constrained composites ($M > 2.0$)	Single composite solution Poorly constrained single event solution Single event solution for $M < .5$ event
Wellbore Breakout	Ten or more distinct breakout zones in a single well with $s_0 \leq 12°$ and/or combined length >300 m Average of breakouts in two or more wells in close geographic proximity with combined length > 300 m and $s_0 \leq 12°$	At least six distinct breakout zones in a single well with $s_0 \leq 20°$ and/or combined length > 100 m	At least four distinct breakouts with $s_0 < 25°$ and/or combined length > 30 m	Less than four consistently oriented breakouts or <30 m combined length in a single well Breakouts in a single well with $s_0 \geq 25°$
Hydraulic Fracture	Four or more hydrofrac orientations in single well with $s_0 \leq 12°$, depth > 300 m Average of hydrofrac orientations for two or more wells in close geographic proximity, $s_0 \leq 12°$	Three or more hydrofrac orientations in a single well with $s_0 < 20°$ Hydrofrac orientations in a single well with $20° < s_0 < 25°$	Hydrofrac orientations in a single well with $20° < s_0 < 25°$. Distinct hydrofrac orientation change with depth, deepest measurements assumed valid One or two hydrofrac orientations in a single well	Single hydrofrac measurement at < 100 m depth

(*Continued*)

Table 12–1. (*Continued*)

	A	B	C	D
Petal Centerline Fracture			Mean orientation of fractures in a single well with s_0<20°	
Overcore	Average of consistent ($s_0 \leq 12°$) measurements in two or more boreholes extending more than two excavation radii from the excavation wall, and far from any known local disturbances, depth > 300 m	Multiple consistent (s_0 <20°) measurements in one or more boreholes extending more than two excavation radii from excavation well, depth >100 m	Average of multiple measurements made near surface (depth > 5–10 m) at two or more localities in close proximity with s_0 â 25° Multiple measurements at depth >100 m with 20° < s_0 < 25°	All near-surface measurements with s_0 > 15°, depth < 5 m All single measurements at depth Multiple measurements at depth with s_0 > 25°
Fault Slip	Inversion of fault-slip data for best-fitting mean deviatoric stress tensor using Quaternary-age faults	Slip direction on fault plane, based on mean fault attitude and multiple observations of the slip vector. Inferred maximum stress at 30° to fault	Attitude of fault and primary sense of slip known, no actual slip vector	Offset coreholes Quarry pop-ups Postglacial surface fault offsets
Volcanic Vent Alignment*	Five or more Quaternary vent alignments or "parallel" dikes with s_0 <12°	Three or more Quaternary vent alignments or "parallel" dikes with s_0 <20°	Single well-exposed Quaternary dike Single alignment with at least five vents	Volcanic alignment inferred from less than five vents

s_0 = standard deviation.
*Volcanic alignments must be based, in general, on five or more vents or cinder cones. Dikes must not be intruding a regional joint set.

ple, have a higher quality if they come from boreholes drilled in association with underground mining rather than from boreholes drilled at the surface.

Stress orientation statistics are at the heart of any question about whether any one technique is more precise than another. For horizontal stress orientation data there is no interpretive distinction between stress directions π (180°) apart. All horizontal stress orientation data are projected onto the interval (θ_i=) 0 to π. The mean and standard deviation of data on orientation of S_H (i.e., θ_i) are easily computed using the statistical approach of Mardia (1972), provided the data are transformed to the interval 0 to 2π (i.e., $\theta_i' = 2\theta_i$). If each orientation datum is considered to be θ_i' in a set of n data, then the mean (M) of the data set is

$$M = \frac{1}{2} \tan^{-1} \frac{C'}{S'} \quad (12\text{–}1)$$

where

$$C' = \frac{1}{n} \sum \cos \theta_i' \quad \text{and} \quad S' = \frac{1}{n} \sum \sin \theta_i' \quad (12\text{–}2)$$

summing from 1 to n. The standard deviation (s_o) of this data set is

$$s_o = \frac{1}{2} \sqrt{-2 \log_e (1 - S_o')} \quad (12\text{–}3)$$

where

$$S_o' = 1 - R' \quad \text{and} \quad R' = \sqrt{C'^2 + S'^2}\,. \quad (12\text{–}4)$$

Precision, not accuracy, is judged by establishing an s_o for the set of orientation data in question. Typical values for s_o vary among stress measurement techniques as is illustrated using data from the Appalachian Basin and other sites (table 12–2). A relatively small s_o is characteristic of data sets established by sampling within an intact volume of rock not cut by bedding discontinuities, joints, shear fractures, or faults. The uncertainty inherent in overcoring is illustrated using data from intact volumes of rock.[1] From nine doorstopper measurements taken at depths less than two meters in Tully Limestone at Ludlowville, New York, s_o is 8°, one of the smallest observed for overcoring stress data anywhere (Engelder and Geiser, 1984). An s_o of 10° is found in overcoring measurements made a meter below the floor of a 20 m deep quarry in Algerie granite at East Otis, Massachusetts (Engelder, 1984). Sampling within a larger volume of rock cut by joints, fractures, and faults invariably leads to stress data with a larger s_o. For example, a set of near-surface overcoring data was taken by Dames and Moore (1978) who used the U.S.B.M. borehole deformation gauge to measure stress in the vicinity of the Nine Mile Nuclear Reactor near Oswego, New York. In all, seventy-four measurements were taken in six different boreholes drilled to 33 m within 200 m of each other. Using all seventy-

Table 12–2. The Orientation of S_H

Location (Details)	# Data	Mean S_H (Clockwise from N)	Standard Deviation (s_o)	Reference
Niagara Escarpment (Linear Re-entrants— N.Y., Ont.)	15	62°	5°	Gross and Engelder, 1991
Ludlowville, N.Y. (Strain Relaxation— Tully Limestone	9	4°	8°	Engelder and Geiser, 1984
East Otis, Mass. (Strain Relaxation — Algerie Granite)	9	64°	10°	Engelder, 1984
Midcontinent U.S.A. (Hydraulic Fracture— Ill., Mich., Wis.)	7	59°	10°	Plumb and Cox, 1987
Appalachian Basin (Hydraulic Fracture— N.Y., Penn., Ohio, W. Va.)	10	65°	15°	Plumb and Cox, 1987
Appalachian Basin (Breakouts— N.Y., Penn., Ohio, W. Va.)	13	58°	18°	Plumb and Cox, 1987
South Canisteo, N.Y. (Hydraulic Fracture— Appleton Well)	19	78°	23°	Evans and Engelder, 1986
South Canisteo, N.Y. (Hydraulic Fracture— Wilkins Well)	29	66°	24°	Evans and Engelder, 1986
Nine Mile Point, N.Y. (Overcoring— USBM Gauge)	74	76°	28°	Dames and Moore, 1978
New England, U.S.A. (Overcoring— USBM Gauge; Doorstopper)	19	44°	29°	Plumb et al., 1984
Mojave Block, Calif. (Overcoring— Doorstopper— locations 5,6,7,8)	8	50°	29°	Sbar et al., 1979
(Overcoring— USBM Gauge— locations 3,8,9,11)				Tullis, 1981

(*Continued*)

Table 12–2. (*Continued*)

Location (Details)	# Data	Mean S_H (Clockwise from N)	Standard Deviation (s_o)	Reference
San Andreas Fault (Hydraulic Fracture—XTLR, MOJ1, MOJ2)	8 5	–35° –21°	31° 13°	(all data) (higher quality data) Zoback et al., 1980
Falkenberg, Germany (Hydraulic Fracture—NB-1; NB-3; HB-4a)	16	–87°	31°	Rummel et al., 1981
Iceland (Hydraulic Fracture)	13	59°	32°	Haimson and Rummel, 1982
San Andreas Fault (Overcoring—USBM Gauge-Lewis Ranch site)	55	–26°	46°	Sbar et al., 1984

four tests without regard to depth or borehole, the mean for S_H is oriented at N76°E with an s_o of 28°. The U.S.B.M. borehole deformation gauge was used in adjacent holes (separated by less than 5 m) drilled to 29 m within 400 m of the San Andreas fault near Palmdale, California (Sbar et al., 1984). Fifty-five tests indicate that the mean for S_H is oriented at N26°W with an s_o of 46° which is larger than most stress data sets.

The issue of accuracy versus precision is illustrated using three strain-relaxation data sets presented above. While the Tully Limestone data (Engelder and Geiser, 1984) show the greatest precision for strain-relaxation measurements, these data are an accurate measure of Alleghanian residual stress but not an accurate measure of the contemporary tectonic stress field which is well known for the Appalachian Basin (Plumb and Cox, 1987). In contrast, the data from the Algerie granite are both precise and accurate measurements of the contemporary tectonic stress field. The strain-relaxation measurements from near Oswego, New York, are less precise but just as accurate as the Algerie granite data in measuring the contemporary tectonic stress field. From this exercise it is not fair to conclude that the accuracy of all near-surface measurements in general is questionable.

Like strain-relaxation data in fractured rock, hydraulic fracture orientation data invariably have a larger s_o than close-spaced overcoring measurements taken within intact volumes of rock. During a hydraulic fracture experiment at South Canisteo, New York, orientation data were taken using televiewer images of 12 (19 half cracks) hydraulic fracture tests between depths of 250 and

1000 m in the Appleton well and 27 (49 half cracks) tests between depths of 200 and 1037 m in the Wilkins well (Evans and Engelder, 1986). The Wilkins well has a mean orientation of N66°E and the Appleton has a mean orientation of N78°E with s_o being 24° and 23°, respectively. With orientation data showing an s_o of about 24°, the difference in orientation between mean orientation for S_H in the Appleton and Wilkins wells is probably not significant. The data from these wells have two of the smallest s_o of any set of hydraulic fractures taken from a single borehole with ten or more measurements (table 12–2). Other single-well data of more than ten measurements per hole include measurements from Iceland where s_o is 32° (Haimson and Rummel, 1982).

The largest single-well orientation data sets come from borehole breakout studies. For example, 38,000 breakout observations within the Cajon Pass scientific well show an s_o of 19° (M. D. Zoback, 1989, personal communication). In general, quality A data ($s_o \leq 12°$) from single boreholes are rare unless stress is sampled within relatively small volumes of intact rock.

The orientation of the contemporary tectonic stress field can be mapped using brittle fractures of neotectonic origin (Hancock and Engelder, 1989). For example, Gross and Engelder's (1991) measurement of the contemporary tectonic stress field at the northern end of the Appalachian Basin based on lineaments along the Niagara Escarpment appears to be both accurate and precise (fifteen data points give $s_o = 5°$). Again, this is an example of near-surface data which reflect the orientation of the contemporary tectonic stress field.

Regional Compilations

Even with the guidelines of a quality control table the process of assessing regional and global stress fields is still subjective. Subjectivity starts with the decision on whether to combine data from adjacent wells for statistical evaluation. An additional tier of subjectivity is added when analyzing data from larger regions such as the Appalachian Basin or the Mojave portion of the San Andreas fault. These two regions represent extremes in contemporary tectonic activity and complexity in stress orientation data. For regional compilations, all tests within individual wells are averaged and treated as a single datum.

Appalachian Basin

Obert's (1962) overcoring measurements in a limestone mine at Barberton, Ohio, were the first stress data from the Appalachian Basin to indicate that S_H is approximately E-W. Early hydraulic fracture data from the Bradford oil fields of western Pennsylvania and New York suggested that S_H is about N70°E (Haimson and Stahl, 1969). Shortly after the early reports of hydraulic fracture tests, Sbar and Sykes (1973) pointed out that there is a significant amount of data suggesting that S_H is roughly ENE to E-W in the Appalachian Basin and surrounding regions. These data begged the question, Just how consistent in orientation and magnitude is the lithospheric stress field over the region the size of the Appalachian Basin? Presently, in the Appalachian Basin, including the states of New York, Pennsylvania, Ohio, and West Virginia, there are ten locations with orientation data based on one to twenty-seven hydraulic fracture tests per well (Plumb and Cox, 1987). The mean for S_H is N65°E and s_o is 15° (table 2–12). This compares with thirteen data on S_H from borehole breakouts showing a mean S_H at N58°E and an s_o of 18°. Ironically, a regional data set composed of a mean orientation datum from several wells appears to have smaller s_o than s_o for individual data sets within single wells of the Appalachian Basin. A relatively small s_o suggests that the lithospheric stress field is rather uniform throughout the Appalachian Basin. Such is not the case for the Mojave portion of the San Andreas fault.

THE MOJAVE PORTION OF THE SAN ANDREAS FAULT, CALIFORNIA

Frictional sliding of rocks generates a significant amount of heat as indicated by laboratory experiments showing temperatures as high as 1100°C (Teufel, 1976) and local melting (Friedman et al., 1974). Heat flow measurement across the San Andreas fault in California indicate that frictional heating along the main trace of the fault is minimal at all points between Cape Mendocino and San Bernardino (Lachenbruch and Sass, 1980). This lack of frictional heating as inferred from heat flow data along the San Andreas fault suggests that depth-averaged frictional shear stress probably does not exceed 20 MPa in the upper 14 km of the San Andreas fault zone, a value much too low for conventional laboratory frictional sliding. For conventional frictional sliding along the San Andreas fault, the right-lateral shear stress should approach 30 MPa at a depth of 3.5 km with a depth-averaged shear stress of 60 MPa. Reconciling frictional sliding on the San Andreas fault with lack of heat flow has given rise to the so-called "frictional stress-heat flow paradox." Stress measurements adjacent to the San Andreas fault give one of the best tools for a resolution of this paradox.

A sizable collection of stress data comes from the vicinity of the Mojave portion of the San Andreas fault north of Los Angeles (fig. 12–2). This section of the fault includes significant bends in the vicinity of both Cajon and Tejon passes. In order to obtain a stress profile normal to the San Andreas fault, the U.S.G.S. drilled a series of boreholes for hydraulic fracture measurements in the vicinity of Palmdale (Zoback et al., 1980; Hickman et al., 1981; Stock and Healy, 1988). Concurrently near-surface stress measurements were attempted using the doorstopper (Sbar et al., 1979) and the U.S.B.M. borehole deformation gauge (Tullis, 1981; Sbar et al., 1984). The deepest stress measurements in the Mojave portion of the San Andreas fault are credited to the U.S. continental

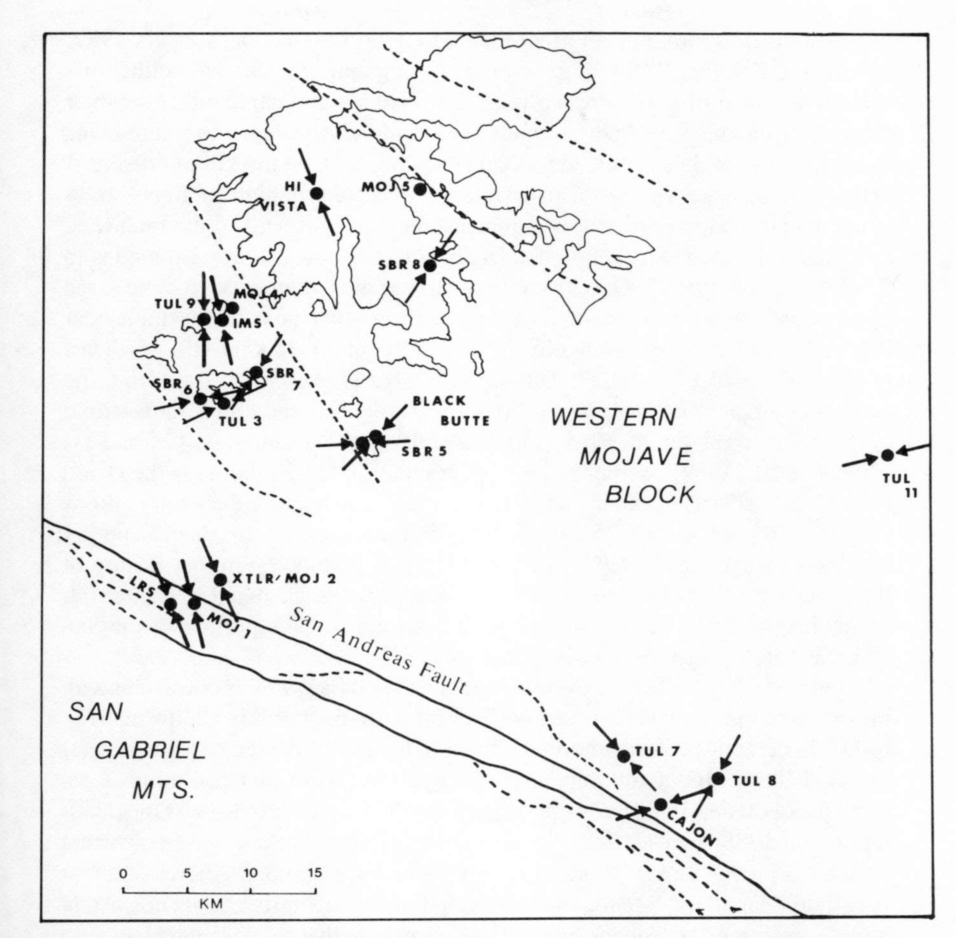

Fig. 12–2. Location of stress measurements in the vicinity of the big bend area of the San Andreas fault north of Los Angeles, California. Sites for hydraulic fracture measurements include: MOJ-4, MOJ-5, XTLR, and MOJ-1 (Zoback et al., 1980); BLACK BUTTE (Stock and Healy, 1989); HI VISTA (Hickman et al., 1981); and CAJON (Zoback and Healy, 1988). Sites for U.S.B.M. borehole deformation gauge measurements are designated TUL-# (Tullis, 1981); LRS and IMS (Sbar et al., 1984). Sites for doorstopper measurements are designated SBR-# (Sbar et al., 1979). Orientation data for TUL-1 and TUL-10 (Tullis, 1981) plus SBR-1 and SBR-2 (Sbar et al., 1979) are omitted in favor of higher quality data showing nearly the same orientation (i.e., LRS, MOJ-1, and XTLR). Orientation data for MOJ-4 (Zoback et al., 1980), and SBR-3 and SBR-4 (Sbar et al., 1979) omitted because of poor-quality data.

scientific drilling program in a borehole at Cajon Pass (Zoback et al., 1987; Vernik and Zoback, 1991). In general, measurements suggest that neither orientation nor magnitude of stress is related in a simple manner to either depth or distance from the San Andreas fault. The hydraulic fracture measurements southeast of Palmdale (XTLR and MOJ-1) show that, in the top km of crust next to the fault, S_H is west of north, an orientation expected for high frictional stress along the fault. However, other deep measurements northeast of the fault (i.e., CAJON and BLACK BUTTE) show that S_H is fault normal (i.e., at right angles to S_H at XTLR and MOJ-1). Data from near-surface measurements give the same impression; the orientation of S_H changes from NNW next to the fault near Palmdale to a variety of orientations (NNW to NE) at 20–30 km to the northeast of the fault (Sbar et al., 1979; Tullis, 1981; Sbar et al., 1984). Data from the Continental Scientific Drilling well (CAJON) shows that the ENE S_H stress field may extend right up the San Andreas fault in the vicinity of Cajon Pass (Zoback et al., 1987). While the average orientation for breakouts in the Cajon Pass well is N57°E (Shimar and Zoback, 1989) some individual orientations vary locally more than ± 35°, particularly in the interval between 1.7 and 2.1 km (Zoback and Healy, 1988). In general, stress orientation in the vicinity of the Mojave portion of the San Andreas is complex (Stock and Healy, 1988). It seems inappropriate to combine all data from the Mojave portion of the San Andreas for the purpose of a statistical analysis.

Fault normal S_H indicated by data from the Cajon Pass well is consistent with the orientation of the tectonic stress field further to the north in California (fig. 6–11). In central California borehole breakouts in wells drilled to 5 km suggest that S_H is ENE (Mount and Suppe, 1987). Likewise, fault-plane solutions from earthquakes at depths of 5–15 km suggest that S_H is oriented NE-SW (Oppenheimer et al., 1988; Jones, 1988). Trilateration data near Parkfield are consistent with a 6.17 ± 1.7 mm/yr shortening perpendicular to the San Andreas fault at Parkfield (Harris and Segall, 1987). Given the large quantity of data consistent with S_H oriented NE, Zoback et al. (1987) conclude that S_H does not change in orientation throughout the thickness of the seismogenic portion of the lithosphere in central California. This orientation is consistent with many geological structures which strike parallel to the San Andreas fault in the Coast Ranges/Great Valley area and the Los Angeles Basin (Hauksson, 1990). The significance of a fault normal S_H is that the San Andreas fault will slip at shear stresses much less than required for laboratory frictional sliding of rock which increases with normal stress. Hickman (1991) argues that because shear stress along most active faults is high, the low shear stress along the San Andreas fault indicates either a low-strength fault gouge or elevated pore pressures. Indeed, some experiments on clay fault gouge suggest that the strength of the San Andreas fault may be quite low (~ 10 MPa) and independent of normal stress (Wang et al., 1980). However, local pockets of higher shear stress (i.e., S_H = NNW) may indicate that the San Andreas fault is not uniformly weak.

Stress Discontinuities as a Function of Depth

Stress discontinuities of two general types are found in the top 5 km of the lithosphere. The first type is indicated by abrupt changes in stress magnitude which are not consistent with a linear increase in stress with depth. Examples are found in the crystalline rocks in Iceland (Haimson and Rummel, 1982), under the Illinois Basin (Haimson and Doe, 1983), and on more than one scale in the Devonian sediments of the Appalachian Plateau (Evans et al., 1989). In each of these cases, the orientation of the stress field does not change appreciably throughout the well. The second type of discontinuity is a systematic change in orientation of the stress field at some depth. One example of this behavior is found at Darlington, Ontario, where an NQ size test hole was drilled through 220 m of Ordovician Limestone and into crystalline basement (Haimson and Lee, 1980). In the Ordovician Limestone the orientation of S_H is about N70°E, a measurement consistent with the North American contemporary tectonic stress field (fig. 12–3). Horizontal stresses in crystalline basement

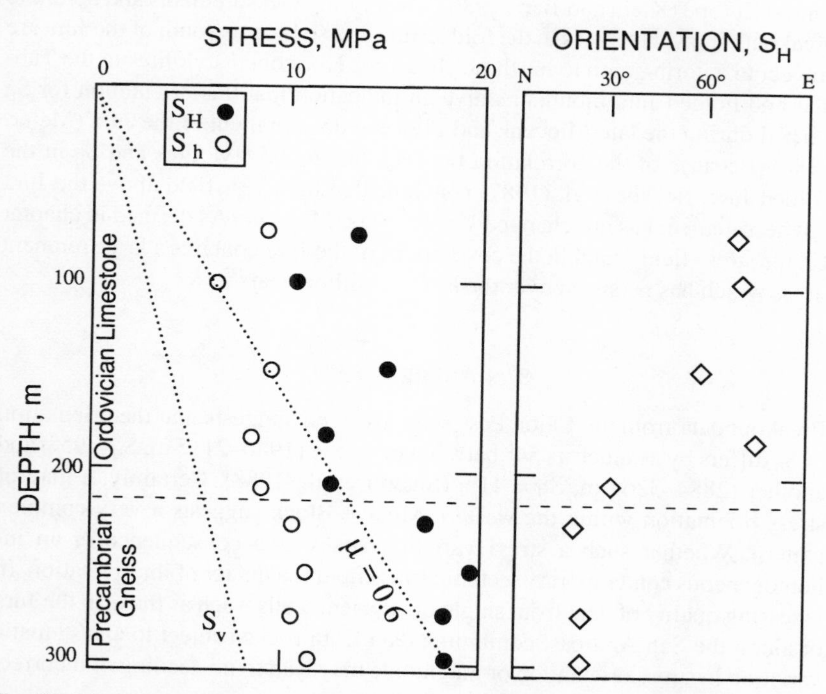

Fig. 12–3. Depth variation of stress magnitudes and S_H directions at Darlington, Ontario, Canada (adapted from Haimson and Lee, 1980).

abruptly jump in magnitude with S_H rotating to about N23°E. This is the odd example where the orientation of S_H from deeper measurements does not conform with the well-established North American platewide trend. Nearby geological features of the same orientation (i.e., N20°-30°E) are the large fracture lineaments dating from the early Paleozoic flexural extension of crystalline rocks of the Adirondack uplift (Bradley and Kidd, 1991). But, even in the Adirondacks most stress measurements indicate that S_H is ENE.

Jura Mountains

A larger-scale orientation discontinuity occurs on the eastern end of the Jura Mountains, Switzerland, where stress in the sedimentary cover is decoupled from stress in the crystalline basement (Becker et al., 1987). In several wells where breakouts were measured in both cover rocks and crystalline basement, an orientation discontinuity was found at a depth of about 300 m. The zone of decoupling is the Triassic salt which permitted the detachment of the folded Jura from basement rocks. In this example the orientation of S_H in the crystalline basement (NNW) agrees with that orientation found throughout most of central Europe (Klein and Barr, 1986). Overcoring measurements and borehole breakouts show that stress in the folded Jura and sediments south of the Jura are subject to a north-south to north-northeast S_H. Horizontal stylolites in the Tabular and Folded Jura mountains give an indication that this orientation for S_H existed during the late Miocene and Pliocene development of the Jura (Meier, 1984). Because of the correlation between horizontal stylolites and S_H in the Folded Jura, Becker et al. (1987) conclude that the stress field above the Jura salt décollement has not changed since the late Miocene. As defined in chapter 10, the stress field found in the cover rocks of the Jura contains a large remnant stress which has persisted over the last ten million years.[2]

San Andreas Fault

Breakout data from the Cajon Pass scientific well suggests that the orientation of S_H differs by as much as 54° between one zone (1946–2115 m, $S_H = 95°$) and another (2884–3264 m, $S_H = 41°$) (Shamir et al., 1988). Certainly, a map of stress orientation within the western Mojave Block suggests a very complex pattern. Whether such a stress variation arises as a consequence of an inhomogeneous contemporary tectonic stress field is a matter of interpretation. In assessing quality of data from single or adjacent wells, such as those in the Jura or along the San Andreas, combining data from rocks subject to a systematic variation in stress orientation or magnitude may lead to misleading or incorrect statistical inferences.

Coupling between the Near-Surface and Deep Stress Fields

Ideas vary concerning whether or not earth stress is a continuous function of depth starting right at the surface. Because surface outcrops are often highly fractured, one hypothesis is that fractures decouple near-surface rocks from stresses below, leaving only components of thermal and residual stress. Here the near-surface is denoted by a top layer of the crust which is most likely to develop sheet fractures; this layer is usually on the order of 10–20 m thick. The decoupling hypothesis is supported by near-surface overcoring measurements where a signal from the contemporary tectonic stress field is not observed during strain-relaxation measurements.

North America

North American examples of near-surface decoupling are found in rocks of the Appalachian Plateau including the Medina Sandstone (Engelder, 1979a), Onondaga Limestone (Engelder, 1979a) and Tully Limestone (Engelder and Geiser, 1984), where strain-relaxation correlates with a component of residual stress dating back to the Alleghanian Orogeny of the Appalachian Mountains. However, there are also many examples of stress measurements where a component of deep lithospheric stress (i.e., the contemporary tectonic stress field of Sbar and Sykes, 1973) is found in the near-surface. Included in this list are measurements in the Algerie granite, Massachusetts (Engelder, 1984), limestones of France (Froidevaux et al., 1980), and the Punchbowl Sandstone, California (Sbar et al., 1984). In the case of the Algerie granite and the Punchbowl Sandstone, very few fractures are found in the near-surface. In brief, outcrops coupled with the deeper lithospheric stress field are common.

Experience has shown that given a dozen or more measurements in the top 30 m of the crust, the orientation of the deep lithospheric stress is accurately defined. Such definition arises because the complications caused by local fracturing plus residual and thermal stress contribute random orientation data so that a weak preferred orientation becomes apparent when mixing data from coupled and decoupled outcrops. Near-surface overcoring data from New England illustrate as well as any the effectiveness of near-surface measurements in detecting the regional contemporary tectonic stress field. Within a few years of the first compilation of stress orientation by Sbar and Sykes (1973), geologists recognized that the entire midcontinent of the United States was part of a continent-scale stress province (Haimson, 1978). However, New England was difficult to characterize in terms of this uniform stress field. Because the first information on stress in New England was based on shallow overcoring measurements which appeared to deviate from the ENE trend of S_H (Hooker and Johnson, 1969), Sbar and Sykes (1973) did not include New England in the large ENE stress province of eastern North America. Citing offset boreholes as evidence (e.g., Block et al., 1979), Zoback and Zoback (1980) also concluded that New England was a stress province distinct from the region immediately to the west. In the early 1980s microearthquake networks were used to detect events in New England from which fault-plane solutions could be constructed. Initial results, based mainly on earthquakes along the Ramapo fault in New York and southern New England, showed a mixture of orientations for P-axes (Yang and Aggarwal, 1981; Pulli and Toksoz, 1981). By this time a compilation of several dozen overcoring measurements at nineteen sites throughout the region (fig. 12–4) suggested that New England is, indeed, part of the mid-continent stress field (Plumb et al., 1984a). Reanalysis of focal mechanisms from the Ramapo fault system (Seborowski et al., 1982) and throughout New England (Gephart and Forsyth, 1985) corroborated Plumb's et al. (1984a) con-

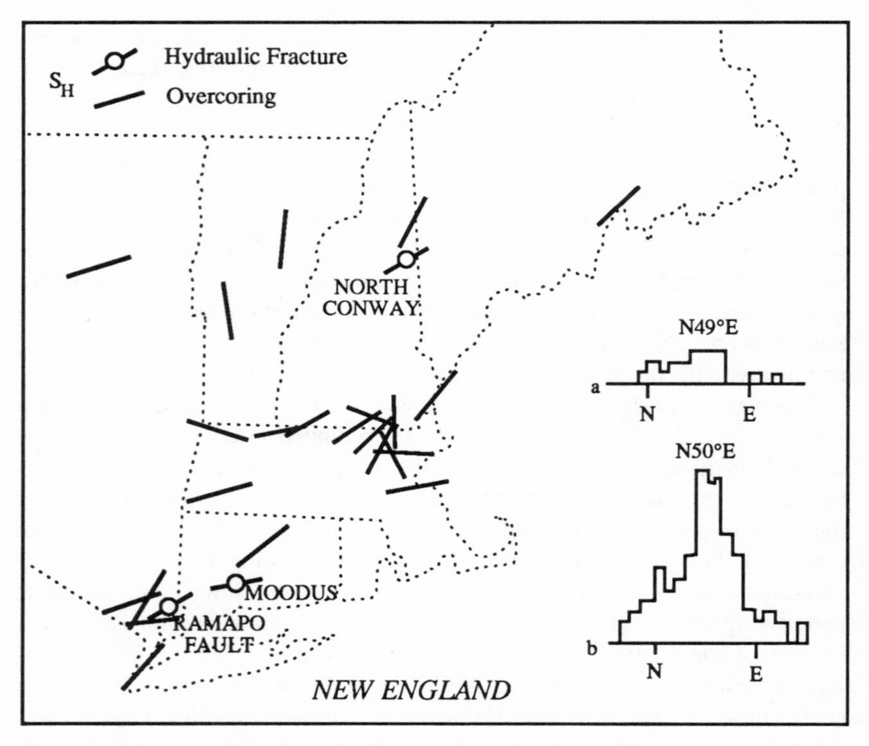

Fig. 12–4. Map of near-surface stress measurements in eastern New York and New England (adapted from Plumb et al., 1984a). Inset histograms are a plot of site average orientations for S_H (*top*) and a compilation of individual measurements of S_H (*bottom*).

clusion. Deep hydraulic fracture measurements at North Conway (Evans et al., 1988), along the Ramapo fault (Zoback, 1986), and in the vicinity of the Moodus microearthquake swarm, Connecticut (Baumgärtner and Zoback, 1989), also confirm that S_H is oriented about ENE for most of New England. Despite the fact that most of the data in figure 12–4 is class D quality, the average orientation of S_H indicated by the data compilation is that of the deep lithospheric stress field.

EUROPE

Coupling of the near-surface to the deep lithosphere is further verified with data from Europe. Europe is characterized by a persistent orientation of S_H from the Alpine foreland to the British Isles, North Sea, and Baltic Shield (fig. 12–5). A compilation of data from overcoring, hydraulic fracturing, borehole breakouts, and flatjack tests shows that S_H on average is about NW-SE (Klein and Barr, 1986). Until recent hydraulic fracture measurements by Rummel et al. (1983) and borehole breakout analyses by Klein and Barr (1986) the orientation of S_H was inferred largely from near-surface stress measurements (overcoring and flatjack tests) and fault-plane solutions (Hast, 1974; Ranalli and Chandler, 1975; Greiner and Illies, 1977). Recent data from deep wells did nothing to change the inferences about the orientation of stress in the deep lithosphere based on near-surface measurements. Through 1966 Hast (1974) had measured earth stress at twenty-nine sites in the Fennoscandian Shield at depths varying from a minimum of 6 m to a maximum of 880 m. Although Hast (1974) does not give the details for calculation of stress orientation at each data point, it is assumed that many, if not all, of these data represent some average of several measurements. A rose diagram for Hast's measurements throughout Fennoscandia shows a cluster for S_H between N-S and N50°W (fig. 12–6). The scatter from Fennoscandia is characteristic for other orientation data sets comparable in size, as is orientation data for S_H from hydraulic fracture tests in Germany (Rummel et al., 1983).

Other examples of near-surface coupling include doorstopper (chap. 7) data from quarries and mines at ten locations in central Europe (Greiner, 1975; Greiner and Illies, 1977). At each location, more than sixty measurements were taken at depths up to 10 m into the quarry floor or 15 m into the wall of a tunnel. The data at each site were averaged and plotted in figure 12–5. The general NNW-SSE trend in S_H for overcoring data is the same as indicated by earthquake fault-plane solutions (Ahorner, 1975). Flatjacks (chap. 8) were used in an attempt to measure contemporary tectonic stress in eight limestone quarries in France (Froidevaux et al., 1980). All eight quarries are subject to a NNW S_H which is between 1 and 5 MPa (fig. 12–5). In one of the quarries doorstoppers were overcored to show that S_H was indeed in the NNW-SSE orientation. Like their counterparts from New England, these measurements provide further ev-

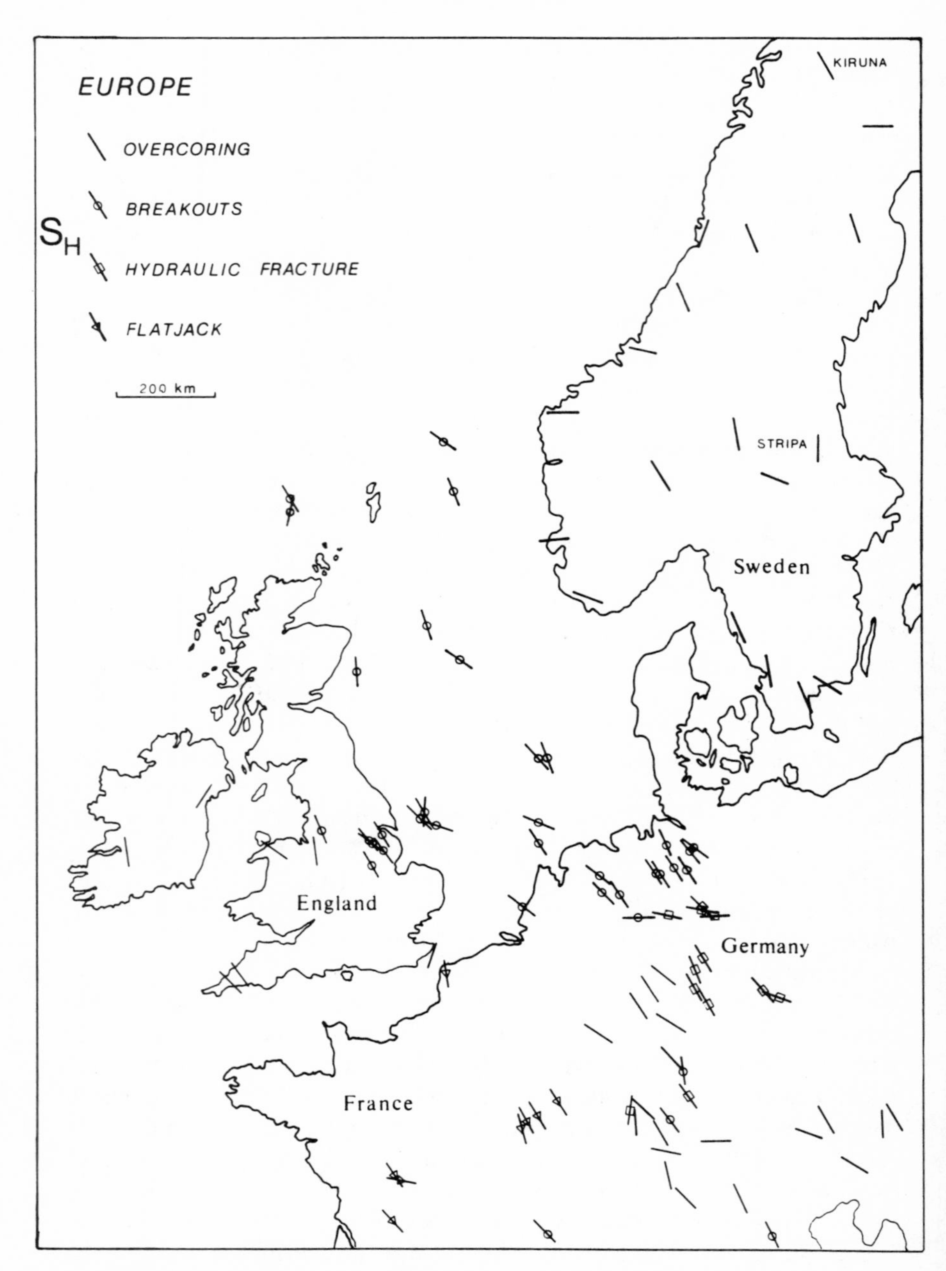

Fig. 12–5. Map of the orientation of S_H for the northwestern portion of the Eurasian plate. Data based on overcoring, borehole breakouts, hydraulic fracturing, and flatjack tests. Data compiled by Klein and Barr (1986). Klein and Barr's data points 18, 27, and 69 omitted because the orientation of S_H was inconsistent with adjacent measurements. Six overcoring measurements in Sweden by Hast (1974) were added.

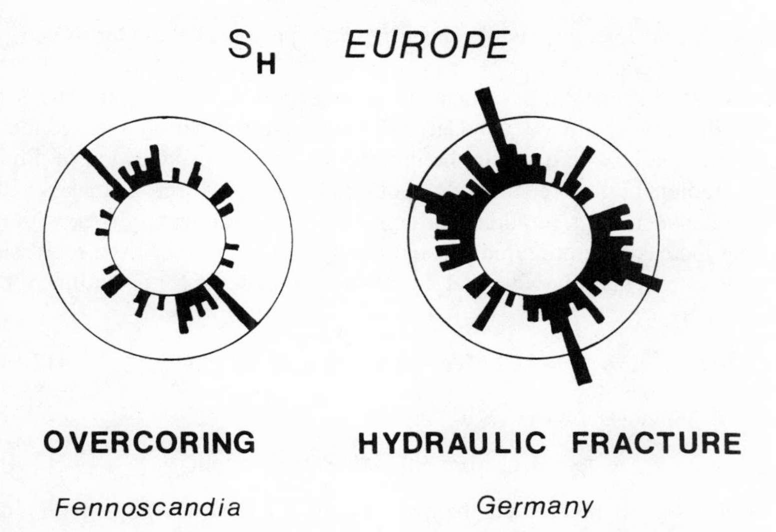

Fig. 12–6. Rose diagrams for the orientation of S_H from 29 data points from Hast (1974) and 69 data points from Rummel et al. (1983)

idence that the orientation of contemporary tectonic stress appears in the near-surface.

Extrapolation of Stress Magnitude with Depth in the Lithosphere

In considering the behavior of stress as a function of depth there are two major issues. The first concerns the extrapolation of stress data to depths in the frictional-slip regime below which measurements are possible. This issue boils down to the question of whether or not stress magnitude increases as a linear function of depth as the relationship between shear stress and normal stress in a linear friction law might suggest. If the upper tier of the ductile-flow regime is encountered, then stress will not increase linearly. The second issue concerns whether shear stress approaches the theoretical frictional strength of the lithosphere everywhere or just within tectonically active areas.

There are a number of ways of expressing stress as a function of depth in the lithosphere. One method is to plot principal stress components which, of course, increase with depth (e.g., Haimson, 1978). Often these plots show that the gradient for S_v with depth is greater than the gradient for either S_H or S_h. Single curves are constructed by plotting mean stress, $[(S_H + S_h + S_v)/3]$, with depth (e.g., Stephansson et al., 1986) or average horizontal stress, $[(S_H + S_h)/2]$, with depth (e.g., Herget, 1988). McGarr (1980) presents data in terms of τ_m as a function of depth.

There is still no final statement on whether any expression for stress increases linearly with depth. McGarr (1980) concludes that on average the τ_m increases linearly with depth in the upper 5 km, with no suggestion of diminishing gradient at the deeper levels of observations. However, he makes a distinction between the τ_m gradient for crystalline and sedimentary rock with the gradient increasing more rapidly in hard rock. McGarr (1980) fits a regression line to his stress data showing that τ_m increases with depth for crystalline rocks according to

$$\tau_m = 5.67 \text{ MPa} + 6.37 \text{ MPa/km} \cdot \text{depth} \qquad (12\text{–}5)$$

and for sedimentary rocks according to

$$\tau_m = 3.11 \text{ MPa} + 3.90 \text{ MPa/km} \cdot \text{depth}. \qquad (12\text{–}6)$$

These stress gradients fall well below that required for frictional slip on faults and fractures with $\mu_s = 0.85$. In contrast to McGarr's analysis, Herget (1986) suggests that the average horizontal stress within the Canadian Shield increases at a higher rate in the top 800 m than at depths below. However, he also suggests that there are locations with extremely high stress where the average horizontal stress increases linearly from the surface. We will return to the question of extrapolation to depth after introducing a means of comparing earth stress with the frictional strength of rocks.

Assuming that the upper crust is fractured, its theoretical strength is based on the frictional strength of rock (Brace and Kohlstedt, 1980). The idea is that joints and fault zones pervade the upper crust to the extent that slip on most favorably oriented discontinuities acts to relieve the buildup of earth stress (chap. 3). As friction is dependent on normal stress, the frictional strength of joints and faults increases with depth. The limit to frictional strength on favorably oriented fault-planes is given in terms of effective principal stresses by rewriting equation 3–30

$$\sigma_1 - P_p = (\sigma_3 - P_p)\left[\sqrt{(\mu_s^2 + 1)} + \mu_s\right]^2 \qquad (12\text{–}7)$$

(Jaeger and Cook, 1969; McGarr et al., 1982). For favorably oriented planes with $\mu_s = 0.85$, frictional slip will take place if $\sigma_1 \approx 5\sigma_3$. If $\mu_s = 0.6$, then slip will take place at $\sigma_1 \approx 3\sigma_3$.

An interesting objective in any compilation of vertical stress profiles is to compare the magnitude of the stress with these empirical limits for frictional strength. A plot of earth stress relative to the frictional sliding envelope based on equation 12–7 gives an indication of crustal stability. Zoback and Healy (1984) examined sets of data on stress in areas of active faulting including normal faulting (the Nevada test site [Stock et al., 1985]), strike-slip faulting (Rangely, Colorado [Raleigh et al., 1976]), and reverse faulting (Punchbowl area, California [Zoback et al., 1980]). In each case Zoback and Healy (1984) conclude that τ_m correlates closely with the frictional strength of active faults in the region, assuming that these faults have frictional properties characteristic of laboratory samples.

To examine whether Zoback and Healy's (1984) observation is generally applicable to the upper lithosphere, stress data from several geological settings are extrapolated to depths of 2.5 km (fig. 12–7). To construct a general plot containing data from both normal and thrust-faulting regimes, the stress state is represented by the octahedral shear stress

$$\tau_{oct} = \frac{1}{3}\sqrt{(\sigma_1 - \sigma_2)^2 + (\sigma_2 - \sigma_3)^2 + (\sigma_3 - \sigma_1)^2}. \qquad (12\text{–}8)$$

τ_{oct} is the shear stress across the plane whose normal is equally inclined to all three principal stress axes. The normal stress across this plane is the mean stress, σ_m. Although frictional strength is measured in terms of the two extreme principal stresses, τ_{oct} gives a better sense for the tendency for frictional slip on faults that are not most favorably oriented for slip. Assuming that the τ is the same for the following cases, τ_{oct} is larger for $\sigma_1 > \sigma_2 = \sigma_3$, than for $\sigma_1 > \sigma_2 > \sigma_3$. In figure 12–7 the frictional envelopes ($\mu_s = 0.6$) for τ_{oct} were minimized by assuming that σ_2 was the average of $\sigma_1 + \sigma_3$. Each stress line is constructed by calculating τ_{oct} for stress data at each depth where S_H and S_h were measured and then finding the best-fit line to the entire data set. Correlation coefficients vary anywhere between 0.95 for data from Cornwall, England (Pine et al., 1983) to as low as 0.17 for data from the Illinois Deep Hole (Haimson and Doe, 1983). A high correlation coefficient for data taken over a depth range of 2 km gives some confidence that τ_{oct} increases as a linear function of depth. In some cases poor correlation coefficients arise because of heterogeneities within the rock. For example, the Illinois Deep Hole intersects a zone of highly fractured rock subjected to lower τ_{oct} (Haimson and Doe, 1983). Measurements such as those at Cornwall, England, support McGarr's (1980) conclusion that linear extrapolation to depths of 5 km is reasonable.

The highest values for τ_{oct} come from crystalline rocks of Cornwall, England; the crystalline rocks of the Appalachian Mountains (Moodus, Connecticut, and Ramapo fault, New York); the western Mojave Block (Black Butte, California); and the Canadian Shield. The lowest values come from sedimentary rocks of the Appalachian Plateau (Auburn and South Canisteo) and the Michigan Basin. In effect, this result is the same as McGarr's (1980) with crystalline rocks being subject to a higher shear stress than sedimentary rocks. In general, stress appears to increase most rapidly in areas of active faulting such as Moodus, Connecticut, and near the San Andreas fault (Black Butte and the Punchbowl area). The least increase of stress with depth is found in the stable continental interior of North America (Canadian Shield,[3] Illinois Basin, and Michigan Basin). Stress in South Africa increases at a rate intermediate between the extremes (table 12–3).

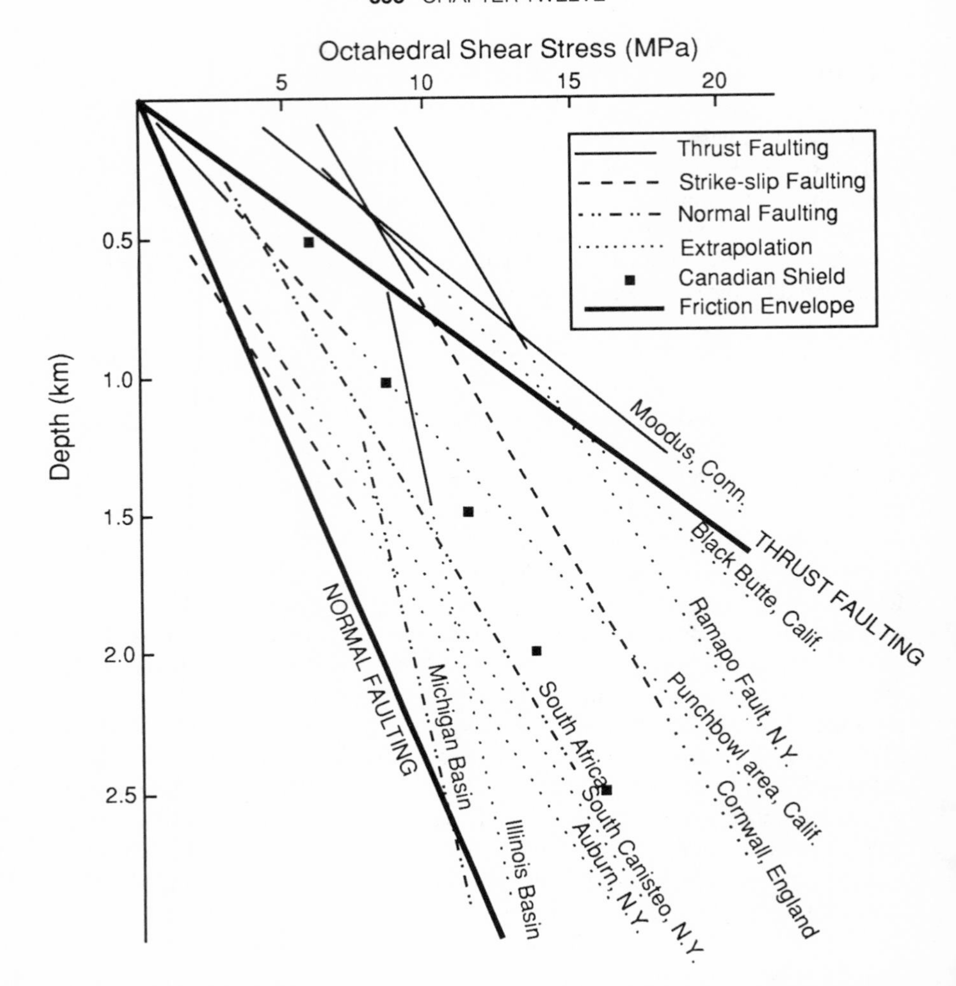

Fig. 12–7. Octahedral shear stress (τ_{oct}) as a function of depth for frictional sliding assuming hydrostatic P_p and $\mu_s = 0.6$ for both thrust faulting and normal faulting (thick dashed lines). τ_{oct} as a function of depth for stress data. Each plot shows depth range of measurements, a linear extrapolation to depth, and the Andersonian stress state indicated by the relative magnitude of the three principal stresses. The data are from: Moodus, Connecticut (Baumgärtner and Zoback, 1989); Black Butte, California (Stock and Healy, 1989); Ramapo fault, New York (Zoback, 1986); Cornwall, England (Pine et al., 1983); the Punchbowl area, California (XTLR of Zoback et al., 1980); Canadian Shield (Herget, 1986); South Canisteo, New York (Evans and Engelder, 1986); South Africa (Gay, 1977); Auburn, New York (Hickman et al., 1985); the Illinois Basin (Haimson and Doe, 1983); and the Michigan Basin (Haimson, 1978).

Table 12–3. Octahedral Shear Stress

Location	# of Data	Deepest Datum (m)	Gradient τ_{oct} (MPa/km)	Correlation Coefficient	Reference
Lab Friction	$\mu_s = .85$	—	25.1	—	Byerlee, 1978
Lab Friction	$\mu_s = .6$	—	14.9	—	Byerlee, 1978
Moodus, Conn.	7	1284	16.1	0.95	Baumgärtner & Zoback, 1989
Black Butte, Calif.	6	635	9.4	0.63	Stock & Healy, 1989
Nakaminato, Japan	9	434	9.0	0.86	Tsukahara, 1983
Punchbowl, Calif.	15	849	8.4	0.96	Zoback et al., 1980
Okabe, Japan	16	429	7.6	0.89	Tsukahara, 1983
Wilkins (Bot.), NY	13	1037	6.3	0.90	Evans & Engelder, 1986
Auburn, NY	4	1482	6.1	0.98	Hickman, et al., 1985
Kent Cliffs, NY	6	864	5.7	0.64	Zoback, 1986
South Africa	12	2500	5.7	0.71	Gay, 1977
Cornwall, England	13	2215	5.5	0.95	Pine & Kwakwa, 1989
Futtsu, Japan	10	437	5.5	0.97	Tsukahara, 1983
South Carolina	10	961	5.5	0.73	Zoback & Hickman, 1982
Iceland	24	576	4.9	0.61	Haimson & Rummel, 1983
Michigan Basin	4	5110	2.1	0.94	Haimson, 1978
SW-German Block	49	448	2.6	0.60	Rummel et al., 1982
Illinois UPH-3	13	1449	1.9	0.17	Haimson & Doe, 1983

In figure 12–7, fault stability is indicated by the position of the stress-depth curve relative to the friction envelopes. For example, fault slip is not predicted for the Illinois Basin showing a thrust-faulting stress regime because all data fall below the friction line for thrust faulting. Thrust faulting is predicted in the vicinity of Moodus, Connecticut, an area where microearthquakes are common (Ebel et al., 1982). τ_{oct} in the sedimentary rocks of the Appalachian Basin suggest that most if not all faults are stable under a strike-slip stress state. Indeed, earthquakes are relatively rare within the Appalachian Basin. Normal faults in the Michigan Basin are stable at depths below 2.5 km.

In a compilation of stress as a function of depth McGarr and Gay (1978) identified the Canadian Shield as an area that has relatively high horizontal stresses. In the Canadian Shield it is common to find that the S_h far exceeds the weight of overburden (Herget, 1986). The extreme opposite is South Africa where horizontal stresses were consistently below the weight of overburden (Gay, 1977). Although there is a significant difference in τ_{oct} at depths less than 2 km, the stress-depth curves converge below this depth (table 12–3).

Extrapolation of stress to great depth next to the Mojave portion of the San Andreas fault is still a subject of debate. Figure 12–7 suggests that τ_{oct} is close to the frictional strength of rock with a hydrostatic pore pressure at Black Butte and in the Punchbowl area of the San Andreas fault (Zoback et al., 1980; Stock and Healy, 1988). It seems common that in the vicinity of active fault zones τ_{oct} approaches the frictional strength of rock (Zoback and Healy, 1984). However, these results are not compatible with the heat flow measurements of Lachenbruch and Sass (1980) and recent measurements by the continental scientific drilling borehole at Cajon Pass which suggests that at 2 km the shear stress is about 5 MPa or about a factor of three lower than that necessary for frictional sliding at hydrostatic pore pressure (Zoback et al., 1987). Low shear stress is compatible with the fault normal orientation of S_H as indicted by the borehole breakout measurements throughout California (Mount and Suppe, 1987). One solution to the stress-heat flow paradox for the San Andreas fault is that conventional laboratory friction data overestimates the strength of most faults at depth (Raleigh and Evernden, 1981). But, the more likely explanation is that portions of the San Andreas fault zone contain exceptionally weak fault gouge or abnormal pore pressures (Hickman, 1991).

Platewide Compilations

Isolated measurements are useful for engineering applications, but when compiled on a regional or platewide basis the result is a powerful data set which constrains hypotheses concerning tectonic processes. Interpretations of early compilations were a bit naïve largely because the constraints of plate tectonics theory had yet to be formulated. In a compilation of data from the Fennoscandian Shield, Hast (1967) observed that $S_H + S_h$ increases linearly with depth and thus concluded that "they behave . . . as a hydrostatic pressure." Hast (1973) reasoned that this linear trend developed because, when horizontal stresses exceeded an upper limit defined by this trend, the bedrock becomes unstable and upward buckling occurs. Although the details of this mechanism were incorrect, Hast's prediction of an upper limit was valid. The frictional strength of joints and faults limit the magnitude of horizontal stress in the frictional-slip regime of the lithosphere. In an early compilation of North American measurements Hooker and Johnson (1969) suggested that S_H was aligned with the tectonic grain of the Appalachian-Ouachita trend. As an understanding of plate tectonics developed, it became clear that S_H for eastern North America was aligned with midocean spreading (Sbar and Sykes, 1973). In places along the Appalachian Mountains its tectonic grain is subparallel to the present spreading direction of the MidAtlantic Ridge, thus Hooker and Johnson's interpretation was reasonable under the circumstances.

Examples of recent continentwide compilations include Zoback and Zoback's (1980; 1989) stress map of the United States (fig. 12–8); Fordjor's et al. (1983) map of Canada (fig. 6–5); Illies' et al. (1981) and Klein's and Barr's (1986) maps of Europe; and Lambeck's et al. (1984) map of Australia. These maps show the orientation of S_H or S_h for as many as four hundred data points (Zoback and Zoback, 1989). Each data point may be, for example, a single fault-plane solution for an earthquake with $M > 3.5$, the average of several hydraulic fracture stress measurements in a single well, or some other measure of stress. Regional stress maps indicate that in some areas there is a remarkable correlation between the orientation of local geological structures and the stress field, whereas in other areas there is no such correlation in orientation. The continental United States contains examples of both situations (fig. 12–8). In the eastern United States the midcontinental stress field does not twist and turn with the sinuous trend of the Appalachian Mountains, yet in the western United States stress provinces and geological provinces are one and the same where younger geological structures are likely to correlate with contemporary tectonic stress fields.

North America Plate

More details are known about stress in the North American lithosphere than in any other lithospheric plate. Zoback and Zoback (1980; 1989) have divided the continental United States into eight stress provinces based on over four hundred higher-quality stress measurements (fig. 12–9). The largest of these, the *Mid-plate stress province*, extends from the Great Plains to the oceanic lithosphere of the western Atlantic Ocean and is characterized by the east-northeast to east-west S_H first documented in Sbar and Sykes's (1973) compilation (Zoback et al., 1986). Active listric normal faulting in the southeastern United

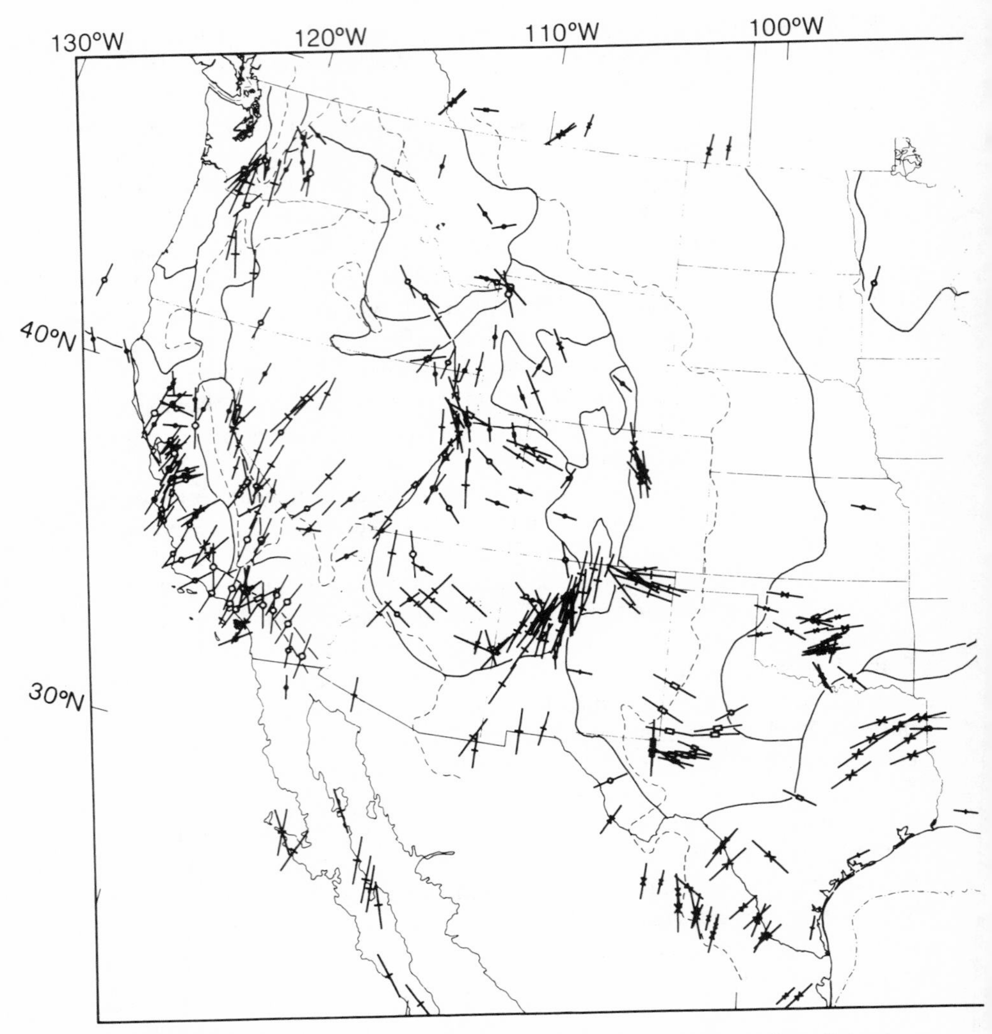

Fig. 12–8. Map of the orientation of S_H for the continental United States (from Zoback and Zoback, 1989).

States defines the *Gulf Coast stress province.* Second largest of the provinces is the *Cordilleran Extensional stress province* which covers much of the western third of the continental United States from Mexico to Idaho and Montana. Because this area is dominated by "basin and range" normal faulting, Zoback and Zoback (1989) identify the range of this province based on west-northwest to east-northeast trending S_h. Most of the Cordilleran extension stress province is characterized by a broad zone of high regional elevation and high heat flow

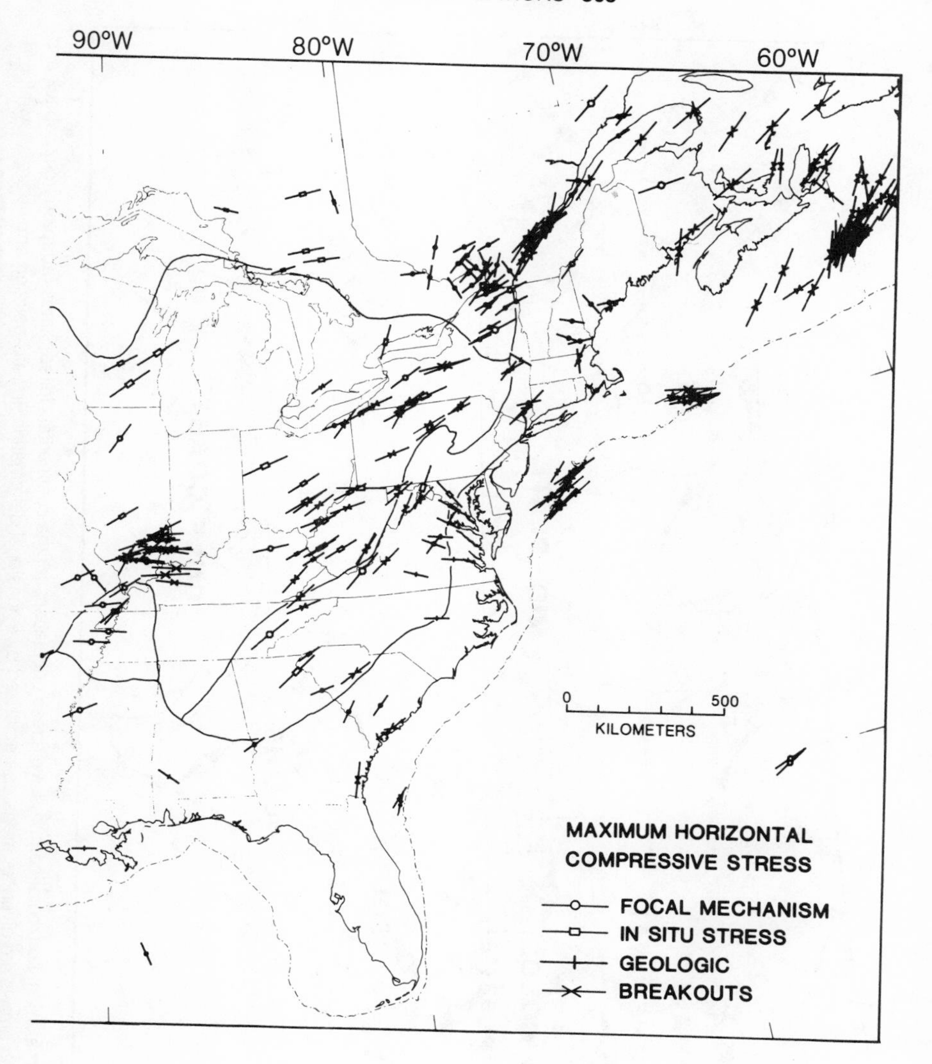

(Eaton, 1979). Inside the Cordilleran Extension province is the *Colorado Plateau stress province* which is characterized by west-northwest S_H. Forming a boundary between the Mid-plate and the Cordilleran stress provinces is the *Southern Great Plains stress province*. Geological characteristics such as basaltic volcanism and north-northeast extension distinguish this province. Three stress provinces on the west coast of the United States include the *San Andreas*, the *Pacific northwest*, and the *Cascade Convergence* provinces. In the San An-

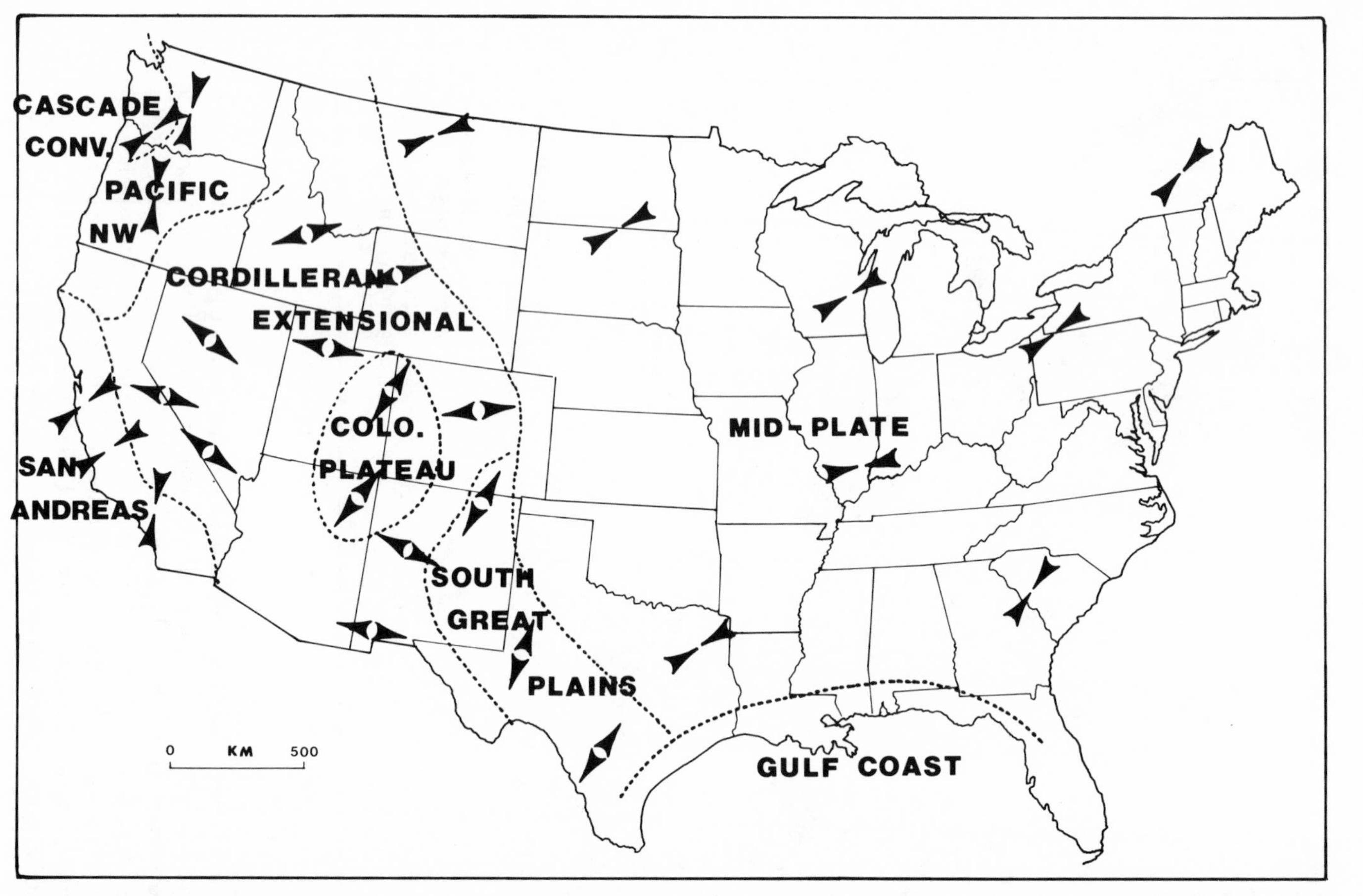

Fig. 12–9. Generalized map of maximum horizontal compressive stress orientations for the continental United States (adapted from Zoback and Zoback, 1980). Outward-pointing arrows are given for areas characterized by extensional deformation. Inward-pointing arrows are shown for regions dominated by compressional tectonism (thrust and strike-slip faulting). Stress provinces are delineated by the dashed lines.

dreas stress province, the province at the boundary between the North American and Pacific lithospheric plates, S_H is roughly normal to the San Andreas fault. The Cascade Convergence zone is characterized by northeast S_H, whereas the Pacific Northwest is characterized by north-south S_H. Canada east of the Rocky Mountain front appears to be part of the same Mid-plate stress province seen in the eastern United States.

From this compilation we conclude that platewide stress fields are interrupted by discontinuities so that small portions of some plates have unique stress provinces. Such provinces are found in western North America (Zoback and Zoback, 1989). This complicated pattern of stress provinces is associated with either subduction or the development of the San Andreas fault system. The latter started developing about 30 million years ago with subduction of the East Pacific Rise. During subduction calc-alkaline magmatism and extension took place in the Basin and Range, Rio Grande rift, and areas near the Snake River plain (Eaton, 1979). This history of subduction and magmatism changed the lithosphere and upper mantle structure in such a manner that even today the state of stress in these regions does not conform with the great North American stress province east of the Great Plains. Zoback and Zoback (1989) suggest that the modern state of stress throughout the western United States is influenced, in a complex manner, by lateral variations in upper mantle structure which are related to neo-San Andreas fault tectonic activity.

World Stress Map

Thirty scientists from nineteen countries collaborated in the World Stress Map project to compile a global database of contemporary in situ stresses in the upper lithosphere (Zoback et al., 1989). Presently there are over 3500 entries in the global database with 3142 data having a quality of C or better. These latter data are plotted on figure 12–10 which supplements earlier global compilations by Ranalli and Chandler (1975) and Richardson et al. (1979). The distribution of data on the World Stress Map includes 50% earthquake focal mechanisms, 31% borehole breakouts, 7% overcoring measurements, 5% fault-slip data, 4% volcanic alignments, and 3% hydraulic fracturing measurements.

Several regional patterns of stress orientation emerge from figure 12–10 (Zoback et al., 1989). Most midplate regions fall within compressive stress regimes where σ_1 is horizontal and faulting is characterized by a component of reverse slip. Most continental areas subject to extensional stress, where faulting is characterized by a component of normal slip, are regions of anomalously high elevation. Broad regions of the lithosphere are subject to a nearly uniform-oriented stress field. In the case of the NE to ENE compression of central and eastern North America the areas is 2×10^7 km^2. These general observations place constraints on the origin of stress in the lithosphere based on plate tectonic processes as discussed in chapter 13.

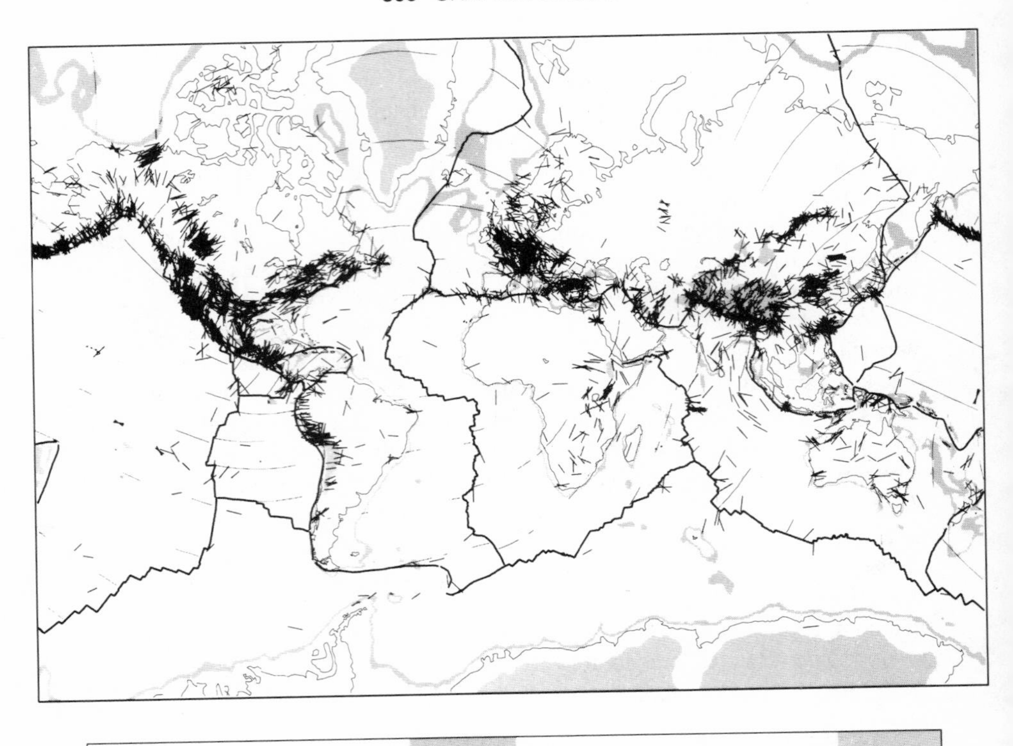

Fig. 12–10. World Stress Map, transverse mercator projection (from Zoback 1992). S_H orientations are plotted on a base of average topography. Lightly dashed lines correspond to absolute velocity trajectories determined using AM1–2 poles of Minster and Jordan (1978).

Concluding Remarks

Extrapolation of stress to depth in the lithosphere suggests that tectonically active areas of the upper lithosphere are near a state of frictional slip. Sedimentary basins are likely to have lower shear stresses relative to crystalline rocks. Apparently, stresses consistent with the upper tier of the ductile-flow regime are more common in sedimentary portions of the lithosphere. Stress data suggest that the interiors of continents are stable (i.e., subject to the crack-propagation or ductile-flow stress regimes but not the frictional-slip stress regime). Finally, platewide lithospheric stress fields are commonly interrupted by regional stress provinces.

— 13 —

Sources of Stress in the Lithosphere

Stress differences of at least 1.5×10^9 dynes/cm^2 (150 MPa) must exist within the outermost 50 km of the earth.

—Harold Jeffreys (1962) on considering the strength of the earth's crust on the basis of the stress necessary for the support of high mountains

The theory of plate tectonics states that major lithospheric plates, about a dozen in number, consist of both crust and upper mantle moving over an asthenosphere composed of ultramafic rocks. *Plate boundaries* are of four types including transform faults where plates slide past one another, midocean ridges where new material is added to plates, subduction zones where older portions of plates are destroyed, and basal surfaces which accommodate the viscous shear between the lithosphere and the asthenosphere. Ever since A. L. Wegener and F. B. Taylor published their ideas on continental drift (i.e., plate tectonics), earth scientists have wondered about the origin of forces responsible for moving the lithospheric plates (Menard, 1986). An understanding of the sources for stress in the lithosphere is, in part, linked to an understanding of the forces driving plate tectonics.

Stresses in the lithosphere arise from mechanisms operating on three scales: platewide, regional, and local. *Platewide* stress fields are found in North America where S_H is oriented approximately ENE over an area of more than 10^7 km^2; in South America where S_H is E-W, and in northwestern Europe where S_H is NNW (chap. 12). Platewide stress fields are generated when forces from convection and buoyancy within the lithosphere and asthenosphere of the earth are opposed by tractions either within the lithosphere or at the boundaries of lithospheric plates including the base of each plate. The notion that plate-boundary forces dominate the stress field of some plate interiors arises from the parallelism between the direction of S_H and the absolute plate velocity. According to the definition in chapter 1, these forces are responsible for the contemporary tectonic stress field. On a much smaller scale, *local stresses* are gravitationally, mechanically, or thermally induced with the scale of the source any where from microscopic (i.e., residual stresses) to the thermal transient associated with the diurnal or annual heating of near-surface rocks up to that of local topography. Somewhere between the scale of platewide and local stresses are *regional stresses* set up by the redistribution of mass within the lithospheric plates or loading at the edges of lithospheric plates. Each of the several stress provinces of the western United States (i.e., fig. 12–9) defines a regional stress field. In general, tectonic stresses arise from either regional or platewide conditions. The platewide stresses associated with the driving mechanisms of plate tectonics are *renewable,* whereas most of the regional stresses are *nonrenewable* (Bott, 1982). Renewable stresses are those which persist through the reapplication of body forces or surface tractions even though strain energy is dissipated. Nonrenewable stresses are relieved by processes such as transient creep or brittle fracture.

Mechanisms Responsible for Platewide Stress Fields

The ultimate source of lithospheric stress is thermal cooling of the earth which leads to density differences and the concomitant buoyancy forces that act on both the asthenosphere and lithosphere. Tomography surveys of the asthenosphere reveal sizable lateral variations in seismic velocities which are interpreted in terms of density heterogeneities providing the driving forces for mantle convection (Vigny et al., 1991). These gravitationally induced body forces are in large part responsible for the motion of the lithospheric plates. Tractions in the form of frictional forces and viscous drag act on the boundaries of lithospheric plates to drive or resist the motion of the plates. There is a general agreement that the lithospheric-scale stress fields are a manifestation of the forces causing lithospheric-plate motion (Forsyth and Uyeda, 1975; Richardson et al., 1976; Chapple and Tullis, 1977). Stresses within the lithospheric plates reflect the pushes and pulls developed by body forces and surface tractions associated with thermally induced gravitational instabilities.

Platewide stress fields arise from the balance of forces driving and resisting plate motion; the balance is among gravitational driving forces at trenches and ridges, resistive tractions such as frictional forces at transform boundaries and collisional zones between plates, and viscous tractions between the lithosphere and asthenosphere (Forsyth and Uyeda, 1975; Chapple and Tullis, 1977; Harper, 1989). As conceived by Forsyth and Uyeda (1975) there are at least eight plausible forces acting on lithospheric plates (fig. 13–1). The viscous coupling between the asthenosphere and lithosphere is called *mantle drag* ($\mathbf{F}_{MD}$) as suggested as early as 1929 (Holmes, 1929). $\mathbf{F}_{MD}$ is either a driving force or resisting force depending on the direction of active flow of thermal convection in the asthenosphere (Chase, 1979). If there is no coherence to flow in the asthenosphere, then the asthenosphere will generally act as a drag on lithospheric motion. Continental lithosphere may have an additional drag force because of the difference in the rheological properties of the asthenosphere under the

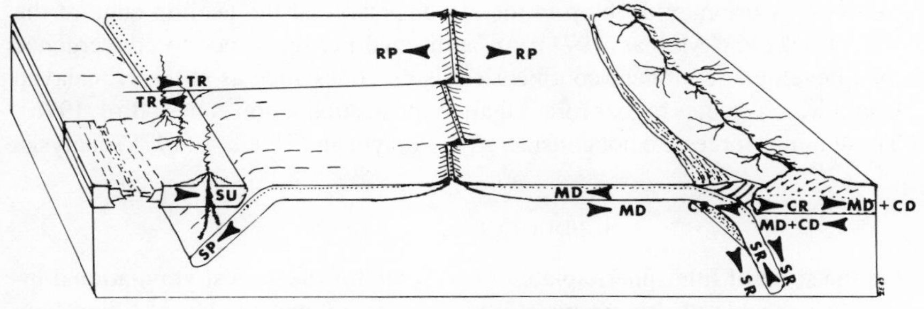

Fig. 13–1. Possible forces acting on lithospheric plates. The forces are defined in the text (adapted from Forsyth and Uyeda, 1975).

oceans and continents (Jordan, 1975). This force is called *continental drag* ($\mathbf{F}_{CD}$). Drag on the continental lithosphere is a combination of two components ($\mathbf{F}_{CD} + \mathbf{F}_{MD}$). At spreading centers along midoceanic ridges the lithosphere slides off of elevated ridges as a consequence of gravitational forces. The body force developed along elevated ridges is known as *ridge push* ($\mathbf{F}_{RP}$) because, in a sense, it acts to push the plates apart (McKenzie, 1972). Transform boundaries develop frictional forces (*transform-fault resistance*) which resist plate motion ($\mathbf{F}_{TR}$) (Hanks, 1977). Several forces act at convergent margins and subduction zones (McKenzie, 1969b). The higher density of the cool, downward-moving slab in the subduction zone sets up a negative buoyancy force known as *slab pull* ($\mathbf{F}_{SP}$). A significant component of slab pull is derived from the elevation of the major phase changes in the cooler subducting lithosphere (Smith and Toksöz, 1972). Drag of the subducting lithosphere on the asthenosphere is a resistive force called *slab resistance* ($\mathbf{F}_{SR}$). Over the subduction zone the two colliding plates grind together setting up resistive forces due to friction and called *colliding resistance* ($\mathbf{F}_{CR}$). Finally, a force, which Elsasser (1971) calls *suction*, draws plates together at the trench ($\mathbf{F}_{SU}$). Possible sources for this "suction" force include the seaward migration of trenches to create a gap between plates, a secondary convection cell above the downgoing slab, upward migration and injection of hot mantle material in the overriding plate, and spreading in a marginal sea (Forsyth and Uyeda, 1975; Hsui, 1988). The absolute magnitude of these driving and resistive forces are not measured directly. However, several modeling studies have used intraplate stress fields to constrain the forces driving and resisting plate motions (e.g., Solomon et al., 1975; Richardson and Cox, 1984; Jurdy and Stefanick, 1991).

In addition to the eight major forces listed by Forsyth and Uyeda (1975) other forces may act to some degree. Although midocean ridges seem to exert a net compression on the adjacent lithosphere, the asthenosphere may resist spreading at the ridge axis (Sleep and Biehler, 1970). In subducting slabs an excess pressure may develop in the asthenosphere at the leading edge of the slab (Isacks and Molnar, 1971). An additional net resistance to convergence may develop at continent-continent collision zones such as at the Himalayan Chain which creates higher forces than found at subduction zones (Bird, 1976). These minor forces are not considered in Forsyth and Uyeda's (1975) analysis.

Subduction Zone Forces

On the scale of lithospheric plates a candidate for the largest gravitational instability capable of driving plate motion is the subducted slab of lithosphere projecting down into the mantle. The largest σ_d anywhere in the lithosphere is likely to develop within the subduction zone if the subducting slab is gravitationally unstable. The net downward buoyancy of slab pull arises from two gravitational forces: the excess density of the cool subducted lithosphere and the weight associated with the reduced depth of the olivine-spinel phase transformation in the cooler lithosphere. Calculation of the slab-pull force is possible because the temperature distribution in the subducting lithosphere is reasonably well constrained (Toksöz et al., 1973). A pulling force per unit length of lithosphere is approximately $3\text{–}5 \times 10^{13}$ Nm^{-1} and this gives a component of extension as much as 600–1000 MPa depending on the thickness of the subducted lithosphere (Turcotte and Schubert, 1982). Resisting the downward motion of the subducted lithosphere are viscous tractions that develop along the base and top of the plate from shear of the mantle through which the lithospheric plate is plunging. The magnitude of the forces acting on the subducting slab is inferred from the velocity of the slab descending into the mantle (Forsyth and Uyeda, 1975). If the slab velocity is lower than terminal velocity, then slab pull will exceed slab resistance and a net slab-parallel extension should develop in the top part of the slab. On the other hand, if the slab velocity exceeds terminal velocity, then slab resistance will exceed slab pull, which would place the top part of the subducted slab into slab-parallel compression. Earthquake fault-plane solutions give an indication that down-dip extension is common in inclined seismic zones under several trenches in the circum-Pacific region, which suggests that slab pull exceeds slab resistance (Kanamori, 1971). However, this small net slab pull is nearly equal to the frictional resistance at the boundary between two lithospheric plates involved in the subduction zone. Based on low stresses (i.e., low level of seismicity), and earthquakes which do not show a consistent extensional direction indicating a dominance of pull in oceanic plates near trenches, it appears that the large forces developed within subducting slabs are locally balanced by tractions on the descending plates (e.g., Toksöz et al., 1973; Richter and McKenzie, 1978; Turcotte and Schubert, 1982). The net force downward is just large enough to keep the plates subducting at a velocity of about 10 cm/yr, which is close to the terminal velocity.

The actual state of stress at subduction plate boundaries is a complex superposition of several stress systems. In modeling this boundary, Bott et al. (1989) recognize five main types of deviatoric stresses: (1) local compression from the downpull of the sinking slab; (2) regional horizontal tension from slab pull and trench suction; (3) local tension in regions of thicker crust; (4) local backarc tension; and (5) bending stress produced by flexure of the lithosphere at the trench. Superimposed on these stress systems are thermal stresses and ridge-push stresses. Other recent papers discussing plate boundary forces include Wortel and Vlaar (1988), Vassiliou and Hager (1988), and Lay et al. (1989).

Transform Faults and Collisional Resistance

Resistance to plate motion takes several forms including shear along large transform faults and frictional resistance when continental portions of plates collide. The largest resistance on transform faults arises from mismatches (asperities) in the shapes of plate boundaries. The big bend in the San Andreas fault just north of Los Angeles is such an asperity. Shear stress along the San Andreas fault is not well known although both heat flow data as well as recent stress measurements suggest that it is less than 10 MPa (Lauchenbruch and Thompson, 1972; Zoback et al., 1987). Evidence for the relatively low tractions along transform faults also comes from the observation that plate velocities are not dependent on the length of bounding transform faults (Forsyth and Uyeda, 1975). Forces at transform boundaries are apparently not important for setting up platewide stress fields within the lithosphere. Likewise, collisional resistance plays a limited role in contributing to platewide stress fields (Richardson, 1991).

Drag Forces

Viscous forces on the base of the plate develop from shear of the convecting mantle beneath the lithospheric plate. For uniform tractions under lithospheric plates, the flow pattern of the mantle must have consistent streamlines over appreciable areas at the base of the lithosphere. These viscous forces will either retard or enhance the motion of the plate depending on the direction of convection. Because of the relatively low viscosities of the asthenosphere, the shear stress on the base of the lithosphere is probably less than 1 MPa (Schubert et al., 1978) and maybe on the order of 10^{-2} MPa (Richardson et al., 1979). If convection cells drive lithospheric plates, the size of the convection cell under the lithosphere has a significant affect on the magnitude of σ_d which may develop by this mechanism. A shear stress from viscous drag, τ, acting over a lateral distance, d, on the base of the lithosphere causes a horizontal normal stress on a cross section of the lithosphere of

$$\sigma_n = \frac{\tau d}{z_l} \qquad (13\text{–}1)$$

where z_l is the lithosphere thickness (Bott and Kusznir, 1984). Over a convection cell a component of extension develops above the rising current and a component of compression develops above the sinking current. The difference between extension ($S_v - S_h$) and compression ($S_H - S_v$) on opposite sides of the convection cell means that maximum σ_d will be half that of equation 13–1. Assuming convection cells with d = 4000 km and z_l = 100 km and an initial lithostatic state of stress, the asthenosphere drag would give rise to maximum σ_d on the order of 20 MPa. Of course, smaller cells would give rise to smaller σ_d.

Based on the distribution of contemporary tectonic stress and earthquakes, Zoback and Zoback (1989) infer that mantle drag on the lithosphere of North America is small. If mantle drag is a driving mechanism, then a drag-related compression should be most pronounced along the leading edge of the North American lithosphere.[1] If seismic activity varies directly as a function of stress, then a basal-drag related stress would cause the highest seismic activity at the leading edge of the lithosphere. In North America the opposite is the case: the distribution of seismic activity increases from west to east. Furthermore, the Great Plains area of North America is subjected to a north-south extensional stress regime which is not compatible with the basal-drag model. Finally, stress magnitude across eastern North America appears quite uniform, a distribution more compatible with a ridge push model for stress generation.

Ridge Push

Ridge push arises from the elevation of the midoceanic ridges and is manifested as a form of gravitational sliding (Lister, 1975; Dahlen, 1981). A general rule for midocean ridges is that the depth to adjacent ocean floor increased with the square root of the distance from the midocean ridge (Parsons and Sclater, 1977). This increasing depth of the ocean floor is explained by the cooling of the oceanic lithosphere as it gradually ages while spreading away from the midocean ridges. Age of the ocean lithosphere correlates directly with distance from the midocean ridges depending only on the rate of spreading from the ridges. As the lithosphere cools while moving away from the midocean ridge it becomes denser. It is assumed that during the initial stages of development, the thin lithosphere is so weak that isostatic compensation is local. Because isostatic compensation is maintained at some depth in the mantle the denser lithosphere tends to sink lower into the mantle (fig. 13–2). Thus by the principal of isostasy the denser and thicker lithosphere has to sit lower in the mantle in order for a thicker column of lightweight sea water above to compensate for its extra weight relative to younger and thinner lithosphere. The formula for the

FORCES ACTING AT A MIDOCEAN RIDGE

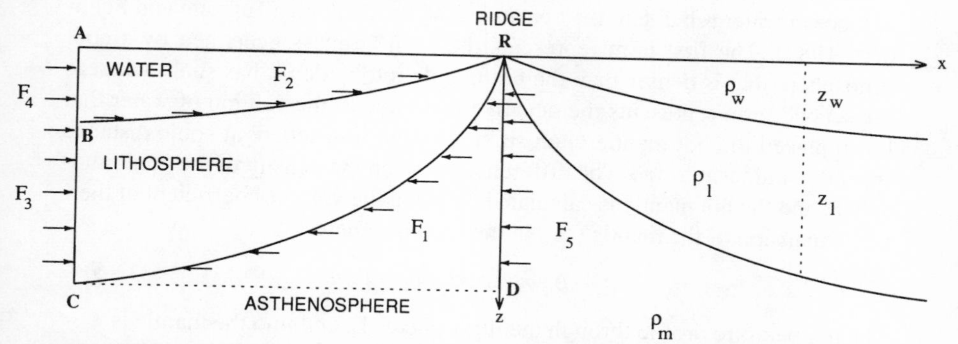

Fig. 13–2. A cross section through a midocean ridge showing the gradual depression of the sea floor with increasing age of lithosphere. The principal of isostasy requires the oceans to deepen with age to offset an increase in density of the lithosphere caused by the thermal contraction on cooling. This cross section through a midocean ridge also shows the horizontal forces acting on a section of the ocean, lithosphere, and mantle at an oceanic ridge (adapted from Turcotte and Schubert, 1982).

depth of the ocean floor as a function of age or distance from the ridge is derived by assuming that the mass per unit area in vertical column extending from the surface to the base of the lithosphere is the same for columns of all ages (Turcotte and Schubert, 1982).

The following derivation of lithospheric plate stresses by the ridge-push mechanism is taken from Turcotte and Schubert (1982). The mass per unit area (M) in a column of oceanic lithosphere of any age is

$$M = \int_0^{z_l} \rho_l dz + z_w \rho_w \qquad (13\text{–}2)$$

where z_l is the thickness of the lithosphere, z_w is the depth of water to the ocean floor, ρ_l is the density of the lithosphere, ρ_m is the density of the deep mantle, and ρ_w is the density of water. At the ridge crest, $\rho_l = \rho_m$, and the mass of a column of vertical height $z_w + z_l$ is $\rho_m(z_w + z_l)$. For this example $z_w = 0$ at the ridge crest. If the depth of isostatic compensation is taken as $z_w + z_l$, then isostasy requires that

$$\rho_m(z_l + z_w) = \int_0^{z_l} \rho_l dz + z_w \rho_w \qquad (13\text{–}3)$$

where the mass of the deep mantle at the mid ocean ridge equals the mass of the oceanic lithosphere plus water at some point off the ridge axis. By rearranging terms

$$\int_0^{z_l} (\rho_l - \rho_m) dz + z_w(\rho_w - \rho_m) = 0. \qquad (13\text{–}4)$$

Equation 13–4 is understood by comparing the density of the cool lithosphere and ocean water relative to the density of the deep mantle (Turcotte and Schubert, 1982). The first term represents the positive mass generated by a cool lithosphere that is denser than the hot mantle into which it has sunk, whereas the second term represents the negative mass due to the column of water that has replaced the hot mantle upon sinking of the lithosphere at some distance from the midocean ridge. The difference between the density of the cool lithosphere and the hot mantle is calculated by using the volume coefficient of thermal expansion of the mantle (α_m) in the asthenosphere

$$\rho_l - \rho_m = -\rho_m \alpha_m (T_l - T_m). \qquad (13\text{–}5)$$

The temperature profile through the lithosphere, T_l, and into the mantle is

$$\frac{T_l - T_s}{T_m - T_s} = 1 - \operatorname{erfc}\left(\frac{z}{2}\sqrt{\frac{u}{\kappa x}}\right) \qquad (13\text{–}6)$$

where erfc(z) is the complementary error function, T_s is the temperature at the seawater-lithosphere interface, T_m is the temperature of the mantle, u is the velocity of spreading from the midocean ridge, x is distance from the midocean ridge, z is depth in the lithosphere, and κ is thermal diffusivity of the lithosphere. Equation 13–6 indicates that at the midocean ridge the surface temperature is equal to the temperature of the asthenosphere, whereas the surface of the lithosphere cools rapidly as the lithosphere spreads from the midocean ridge. Substituting the temperature profile into the equation for volume coefficient of thermal expansion (13–5) and then substituting the result into the equation for isostasy (13–4) we obtain

$$z_w(\rho_m - \rho_w) = \rho_m \alpha_m (T_m - T_S) \int_0^{z_l} \operatorname{erfc}\left(\frac{z}{2}\sqrt{\frac{u}{\kappa x}}\right) dz. \qquad (13\text{–}7)$$

Turcotte and Schubert (1982) rewrite this equation using the definite integral of the erfc to obtain

$$z_w = \frac{2\rho_m \alpha_m (T_m - T_s)}{(\rho_m - \rho_w)} \sqrt{\frac{\kappa x}{\pi u}}. \qquad (13\text{–}8)$$

Equation 13–8 predicts that the depth of the ocean increases with the square root of the distance from the ridge or the square root of the age according to Parsons and Sclater (1977).

From these equations the gravitational forces causing ridge push are derived using a force-balance approach on a unit cross section (fig. 13–2). The analysis proceeds assuming that the mantle acts like a fluid where the elevation of the ridge crest develops a pressure head as the deep mantle ascends and then spreads horizontally away from the ridge crest. All forces are referenced to the ridge crest where z = 0. The integrated horizontal force on the base of the lithosphere is F_l (fig. 13–2). The net horizontal force of the mantle, a fluid, on a cross section centered on the ridge crest (RD) must equal the net horizontal force on the base of the lithosphere (F_l). So, the force beneath the ridge crest is

$$F_5 = F_l = \int_0^{z_w + z_l} \rho_m g z \, dz \qquad (13\text{–}9)$$

where g is the gravity constant. Turcotte and Schubert (1982) rewrite equation 13–9 as

$$F_l = g \int_0^{z_w} \rho_m z \, dz + g \int_0^{z_l} \rho_m (z_w + \bar{z}) \, d\bar{z} \qquad (13\text{–}10)$$

where $\bar{z} = z - z_w$. The integrated force on the upper surface of the lithosphere is the pressure due to the water over lithosphere that has sunk into the mantle and is equal to the net force across AB

$$F_2 = F_4 = \int_0^{z_w} \rho_w g z \, dz. \qquad (13\text{–}11)$$

The horizontal force acting on the section of the lithosphere BC is the integral of the pressure in the lithosphere P_l across BC

$$F_3 = \int_0^{z_l} P_l d\bar{z} \qquad (13\text{–}12)$$

where P_l is given by

$$P_l = \rho_w g z_w + \int_0^{\bar{z}} \rho_l g \, d\bar{z}' \qquad (13\text{–}13)$$

where ρ_l is the density of the lithosphere and $\bar{z}'$ recognizes a change in the z coordinate system for the integration. This calculation assumes that the lithosphere is in a state of lithostatic stress. Substituting back into the equation for the force acting on a cross section of the lithosphere

$$F_3 = \int_0^{z_l} \left(\rho_w g z_w + \int_0^{\bar{z}} \rho_l g \, d\bar{z}' \right) d\bar{z}. \qquad (13\text{–}14)$$

The net horizontal force, F_{RP}, on the lithosphere near the midocean ridge is then found by combining equations 13–10, 13–11, and 13–14

$$F_{RP} = F_l - F_2 - F_3 = g \int_0^{z_w} (\rho_m - \rho_w) z \, dz$$

$$+ g \int_0^{z_l} \left\{ (\rho_m - \rho_w) z_w + \rho_m \bar{z} + \int_0^{\bar{z}} \rho_l \, d\bar{z}' \right\} d\bar{z}. \qquad (13\text{–}15)$$

Turcotte and Schubert (1982) substitute the equation for volume thermal expansion (13–5) and the equation for the temperature profile through the lithosphere (13–6) to arrive at an equation for ridge push in terms of the the distance of the lithosphere from the ridge crest

$$F_{RP} = g\rho_m\alpha_m(T_m - T_s)\left(1 + \frac{2\rho_m\alpha_m(T_m - T_s)}{\pi(\rho_m - \rho_w)}\right)\frac{\kappa x}{u}. \qquad (13\text{–}16)$$

Forces due to midocean ridges and the associated thickening of the oceanic lithosphere are then modeled in terms of density, bathymetry, and thickness of the young oceanic lithosphere where the latter two parameters are used to derive a temperature structure of the lithosphere. Calculation of this force[2] indicates that it is about 4×10^{12} Nm^{-1} so that if it is supported across a full 100 km thick lithosphere the equivalent σ_n above lithostatic stress is 40 MPa. If on the other hand, this ridge-push force is supported in the upper and stronger portions of the lithospheric plate then σ_d could grow as high as 200 MPa across, say, a 20 km thick plate.

The ridge-push model for driving plate motion is evaluated using earthquake fault-plane solutions to define the state of stress on either side of midocean ridges. Early observations of midocean earthquakes showed that oceanic lithosphere older than about 35 Ma was subject to compression (Mendiguren, 1971). Later studies pointed out that S_H was oriented roughly normal to the ridge crests (Sykes and Sbar, 1974). A tabulation of earthquake fault-plane solutions shows that normal faulting is common in the younger oceanic lithosphere near ridge crests, suggesting that an extensional stress is found in the vicinity of midocean ridges (Wiens and Stein, 1985). Thermal contraction is parallel to the ridge crest for young lithosphere. A transition from mixed extensional and compressive mechanisms in lithosphere less than 35 Ma old to purely compressional mechanisms at 35 Ma suggests the ridge-push mechanism dominates the stable interior of lithospheric plates only after the new oceanic lithosphere has cooled and contracted for 35 Ma (Bergman and Solomon, 1984). Furthermore, the ridge push forces appear to be very small near the ridge crests as expected from equation 13–15, where $F_{RP} = 0$ at $x = 0$ (Wiens and Stein, 1984).

The relative importance of ridge-push as a driving mechanism for motion of the North American lithosphere is suggested by the fact that the North American plate lacks any attached subducting slab. Furthermore, the contemporary tectonic stress field within the lithosphere of the North American plate has the wrong orientation for slab pull. Ridge push and slab pull have much different effects on the stress field within the interior of the plates. A compressional force from ridge push, as manifest by the orientation of S_H, develops as some integrated average of all ridges contributing to ridge push but is oriented roughly normal to the nearest ridge axis, whereas an extensional force from slab pull is normal to the subduction zone. On the basis of this geometric argument, the ridge-push model is given higher credibility for generation of major elements of the present stress field in the lithosphere of North America (Zoback et al., 1986). A gravitationally induced stress field associated with ridge push would impose an S_H parallel to the transform faults of the ridge and there is a strong correlation between the stress orientations in the eastern United States (S_H = ENE) and the overall North Atlantic transform fault orientations. Such a correlation is seen in a plot of transform fault orientation compiled on an equal-area projection centered on the Adirondack Mountain region of New York (fig. 13–3). Furthermore, the trajectory of the Yellowstone Hot Spot, Wyoming, correlates well on the same plot indicating the absolute motion of the North American lithosphere relative to the asthenosphere. The correlation among the transform orientations, stress orientations (i.e., S_H = ENE), and North American plate motions strongly suggests that the source for the contemporary tectonic stress in eastern North America is the gravitationally induced stress developed at the Mid Atlantic Ridge.

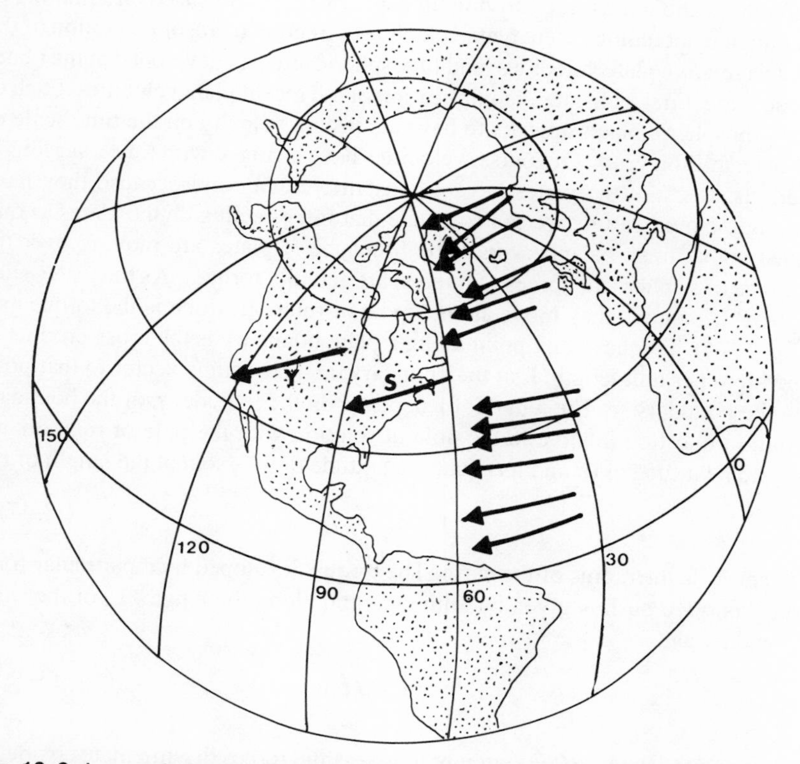

Fig. 13–3. A map of transform faults along the North Atlantic midocean ridge showing their orientation relative to North America (adapted from Wise, personal communication, 1984). S represents the orientation of S_H throughout eastern North America. Y represents the track of the Yellowstone hot spot.

Continental-size stress fields such as found in North America arise because the plates act as a stress guide absorbing the S_H developed at the ridges by the ridge-push or basal drag mechanisms. Furthermore, the thermally induced driving mechanisms assure that these large stress fields are maintained without relaxing; basically the source for the stress is renewable (Bott and Kusznir, 1984). The homogeneous stress field in the central and eastern United States is apparently not influenced by the relatively low resistive shear forces on the San Andreas transform because shear stresses generated by the San Andreas system are probably not transmitted through the thermally weakened, actively deforming western Cordillera (Zoback and Zoback, 1989).

Constraints on the Forces Driving Plate Tectonics

Inferences about the forces driving lithospheric plates are based on a number of parameters including recent plate velocities, direction of absolute motion of the plates, relative plate velocities, and lithospheric stresses developed at the plate-wide scale. First, consider the implication of the recent plate velocities. Each of the lithospheric plates appears to have a constant velocity on the time scale of about a million years. Of course, velocities have changed with time over longer periods. Because the plates have a constant velocity and because they have negligible momentum, we can conclude that forces acting on the plates as outlined above must balance. Since the lithospheric plates are moving over the surface of a sphere the forces of interest are actually torques. As the plates slide across the sphere, they move about a pole of rotation which is the torque axis (fig. 13–4). Torque at any point along a plate boundary is the cross product of the force per unit length, **f**, at the boundary, and the radius vector to that point on the boundary, **r**. The length of **r** depends on the latitude, γ, of the boundary point in question relative to the pole of rotation with the pole of rotation, assigned a latitude of 0° and its equator a latitude of 90°, so that the length of **r** is

$$r = R_e \sin \gamma \tag{13–17}$$

where R_e is the radius of the earth. The torque developed by a particular force such as ridge push is given by a line integral along the length, L, of the ridge boundary as

$$\mathbf{T}_r = \int_L \mathbf{r} \times \mathbf{f} \, dl. \tag{13–18}$$

Forsyth and Uyeda (1975) attempt to assess the forces driving plates by determining the relative size of each force which would minimize the components of net torque acting on each plate. Their calculations suggest that the driving forces are on the same order as discussed above.

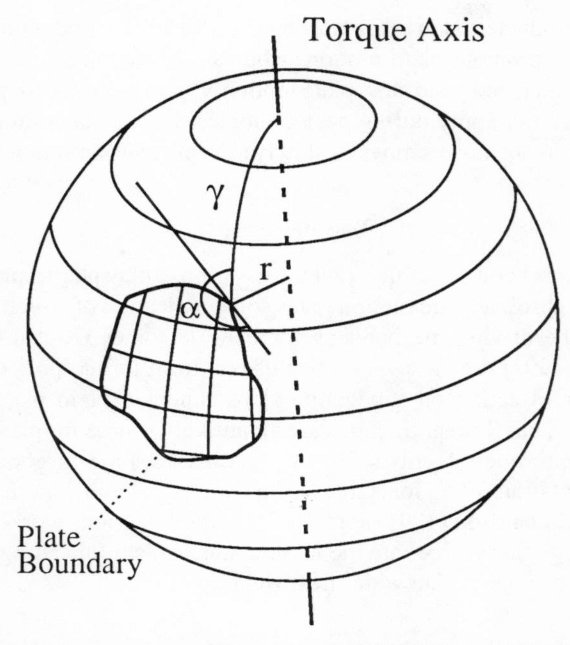

Fig. 13–4. Schematic of the motion of a lithospheric plate relative to its torque axis (adapted from Forsyth and Uyeda, 1975). r is the distance to the torque axis, γ is the colatitude of the position of the plate boundary, and α is the angle between the strike of the boundary and the azimuth to the pole of the torque axis.

Relative Plate Velocities

The relative velocities of lithospheric plates is another constraint for the various forces acting on lithospheric plates. The motion of plates with respect to underlying mantle is termed the "absolute" motion of plates. Information on the velocity vector of the plates is derived from paleomagnetic considerations (Jurdy, 1978; Gordon et al., 1978; Engebretson et al., 1985) and the assumption that hot-spots have a fixed position in the mantle (Morgan, 1972). There are several relationships between absolute plate motion and the lithospheric plates which have a bearing on predicting the relative importance of the forces driving plate motion. It is assumed that those plates subject to larger gravitational forces will have higher velocities, whereas those plates subject to larger resistive forces will move relatively slower. Indeed, plate velocity varies inversely with the area of continental lithosphere (Minster et al., 1974) and varies directly with the percentage of plate boundary connected with subducting slabs (Forsyth and Uyeda, 1975). Plate velocity is proportional to the sum of percentages of plate boundary connected to ridges and downgoing slabs (Aggarwal, 1978). There is also a correlation between the absolute plate velocities and the age of the subducted slabs (Carlson et al., 1983a,b). Subducting slabs and ridges tend to promote plate motion, whereas the deep roots of continental lithosphere tend to drag and slow plate motion. However, rules for present plate velocities may not apply during past geological history as dominantly continental plates do not always move as slowly as at present (Solomon et al., 1977).

Torque Poles

The correlation between torque poles for various lithospheric plates and the directions of absolute plate motion gives some indication of which forces actually act as the driving mechanisms for plate tectonics (Richardson, 1991). Richardson (1991) draws several conclusions from torque pole calculations. First, the forces acting on subducting slabs cannot explain platewide stress fields. Second, basal shear tractions are an unlikely source for platewide stress fields. Third, torque directions for ridge-push forces are in good agreement with the orientations of S_H for stable North America, South America, and western Europe. Richardson (1991) agrees with several previous workers who conclude that ridge-push forces are the favored mechanism for setting up the long wavelength features of platewide stress fields.

The Origin of Regional Stress Fields

Variations in Thickness or Density of the Lithosphere

Artyushkov (1973) suggests that regional tectonic stresses arise as an indirect consequence of density differences in the lithosphere. For isostatic equilibrium, a thick but weak crust requires a root area surrounded by a denser mantle. In this case compensation is local as envisioned by Airy where the total mass above a compensation depth is constant. An example of local compensation is found at the midocean ridges (fig. 13–2). Density differences give root areas a positive potential gravitational energy by Archimedes principle, which tends to lift the root areas up out of the mantle. The vertical component of the buoyancy force is balanced by the weight of the uplifted crust; however, the horizontal component of buoyancy is unbalanced. Density differences develop both at the edge of continental crust as well as in the vicinity of crustal bulges such as mountain belts within continental crust (fig. 13–5). Bott and Dean (1972) point out that at passive margins unbalanced buoyancy forces cause larger compressive stress in oceanic crust relative to adjacent continental crust. Weight developed within the isostatically compensated uplift will cause the uplifted area to spread sideways in order to diminish the positive potential energy associated with the uplifted area and the root zone (fig. 13–5b). Such spreading can induce extensional stresses beneath high plateaus (England and Houseman, 1989). With Airy compensation the σ_d arising from the density differences are the

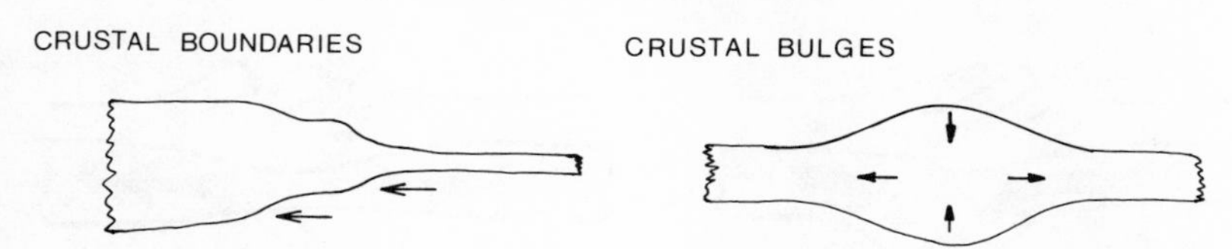

Fig. 13–5. Crustal structure in the regions of an isostatically compensated uplift where density contrasts give rise to stress differences. Arrows indicate the forces arising from density contrasts.

same order of magnitude as the load (McNutt, 1980). In the case of variation in crustal thicknesses, the vertical stress arising from the topographic relief, h, constitutes the load so that

$$\sigma_d = \rho g h. \tag{13–19}$$

One kilometer of topography will give rise to a stress difference of about 25 MPa according to the Artyushkov theory.

Lithospheric thickness inhomogeneities can cause elevated regions of the crust to shorten or lengthen depending on the thermal structure of the underlying lithosphere (Fleitout and Froidevaux, 1982; 1983). A dense cold lithospheric root has a mass excess which causes net compressive stresses in the root. These compressive stresses are reflected by compressional tectonics and thickening of the upper crust (fig. 13–6). By Artyushkov's analysis the thickened upper crust will eventually spread due to its mass deficiency. If the dense cold lithospheric root breaks away and settles into the mantle, hot mantle fills the void. This hot material causes a mass deficiency which leads to spreading of the lithosphere and extensional tectonics in the upper crust. Stresses associated with mass anomalies in the lower crust are given by

$$\sigma_d = \frac{\Delta m g z}{z_l} \tag{13–20}$$

where Δm is the difference in mass between the cold lithosphere and hot mantle, z is the depth to the mass anomaly, and z_l is the thickness of the lithosphere. This equation suggests that deeper mass anomalies have larger effects.

The effective thickness of the load-bearing portion of the lithosphere may vary in time because of stress decay by thermally activated creep in the lower portion of the lithosphere (Kusznir, 1982). If the lithosphere is subject to a uniform lateral load such as is developed from the ridge push mechanism, then the stress normal to the ridge varies according to the thickness of the lithosphere. Thinning of the load-bearing portion of the lithosphere by thermally activated creep redistributes the load so that stress in the lithosphere is amplified. A pos-

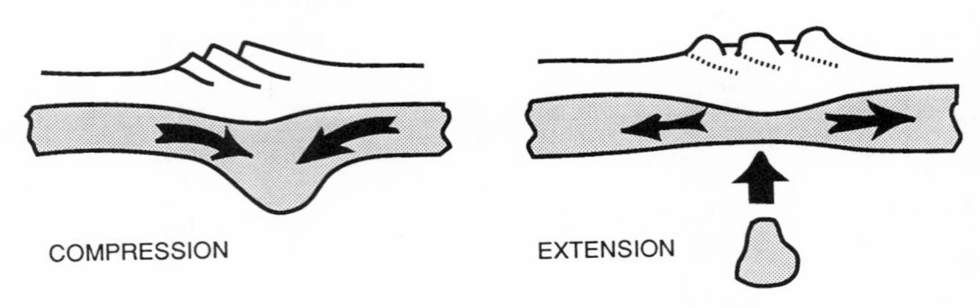

Fig. 13–6. The time evolution of the stresses associated with the lithospheric thickening process (adapted from Fleitout and Froidevaux, 1982). A dense, cold lithospheric root generates compression in the upper crust ($S_H > S_V$). After a cold blob detaches, rapid uplift causes extension in the upper crust ($S_H < S_V$).

sible example of stress amplification is found in the Basin and Range province of the western United States where a lithosphere thinning from 150 to 20 km in 100 Ma may have amplified an initial normal stress of 10 MPa to 75 MPa (Kusznir, 1982).

Thermal Stresses

Several cycles of heating and cooling have affected much of the continental lithosphere of the earth. The thermal history of most oceanic lithosphere is unusually simpler, yet thermal stresses develop there also. During the cooling of oceanic lithosphere at midocean ridges plane-strain conditions lead to a large component of thermal stress parallel to the ridge. Upon cooling, the oceanic lithosphere may contract to develop thermal stresses of up to 400 MPa (Turcotte, 1974a). As these cooling stresses are tensile, they are probably dissipated by brittle fracture long before they can build to this magnitude. With deviatoric tension in this orientation transform faults could have developed as thermal cooling cracks. This hypothesis is supported by the observation that there is a direct correlation between spacing of transform faults at midocean ridges and the spreading rates of those ridges (Sandwell, 1986). Thermal stresses thus appear to control the spacing between transform faults in oceanic lithosphere.

Thermal stresses will develop anytime materials of different thermal expansivities are welded together as was illustrated in the discussion of residual stress (chap. 10). Residual stresses from thermal contraction are likely on all scales up to the size of plutons (Savage, 1978). Thermal stresses are also associated with temperature changes during sedimentation. Finally, thermal stresses develop during the heating of subducting oceanic lithosphere (House and Jacob, 1982). Heating may cause a downdip compression in the uppermost region of the subducting lithosphere, whereas an underlying region develops downdip extension. This model, which allows the generation of deviatoric tension on the order of 500 MPa, is one explanation for double seismic zones seen in the western Pacific.

MEMBRANE STRESSES

Lithospheric plates must conform to the geoid as they move over the surface of the earth. However, the geoid is an oblate ellipsoid with its axis through the earth's poles. As a consequence, the curvature of any plate drifting north/south must change to conform to the curvature of the geoid. Membrane stresses of as much as 100 MPa may develop in the lithosphere as a consequence of drift in the north/south direction (Turcotte, 1974b). To date there is no compelling evidence that these stresses exist.

TIDAL STRESSES

The oscillatory strain associated with earth tides leads to a stress on the order of 10^{-3} MPa. One popular notion is that these small stresses may superimpose existing stresses to trigger earthquakes on faults loaded to the point of failure (e.g., Bott and Kusznir, 1984).

Lithospheric Flexure

The Artyushkov (1973) model for tectonic stress in a weak crust suggests that the lithosphere is not capable of supporting stress differences greater than the loads generated by topographic relief. In this case topography is locally compensated. However, early perceptions concerning the relative strength of the earth's outer layer came from an estimate of σ_d required to support topographic highs which were uncompensated. Some models assumed that if the earth's topography is uncompensated, σ_d of up to 60 MPa is required to support topography in Africa and North America (Darwin, 1882; Watts et al., 1980) Even larger σ_d is required to support uncompensated topographic highs in the Himalayas (Jeffreys, 1962). Later studies of gravity anomalies and surface deformation suggested that lithosphere is, indeed, capable of maintaining long-term loads (Watts et al., 1975). An important and long-standing debate on earth stress concerns the magnitude of σ_d in the lithosphere as a consequence of its ability to support long-term loads. One side of the debate is carried by those who conclude that high stress differences are maintained during crustal deformation such as found at the bending of the lithosphere near subduction zones (e.g., Hanks, 1977). The other side is carried by those who point out that laboratory creep experiments indicate that rocks might relax under the long-term loads (e.g., Carter, 1976).

Long-term loads are generated during the redistribution of mass within and on the lithosphere by the following processes (fig. 13–7). During orogenic events mass is forced laterally or vertically as a consequence of the impact between two lithospheric plates. Usually the lithosphere thickens through overthrusting, pervasive volume-constant strain with horizontal shortening and vertical lengthening, and the movement of magma by intrusion and extrusion. If the orogenic event has come to a halt then this mechanism for mass distribution is considered inactive. The Bouguer gravity anomaly across the Appalachian Mountains is explained as the result of loading of the continental lithosphere by surface thrust sheets and a much more massive body buried in the hinterland (Karner and Watts, 1983). Volcanism associated with hot spots can load the crust as is indicated by the flexure of the lithosphere under the Hawaiian Islands and other seamounts (Watts and Ribe, 1984). Other features causing long-term loads (> 10^6 years) on oceanic lithosphere include seamount chains (Walcott, 1970a), uplifted seamounts (McNutt and Menard, 1978), oceanic islands (Watts et al., 1975), and island arc-trench systems (Hanks, 1971). Mass redistribution occurs with the intrusion of high-density magmas commonly associated with the initial stages of rifting (Larson et al., 1986). Ocean basins may form if the rifting event evolves to maturity. The process of eroding mountain chains and depositing sediments in adjacent basins amounts to a very efficient mechanism for the redistribution of mass on the lithosphere (Walcott, 1972; Watts et al., 1982; Cloetingh et al., 1984). Glacial advances and retreats shift the mass of the lithospheric plates in a relatively short time span (Walcott, 1970b; Hasegawa et al., 1985). Some earthquakes in the area of Baffin Bay and the Labrador Sea have locations and fault-plane solutions consistent with the stress state predicted from glacial unloading (Stein et al., 1979). Loading or unloading of the crust produces an elastic deformation through uniaxial strain

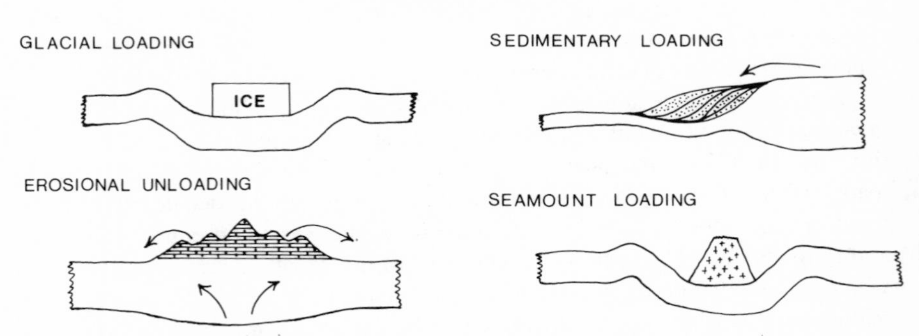

Fig. 13–7. Various mechanisms causing lithospheric flexure by loading and unloading.

deformation that may lead to the generation of intermediate scale stresses (Haxby and Turcotte, 1976). However, McGarr (1988) argues that this effect is usually overestimated. In the elastic lithosphere large σ_d developed by redistribution of mass does not contribute to plate motion because the accompanying elastic deformations are always small.

The shape of features such as the trench and outer rise or seamount depressions as measured by bathymetry matches the free air gravity anomaly across these features (Watts and Talwani, 1974). The correlation between gravity and topography suggests that lithospheric flexure is dynamic, supported by stresses in the lithosphere (Chapple and Forsyth, 1979). Models for the lithosphere suggest that its response to long-term loads is similar to a thin elastic plate (i.e., the lithosphere) overlying a fluid (i.e., the asthenosphere) (Watts et al., 1975, 1980; Turcotte and Schubert, 1982). If this model applies, compensation is distributed over a sizable region. Assuming that the lithosphere is an elastic plate, calculation of stress within the lithosphere is based on the bending of an elastic plate whose thickness is t (fig. 13–8). As the plate bends (i.e., folds), the upper portion of the plate is stretched, whereas the lower portion of the plate is shortened. The neutral surface, the plane between the stretched and shortened portions of the plate, is folded without changing length. The neutral fiber is that curve made by the intersection of the neutral surface and a plane cutting normal to the fold axis. If the neutral fiber is assigned a length, l, then a curve a distance z above the neutral fiber has the length, l + Δl. The strain parallel to the neutral fiber is proportional to the radius of curvature, R, of the elastic plate

$$\varepsilon_x = \frac{\Delta l}{l} = \frac{z}{R}. \qquad (13\text{–}21)$$

ELASTIC BUCKLE

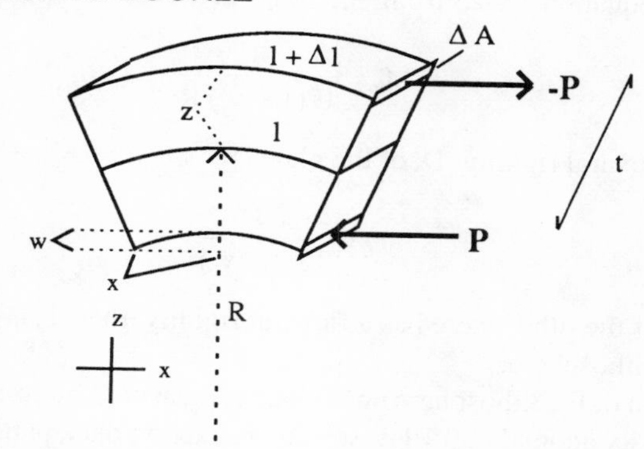

Fig. 13–8. Bending of an elastic plate (thickness, t, and radius of curvature, R). Such bending gives rise to compressional (**P**) and extensional (–**P**) forces which increase as a function of distance from the neutral fiber. The plate's neutral fiber has a length l.

If w is the deflection of the elastic plate in the direction z at a distance, x, from the origin, then

$$\frac{1}{R} = \frac{d^2w}{dx^2}. \tag{13–22}$$

By Hooke's Law a force per unit area on a small strip at z from the neutral fiber is

$$\sigma_x = \frac{\Delta P}{\Delta A} = \frac{Ez}{(1 - \nu^2)R}. \tag{13–23}$$

As the plate bends the force, **P**, per unit area, ΔA on a cross section of the plate increases with decreasing radius of curvature. Under static conditions for an elastic plate with a slight bend the normal stress (σ_x) within the elastic plate increases linearly with distance from the neutral fiber as a consequence of Hooke's Law. If each cross-sectional element has a width approaching zero, then $P = \sigma_x dz$. This force, P, exerts a torque about the neutral fiber of the elastic plate given by $\sigma_x z dz$. The bending moment on an elastic plate is defined as the integrated torque about the neutral fiber

$$M_b = \int_{-\frac{t}{2}}^{\frac{t}{2}} \sigma_x z \, dz \tag{13–24}$$

where t is the thickness of the elastic plate. Substituting equation 13–23 into 13–24, we derive

$$M_b = \frac{E}{(1 - \nu^2)R} \int_{-\frac{t}{2}}^{\frac{t}{2}} z^2 \, dz. \tag{13–25}$$

Integrating equation 13–25 from –t/2 to t/2 gives

$$M_b = \frac{Et^3}{12(1 - \nu^2)\, R} \tag{13–26}$$

where the flexural rigidity, D, of the plate is

$$D = \frac{ET_e^3}{12(1 - \nu^2)} \tag{13–27}$$

and $t = T_e$. If the lithosphere has a flexural rigidity, then T_e is the elastic thickness of the lithosphere.

Deflection of the lithosphere under load is measured by comparing observed free air gravity anomaly profiles, seismic refraction data, patterns of uplift and subsidence, or elevation and bathymetry data (Watts et al., 1980). To calculate bending stresses within the lithosphere, flexural rigidity, D, is required. In practice, D is estimated using elastic plate theory (Turcotte and Schubert, 1982).

Profiles of computed gravity anomalies based on elastic plates with various values of D are matched to observed free-air anomalies to identify the appropriate D for a particular lithostatic load such as a seamount (Watts et al., 1975). Flexural rigidity ranges from 7×10^{19} to 1×10^{24} N/m for oceanic lithosphere (Walcott, 1970c; Watts et al., 1975) and 5×10^{20} to 4×10^{24} N/m for continental lithosphere (Haxby et al., 1976; Cochran, 1980). If we know the flexural rigidity and radius of curvature of the lithospheric plate we can solve equation 13–26 for the bending moment. Bending stresses are found by integrating equation 13–24 where we use equation 13–27 to solve for T_e. The magnitude of the bending stresses within oceanic lithosphere is illustrated by considering the depression and free-air gravity anomaly that has developed as a consequence of loading by the Hawaiian ridge near Oahu (Watts et. al., 1980). Based on the elastic flexure model, the T_e beneath the Hawaiian ridge is about 30 km. Under Oahu the lithosphere takes the shape of a syncline with maximum compression at the top of the plate where bending stresses are on the order of 700 MPa. According to Goetze and Evans (1979) oceanic lithosphere at a depth of 10 km can sustain a horizontal compression of this magnitude. Kirby (1983) suggests that the strength of olivine in oceanic lithosphere can support σ_d in excess of 1500 MPa at a depth of 40 km. A significant consequence of this elastic plate model for the lithosphere is that large bending stresses (150–700 MPa) are developed to accommodate the measured flexure of the lithosphere. A regional horizontal deviatoric compression (i.e., a tectonic stress) of up to several hundred MPa over the thickness of the lithosphere is associated with oceanic plates as they bend downward into subduction zones (Hanks, 1971).

Support for the hypothesis that the lithosphere has a long-term strength comes from one critical observation. The shape of the depression or bend caused by loading is proportional to the difference in age between the lithosphere and the load (Watts et al., 1980). All oceanic lithosphere starts relatively warm and quite thin at the midocean ridges. As the lithosphere matures by sliding away from the ridges, it cools and thickens. One consequence of this cooling is that the lithosphere's strength increases. This added strength is reflected in a thicker elastic layer and, hence, greater flexural rigidity (fig. 13–9). A sharp bend develops in response to loads applied when the lithosphere is very thin. Loads on thin ocean crust are found near midocean ridge crests including the East Pacific rise crest (Cochran, 1979), the Juan de Fuca rise crest (McNutt, 1979), and the Mid Atlantic Ridge crest (McKenzie and Bowin, 1976). In contrast, the oldest oceanic lithosphere responds to loads by deflecting downward with a gentle bend having a larger wavelength. Loading of the oldest lithosphere occurs at trenches such as the Mariana, Bonin, and Kuril (Watts, 1978). The effect of loading of oceanic lithosphere of intermediate age is seen around seamounts such as the Hawaiian-Emperor seamount chain and Great Meteor seamounts. With knowledge of flexural rigidity, we can calculate the thickness of the elastic lithosphere from equation 13–27. Assuming $E = 10^{11}$ MPa, and

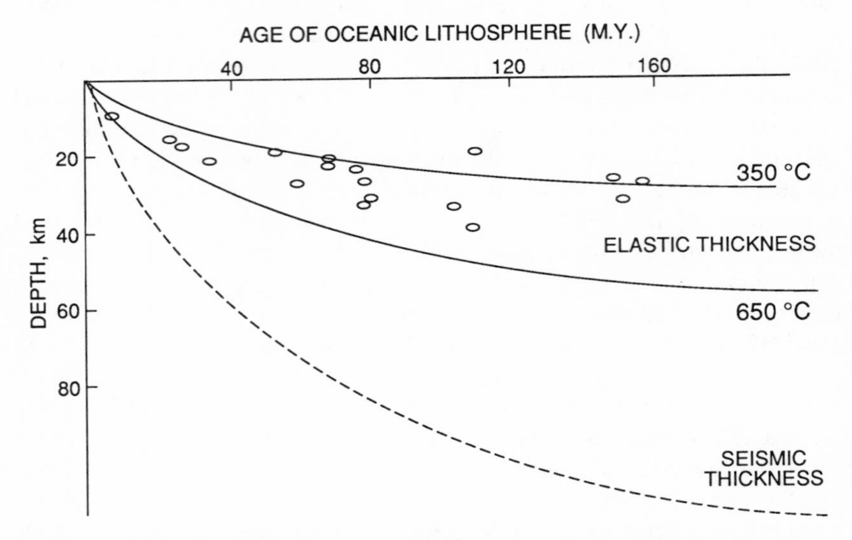

Fig. 13–9. Data points indicating the effective elastic thickness of the oceanic lithosphere versus age of the lithosphere at the time of loading (adapted from Watts et al., 1980). The solid lines are the 350°C and the 650°C oceanic isotherms based on the cooling plate model of Parsons and Sclater (1977). The dashed line is based on surface wave studies from Leeds et al. (1974) showing the seismic thickness of the oceanic crust.

$\nu = .25$, estimates of flexural rigidity imply that T_e of the oceanic lithosphere is between 2 and 54 km and that for the continental lithosphere the T_e is between 2 and 100 km (Watts et al., 1980). Studies of the continental lithosphere also indicate that flexural rigidity increases with thermal age but is significantly reduced by inelastic deformation at large strains (McNutt et al., 1988; Suber et al., 1989).

Another technique for computing the thickness of "strong" oceanic lithosphere is based on surface wave studies. Surface waves load the crust for very short intervals (< 150 seconds). Under loads of short duration the oceanic lithosphere appears two to three times thicker than is computed for load with a duration in excess of 10^6 years. The thickness of the lithosphere associated with short-term loads is labeled the "seismic" thickness in contrast to the 'elastic' thickness developed under longer-term loads. Figure 13–9 shows the seismic thickness of the oceanic lithosphere taken mainly from the surface wave studies of Leeds et al. (1974).

The shape forced on the lithosphere by a load depends on the rheology of the plate (Forsyth, 1980). Three general rheological models for the deformation of

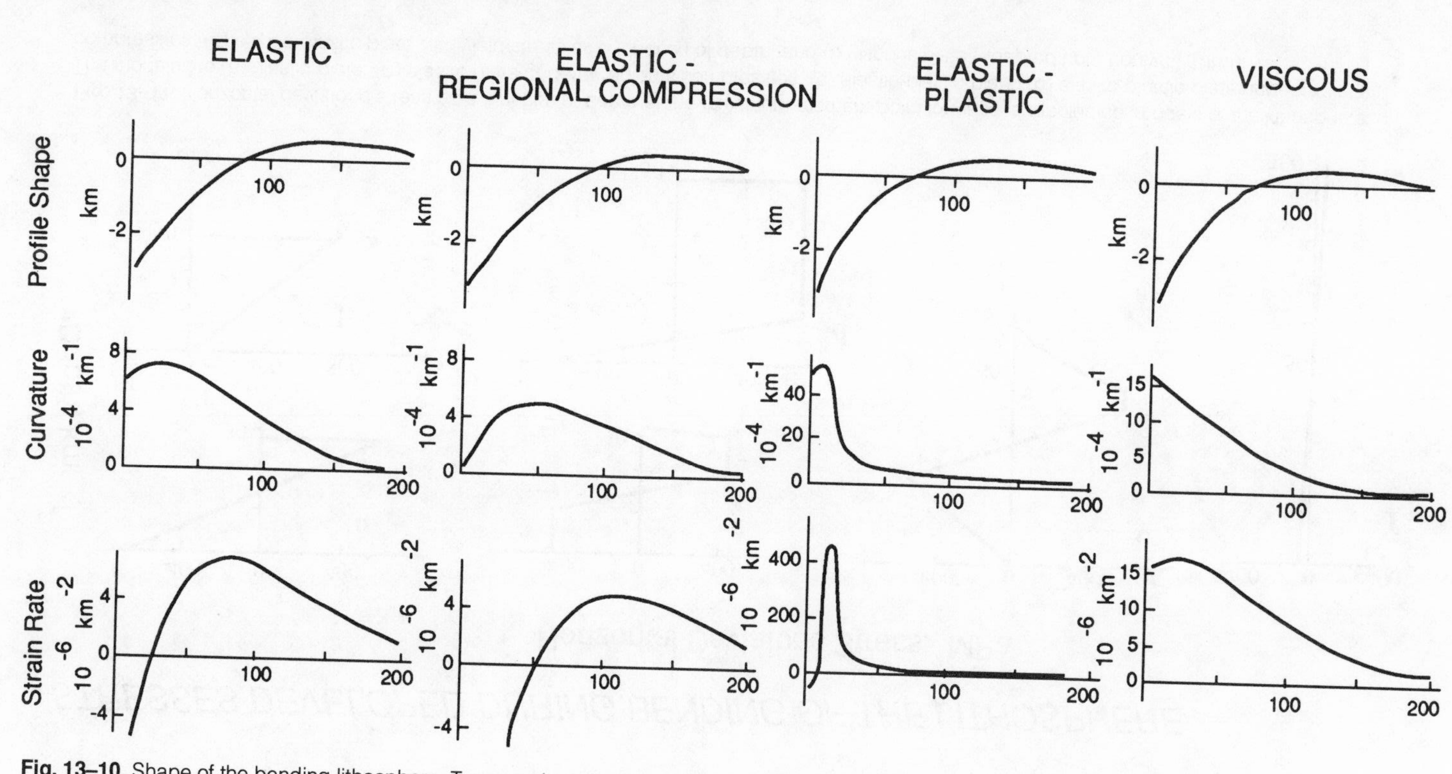

Fig. 13–10. Shape of the bending lithosphere. Top row gives the topography, middle row gives the curvature, and bottom row gives the rate of extension, all as a function of distance from the trench in km. The columns are four different rheological models for the plate bending into a subduction zone (adapted from Forsyth, 1980).

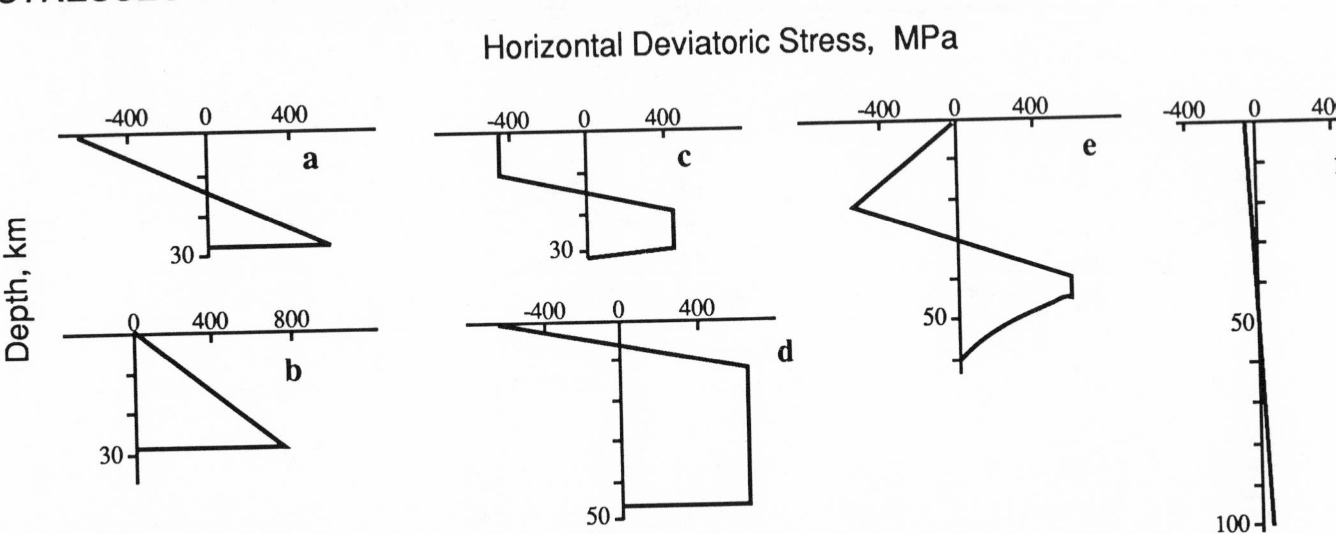

Fig. 13–11. Horizontal deviatoric stress as a function of depth at the maximum-moment point for various rheological models of the lithosphere. The models are (a) elastic plate; (b) elastic plate under a regional compression; (c) elastic-plastic plate; (d) elastic-plastic plate under regional compression; (e) elastic-plastic plate with yield stress as a function of depth; and (f) viscous plate (adapted from Forsyth, 1980).

the lithosphere through an outer rise and into a subduction zone include: (1) a uniform elastic plate (Caldwell et al., 1976); (2) a uniform plate with elastic-perfectly plastic rheology (Turcotte et al., 1978; and (3) a uniform viscous plate with no regional stress (Forsyth, 1980). Variations on these general models include: a uniform elastic plate under regional compression (Hanks, 1971); a uniform plate with elastic-perfectly plastic rheology under regional compression (McAdoo et al., 1978); and a plate with elastic-perfectly plastic rheology in which yield stress is a function of depth (Chapple and Forsyth, 1979). A test for these three models is based on the shape profile and curvature of various outer rise and trench systems (fig. 13–10). The stress profile as a function of depth for each of these models (i.e., fig. 13–11) must be consistent in detail with the predictions of experimental rock mechanics. Forsyth (1980) points out that the elastic-plastic model has the best fit to the shape profile. Other support comes from earthquake data where normal fault earthquakes are known to concentrate in the vicinity of the seaward side of the trench axis. These earthquakes are attributed to bending of the lithospheric plate in tension above a neutral fiber. The derivative of the curvature with respect to distance across the outer rise-trench system (d^3w/dx^3) describes where the deformation is achieved during subduction of oceanic lithosphere (fig. 13–10). The third derivative of topography for the elastic-perfectly plastic model is maximum in the vicinity of the concentration of normal fault earthquakes. In contrast, the third derivative of topography for the elastic and viscous models shows a broad peak which does not center on the known distribution of normal fault earthquakes in the outer rise-trench system. Once maximum bending is achieved no more normal faulting will take place. The elastic-perfectly plastic model with yield stress as a function of depth is most consistent with laboratory data on rock strength (Forsyth, 1980). This model includes the characteristic that the upper 20 km of the plate obey the Coulomb failure law for frictional sliding. Only at depths of 40–50 km is the plastic yield criterion exceeded for bending plates in compression.

Concluding Remarks

Plate tectonics theory states that the lithosphere of the earth (crust plus upper mantle) moves over the asthenosphere (middle and lower mantle). This process is believed to be driven by mechanisms including gravitational forces developed as thermal cooling of oceanic lithosphere takes place next to midoceanic ridges. Even larger gravitational forces are developed within the subducting lithosphere. Motion of the lithosphere is resisted by friction along strike-slip plate boundaries and at subduction zones. Viscous drag also occurs at the base of lithospheric plates. Because the forces at subduction zones appear to balance, platewide stress fields apparently arise from the gravitational forces at midocean ridges. However, the highest stress differences are apparently associated with the bending of the lithospheric plates with long-term strengths. The driving forces for plate tectonics are dissipated either by faulting in the frictional-slip regime of the lithosphere or by creep in the ductile-flow regime of the lithosphere.

Epilogue

During the past twenty years, data from direct measurement of stress in the lithosphere accumulated at an unprecedented rate. The interpretation of large quantities of in situ stress data allowed the general understanding of earth stress as expressed in this monograph. Rapid progress in the accumulation of in situ data depended on the congruence of three elements: (1) the confirmation of continental drift and with it, a global context for data on stress in the lithosphere; (2) the seminal work of many scientists having seemingly unrelated specialities including miners and tunnelers, field geologists, quarry operators, petroleum engineers, seismologists, laboratory experimentalists, and computer modelers; and (3) a significant surge of funding from both the U.S. federal government and private industry. Although this monograph deals with the technical aspects of lithospheric stress, funding was overwhelmingly important for advances over the past twenty years and, therefore, deserves some attention.

Briefly, the surge in funding for earth stress measurements in the United States was stimulated by two major socioeconomic forces: the search for domestic oil brought on by the 1973 oil embargo; and a sudden public interest in earthquake prediction. One by-product of the intensified exploration for oil during the 1970s was major funding for research in areas such as massive hydraulic fracturing, borehole logging, and core fabric analyses. Department of Energy–sponsored projects, the geothermal energy program at Los Alamos, the oil shale program in Colorado, and the gas-shales program in the Appalachian Basin, all had components dealing with various aspects of earth stress. Another sector of the energy industry involved nuclear power plants which were considered vulnerable to earthquake shaking. The Nuclear Regulatory Commission required the careful analysis of seismic potential in the vicinity of reactors.

A sudden public interest in earthquake prediction sprung from two unrelated events. First, Aggarwal et al. (1975) predicted an earthquake in the Adirondack Mountains of New York. The physical basis for this prediction was the dilatancy-diffusion model for the earthquake cycle developed by Nur (1972) and applied by Scholz et al. (1973). Shortly afterward, public attention on earthquakes increased with the announcement that geodetic data showed a rapid uplift of the western Mojave Desert over a period of less than ten years (Castle et al., 1976). Many worried that this uplift, the Palmdale Bulge, was the forerunner of a major earthquake along that portion of the San Andreas fault which last ruptured in 1857. Funds were allocated by the United States Congress to the U.S.G.S. for extensive investigations into earthquake generation and prediction with particular focus on the Palmdale Bulge. With the large influx of funds for earthquake prediction, the U.S.G.S. organized the necessary equipment for drilling dedicated holes to more than 1 km for stress measurements and related tests (Zoback et al., 1980). Concurrently, with support for hydraulic fracture measurements from the National Science Foundation, Haimson (1977) demonstrated the merits of a systematic study of earth stress on a regional basis. Subsequently, regional surveys of earth stress have answered many of the questions concerning the behavior of stress in the top km of the crust. Finally, the next frontier for earth stress measurements is in the depth range of 3–10 km with the future for these measurements tied directly to federal funding for continental scientific drilling programs.

Notes

1 Basic Concepts

1. The state of stress is a statement about orientation and/or magnitude of stress within the lithosphere.

2. Although they are treated as principal stresses, S_v, S_H, and S_h have different relative magnitudes depending on the geological setting and, therefore, have no a priori relationship to the principal compressive stresses designated as $\sigma_1 > \sigma_2 > \sigma_3$.

3. This book treats compressive stresses as positive because an overwhelming majority of rocks within the earth are subjected to compressive stresses. If compression is positive, then contractional strain is positive and extensional strain is negative. These sign conventions are confusing because the engineers treat compressive stress as negative. Furthermore, structural geologists treat finite extensional strain as positive largely because the very large elongation found in ductile shear zones is most conveniently managed as a positive natural logarithm (Ramsay and Huber, 1983) and, therefore, finite contractional strain is negative. In dealing with elastic strains associated with earth stress the sign convention is of little consequence because the strains in question are always very small.

4. The coordinate system used from here on has z normal to the earth's surface and x plus y parallel to the earth's surface.

5. Note that P_p is given a positive sign because of the poroelastic effect. This sign convention is maintained even when P_i (i.e., P_p within a crack) exerts a tensile force to cause crack propagation.

6. Note that, while opening-mode crack propagation is an important element of shear rupture, shear crack propagation is also another element (e.g. Cox and Scholz, 1988). In this book, reference to the crack propagation regime implies that the state of stress favors just opening mode crack propagation.

2 Stress in the Crack-Propagation Regime

1. In subsequent discussion the term, crack, is used to mean any geologic fracture subject to opening mode loading unless otherwise stated.

2. The crack tip in thin plates of metal is treated in plane stress because there is no lateral constrain to prevent displacement in the x direction.

3. K_I is properly defined using equation 2–17. Students often make the mistake of rearranging equation 2–16 to solve for K_I.

4. P_i is the fluid pressure necessary to initiate joint propagation and P_b is the fluid pressure necessary to initiate the propagation of commercial hydraulic fractures.

5. Scott et al. (1991) found K_{Ic} for the Ithaca siltstone to be on the order of 2.5 MP A.M.$^{1/2}$.

6. As defined in chapter 8, a stressmeter is any device which permits the direct readout of stress without knowledge of the elastic properties of the rock in which the device is imbedded.

7. At depths greater than 6 km in the Gulf of Mexico, lithostatic P_p is sometimes found indicating that at that depth $S_h \geq S_v$.

8. These calculations give no information on the absolute value of either P_i or S_h.

9. Stress trajectories are trend lines connecting stress orientation data on a map or cross section.

10. Cross-fold joints are those vertical cracks whose normal is roughly parallel to the fold axes of the Appalachian Plateau (fig. 2–14).

11. The Allegheny Front, a boundary between the Appalachian Valley and Ridge and Plateau, is largely controlled by the southeastern edge of the Silurian salt basin where décollement faulting climbed up from the Cambrian shales into the salt beds. Low strength of the salt changed the character of the Appalachian foreland tectonics from duplex structures of the Valley and Ridge to layer parallel shortening of the Appalachian Plateau (Davis and Engelder, 1987).

3 Stress in the Shear-Rupture and Frictional-Slip Regimes

1. It is appropriate to refer to reactivated joints as either shear fractures or minor faults.

2. On the Mohr circle, σ_n and τ acting on a particular plane are found at the angle $2\theta°$ measured counterclockwise from the σ_n axis.

3. Assume that the host rock has a bulk density of 2.5 g/cc.

4. In general, post-glacial thrust faults have larger throws than minor faulting associated with post-glacial popups. However, the distinction between these structures may be quite arbitrary.

4 Stress in the Ductile-Flow Regime

1. Pressure solution is the popular term used for stress solution. The adjective, stress, is substituted for pressure whenever enhanced dissolution of a solid is a consequence of stress in the solid and not pressure in the solution surrounding the solid.

2. Stress was much higher in the Onondaga Limestone and other units which served as a stress guide and faulted during layer parallel shortening of the Appalachian Plateau (e.g., Gwinn, 1964).

5 Hydraulic Fracture

1. Fracture pressure gradient (instantaneous shut-in pressure divided by depth) is another term used in the petroleum literature to express a normalized parting pressure (An-

derson et al., 1973). When expressed as psi/ft, the number for fracture pressure gradient is approximately the lifting factor. Both the intake pressure at the depth of the pressure parting and instantaneous shut-in pressure are equivalent to the σ_3 as discussed later in this chapter.

2. Until the extensive use of borehole televiewers during the past ten years, the impression packer was the primary tool for detecting the orientation of hydraulic fractures.

3. By strict definition the term, hydraulic fracture, is a misnomer for hydraulic crack. However, because this term is so well entrenched in the literature, this book will continue to use it.

4. The borehole televiewer is discussed in detail in chapter 6.

5. The Devonian section in the vicinity of South Canisteo is an exception to this rule.

6. This list is largely a compilation of issues discussed at the Second International Workshop on Hydraulic Fracture Stress Measurements, June 15–18, 1988, in Minneapolis, Minnesota.

6 Borehole and Core Logging

1. The tool also measures the borehole inclination and the orientation of the tool.

2. Later work suggests that wellbore failure probably does not occur at the wellbore surface in sedimentary rocks (Plumb, 1989).

3. Dark bands can also arise if the BHTV tool is not centered in the borehole.

4. This analysis assumes that ϕ_b is the width of the failed region. The analysis may apply only for the case where $\sigma_\theta = \sigma_1$ and $\sigma_r = \sigma_3$.

7 Strain-Relaxation Measurements

1. Note that expansion is negative.

2. A pilothole is a small hole drilled prior to a borehole for the purpose of installing a strain cell or stressmeter.

3. Some commercial strain gauge rosettes come with a fourth component. The fourth and redundant component allows an evaluation of measurement error within one sample.

4. The gauge length of a strain gauge is equivalent to the distance between two bench marks in the Hoover Dam experiment (fig. 7–1).

5. The Beckley coalbed is located within the New River Formation of the Pennsylvanian Pottsville Group in West Virginia.

8 Stressmeters and Crack Flexure

1. The active element is that part of the stressmeter which reacts against the deforming wall of a borehole or crack.

2. Unlike a true crack with a sharp tip, the flatjack slot has a rounded tip.

3. Again, this is not strictly a stressmeter because the fluid-filled slot is highly deformable, and thus the gauge does not have as high a modulus as the surrounding rock.

9 Microcrack-Related Phenomena

1. Mason of Redstone, New Hampshire, and Thunburg of Milford, New Hampshire, worked in some of New England's most famous granite quarries during the middle part of the twentieth century. They say that the discriminating eye of a quarryman can identify the nature of a familiar granite by a visual inspection of various surfaces. Confirmation of the nature comes from the feel or touch of the granite by simply running fingers lightly over a freshly cut surface; cracks running parallel to the nature are generally smoother.

2. σ_m or $\delta\sigma_i$ are similar to the lithostatic stress (S_v) or confining pressure (σ_3) and differential stress ($\sigma_d = \sigma_1 - \sigma_3$), respectively.

3. The direct model of Blanton (1983) requires a large number of input data and assumes a constant Poisson's ratio which makes accuracy uncertain in some cases. To compensate for these difficulties Warpinski and Teufel (1989b) have developed a strain-history model which requires fitting a theoretical model to the measured strain history.

4. In the laboratory, confining pressure is a hydrostatic pressure.

5. Stress data from UPH-3 are found in figure 5–14.

6. The actual orientation of the sample and, hence, its rift plane, is unknown.

10 Residual and Remnant Stresses

1. Concrete held in compression has a much higher tensile strength than relaxed concrete.

2. Adjacent portions of the rock must be the same size and shape for this rule to hold.

3. Note that the sign of locked-in stress in the grains is compressive, whereas the sign of the locked-in stress in the prestressed concrete is tensile.

4. Note that remnant stresses refer to boundary tractions and not intergranular stresses.

5. Fifteen years after this observation, it still remains my most exciting discovery.

6. First order, hydrofracture measurements, particularly the ISIP, are not affected by the presence of grain-scale residual stresses.

12 Data Compilations

1. Intact volumes larger than a few m^3 are rare.

2. Rocks above the salt décollement in the Appalachian Basin also show a remnant stress (Evans, 1989).

3. The stress-depth curve for the Canadian Shield was constructed by using Herget's

(1986) data for S_H/S_v and S_h/S_v at the arbitrary depths of 1 km and 2 km (fig. 3, 1986). S_v is assumed to increase at 25 MPa/km.

13 Sources of Stress in the Lithosphere

1. Because the western Cordillera is thermally weakened and actively deforming, the leading edge of North America is taken as the southwest margin of the Great Plains.

2. Note that ridge-push forces are a full order of magnitude less than the negative buoyancy forces of slab pull.

References

Abou-sayed, A.S., Brechtel, C.E., and Clifton, R.J., 1978, In situ stress determination by hydrofracturing: A fracture mechanics approach: *Journal of Geophysical Research,* 83:2851–62.

Adams, F.D., 1910, An experimental investigation into the action of differential pressures on certain minerals and rock, employing the process suggested by Professor Kick: *Journal of Geology,* 18:489–525.

Adams, F.D., and Bancroft, J.A., 1917, On the amount of internal friction developed in rocks: *Journal of Geology,* 25:597–637.

Adams, J., 1982, Stress-relief buckles in the McFarland quarry, Ottawa: *Canadian Journal of Earth Science,* 19:1882–87.

Adams, L.H., and Williamson, E.D., 1923, On the compressibility of minerals and rocks at high pressures: *Journal of Franklin Institute,* 195:475–529.

Agapito, J.F.T., Mitchell, S.J., Hardy, M.P., and Aggson, J.R., 1980, A study of ground control Problems in coal mines with high horizontal stresses: in Summers, D.A., ed. *Rock Mechanics, A State of the Art:* Proceedings of the 21st Symposium on Rock Mechanics, Rolla, Missouri, Univ. of Missouri, pp. 320–25.

Aggarwal, Y.P., 1978, Scaling laws for global plate motions: consequences for the driving mechanism, interplate stresses and mantle viscosity: *EOS,* 59:1202.

Aggarwal, Y.P., Sykes, L.R., Simpson, D.W., and Richards, P.G., 1975, Spatial and temporal variations in the ts/tp and in P-wave residuals at Blue Mountain Lake: application to earthquake prediction: *Journal of Geophysical Research,* 80:718–32.

Ahorner, L., 1975, Present-day stress field and seismotectonic block movements along major fault zones in Central Europe: *Tectonophysics,* 29:233–49.

Aki, K., 1966, Generation and propagation of G waves from the Niigate earthquake of June 16, 1964. Part 2. Estimation of earthquake moment, released energy, and stress-strain drop from the G wave spectrum: *Bulletin of the Earthquake Research Institute,* Tokyo University, 44:73–88.

———, 1972, Scaling law of earthquake source time-function: *Geophysical Journal of the Royal Astronomical Society,* 72:3–25.

Aleksandrowski, P., 1985, Graphical determination of principal stress directions for slickenside lineation populations: An attempt to modify Arthaud's method: *Journal of Structural Geology,* 7:73–82.

Alexander, L.G., 1960, Field and laboratory tests in rock mechanics: Proceedings of the Third Australian-New Zealand Conference on Soil Mechanics and Foundation Engineering, pp. 161–68.

———, 1983, Note on effects of infiltration on the criterion for breakdown pressure in hydraulic fracturing stress measurements: in Zoback, M.D., and Haimson, B.C., eds., *Hydraulic Fracturing Stress Measurements,* Washington, D.C., National Academy Press, D.C. pp. 143–48.

Alvarez, W., Engelder, T., and Lowrie, W., 1976, Formation of spaced cleavage and folds in brittle limestone by dissolution: *Geology,* 4:698–701.

Anderson, E.M., 1905, Dynamics of faulting: *Transactions Edinburgh Geological Society,* 8:387–402.

———, 1942, *The dynamics of faulting*: 1st ed., Edinburgh, Oliver and Boyd.

———, 1951, *The dynamics of faulting and dyke formation with application to Britain:* 2d ed., Edinburgh, Oliver and Boyd.

Anderson, O.L., and Grew, P.C., 1977, Stress corrosion theory of crack propagation with applications to geophysics: *Reviews of Geophysics and Space Physics,* 15:77–104.

Anderson, R.A., Ingram, D.S., and Zanier, A.M., 1973, Determining fracture pressure gradients from well logs: *Journal of Petroleum Technology,* 26:1259–68.

Anderson, T.O. and Stahl, E.J., 1967, A study of induced fracturing using an instrumental approach: *Journal of Petroleum Technology,* 19:261–67.

Angelier, J, 1979, Determination of the mean principal stress from a given fault population: *Tectonophysics,* 56:T17–T26.

Angelier, J., and Goguel, J., 1979, Sur une méthode simple de détermination des axes principaus des contraintes pour une population de failles: *Paris Académie de Science Comptes Rendus,* 288:307–10.

Angelier, J., and Mechler, P., 1977, Sur une methode graphique de recherche des contraintes principales egalment utilisable en tectonique et en seismolgie: La methode des diedres droits: *Bulletin de Societie Geologique de France,* 19:1309–18.

Angelier, J., Colletta, B., and Anderson, R.E., 1985, Neogene paleostress changes in the Basin and Range: A case study at Hoover Dam, Nevada-Arizona: *Geological Society of America Bulletin,* 96:347–61.

Angelier, J., Colletta, B., Chorowicz, J., Ortlieb, L., and Rangin, C., 1981a, Fault tectonics of the Baja California peninsula and the opening of the Sea of Cortez, Mexico: *Journal of Structural Geology,* 3:347–57.

Angelier, J., Dumont, J.F., Karamenderesi, H., Poisson, A., Simsek, S., and Uysal, S., 1981b, Analysis of fault mechanisms and expansion of southwestern Anatolia since the late Miocene: *Tectonophysics,* 75:T1–T9.

Angelier, J., Terentola, A., Valette, B., and Manoussis, S., 1982, Inversion of field data in fault tectonics to obtain the regional stress: I. Single phase fault populations: A new method of computing the stress tensor: *Geophysical Journal of the Royal Astronomical Society,* 69:607–21.

Angus, S., Armstrong, B., and de Reuck, K.M., 1978, International thermodynamic tables of the fluid state—Methane: *International union of pure and applied chemistry thermodynamics tables project,* v. 5, New York, Pergamon Press.

Arabasz, W.J., Smith, R.B., and Richins, W.B., 1979, Earthquake studies along the Wasatch Front, Utah: Network monitoring seismicity, and seismic hazards, in Arabasz, W.J., Smith, R.B., and Richins, W.B., eds., *Earthquake Studies in Utah, 1850 to 1978,* Salt Lake City, University of Utah, pp. 253–86.

Armijo, R., Carey, E., and Cisternas, A., 1982, The inverse problem in microtectonics and the separation of tectonic phases: *Tectonophysics,* 82:145–60.

Artyushkov, E.V., 1973, Stresses in the lithosphere caused by crustal thickness inhomogeneities: *Journal of Geophysical Research,* 78:7675–7708.

Atkinson, B.K., 1984, Subcritical crack growth in geological materials: *Journal of Geophysical Research,* 89:4077–4114.

Ault, C.H., Harper, D., Smith, C.R., and Wright, M.A., 1985, *Faulting and jointing in and near surface mines of sourthwestern Indiana:* U.S. Nuclear Regulatory Commission NUREG/CR-4117.

Avé Lallemant, H.G., 1978, Experimental deformation of diopside and websterite: *Tectonophysics,* 48:1–27.

———, 1985, Subgrain rotation and dynamic recrystallization of olivine, upper mantle diapirism, and extension of the Basin and Range Province: *Tectonophysics,* 119:89–117.

Avé Lallemant, H.G., Mercier, J-C., Carter, N.L., and Ross, J.V., 1980, Rheology of the upper mantle: Inferences from peridotite xenoliths: *Tectonophysics,* 70:85–113.

Aydin, A., and Johnson, A.M., 1978, Development of faults as zones of deformation bands and as slip surfaces in sandstone: *Pure and Applied Geophysics,* 116:931–42.

Babcock, E.A., 1978, Measurement of subsurface fractures from dipmeter logs: *American Association of Petroleum Geologists Bulletin,* 62:1111–26.

Badgley, P.C., 1965, *Structural and Tectonic Principals:* New York, Harper & Row.

Bahadur, S., 1972, In situ stress measurements in the Black Hills, South Dakota [Ph.D. thesis]: Rapid City, South, Dakota, South Dakota School of Mines and Technology.

Bahat, D., 1979, Theoretical considerations on mechanical parameters of joint surfaces based on studies on ceramics: *Geological Magazine,* 116:81–92.

Bahat, D. and Engelder, T., 1984, Surface morphology on cross-fold joints of the Appalachian Plateau, New York and Pennsylvania: *Tectonophysics,* 104:299–313.

Barker, C., 1972, Aquathermal pressuring—Role of temperature in development of abnormal-pressure zones: *American Association of Petroleum Geologists Bulletin,* 56:2068–71.

Barton, C.A., and Zoback, M.D., 1988, Determination of in situ stress orientation from borehole guided waves: *Journal of Geophysical Research,* 93:7834–44.

Barton, C.A., Zoback, M.D., and Burns, K.L., 1988, In situ stress orientation and magnitude at the Fenton geothermal site, New Mexico, determined from wellbore breakouts: *Geophysical Research Letters,* 15:467–70.

Barton, N., 1976, The shear strength of rock and rock joints: *International Journal of Rock Mechanics and Mining Sciences,* 13:255–79.

Baumgärtner, J., and Zoback, M.D., 1989, Interpretation of hydraulic pressure-time records using interactive analysis methods: *International Journal of Rock Mechanics and Mining Sciences,* 26:461–70.

Beach, A., 1975, The geometry of en echelon vein arrays: *Tectonophysics,* 28:245–64.

———, 1977, Vein arrays, hydraulic fractures, and pressure solution structures in a deformed flysch sequence, S.W., England: *Tectonophysics,* 40:201–25.

Becker, A., Blümling, P., and Müller, W.H., 1987, Recent stress field and neotectonics in the Eastern Jura Mountains, Switzerland: *Tectonophysics,* 135:277–88.

Becker, G.F., 1893, Finite homogeneous strain, flow and rupture of rocks: *Geological Society of America Bulletin,* 4:13–90.

Bell, J.S., and Gough, D.I., 1979, Northeast-southwest compressive stress in Alberta: evidence from oil wells: *Earth and Planetary Science Letters ,* 45:475–82.

———, 1983, The use of borehole breakouts in the study of crustal stress: in Zoback,

M.D., and Haimson, B.C., eds., *Hydraulic Fracturing Stress Measurements,* Washington, D.C, National Academy Press, pp. 201–09.

Bell, T.M., and Etheridge, M.A., 1973, The deformation and recrystallization of quartz in a mylonite zone, central Australia: *Tectonophyiscs,* 32:235–67.

Bergerat, F., 1987, Stress fields in the European platform at the time of Africa-Eurasia collision: *Tectonics,* 6:99–132.

Bergman, E.A., 1986, Intraplate earthquakes and the state of stress in oceanic lithosphere: *Tectonophysics,* 132:1–35.

Bergman, E.A., and Solomon, S.C., 1980, Oceanic intraplate earthquakes: Implications for local and regional intraplate stress: *Journal of Geophysical Research,* 85:5389–5410.

———, 1984, Source mechanisms of earthquakes near mid-ocean ridges from body waveform inversion: implications for the early evolution of oceanic lithosphere: *Journal of Geophysical Research,* 89:11,415–41.

Berry, F.A.F., 1973, High fluid potentials in California Coast Ranges and their tectonic significance: *American Association of Petroleum Geologists Bulletin,* 57:1219–49.

Bethke, C.M., Harrison, W.J., Upson, C., and Altaner, S.P., 1988, Supercomputer analysis of sedimentary basins: *Science,* 239:261–67.

Bielenstein, H.U., and Barron, K., 1971, In situ stresses—A summary of presentations and discussions given in Theme I: Proceedings of the 4th Canadian Symposium on Research in Tectonics and the 7th Canadian Symposium on Rock Mechanics, Edmonton, Alberta.

Bieniawski, Z.T., 1984, *Rock mechanics design in mining and tunneling:* Rotterdam, A.A. Balkema.

Bilham, R., and King, G., 1989, The morphology of strike-slip faults: Examples from the San Andreas fault, California: *Journal of Geophysical Research,* 94:10,204–16.

Biot, M.A., 1941, General theory of three-dimensional consolidation: *Journal of Applied Physics* 12:155–64.

Birch, F., 1960, The velocity of compressional waves in rocks to 10 kilobars: *Journal of Geophysical Research,* 65:1083–1102.

Bird, G.P., 1976, Thermal and mechanical evolution of continental convergence zones: Zagros and Himalayas [Ph.D. thesis]: Cambridge, Mass., Massachusetts Institute of Technology.

Bird, J.E., Mukherjee, A.K., and Dorn, J.F., 1969, Correlations between high temperature creep behavior and structure: in D.G. Brandon and R. Rosen, eds., *Quantitative Relation Between Properties and Microstructure, Proceedings of an International Conference,* Haifa, Israel, pp. 255–342.

Bishop, J.F.W., and Hill, R., 1951, A theory of the plastic distortion of a polycrystalline aggregate under combined stresses: *Philosophical Magazine,* 42:414–27.

Bjarnason, B., Leijon, B., and Stephansson, O., 1986, The Bolmen project—rock stress measurements using hydraulic fracturing and overcoring techniques: Research Report BeFo 160:1/86, Swedish Rock Engineering Research Foundation, Stockholm.

Bjarnason, B., Ljunggren, C., and Stephansson, O., 1989, New developments in hydrofracturing stress measurements at Luleå University of Technology: *International Journal of Rock Mechanics and Mining Sciences,* 26:579–86.

Blackwood, R.L., 1978, Diagnostic stress-relief curves in stress measurement by overcoring: *International Journal of Rock Mechanics and Mining Sciences,* 15:205–9.

Blanton, T.L., 1983, The relation between recovery deformation and in situ stress magnitudes: SPE 11624, presented at 1983 SPE/DOE Symposium on Low Permeability Gas Reservoirs, Denver, Colorado, March 14–16, 1983.

Blanton, T.L., and Teufel, L.W., 1983, A field test of the strain recovery method of stress determination in Devonian Shales: SPE 12304, presented at the eastern Regional SPE Meeting, Champion, PA., November 9–11, 1983.

Blanton, T.L., Dischler, S.A., and Patti, N.C., 1982, Mechanical properties of Devonian shales from the Appalachian Basin: Final report to U.S.D.O.E., Morgantown Energy Technology Center, Morgantown, W.V.

Blenkinsop, T.G., and Drury, M.R., 1988, Stress estimates and fault history from quartz microstructures: *Journal of Structural Geology,* 10:673–84.

Blès. J.L. and Feuga, B., 1986, *The Fracture of Rocks:* Elsevier, Amsterdam.

Block, J.W., Clement, R.C., Lew, L.R., and deBoer, J., 1979, Recent thrust faulting in southeastern Connecticut: *Geology,* 7:79–82.

Boland, J.N., and Tullis, T.E., 1986, Deformation behavior of wet and dry clinopyroxenite in the brittle to ductile transition region: in Hobbs, B.E., and Heard, H.C., eds. *Mineral and Rock Deformation: Laboratory Studies:* Geophysical Monograph 36, American Geophysical Union, pp. 35–50.

Bonilla, M.G., and Buchanan, J.M., 1970, Interim report on worldwide historic surface faulting: U.S., Geological Survey Open-File Report, Washington, D.C.

Bonnechere, F., 1967, A comparative study in in situ rock stress measurement techniques [M.Sc. thesis]: Minneapolis, Minnesota, University of Minnesota.

Borg, I., and Turner, F.J., 1953, Deformation of Yule Marble, Part VI: *Geological Society of America Bulletin,* 64:1343–52.

Bott, M.H.P., 1959, The mechanics of oblique slip faulting: *Geological Magazine,* 46:109–17.

Bott, M.H.P., 1982, *The Interior of The Earth: Its Structure, Constitution and Evolution:* London, Edward Arnold.

Bott, M.H.P., and Dean, D.S., 1972, Stress systems at young continental margins: *Nature,* 235:23–25.

Bott, M.H.P., and Kusznir, N.J., 1984, The origin of tectonic stress in the lithosphere: *Tectonophysics,* 105:1–13.

Bott, M.H.P., Waghom, G.D., and Whittaker, A., 1989, Plate boundary forces at subduction zones and trench-arc compression: *Tectonophysics,* 170:1–15.

Bowden, F.P., and Tabor, D., 1964, *The friction and lubrication of solids, Part II:* London, Clarendon Press.

Brace, W.F., 1971, Micromechanics in rock systems: in M. Te'eni ed., *Structure, Solid Mechanics, and Engineering Design,* London, Wiley-Interscience, pp. 187–204.

Brace, W.F., and Bombolakis, E.G., 1963, A note on brittle crack growth in compression: *Journal of Geophysical Research,* 68:3709–13.

Brace, W.F., and Byerlee, J.D., 1966, Stick-slip as a mechanism for earthquakes: *Science,* 153:990–92.

———, 1970, California earthquakes: Why only shallow focus?: *Science,* 168:1573–75.

Brace, W.F., and Kohlstedt, D.L., 1980, Limits on lithospheric stress imposed by laboratory experiments: *Journal of Geophysical Research,* 85:6248–52.

Brace, W.F., and Martin, R.J., 1968, A test of the law of effective stress for crystalline rocks of low porosity: *International Journal of Rock Mechanics and Mining Sciences,* 5:415–26.

Brace, W.F., and Walsh, J.B., 1962, Some direct measurements of the surface energy of quartz and orthoclase: *American Mineralogist,* 47:1111–22.

Brace, W.F., Paulding, B., and Scholz, C.H., 1966, Dilatancy in the fracture of crystalline rocks: *Journal of Geophysical Research,* 71:3939–54.

Bradley, D.C., and Kidd W.S.F., 1991, Flexural extension of the upper continental crust in collisional foredeeps: *Geological Society of America Bulletin* 103:1416–38.

Bredehoeft, J. D., Wolff, R.G., Keys, W.S., and Schuter, E., 1976, Hydraulic fracturing to determine the regional in situ stress field, Piceance Basin, Colorado: *Geological Society of America Bulletin,* 87:250–58.

Bretz, J.H., 1942, Vadose and phreatic features of limestone caverns: *Journal of Geology,* 50:675–811.

Briegel, U., and Goetze, C., 1978, Estimates of differential stress recorded in the dislocation structure of Lochseiten limestone (Switzerland): *Tectonophysics,* 48:61–76.

Broek, D., 1987, *Elementary Engineering Fracture Mechanics:* Boston, Massachusetts, Martinus Nijhoff.

Brown, A., 1973, In-situ strain measurement by photoelastic bar gauges, 1. The method: *Tectonophysics,* 19:383–97.

———, 1974, Photoelastic measurements of recoverable strain at four sites: *Tectonophysics,* 21:135–64.

Brown, E.T. and Hoek, E., 1978, Trends in relationships between measured in situ stresses and depth, *International Journal of Rock Mechanics and Mining Sciences:* 15:211–15.

Brown, M.W., 1990, Tunnel drilling, old as Babylon, becomes much safer: *The New York Times,* No. 48437, 140:16.

Brune, J.N., 1961, Radiation pattern of Rayleigh waves from the southeast Alaska earthquake of July 10, 1958: *Publication of Dominion Observatory,* Ottawa, 4:373–83.

———, 1970, Tectonic stress and the spectra of seismic shear waves from earthquakes: *Journal of Geophysical Research,* 75:4997–5009.

Brune, J.N., and Allen, C.R., 1967, A low stress-drop, low-magnitude earthquake with surface faulting, the Imperial, California, earthquake of March 4, 1966: *Bulletin Seismological Society of America,* 57:501–14.

Bruner, W.M., 1984, Crack growth during unroofing of crustal rocks: Effects on thermoelastic behavior and near-surface stresses: *Journal of Geophysical Research,* 89:4167–84.

Bucher, W.H., 1920, The mechanical interpretation of joints: *Journal of Geology,* 28:707–30.

Buen, B., 1977, Residualspenninger i Fjell. *Rapport fra Geologisk Institutt,* Norges Tekniske Hogskole, Universitetet i Trondheim, 147 p.

Bullen, K.E., 1953, *An Introduction to the Theory of Seismology:* 2d ed., Cambridge, Cambridge University Press.

Burg, J.P., and Laurent, P., 1978, Strain analysis of a shear zone in a granodiorite: *Tectonophysics,* 47:15–42.

Burlet, D., Cornet, F.H., and Feuga, B., 1989, Evaluation of the HTPF method of stress determination in two kinds of rock: *International Journal of Rock Mechanics and Mining Sciences,* 26:673–80.

Byerlee, J.D., 1978, Friction in rocks: *Pure and Applied Geophysics,* 116:615–26.

Byerlee, J.D., and Brace, W.F., 1968, Stick-slip, stable sliding, and earthquakes—Effect of rock type, pressure, strain rate, and stiffness: *Journal of Geophysical Research,* 73:6031–37.

Byerly, P., 1926, The Montana earthquake of June 28, 1925: *Bulletin Seismological Society of America,* 16:209–63.

Caldwell, J.G., Haxby, W.F., Karig, D.E., and Turcotte, D.L., 1976, On the applicability of a universal elastic trench profile: *Earth and Planetary Science Letters,* 31: 239–46.

Cannaday, F.X., and Leo, G.M., 1966, Piezoelectric pulsing equipment for sonic velocity measurements in rock samples from laboratory size to minor pillars: Reports of Investigation 6810, U.S. Bureau of Mines, Washington, D.C.

Carey, E., and Brunier, B., 1974, Analyse théorique et numérique d'un modèle mécanique élémentaire appliqué à l'étude d'une population de failles: *Paris Académie de Science Comptes Rendus,* 279:891–94.

Carlson, R.L., Hilde, T.W.C., and Uyeda, S., 1983a, The driving mechanism of plate tectonics: Relation to age of the lithosphere at trenches: *Geophysical Research Letters,* 10:297–300.

———, 1983b, Correction to "The driving mechanism of plate tectonics: Relation to age of the lithosphere at trenches": *Geophysical Research Letters,* 10:987.

Carlson, S.R., and Wang, H.F., 1986, Microcrack porosity and in situ stress in Illinois borehole UPH-3: *Journal of Geophysical Research,* 91:10,421–28.

Carslaw, H.S., and Jaeger, J.C., 1959, *Conduction of heat in solids:* 2d ed., Oxford, Clarendon Press.

Carter, N.L., 1976, Steady state flow of rocks: *Reviews of Geophysics and Space Physics,* 14:301–60.

Carter, N.L., and Avé Lallemant, H.G., 1970, High temperature flow of dunite and peridotite: *Geological Society of America Bulletin,* 81:2181–2202.

Carter, N.L., and Friedman, M., 1965, Dynamic analysis of deformed quartz and calcite from the Dry Creek Ridge Anticline, Montana: *American Journal of Science,* 263:747–85.

Carter, N.L., and Kirby, S.H., 1978, Transient creep and semibrittle behavior of crystalline rocks: *Pure and Applied Geophysics,* 116:807–39.

Carter, N.L., and Raleigh, C.B., 1969, Principal stress directions from plastic flow in crystals: *Geological Society of America Bulletin,* 80:1231–64.

Carter, N.L., and Tsenn, M.C., 1987, Flow properties of continental lithosphere: *Tectonophysics,* 136:27–63.

Carter, N.L., Christie, J.M., and Griggs, D.T., 1964, Experimental deformation and recrystallization of quartz: *Journal of Geology,* 72:687–733.

Carter, N.L., Hansen, F.D., and Senseny, P.E., 1982, Stress magnitudes in natural rock salt: *Journal of Geophysical Research,* 87:9289–9300.

Castle, R.O., Church, J.P., and Elliott, 1976, Aseismic uplift in Southern California: *Science,* 192:251–53.

Célérier, B., 1988, How much does slip on a reactivated fault plane constrain the stress tensor?: *Tectonics,* 7:1257–78.

Chandra, U., 1977, Earthquakes of peninsular India—A seismotectonic study: *Bulletin of the Seismological Society of America,* 67:1387–1413.

Chapple, W.M., and Forsyth, D.W., 1979, Earthquakes and the bending of plates at trenches: *Journal of Geophysical Research,* 84:6729–49.

Chapple, W.M., and Spang, J.H., 1974, Significance of layer-parallel slip during folding of layered sedimentary rocks: *Geological Society of America Bulletin,* 85:1523–34.

Chapple, W.M., and Tullis, T.E., 1977, Evaluation of the forces that drive the plates: *Journal of Geophysical Research,* 82:1969–84.

Chase, C.G., 1979, Asthenospheric counterflow: A kinematic model: *Geophysical Journal of the Royal Astronomical Society,* 56:1–18.

Chen, W-P. and Molnar, P., 1983, Focal depths of intracontinental and intraplate earthquakes and their implications for the thermal and mechanical properties of the lithosphere: *Journal of Geophysical Research,* 88:4183–4214.

Cheng, C.H., and Toksoz, M.N., 1984, Generation, propagation, and analysis of tube waves in a borehole: in M.N., Toksoz and R.R. Stewart, eds., *Vertical Seismic Profiling,* Part B, Advanced Concepts, Geophysical Press.

Chinnery, M.A., 1964, The strength of the earth's crust under horizontal shear stress: *Journal of Geophysical Research,* 69: 2085–89.

Chopra, P.N., and Paterson, M.S., 1981, The experimental deformation of dunite: *Tectonophysics,* 78:453–73.

———, 1984, The role of water in the deformation of dunite: *Journal of Geophysical Research,* 89:7861–76.

Christie, J.M., and Ord, A., 1980, Flow stress from microstructures of mylonites: *example and current assessment: Journal of Geophysical Research,* 85:6253–62.

Clark, B.R., 1981, Stress anomaly accompanying the 1979 Lytle Creek Earthquake: Implications for earthquake prediction: *Science,* 211:51–53.

Cleary, M.P., Crockett, J.I., Martinez, J.I., and Narendran, V.M., 1983, Surface integral schemes for fluid flow and induced stresses around fractures in underground reservoirs: SPE Paper 11632, Symposium on Low Permeability Gas Reservoirs, Society of Petroleum Engineers, Dept. of Energy, Denver, Colorado.

Cliffs Minerals, Inc., 1982, Analysis of the Devonian Shales in the Appalachian Basin, U.S. Department of Energy contract DE-AS21–80MC14693, Final Report, Washington, D.C., Springfield Clearing House.

Cloetingh, S., and Wortel, R., 1986, Stress in the Indo-Australian plate: *Tectonophysics,* 132:49–67.

Cloetingh, S., Wortel, R., and Vlaar, N.J., 1983, State of stress at passive margins and initiation of subduction zones: *Studies in Continental Margin Geology,* American Association of Petroleum Geologists Memoir 34, pp. 717–23.

———, 1984, Passive margin evolution, initiation of subduction and the Wilson cycle: *Tectonophysics,* 109:147–63.

Coates, D.F., 1964, Residual stress in rocks: in Judd, W.R., ed., *State of Stress in the Earth's Crust,* New York, Elsevier, pp. 679–88.

Coates, D.F., and Yu, Y.S., 1970, A note on the stress concentrations at the end of a cylindrical hole: *International Journal of Rock Mechanics and Mining Sciences,* 7:583–88.

Cochran, J.R., 1979, An analysis of isostasy in the world's ocean, 2, Midocean ridge crests: *Journal of Geophysical Research,* 84:4713–29.

———, 1980, Some remarks on isostasy and the long-term behavior of the continental lithosphere: *Earth and Planetary Science Letters,* 46:266–74.

Conel, J. E., 1962, Studies of the development of fabrics in naturally deformed limestones [Ph.D. thesis]: Pasadena, California, California Institute of Technology.

Conrad, R.E., and Friedman, M., 1976, Microscopic feather fractures in the faulting process: *Tectonophysics,* 33:187–98.

Cooper, R.F., and Kohlstedt, D.L., 1986, Rheology and structure of olivine-basalt partial melts: *Journal of Geophysical Research,* 91:9315–23.

Cornet, F.H., 1986, Stress determination from hydraulic tests on preexisting fractures—The H.T.P.F. method: in Stephansson, O., ed., *Rock Stress and Rock Stress Measurements:* CENTEK Pub., Luleå, Sweden, pp. 301–12.

Cornet, F.H., and Valette, 1984, In situ stress determination from hydraulic injection test data: *Journal of Geophysical Research,* 89:11527–37.

Cox, J.W., 1970, The high resolution dipmeter reveals dip-related borehole and formation characteristics: paper presented at 11th Annual Logging Symposium, Society of Professional Well Logging Analysts, Los Angeles, California.

Cox, S.J.D., and Scholz, C.H., 1988, On the formation and growth of faults: an experimental study: *Journal of Structural Geology,* 10:413–30.

Crampin, S., 1977, A review of the effects of anisotropic layering on the propagation of seismic waves: *Geophysical Journal of the Royal Astronomical Society,* 53:467–96.

———, 1984, Effective anisotropic elastic constants for wave propagation through cracked solids: *Geophysical Journal of the Royal Astronomical Society,* 76:135–45.

Creuels, F.H., and Hermes, J.M., 1956, Measurement of changes in rock-pressure in the vicinity of working face: International Strata Control Conference, Essen, W. Germany.

Crosby, W.O., 1882, On the classification and origin of jointed structures: Proceedings of the Boston Society of Natural History, 22:72–85.

Cyrul, T., 1983, Notes on stress determination in Heterogeneous rocks: *Field Measurements in Geomechanics,* Rotterdam, A.A. Balkema, pp. 59–70.

Dahlen, F.A., 1981, Isostasy and the ambient state of stress in the oceanic lithosphere: *Journal of Geophysical Research,* 86:7801–7.

Dale, T.N., 1923, The commercial granites of New England: *U.S. Geological Survey Bulletin* 738.

Dames and Moore, 1978, Geologic investigation: Nine Mile Point nuclear station unit 2, Doc. 04707–125–19 to Niagara Mohawk Power Corporation, Syracuse, New York.

Darwin, G.H., 1882, On the stresses caused in the interior of the earth by the weight of continents and Mountains: *Philosophical Transactions of the Royal Society,* 173:187–230.

Daubree, A., 1879, *Geologie Experimentale:* Librare des Corps des points et Chausser des Mines et des Telegraphs, Paris.

Davis, D.M., and Engelder, T., 1985, The role of salt in fold-and-thrust belts: *Tectonophysics,* 119:67–88.

de Boer, J.Z., McHone, J.G., Puffer, J.H., Ragland, P.C., and Whittington, D., 1988, Mesozoic and Cenozoic magmatism: in Sheridan, R.E., and Grow, J.A., eds., *The Geology of North America,* vol. I-2, The Atlantic Continental Margin, Geological Society of America, pp. 217–41.

de Boer, R.B., 1977, Pressure solution: theory and experiments: *Tectonophysics,* 39:287–301.

de Bree, P., and Walters, J.V., 1989, Micro/minifrac test procedures and interpretation for in situ stress determination: *International Journal of Rock Mechanics and Mining Sciences,* 26:515–522.

de la Cruz, R.V., 1977, Jack fracturing technique of stress measurement: *Rock Mechanics,* 9:27–42.

———, 1978, Modified borehole jack method for elastic property determination in rocks: *Rock Mechanics,* 10:221–39.

de la Cruz, R.V., and Raleigh, C.B., 1972, Absolute stress measurements at the Rangely Anticline, northwestern Colorado: *International Journal of Rock Mechanics and Mining Sciences,* 9:625–34.

Dean, A.H., and Beatty, R.A., 1968, Rock stress measurements using cylindrical jacks and flatjacks at north Broken Hill Limited: in Radmanovich, M., and Woodcock, J.T., eds., Monograph Series, No. 3, Australian Institute of mining and Metallurgy.

Dean, A.J., 1937, The Lake Copais, Boeotia, Greece: Its drainage and development: *Journal of Institution of Civil Engineering* (London), 5:287.

Delaney, P.T., and Pollard, D.D., 1981, U.S. Geological Survey Professional Paper 1202.

Delaney, P.T., Pollard, D.D., Ziony, J.I., and McKee E.H., 1986, Field relations between dikes and joints: Emplacement processes and Paleostress Analysis: *Journal of Geophysical Research,* 91:4920–38.

Detournay, E., and Roegiers, J., 1986, Comment on "Well bore breakouts and in situ stress" by M.D., Zoback, D. Moos, L. Mastin, and R.N. Anderson: *Journal of Geophysical Research,* 91:14161–62.

Detournay, E., Roegiers, J-C., and Cheng, A., 1987, Some new examples of poroelastic effects in rock mechanics: in *Rock Mechanics,* 28th U.S. Symposium on Rock Mechanics, Rotterdam, A.A. Balkema, pp. 575–84.

Dey, T.N., and Brown, D.W., 1986., Stress measurements in a deep granitic rock mass using hydraulic fracturing and differential strain curve analysis: in Stephansson, O., ed., *Rock Stress and Rock Stress Measurements:* CENTEK Pub., Luleå, Sweden, pp. 351–57.

Dickey P.A., and Andersen, K.H., 1950, The behavior of water-input wells: API., *Secondary Recovery of Oil in the United States,* 2d ed., pp. 317–41.

Dickinson, G., 1953, Geological aspects of abnormal reservoir pressures in Gulf Coast Louisiana: *American Association of Petroleum Geologists Bulletin,* 37:410–32.

Dieterich, J.H., 1972, Time-dependent friction in rocks: *Journal of Geophysical Research,* 77:3690–97.

———, 1978, Preseismic fault slip and earthquake prediction: *Journal of Geophysical Research,* 83:3940–47.

———, 1979, Modeling rock friction, 1. Experimental results and constitutive equations: *Journal of Geophysical Research,* 84:2161–68.

Dietrich, D., and Song, H., 1984, Calcite fabrics in a natural shear environment, the Helvetic nappes of western Switzerland: *Journal of Structural Geology,* 6:19–32.

Dunning, J.D., Petrovski, D., Schuyler, J., and Owens, A., 1984, The effects of aqueous chemical environments on crack propagation in quartz: *Journal of Geophysical Research,* 89:4115–24.

Durham, W.G., and Goetze, C., 1977, Plastic flow of oriented single crystals of olivine, 1. Mechanical data: *Journal of Geophysical Research,* 82:5737–53.

Durney, D.W., 1972, Solution-transfer, an important geological deformation mechanism: *Nature,* 235:315–17.

———, 1978, Early theories and hypotheses on pressure-solution-redeposition: *Geology,* 6:369–72.

Dyer, R., 1988, Using joint interactions to estimate paleostress ratios: *Journal of Structural Geology,* 10:685–99.

Dyke, C.G., 1989, Core discing: Its potential as an indicator of principal in situ stress directions: in Maury, V., and Fourmaintraux, D., eds., *Rock at Great Depth,* Rotterdam, A.A. Balkema, pp. 1057–64.

Eaton, G.P., 1979, Regional geophysics, Cenozoic tectonics, and geologic resources of the Basin and Range province and adjoining regions: in Newman, G.W., and Goode, H.D., eds., *1979 Basin and Range Symposium:* Rocky Mountain Association of Geologists and Utah Geological Association, pp. 11–39.

Ebel, J.E., Vudler, V., and Celata, M., 1982, The 1981 microearthquake swarm near Moodus, Connecticut: *Geophysical Research Letters,* 9:397–400.

Eisbacher, G.H., and Bielenstein, H.V., 1971, Elastic strain recovery in Proterozoic rocks near Elliot Lake, Ontario: *Journal of Geophysical Research ,* 76:2012–21.

Ekstrom, M.P., Dahan, C.A., Chen, M-Y., Lloyd, P.M., and Rossi, D.J., 1987, Formation imaging with microelectrical scanning arrays: *Log Analyst,* 28:294–306.

Elliott, D., 1973, Diffusion flow laws in metamorphic rocks: *Geological Society of America,* 84:2645–64.

Elsasser, W.M., 1971, Sea-floor spreading as thermal convection: *Journal of Geophysical Research,* 76:1101–12.

Emery, C.L., 1961, The measurement of strains in mine rock: *International Symposium on Mining Resources,* vol. 2, New York, Pergamon Press.

Enever, J., 1988, Ten years experience with hydraulic fracture stress measurements in Australia: in Haimson, B.C., Rogiers, J.C., and Zoback, M.D., eds., *Proceedings of the 2nd International Workshop on Hydraulic Fracture Stress Measurements,* June 15–18, 1988, Minneapolis, Minn., Geological Engineering Program, Univ. of Wisconsin, Madison, WI.

Enever, J.R., and Wooltorton, B.A., 1983, Experience with hydraulic fracturing as a means of estimating in situ stress in Australian coal basin sediments: in Zoback, M.D., and Haimson, B.C., eds., *Hydraulic Fracturing Stress Measurements,* Washington, D.C., National Academy Press, pp. 28–43.

Engebretson, D.C., Cox, A., and Gordon, R.G, 1985, Relative Motions between Oceanic and Continental Plates in the Pacific Basin: *Geological Society of America Special Paper 206,* Geological Society of America, Boulder, Colorado.

Engelder, T., 1974, Cataclasis and the generation of fault gouge: *Geological Society of America Bulletin,* 85:1515–22.

———, 1979a, The nature of deformation within the outer limits of the central Appalachian foreland fold and thrust belt in New York State: *Tectonophysics,* 55:289–310.

———, 1979b, Mechanisms for strain within the Upper Devonian clastic sequence of the Appalachian Plateau, western New York: *American Journal of Science,* 279:527–42.

Engelder, T., 1981, General characteristics of strain relaxation: A note on sample preparation for large-scale tests: *Geophysical Research Letters,* 8:687–89.

———, 1982a, Is there a genetic relationship between selected regional joints and contemporary stress within the lithosphere of North America?: *Tectonics,* 1:161–177.

———, 1982b, Reply to a comment on "Is there a genetic relationship between selected regional joints and contemporary stress within the lithosphere of North America?" by A.E. Scheidegger: *Tectonics,* 1:465–70.

Engelder, T., 1982c, A natural example of the simultaneous operation of a free-face dissolution and a pressure solution: *Geochimica Cosmochimica Acta.,* 46:69–74.

———, 1984, The time-dependent strain relaxation of Algerie Granite: *International Journal of Rock Mechanics and Mining Sciences,* 21:63–73.

———, 1985, Loading paths to joint propagation during a tectonic cycle: An example from the Appalachian Plateau, U.S.A.: *Journal of Structural Geology,* 7:459–76.

———, 1986, Time-dependent strain relaxation of gneiss from the 1-km deep hole near Kent Cliffs, New York: in Statton, C.T., ed., Kent Cliffs borehole research project: A determination of the magnitude and orientation of tectonic stress in southeastern New York State: Research Report EP 84-27 to Empire State Electric Energy Research Corporation, Albany, New York, Appendix I.

———, 1989, The analysis of pinnate joints in the Mount Desert Island Granite: Implications for post-intrusion kinematics in the coastal volcanic belt, Maine: *Geology,* 17:564–67.

———, 1990, Smoluchowski's dilemma revisited: A note on the fluid-pressure history of the Central Appalachian Fold-Thrust Belt: in Norton, D., and Bredehoeft, J. D., eds., *Studies in Geophysics: The Role of Fluids in Crustal Processes:* National Academy of Science, Washington D.C. pp. 140–47.

Engelder, T., and Engelder, R., 1977, Fossil distortion and décollement tectonics of the Appalachian Plateau: *Geology,* 5:457–60.

Engelder, T., and Geiser, P.A., 1980, On the use of regional joint sets as trajectories of paleostress fields during the development of the Appalachian Plateau, New York: *Journal of Geophysical Research,* 85:6319–41.

———, 1984, Near-surface in situ stress, 4, Residual stress in the Tully limestone, Appalachian Plateau, New York: *Journal of Geophysical Research,* 89:9365–70.

Engelder, T., and Lacazette, A., 1990, Natural Hydraulic Fracturing: in Barton, N., and Stephansson, O., eds., *Rock Joints,* Rotterdam, A.A. Balkema, pp. 35–44.

Engelder, T., and Marshak, S., 1985, Disjunctive cleavage formed at shallow depths in sedimentary rocks: *Journal of Structural Geology,* 7:327–43.

———, 1988, The analysis of data from rock-deformation experiments: in Marshak, S. and Mitra, G., eds., *Basic Methods of Structural Geology,* Englewood Cliffs, New Jersey, Prentice-Hall, pp. 193–212.

Engelder, T., and Oertel, G., 1985, The correlation between undercompaction and tectonic jointing within the Devonian Catskill Delta: *Geology,* 13:863–66.

Engelder, T., and Plumb, R.A., 1984, Changes in in situ ultrasonic properties of rock on strain relaxation: *International Journal of Rock Mechanics and Mining Sciences,* 21:75–82.

Engelder, T., and Sbar, M.L., 1976, Evidence for uniform strain orientation in the Potsdam sandstone, northern New York, from in situ measurements: *Journal of Geophysical Research,* 81:3013–17.

———, 1977, The relationship between in situ strain and outcrop fractures in the Potsdam sandstone, Alexandria Bay, New York: *Pure and Applied Geophysics,* 115:41–55.

———, 1984, Near-surface in situ stress: *Introduction: Journal of Geophysical Research,* 89:9321–22.

Engelder, T., Sbar, M.L., and Kranz, R., 1977, A mechanism for strain relaxation of Barre Granite: Opening of microfractures: *Pure and Applied Geophysics,* 115:27–40.

England, P., and Houseman, G., 1989, Extension during continental convergence, with application to the Tibetan Plateau: *Journal of Geophysical Research,* 94:17561–79.

Epstein, A.G., Epstein, J.B., and Harris, L.D., 1977, Conodont color alteration—an index to organic metamorphism: U.S. Geological Survey Professional Paper 995.

Etchecopar, A., Vasseur, G., and Daignieres, M., 1981, An inverse problem in microtectonics for the determination of stress tensors from fault striation analysis: *Journal of Structural Geology,* 3:51–65.

Evans, K., 1983, Some examples and implications of observed elastic deformation associated with growth of hydraulic fractures in the earth: in Zoback, M.D., and Haimson, B.C., eds., *Hydraulic Fracturing Stress Measurements,* Washington, D.C., National Academy Press, pp. 246–59.

———, 1987, A laboratory study of two straddle-packer systems under simulated hydrofrac stress-measurement conditions: *Journal of Petroleum Technology,* 109:180–90.

———, 1989, Appalachian Stress Study 3: Regional scale stress variations and their relation to structure and contemporary tectonics: *Journal of Geophysical Research,* 94:17619–45.

Evans, K., and Engelder, T., 1986, A study of stress in Devonian Shale: Part I - 3D stress mapping using a wireline minifrac system: Society of Petroleum Engineers Paper No. 15609.

———, 1989, Some problems in estimating horizontal stress magnitudes in "thrust" regimes: *International Journal of Rock Mechanics and Mining Sciences,* 26:647–60.

Evans, K., Engelder, T., and Plumb, R.A., 1989a, Appalachian Stress Study 1: A detailed description of in situ stress variations in Devonian Shales of the Appalachian Plateau: *Journal of Geophysical Research,* 94:1729–54.

Evans, K., Oertel, G., and Engelder, T., 1989b, Appalachian Stress Study 2: Analysis of Devonian shale core: Some implications for the nature of contemporary stress variations and Alleghanian deformation in Devonian rocks: *Journal of Geophysical Research,* 94:1755–70.

Evans, K., Scholz, C.H., and Engelder, T., 1988, An analysis of horizontal fracture initiation during hydrofracture stress measurements in granite at North Conway, New Hampshire: *The Geophysical Journal of the Royal Astronomical Society,* 93:251–64.

Evans, R.H., 1936, The elasticity and plasticity of rocks and artificial stone: *Proceedings of the Leeds Philosophical Literary Society,* 3:145–58.

Faill, R.T., 1973, Kink band folding, Valley and Ridge Province, Pennsylvania: *Geological Society of America Bulletin,* 84:1289–1314.

Feves, M., Simmons, G., and Siegfried, R., 1977, Microcracks in crustal igneous rocks: Physical properties, in Heacock, J.G. ed., *The Earth's Crust:* Geophysical Monograph 20, American Geophysical Union, pp. 95–117.

Fleitout, L., and Froidevaux, C., 1982, Tectonics and topography for a lithosphere containing density heterogeneities: *Tectonics,* 1:21–56.

———, 1983, Tectonic stress in the lithosphere: *Tectonics,* 2:315–24.

Fletcher, J.B., 1982, A comparison between the tectonic stress measured in situ and stress parameters from induced seismicity at Monticello Reservoir, South Carolina: *Journal of Geophysical Research,* 87:6931–44.

Fletcher, J.B., and Sykes, L.R., 1977, Earthquakes related to hydraulic mining and natural seismic activity in western New York State: *Journal of Geophysical Research,* 82:3767–80.

Flugge, W., 1967, *Viscoelasticity:* Waltham, Massachusetts, Blaisdell Publishing.

Fordjor, C.K., Bell, J.S., and Gough, D.I., 1983, Breakouts in Alberta and stress in the North American Plate: *Canadian Journal of Earth Sciences,* 20:1445–55.

Forsyth, D.W., 1980, Comparison of mechanical models of the oceanic lithosphere: *Journal of Geophysical Research,* 85:6364–68.

Forsyth, D.W., and Uyeda, S., 1975, On the relative importance of the driving forces of plate motion: *Geophysical Journal of the Royal Astronomical Society,* 43:163–200.

Friedman, M., 1963, Petrofabric analysis of experimentally deformed calcite-cemented sandstones: *Journal of Geology,* 71:12–37.

———, 1964, Petrofabric techniques for the determination of principal stress directions in rocks: in Judd, W.R., ed., *State of stress in the Earth's Crust,* New York, Elsevier, pp. 451–552.

———, 1967a, Measurement of the state of residual elastic strain in quartzose rocks by means of X-ray diffractometry: *Norelco Reporter,* 14:7–9.

———, 1967b, Description of rocks and rock masses with a view to their physical and mechanical behavior: Proceedings of the First International Congress of the Society Rock Mechanics, 3:181–97.

———, 1972a, Residual elastic strain in rocks: *Tectonophysics,* 15:297–330.

———, 1972b, X-ray analysis of residual elastic strain in quartzose rocks, in Basic and Applied Rock Mechanics: in K.E. Gray, ed., 10th U.S. Symposium on Rock Mechanics, Austin Texas, pp. 573–95.

Friedman, M., and Bur, T.R., 1974, Investigation of the relations among residual strain, fabric, fracture and ultrasonic attenuation and velocity in rocks: *International Journal of Rock Mechanics and Mining Sciences,* 11:221–34.

Friedman, M., and Conger, F.B., 1964, Dynamic interpretation of calcite twin lamellae in a naturally deformed fossil: *Journal of Geology,* 72: 361–68.

Friedman, M., and Heard, H.C., 1974, Principal stress ratios in Cretaceous limestones from the Texas Gulf Coast: *American Association of Petroleum Geologists,* 58:71–78.

Friedman, M., and Logan, J.M., 1970, Influence of residual elastic strain on the orientation of experimental fractures in three quartzose sandstones: *Journal of Geophysical Research,* 75:387–405.

Friedman, M., and Sowers, G.M., 1970, Petrofabrics: a critical review: *Canadian Journal of Earth Sciences,* 7:447–97.

Friedman, M., and Stearns, D.W., 1971, Relations between stresses from calcite twin lamellae and macrofractures: *Geological Society of America Bulletin,* 82:3151–61.

Friedman, M., Handin, H., and Alani, G., 1972, Fracture-surface energy of rocks: *International Journal of Rock Mechanics and Mining Sciences,* 9:757–66.

Friedman, M., Handin, H., Logan, J.M., Min, K.D., and Stearns, D.W., 1976, Experimental drape folds in multilithologic layered specimens: *Geological Society of America Bulletin,* 87:1049–66.

Friedman, M., Logan, J.M., and Rigert, J.A., 1974, Glass indurated quartz gouge in sliding-friction experiments on sandstones: *Geological Society of America Bulletin,* 85:937–42.

Frizzell, V.A., and Zoback, M.L., 1987, Stress orientation determined from fault slip data in Hamel Wash area, Nevada, and its relation to contemporary regional stress field: *Tectonics,* 6:89–98.

Froidevaux, C., Paquin, C., and Souriau, M., 1980, Tectonic stresses in France: In situ measurements with a flatjack: *Journal of Geophysical Research,* 85:6342–46.

Frost, H.J., and Ashby, M.F., 1982, *Deformation mechanism maps: The plasticity and creep of metals and ceramics:* New York, Pergamon Press.

Gallagher, J.J., Friedman, M., Handin, J., and Sowers, G., 1974, Experimental studies relating to microfracture in sandstone: *Tectonophysics,* 21:203–47.

Gay, N.C., 1975, In-situ stress measurements in South Africa: *Tectonophysics,* 29:447–59.

———, 1977, Principal horizontal stresses in Southern Africa: *Pure and Applied Geophysics,* 115:3–10.

Geertsma, J., 1957, The effect of fluid pressure decline on volumetric changes of porous rock: *Transactions AIME,* 210:331–39.

Geiser, P.A., and Engelder, T., 1983, The distribution of layer parallel shortening fabrics in the Appalachian foreland of New York and Pennsylvania: Evidence for two non-coaxial phases of the Alleghanian orogeny: in Hatcher, R.D., Jr., Williams, H., and Zietz, I., eds., *Contributions to the tectonics and geophysics of Mountain chains: Geological Society of America Memoir* 158, pp. 161–75.

Geiser, P.A., 1988, Mechanisms of thrust propagation: Some examples and implications for the analysis of overthrust terranes: *Journal of Structural Geology,* 10:829–46.

Gephart, J.W., 1988, On the use of stress inversion of fault-slip data to infer the frictional strength of rocks: *EOS,* 69:1462.

Gephart, J.W., and Forsyth, D., 1984, An improved method for determining the regional stress tensor using earthquake focal mechanism data: Application to the San Fernando earthquake sequence: *Journal of Geophysical Research,* 89:9305–20.

———, 1985, On the state of stress in New England as determined from earthquake focal mechanisms: *Geology,* 13:70–72.

Gilbert, J.F., and MacDonald, G.J.F., 1960, Free oscillations of the earth, I. Toroidal oscillation: *Journal of Geophysical Research,* 65:675–93.

Gladwin, M.T., and Stacy, F.D., 1973, Ultrasonic pulse velocity as a rock stress sensor: *Tectonophysics,* 21:39–45.

Glover, G. and Sellars, C.M., 1973, Recovery and recrystallization during high temperature deformation of å-iron: *Metallurgical Transactions,* 4:765–75.

Goetze, C., 1975, Sheared lherzolites: from the point of view of rock mechanics: *Geology,* 3:172–73.

Goetze, C., and Evans, B., 1979, Stress and temperature in bending lithosphere as constrained by experimental rock mechanics: *Geophysical Journal of the Royal Astronomical Society,* 59:463–78.

Goodman, R.E., 1980, *Introduction to Rock Mechanics:* New York, J. Wiley and Sons.

Gordon, R.G., Cox, A., and Harter, C.E., 1978, Absolute motion of an individual plate estimated from its ridge and trench boundaries: *Nature,* 274:752–55.

Gough, D.I., and Bell, J.S., 1981, Stress orientations from oil-well fractures in Alberta and Texas: *Canadian Journal of Earth Science,* 18:638–45.

———, 1982, Stress orientation from borehole wall fractures with examples from Colorado, east Texas, and northern Canada: *Canadian Journal of Earth Science,* 19:1358–70.

Gough, D.I., and Gough, W.I., 1970, Stress and deflection in the lithosphere near Lake Kariba: *Geophysical Journal of the Royal Astronomical Society,* 21:65–78.

———, 1987, Stress near the surface of the earth: *Annual Review of Earth and Planetary Sciences,* 15:545–66.

Gregory, E.G., Rundle, T.A., McCabe, W.M., and Kim, K., 1983, In situ stress measurement in a jointed basalt: Proceedings of the Rapid Excavation and Tunnelling Conference, Chicago, 1:42–61.

Greiner, G., 1975, In situ stress measurements in southwest Germany: *Tectonophysics,* 29:265–74.

Greiner, G., and Illies, J.H., 1977, Central Europe: Active or residual tectonic stresses: *Pure and Applied Geophysics,* 115:11–26.

Griffith, A.A., 1920, The phenomena of rupture and flow in solids: *Philosophical Transactions Royal Society of London,* A221:163

———, 1924, Theory of rupture: *Proceedings of the First International Congress of Applied Mechanics,* Delft, pp. 55–63.

Griggs, D.T., 1936, Deformation of rocks under high confining pressures: *Journal of Geology,* 44:541–77.

Griggs, D.T., and Blacic, J.D., 1965, Quartz-anomalous weakness of synthetic crystals: *Science,* 147:292–95.

Griggs, D.T., and Handin, J. (eds.), 1960, Rock Deformation: *Geological Society of America Memoir* 79.

Griswold, G.B., 1963, How to measure rock pressures: *new tools: Engineering and Mining Journal,* 164:90–95.

Gronseth, J.M., 1982, Determination of the instantaneous shut-in pressure from hydraulic fracturing data and its reliability as a measure of the minimum principal stress: in Goodman, R.E., and Heuze, F.E., eds., *Issues in Rock Mechanics: Proceedings of the 23rd U.S. Symposium on Rock Mechanics,* New York, Society of Mining Engineers, pp. 183–89.

Gronseth, J.M., and Detournay, E., 1979, Improved stress determination procedures by hydraulic fracturing: Final Report for the U.S. Geological Survey, Menlo Park, Calif. Contract #14-08-001-16768.

Gronseth, J.M., and Kry, P.R., 1983, Instantaneous shut-in pressure and its relationship to the minimum in situ stress: in Zoback, M.D., and Haimson, B.C., eds., *Hydraulic Fracturing Stress Measurements,* Washington, D.C., National Academy Press, pp. 55–67.

Groshong, R.H., 1972, Strain calculated from twinning in calcite: *Geological Society of America,* Bulletin 83:2025–37.

———, 1975, Strain, fractures, and pressure solution in natural single-layer folds: *Geological Society of America Bulletin,* 86:1363–76.

———, 1988, Low-temperature deformation mechanisms and their interpretation: *Geological Society of America Bulletin,* 100:1329–60.

Groshong, R.H., Teufel, L.W., and Gasteiger, C., 1984, Precision and accuracy of the calcite strain-gauge technique: *Geological Society of America Bulletin,* 95:357–63.

Gross, M., and Engelder, T., 1991, Asymmetric re-entrants in the Niagara Escarpment: A case for neotectonic joints in North America: *Tectonics,* 10:631–41.

Gueguen, Y., and Nicolas, A., 1980, Deformation of mantle rocks: *Annual Review of Earth and Planetary Science,* 8:119–44.

Guenot, A., and Santarelli, F.J., 1988, Borehole stability: A new challenge for an old problem: in Cundall, P. A., Sterling, R.L., and Starfield, A.M., eds., *Key Questions in Rock Mechanics,* Rotterdam, A.A. Balkema, pp. 453–60.

Gwinn, V.E., 1964, Thin-skinned tectonics in the Plateau and northwestern Valley and Ridge Province of the Central Appalachians: *Geological Society of America Bulletin,* 75:863–900.

Hacker, B.R., Yin, A., Christie, J.M., and Snoke, A.W., 1990, Differential stress, strain rate, and temperatures of mylonitization in the Rudy Mountains, Nevada: Implications for the rate and duration of uplift: *Journal of Geophysical Research,* 95:8569–80.

Hafner, W., 1951, Stress distributions and faulting: *Geological Society of America Bulletin,* 62:373–98.

Haimson, B.C., 1968, Hydraulic fracturing in porous and nonporous rock and its potential for determining in situ stresses at great depth [Ph.D. thesis]: Minneapolis, Minnesota, University of Minnesota.

———, 1973, Earthquake related stress at Rangely, Colorado: in Hardy, H.R., and Stefanko, R., eds., *New Horizons in Rock Mechanics:* Proceedings of 14th U.S. Symposium on Rock Mechanics, pp. 689–708.

———, 1974, A simple method for estimating in situ stresses at great depths, field testing and instrumentation of rock: American Society for Testing of Materials, Special Publication 554:156–82.

———, 1977, Crustal stress in the United States as derived from hydrofracturing tests: in Heacock, J.C., ed., *The Earth's Crust:* Geophysical Monograph 20, American Geophysical Union, pp. 576–92.

———, 1978a, The hydrofracturing stress measuring method and recent field results: *International Journal of rock Mechanics and Mining Sciences,* 15:167–78.

———, 1978b, Crustal stress in the Michigan Basin: *Journal of Geophysical Research,* 83:5857–63.

———, 1979, Field measurements and laboratory tests for design of energy storage caverns: in Proceedings of the 4th International Society for Rock Mechanics Congress, Rotterdam, A. A. Balkema, 2:195–201.

———, 1980, Near-surface and deep hydrofracturing stress measurements in Waterloo quartzite: *International Journal of Rock Mechanics and Mining Sciences,* 17:81–88.

———, 1981, Confirmation of hydrofracturing results through comparisons with other stress measurements: in Einstein, H.H., ed., *Rock Mechanics from Resarch to Application:* Proceedings of 22nd U.S. Symposium on Rock Mechanics, Cambridge, Massachusetts, MIT Press, pp. 379–85.

———, 1982, Comparing hydrofracturing deep measurements with overcoring near-surface tests in three quarries in western Ohio: in Goodman, R.E., and Heuze, F.E., eds., *Issues in Rock Mechanics:* Proceedings of the 23rd U.S. Symposium on Rock Mechanics, New York, Society of Mining Engineers, pp. 190–202.

Haimson, B.C., and Doe, T.W., 1983, State of stress, permeability, and fractures in the Precambrian granite of northern Illinois: *Journal of Geophysical Research,* 88:7355–72.

Haimson, B.C., and Fairhurst, C., 1967, Initiation and extension of hydraulic fractures in rock: *Society of Petroleum Engineers Journal,* 7:310–18.

Haimson, B.C., and Lee, C.F., 1980, Hydrofracturing stress determination at Darlington, Ontario: in *Underground Rock Engineering:* Proceedings of the 13th Canadian Symposium on Rock Mechanics, Canadian Inst. Mining and Metallurgy, Special Volume, pp. 42–50.

Haimson, B.C., and Lee, M.Y., 1984, Development of a wireline hydrofracturing technique and its use at a site of induced seismicity: in Dowding, C.H., and Singh M.M., eds., *Rock Mechanics in Productivity and Protection:* Proceedings of the 25th U.S. Symposium of Rock Mechanics, New York, Society of Mining Engineers of AIME, pp. 194–203.

Haimson, B.C., and Rummel, F., 1982, Hydrofracturing stress measurements in the Iceland Research Drilling Project drill hole at Reydarfjordur, Iceland: *Journal of Geophysical Research,* 87:6631–59.

Haimson, B.C., and Stahl, E.J., 1969, Hydraulic fracturing and the extraction of minerals through wells: in Rau, J.L. and Dellwig, L.F., eds., Proceedings of the 3rd Symposium on Salt, Northern Ohio Geological Society, Cleveland, Ohio, pp. 421–32.

Haimson, B.C., and Voight, B., 1977, Crustal stress in Iceland: *Pure and Applied Geophysics,* 115:153–90.

Haimson, B.C., Lacomb, J., Jones, A.H., and Green, S.J., 1974, Deep stress measurements in tuff at the Nevada test site: Proceedings of the 3rd International Society of Rock Mechanics Congress, Denver, 2A:557–62.

Hallbauer, D.K., Wagner, H., and Cook, N.G., 1973, Some observations concerning the microscopic and mechanical behavior of quartzite specimens in stiff, triaxial compression tests: *International Journal of Rock Mechancis and Mining Sciences,* 10:713–26.

Hancock, P.L., and Engelder, T., 1989, Neotectonic Joints: *Geological Society of America Bulletin,* 101:1197–1208.

Handin, J.W., 1971, Studies in Rock Fracture: 12th Quarterly Technical Report—Center for Tectonophysics, Texas A&M Research Foundation, College Station, Texas.

Handin, J.W., and Fairbairn, H.W., 1953, Experimental deformation of Hasmark dolomite: *Geological Society of America Bulletin,* 64:1429–30.

Handin, J.W., and Griggs, D.T., 1951, Deformation of Yule marble: *Part II,* Predicted fabric changes: *Geological Society of America Bulletin,* 62:863–86.

Handin, J.W., and Hager, R.V., 1957, Experimental deformation of sedimentary rocks under confining pressure: Tests at room temperature on dry samples: *American Association of Petroleum Geologists Bulletin,* 41:1–50.

Handin, J.W., Heard, H.C., and Magouirk, J.N., 1967, Effects of the intermediate principal stress on the failure of limestone, dolomite, and glass at different temperatures and strain rates: *Journal of Geophysical Research,* 72:611–40.

Handy, M.R., 1990, The solid-state flow of polymineralic rocks: *Journal of Geophysical Research,* 95:8647–61.

Hanks, T.C., 1971, The Kuril Trench-Hokkaido Rise system: Large shallow earthquakes and simple models of deformation: *Geophysical Journal of the Royal Astronomical Society,* 23:273–89.

———, 1977, Earthquake stress drops, ambient tectonic stresses and stresses that drive plate motions: *Pure and Applied Geophysics,* 115:441–58.

Hanks, T.C., and Raleigh, C.B., 1980, The conference on magnitude of deviatoric stresses in the earth's crust and uppermost mantle: *Journal of Geophysical Research,* 85:6083–85.

Hansen, E.C., and Borg, I.Y., 1962, The dynamic significance of deformation lamellae in quartz of a calcite-cemented sandstone: *American Journal of Science,* 260:321–66.

Hansen, F.D., and Carter, N.L., 1983, Semibrittle creep of dry and wet Westerly granite at 1000 MPa: in Mathewson, C.C., ed., Rock Mechanics—Theory—Experiment—Practice: Proceedings of the 24th U.S. Symposium on Rock Mechanics, College Station, Texas, Texas A&M University Press, pp. 429–47.

Hardcastle, K.C., 1989, Possible paleostress tensor configurations derived from fault-slip data in eastern Vermont and western New Hampshire: *Tectonics,* 8:265–84.

Hardy, H.R., 1981, Applications of acoustic emission techniques to rock and rock structures: A state-of-the-art Review, Acoustic Emissions in Geotechnical Engineering Practice: STP 750 American Society for Testing and Materials.

Harper, J.F., 1989, Forces driving plate tectonics: The use of simple dynamical models: *Reviews of Aquatic Science,* 1:319–36.

Harper, T.R., Appel, G., Pendleton, M.W., Szmanski, J.S., Taylor, R.K., 1979, Swelling strain development in sedimentary rocks in northern New York: *International Journal of Rock Mechanics and Mining Sciences,* 16:271–92.

Harris, R.A., and Segall, P., 1987, Detection of a locked zone at depth on the Parkfield, California, segment of the San Andreas fault: *Journal of Geophysical Research,* 92:7945–62.

Hartzell, S.H., and Brune, J.N., 1977, Source parameters for the January 1975 Brawley-Imperial Valley earthquake swarm: *Pure and Applied Geophysics,* 115:333–55.

Hasegawa, H.S., Adams, J., and Yamazaki, K., 1985, Upper crustal stresses and vertical stress migration in eastern Canada: *Journal of Geophysical Research,* 90:3637–48.

Haskell, N.A., 1963, Radiation patterns of Raleigh waves from a fault of arbitrary dip and direction of motion in a homogeneous medium: *Bulletin of the Seismological Society of America,* 53:619–42.

Hast, N., 1958, The measurement of rock pressure in mines: *Sveriges Geologiska Undersokning,* Series C Arsbok 52, No. 3.

———, 1967, The state of stresses in the upper part of the earth's crust: *Engineering Geology,* 2:5–17.

———, 1973, Global measurements of absolute stress: *Philosophical Transactions,* Royal Society of London, A., 274:409–19.

———, 1974, The state of stress in the upper part of the earth's crust as determined by measurements of absolute rock stress: *Naturwissenschaften,* 61:468–75.

———, 1979, Limits of stress measurements in the earth's crust: *Rock Mechanics,* 11:143–50.

Hauksson, E., 1990, Earthquakes, faulting, and stress in the Los Angeles Basin: *Journal of Geophysical Research,* 95:15,365–94.

Hauksson, E., and Jones, L.M., 1989, The 1987 Whittier Narrows earthquake sequence in Los Angeles, southern California: Seismological and tectonic analysis: *Journal of Geophysical Research,* 94:9569–89.

Hawkes, I., 1968, Theory of the photoelastic biaxial strain gauge: *International Journal of Rock Mechanics and Mining Sciences,* 5:57–63.

Hawkes, I, and Bailey, W.V., 1973, Design, develop, test, and demonstrated low-cost cylindrical stress gauges and associated components capable of measuring change of stress as a function of time in underground coal mines: U.S. Bureau of Mines Contract Report No. 40220050.

Hawkes, I., and Fellers, G.E., 1969, Theory of the determination of the greatest principal stress in a biaxial stress field using photoelastic hollow cylinder inclusions: *International Journal of Rock Mechanics and Mining Sciences,* 6:143–58.

Hawkes, I., and Hooker, V.E., 1974, The vibrating wire stressmeter: Proceedings of the 3rd International Society of Rock Mechanics Congress, Denver, 2A:439–44.

Hawkes, I., and Moxon, S., 1965, The measurement of in situ rock stress, using the photoelastic biaxial gauge with the overcoring technique: *International Journal of Rock Mechanics and Mining Sciences,* 2:405–14.

Hawkes, I., Mellor, M., and Gariepy, S., 1973, Deformation of rocks under uniaxial tension: *International Journal of Rock Mechanics and Mining Sciences,* 10:493–07.

Haxby, W.F., and Turcotte, D.L., 1976, Stresses induced by the addition or removal of overburden and associated thermal effects: *Geology,* 4:181–84.

Haxby, W.F., Turcotte, D.L., and Bird, J.M., 1976, Thermal and mechanical evolution of the Michigan basin: *Tectonophysics,* 36:57–75.

Hayashi, K., and Sakurai, I., 1988, Interpretation of shut-in curves of hydraulic fracturing for tectonic stress measurements: *International Journal of Rock Mechanics and Mining Sciences,* 26:477–82.

Hayashi, M., Kanagawa, T., Hibino, S., Motozima, M., and Kitahara, Y., 1979, Detection of anisotropic geostresses trying by acoustic emission and nonlinear rock mechanics on large excavating caverns: Proceedings of the 4th International Congress on Rock Mechanics, Montreux, 2:211–18.

Healy, J.H., and Zoback, M.D., 1988, Hydraulic fracturing in situ stress measurements to 2.1 km Depth at Cajon Pass, California: *Geophysical Research Letters,* 15:1005–8.

Healy, J.H., Rubey, W.W., Griggs, D.T., and Raleigh, C.B., 1968, The Denver earthquakes: *Science,* 161:1301–10.

Heard, H.C., 1960, Transition from brittle to ductile flow in Solnhofen Limestone as a function of temperature, confining pressure, and interstitial fluid pressure: in Griggs, D., and Handin, J., eds., *Rock Deformation: Geological Society of America Memoir* 79:193–226.

———, 1963, Effect of large changes in strain rate in the experimental deformation of Yule marble: *Journal of Geology,* 71:162–95.

Heard, H.C., and Carter, N.L., 1968, Experimentally induced “natural” intragranular flow in quartz and quartzite: *American Journal of Science,* 266:1–42.

Heck, E.T., 1960, Hydraulic fracturing in light of geologic conditions: *Producers Monthly,* 24:12–19.

Heim, A., 1878, *Untersuchungen über den Mechanismus der Gebirgsbildung:* Basel, Schwabe, vol. 2.

Herget, G., 1986, Changes of ground stresses with depth in the Canadian Shield: in Stephansson, O., ed., *Rock Stress and Rock Stress Measurements:* CENTEK Pub., Luleå, Sweden, pp. 61–68.

———, 1988, Stresses in Rock: Rotterdam, A.A. Balkema.

Herrick, J.P., 1949, *Empire Oil:* New York, Dodd, Mead, & Company.

Herrmann, R.B., 1979, Surface wave focal mechanisms for eastern North American earthquakes with tectonic implications: *Journal of Geophysical Research,* 84:3543–52.

Hickman, S.H., 1990, The strength of active faults in the earth's crust and the high stress/low stress controversy: *EOS,* 71:1579–80.

———, 1991, Stress in the lithosphere and the strength of active faults: *Reviews of Geophysics Supplement:* 759–75.

Hickman, S.H., and Zoback, M.D., 1983, The interpretation of hydraulic fracturing pressure-time data for in situ stress determination: in Zoback, M.D., and Haimson, B.C., eds., *Hydraulic Fracturing Stress Measurements,* Washington, D.C., National Academy Press, pp. 44–54.

Hickman, S.H., Healy, J.H., and Zoback, M.D., 1985, In situ stress, natural fracture distribution, and borehole elongation in the Auburn geothermal well, Auburn, New York: *Journal of Geophysical Research,* 90:5497–5512.

Hickman, S.H., Healy, J.H., Zoback, M.D., and Svitek, J., 1981, Recent in situ stress measurements at depth in the western Mojave Desert: *EOS,* 62:1048.

Hickman, S.H., Zoback, M.D., and Healy, J.H., 1988, Continuation of a Deep Borehole Stress measurement profile near the San Andreas Fault, 1. Hydraulic fracturing stress measurements at Hi Vista, Mojave Desert, California: *Journal of Geophysical Research,* 93:15,183–95.

Higgs, D.V., and Handin, J.W., 1959, Experimental deformation of dolomite single crystals: *Geological Society of America Bulletin,* 70:245–78.

Hodgson, J.H., 1957, Nature of faulting in large earthquakes: *Geological Society of America Bulletin,* 68:611–44.

Hodgson, R.A., 1961, Classification of structures on joint surfaces: *American Journal of Science,* 259:493–502.

Hoek, E., and Bieniawski, Z.T., 1965, Brittle fracture propagation in rock under compression: *International Journal of Fracture Mechanics,* 1:137–55.

Hoek, E., and Brown, E.T., 1980, Empirical strength criterion for rock masses: *Journal of Geotechnical Engineering.* Div., *ASCE,* 106:1013–35.

Holcomb, D.J., and McNamee, M.J., 1984, Displacement gauge for the rock mechanics laboratory: Sandia National Laboratories Report SAND84-0651.

Holmes, A., 1929, A review of the continental drift hypothesis: *Mining Magazine,* 40:205–9, 286–88, 340–47.

Holst, T. B., and Foote, G. R., 1981, Joint orientation in Devonian rocks in the northern portion of the lower peninsula of Michigan: *Geological Society of America Bulletin,* 92:85–93.

Holzhausen, G.R., Branagan, P., Egan, H., and Wilmer, R., 1989, Fracture closure pressure from the free-oscillation measurements during stress testing in complex reservoirs: *International Journal of Rock Mechanics and Mining Sciences,* 26:533–40.

Holzhausen, G.R., and Johnson, A.M., 1979a, Analyses of longitudinal splitting of uniaxially compressed rock cylinders: *International Journal of Rock Mechanics and Mining Sciences,* 16:163–77.

———, 1979b, The concept of residual stress in rock: *Tectonophysics,* 58:237–67.

Honda, H., 1957, The mechanism of the earthquakes: Science Reports, Tohoku University, Series 5, *Geophysics,* 9:1–46.

Hooker V.E., and Bickel, D.L., 1974, Overcoring equipment and techniques used in rock stress determination: Information Circular of the U.S. Bureau of Mines, No. 8618.

Hooker, V.E., and Duvall, W.I., 1971, In situ rock temperature: Stress investigations in rock quarries: Report of Investigations 7589, Washington, D.C., U.S. Bureau of Mines.

Hooker, V.E., and Johnson, C.F., 1969, Near-surface horizontal stresses, including the effects of rock anisotropy: Report of Investigations 7224, Washington, D.C., U.S. Bureau of Mines.

Hooker, V.E., Aggson, J.R., and Bickel, D.L., 1974, Improvements in the three component borehole deformation gauge and overcoring techniques: Report of Investigations 7894, Washington, D.C., U.S. Bureau of Mines.

Hoskins, E.R., 1966, An investigation of the flatjack method of measuring rock stress: *International Journal of Rock Mechanics and Mining Sciences,* 3:249–64.

———, 1967, An investigation of strain relief methods of measuring rock stress: *International Journal of Rock Mechanics and Mining Sciences,* 4:155–64.

Hoskins, E.R., and Bahadur, S., 1971, Measurements of residual strain in the Black Hills: *EOS,* 52:122.

Hoskins, E.R., and Daniells, P.A., 1970, Measurements of residual strain in rocks: *EOS,* 51:826–27.

Hoskins, E.R., and Russell, J.E., 1981, The origin of the measured residual strains in crystalline rocks: in Carter, N.L., Friedman, M., Logan, J.M., and Stearns, D.W., eds., *The Mechanical Behavior of Crustal Rocks,* Geophysical Monograph 24, American Geophysical Union, pp. 187–98.

Hoskins, E.R., Jaeger, J.C., and Rosengren, K.J., 1968, A medium-scale direct friction experiment: *International Journal of Rock Mechanics and Mining Sciences,* 5:143–54.

Hoskins, E.R., Russell, J.E., Beck, K., and Mohrman, D., 1972, In situ and laboratory measurements of residual strain in the coast ranges north of San Francisco Bay: *EOS,* 53:1117.

Hottman, C.E., Smith, J.H., and Purcell, W.R., 1979, Relationship among earth stresses, pore pressure and drilling problems, offshore Gulf of Alaska: *Journal of Petroleum Technology,* 31:1477–84.

House, L.S., and Jacob, K.H., 1982, Thermal stresses in subducting lithosphere can explain double seismic zones: *Nature,* 295:587–89.

Howard, G.C., and Fast, C.R., 1970, *Hydraulic Fracturing:* Monograph Series, Society of Petroleum Engineers of AIME, vol. 2.

Hsui, A.T., 1988, Application of fluid mechanic principles to the study of geodynamic processes at trench-arc-back arc systems: *Pure and Applied Geophysics,* 128:661–81.

Huang, Q., 1988, Computer-based method to separate heterogeneous sets of fault-slip data into subsets: *Journal of Structural Geology,* 10:297–99.

Hubbert, M.K., and Rubey, W.W., 1959, Role of fluid pressures in mechanics of overthrust faulting: I. Mechanics of fluid-filled porous solids and its application to overthrust faulting: *Geological Society of America Bulletin,* 70:115–66.

Hubbert, M.K., and Willis, D.G., 1957, Mechanics of hydraulic fracturing: *Journal of Petroleum Technology,* 9:153–68.

Hudson, J.A., and Cooling, C.M., 1988, In situ rock stresses and their measurement in the U.K.—Part I. The current state of knowledge: *International Journal of Rock Mechanics and Mining Sciences,* 25:363–70.

Hughson, D.R., and Crawford, A.M., 1986, Kaiser effect gauging: A new method for determining the pre-existing in situ stress from an extracted core by acoustic emission: in Stephansson, O., ed., *Rock Stress and Rock Stress Measurements:* CENTEK Pub., Luleå, Sweden, pp. 359–68.

Hunt, J.M., 1990, Generation and Migration of Petroleum from abnormally pressured fluid compartments: *American Association of Petroleum Geologists Bulletin,* 74:1–12.

Hurlow, H.A., 1988, P-T conditions of mylonitization in a Tertiary extensional shear zone, Rudy Mountains-East Humboldt Range, Nevada: Geological Society of America Abstract with Programs, 20:170.

Hustrulid, W., and Leijon, B., 1983, CSIRO strain cell measurements: in In situ stress measurements at the Stripa Mine, Sweden, Technical Information Report No. 44, Swedish-American cooperative Program on Radioactive Waste Storage in Mined Caverns in Crystalline Rock, Lawrence Berkeley Laboratory, California.

Hyett, A.J., Dyke, C.G., and Hudson, J.A., 1986, A critical examination of basic concepts associated with existence and measurements of in situ stress: in Stephansson, O., ed., *Rock Stress and Rock Stress Measurements:* Luleå, Sweden, CENTEK Pub., pp. 687–94.

Illies, J.H., and Baumann, H., and Hoffers, 1981, Stress pattern and strain release in the Alpine Foreland: *Tectonophysics,* 71:157–72.

Inglis, C.E., 1913, Stresses in a plate due to the presence of cracks and sharp corners: *Transactions,* Institute of Naval Architecture, 55:219.

Ingraffea, A.R., 1987, Theory of crack initiation and propagation in rock, in Atkinson, B., ed., *Fracture Mechanics of Rock,* London, Academic Press Inc.

Irwin, G.R., 1957, Analysis of stress and strains near the end of a crack traversing a plate: *Journal of Applied Mechanics,* 24:361–64.

———, 1958, Fracture: in S. Flugge, ed., *Handbuch der Physik,* vol VI, Berlin, Springer-Verlag, pp. 551–90.

Isacks B., and Molnar, P., 1971, Distribution of stresses in the descending lithosphere from a global survey of focal mechanism solutions of mantle earthquakes: *Reviews of Geophysics and Space Physics,* 9:103–74.

Isacks, B., Oliver, J., and Sykes, L.R., 1968, Seismology and the new global tectonics: *Journal of Geophysical Research,* 73:5855–99.

Jacobi, D., 1958, Instrumentation for rock pressure research: *Colliery Engineering,* 25:81–85.

Jaeger, J.C., and Cook, N.G.W., 1963, Pinching-off and disking of rocks: *Journal of Geophysical Research,* 68:1759–65.

———, 1969, *Fundamentals of Rock Mechanics:* London, Methuen & Co.

Jamison, W.R., and Spang, J.H., 1976, Use of calcite to infer differential stress: *Geological Society of America Bulletin,* 87:868–72.

Jeffreys, H., 1962, *The Earth:* 4th ed., England, Cambridge University Press.

Johnson, J.N., et al., 1974, Anisotropic mechanical properties of Kayenta sandstone (mixed company site) for ground motion calculations: Terra Tek Report TR 74-61, Salt Lake City, Utah.

Johnson, R.B., 1961, Patterns and origin of radial dike swarms associated with West

Spanish Peak and Dike Mountain, South Central Colorado: *Geological Society of America Bulletin,* 72:579–90.

Jones, L.M., 1988, Focal mechanisms and the state of stress on the San Andreas fault in southern California: *Journal of Geophysical Research,* 93:8869–91.

Jordan, T.H., 1975, The continental tectosphere: *Reviews of Geophysics and Space Physics,* 13:1–12.

Jurdy, D.M., 1978, An alternative model for early Tertiary absolute plate motions: *Geology,* 6:469–72.

Jurdy, D.M., and Stefanick, M., 1992, Stress observations and driving force models for the South American plate: *Journal of Geophysical Research,* vol. 97 (forthcoming).

Kaiser, J., 1953, Erkenntnisse und Folgerungen aus der Messung von Gerauschen bei Zugbeanspruchung von metallischen Werkstoffen: *Archiv für das Eisenhuttenwesen,* 24:43–45.

Kanagawa, T., Hayashi, M., Kitahara, Y., 1981, Acoustic emission and overcoring methods for measuring tectonic stresses: Proceedings of the International Symposium on Weak Rock, Tokyo, Japan.

Kanagawa, T., Hibino, S., Ishida, T., Hayashi, M., Kitahara, Y., 1986, 4. In situ stress measurements in Japanese Islands: overcoring results from a multielement gauge used at 23 sites: *International Journal of Rock Mechanics and Mining Sciences,* 23:29–39.

Kanamori, H., 1971, Seismological evidence for a lithospheric normal faulting—The Sanriku earthquakes of 1933: *Physics of Earth and Planetary Interiors,* 4:289–300.

Kanamori, H., and Allen, C., 1986, Earthquake repeat time and average stress-drop: in Das, S., Boatwright, J., and Scholz, C., eds., *Earthquake Source Mechanics:* AGU Maurice Ewing Series, 6:227–36.

Kanamori, H., and Anderson, D.L., 1975, Theoretical basis of some empirical relations in seismology: *Bulletin of the Seismological Society of America,* 65:1073–95.

Kappmeyer, J., and Wiltschko, D.V., 1984, Quartz deformation in the Marquette and Republic troughs, Upper Peninsula of Michigan: *Canadian Journal of Earth Science,* 21:793–801.

Karato, S., Paterson, M.S., and Fitz, J.D., 1986, Rheology of synthetic olivine aggregates: Influence of grain size and wear: *Journal of Geophysical Research,* 91:8151–76.

Karig, D.E, and Hou G., 1992, High-stress consolidation experiments and their geological implications: *Journal of Geophysical Research,* 97:289–300.

Karner, G.D., and Watts, A.B., 1983, Gravity anomalies and flexure of the lithosphere at mountain ranges: *Journal of Geophysical Research,* 88:10449–74.

Kasahara, K., 1981, *Earthquake mechanics:* Cambridge, Cambridge University Press.

Kehle, R.O., 1964, Determination of tectonic stresses through analysis of hydraulic well fracturing: *Journal of Geophysical Research,* 69:259–70.

Kilsdonk, B., and Fletcher, R.C., 1989, An analytical model of hanging-wall and footwall deformation at ramps on normal and thrust faults: *Tectonophysics,* 163:153–68.

King, B.C.P., and Vita-Finzi, C., 1981, Active folding in the Algerian earthquake of 10 October, 1980: *Nature,* 292:22–26.

Kirby, S.H., 1980, Tectonic stresses in the lithosphere: Constraints provided by the experimental deformation of rocks: *Journal of Geophysical Research,* 85:6353–63.

———, 1983, Rheology of the lithosphere: *Reviews of Geophysics and Space Physics,* 21:1458–87.

———, 1985, Rock mechanics observations pertinent to the rheology of the continental lithosphere and the localization of strain along shear zones: *Tectonophysics,* 119:1–27.

Kirby, S.H., and Kronenberg, A., 1984, Deformation of clinopyroxenite: Evidence for a transition in flow mechanisms and semibrittle behavior: *Journal of Geophysical Research,* 89:3177–92.

Kirby, S.H., and Raleigh, C.B., 1973, Mechanisms of high-temperature, solid-state flow in minerals and ceramics and their bearing on the creep behavior of the mantle: *Tectonophysics,* 19:165–94.

Kirsch, G., 1898, Die Theorie der Elastizität und die Bedurfnisse der Festigheitslehre: *Zeitschrift des Vereines Deutscher Ingenieure,* 42:797.

Klein, R.J., and Barr, M.V., 1986, Regional state of stress in western Europe: in Stephansson, O., ed., *Rock Stress and Rock Stress Measurements:* CENTEK Pub., Luleå, Sweden, pp. 33–44.

Kleinspehn, K.L., Pershing, J., and Teyssier, C., 1989, Paleostress stratigraphy: A new technique for analyzing tectonic control on sedimentary-basin subsidence: *Geology,* 17:253–56.

Knopoff, L., 1958, Energy release in earthquakes: *Geophysical Journal,* 1:44–52.

Koch, P.S., 1983, Rheology and microstructures of experimentally deformed quartz aggregates [Ph.D. thesis]: Los Angeles, California, University of California.

Kohlbeck, F., and Scheidegger, A.E., 1983, *Application of strain gauges on rock and concrete: Field Measurements in Geomechanics,* A.A. Balkema, Rotterdam, pp. 197–207.

Kohlstedt, D.L., and Weathers, M.S., 1980, Deformation-induced microstructures, paleopiezometers, and differential stresses in deeply eroded fault zones: *Journal of Geophysical Research,* 85:6269–85.

Kohlstedt, D.L., Cooper, R.F., Weathers, M.S., and Bird, J.M., 1979, Paleostress analysis of deformation induced microstructures: Moine thrust zone and Ikertoq shear zone: Analysis of Actual Fault Zones in Bedrock, U.S. Geological Survey Open File Report 79–1239, pp. 394–425.

Komar, C. and Bolyard, T., 1981, Structure/stress-ratio relationships within the Appalachian basin: Map 759-027, U.S. Government Printing Office, Washington, D.C.

Kowallis, B.J., and Wang, H.F., 1983, Microcrack study of granitic cores from Illinois Deep Borehole UPH3: *Journal of Geophysical Research,* 88:7373–80.

Kowallis, B.J., Roeloffs, E.A., and Wang, H.F., 1982, Microcrack studies of basalts from the Iceland Research Drilling Project: *Journal of Geophysical Research,* 87:6650–56.

Kozlovsky, Ye. A. (ed.), 1987, *The Superdeep Well of the Kola Peninsula.* Translated from the original Russian edition, published in 1984, New York, Springer-Verlag.

Kranz, R.L., 1979, Crack-crack and crack-pore interactions in stressed granite: *International Journal of Rock Mechanics and Mining Sciences,* 16:37–47.

———, 1983, Microcracks in rocks: A review: *Tectonophysics,* 100:449–80.

Kranz, R.L., and Scholz, C.H., 1977, Critical dilatant volume of rocks at the onset of tertiary creep: *Journal of Geophysical Research,* 82:4893–98.

Kulander, B.R., and Dean, S.L., 1985, Hackle plume geometry and joint propagation dynamics: in Stephansson, O., ed., *Rock Stress and Rock Stress Measurements:* Luleå, Sweden, CENTEK Pub., pp. 85–94.

Kulander, B.R., Barton, C.C., and Dean, S.L., 1979, The application of fractography to core and outcrop fracture investigations: Report to U.S.D.O.E. Morgantown Energy Technology Center, METC/SP-79/3.

Kulander, B.R., Dean, S.L., and Ward, B.J., 1990, Fractured Core Analysis: Interpretation, Logging, and Use of Natural and Induced Fractures in Core: AAPG Methods in Exploration Series, No. 8, American Association of Petroleum Geologists, Tulsa, Oklahoma.

Kurita, K., and Fujii, N., 1979, Stress memory of crystalline rocks in acoustic emission: *Geophysical Research Letters,* 6:9–12.

Kusznir, N.J., 1982, Lithosphere response to externally and internally derived stresses: A viscoelastic stress guide with amplification: *Geophysical Journal of the Royal Astronomical Society,* 70:399–414.

Lacazette, A., 1991, Natural hydraulic fracturing in the Bald Eagle sandstone in Central Pennsylvania and the Ithaca siltstone at Watkins Glen, New York [Ph.D. thesis]: University Park, Pennsylvania, The Pennsylvania State University.

Lacazette, A., and Engelder, T., 1987, Reducing fluids and the origin of natural fractures in the Bald Eagle Sandstone, Pennsylvania: *Geological Society of America Abstracts with Programs,* 19:737.

———, 1992, Fluid-driven cyclic propagation of a joint in the Ithaca siltstone, Appalachian Basin, New York: in Evans, B. and Wong, T-F., eds., *The Brace Volume on Geological Rock Mechanics:* New York, Academic Press (forthcoming).

Lachenbruch, A.H., and Sass. J.H., 1980, Heat flow and energetics of the San Andreas fault zone: *Journal of Geophysical Research,* 85:6185–6222.

Lachenbruch, A.H., and Thompson, G.A., 1972, Oceanic ridges and transform faults: Their intersection angles and resistance to plate motions: *Earth and Planetary Science Letters,* 15:116–22.

Lambeck, K., McQueen H.W.S., Stephenson, R.A., and Denham, D., 1984, The state of stress within the Australian continent: *Annales Geophysicae,* 2:723–41.

Lang, P.A., Thompson, P.M., and Ng., L.K.W., 1986, The effect of residual stress and drill hole size on the in situ stress determined by overcoring: Proceedings of the International Symposium on Rock Stress and Rock Stress Measurement, Stockholm, Sweden.

Larsen, W., Reilinger, R., and Brown, L., 1986, Evidence of ongoing crustal deformation related to magmatic activity near Socorro, New Mexico: *Journal of Geophysical Research,* 91:6283–92.

Laurent, Ph., Bernard, Ph., Vasseur, G., and Etchecopar, A., 1981, Stress tensor determination from the study of e twins in calcite: A linear programming method: *Tectonophysics,* 78:651–60.

Laurent, Ph., Tourneret, C., and Laborde, O., 1990, Determining deviatoric stress tensors from calcite twins: Applications to Monophased synthetic and natural polycrystals: *Tectonics,* 9:379–89.

Lawn, B.R., and Wilshaw, T.R., 1975, *Fracture of brittle solids:* Cambridge, Cambridge University Press.

Lawrence, R.D., 1970, Stress analysis based on albite twinning of plagioclase feldspars: *Geological Society of America Bulletin,* 81:2507–12.

Lay, T., Astiz, L., Kanamori, H., and Christensen, D.H., 1989, Temporal variation of large intraplate earthquakes in coupled subduction zones: *Physics of Earth and Planetary Interiors,* 54:258–12.

LeConte, J., 1882, Origin of jointed structure in undisturbed clay and marl deposits: *American Journal of Science,* 3rd Series, 23:233–34.

Lee, C.F., and Lo, K.Y., 1976, Rock squeeze study of two deep excavations at Niagara Falls: Proceedings, ASCE Speciality Conference on Rock Engineering, Boulder, Colorado, pp. 116–40.

Lee, F.T., Abel, J.F., and Nichols, T.C., 1976, The relation of geology to stress changes caused by underground excavation in crystalline rocks at Idaho Springs, Colorado: U.S. Geological Survey Professional Paper 965.

Lee, F.T., Miller, D.R., and Nichols, T.C., 1979, The relation of stresses in granite and gneiss near Mount Waldo, Maine, to structure, topography, and rockbursts: in Gray, K.E., ed., Proceedings, 20th U.S. Symposium on Rock Mechanics, Austin, Texas, University of Texas Press, pp. 663–73.

Leeds, A.R., Knopoff, L., and Kausel, E.G., 1974, Variations of an upper mantle structure under the Pacific Ocean: *Science,* 186:141–43.

Leeman, E.R., 1964a, The measurement of stress in rock Parts I–III: *Journal of South African Institute of Mining and Metallurgy,* 65:45–114, 254–84.

———, 1964b, Absolute rock stress measurement using a borehole trepanning stress-relieving technique: Proceedings of the 6th U. S. Symposium on Rock Mechanics, Rolla, Missouri, pp. 407–26.

———, 1968, The determination of the complete state of stress in rock in a single borehole—laboratory and underground measurements: *International Journal of Rock Mechanics and Mining Sciences,* 5:31–56.

———, 1971, The CSIR "Doorstopper" and triaxial rock stress measuring instruments: *Rock Mechanics,* 3:25–50.

Legget, R. F., 1939, *Geology and Engineering:* New York, McGraw-Hill Book Co.

Lehnhoff, T.F., Stefansson, B., Thirumalai, K., and Wintezak, T.M., 1982, The core disking phenomenon and its relation to in situ stress at Hanford: SD-BWI-TI-085, Rockwell Hanford Operations, Washington.

Leijon, B.A., 1983, Rock stress measurements with the LUH-gauge at the near-surface test facility, Hanford Test Site: Research Report 1983:18, Luleå, Sweden, Luleå University of Technology.

Leijon, B.A., and Stillborg, B.L., 1986, A comparative study between two rock stress measurement techniques at Luossavaara Mine: *Rock Mechanics and Rock Engineering,* 19:143–63.

Leith, C.K., 1913, *Structural Geology:* New York, Holt.

Lespinasse, M., and Pecher, A., 1986, Microfracturing and regional stress field: A Study of the preferred orientations of fluid inclusion planes in a granite from the Massif Central, France: *Journal of Structural Geology,* 8:169–80.

Li, Y.-G., Leary, P.C., and Henyey, T.L., 1988, Stress orientation inferred from shear wave splitting in basement rock at Cajon Pass: *Geophysical Research Letters,* 15:997–1000.

Lieurance, R.S., 1932, Stresses in foundation at Boulder Dam: Technical Memo, 346, Bureau of Reclamation, Denver, Colorado.

Lisle, R.J., 1987, Principal stress orientations from faults: An additional constraint: *Annales Tectonicae,* 1:155–58.

Lister, C.R.B., 1975, Gravitational drive on oceanic plates caused by thermal contraction: *Nature,* 257:663–65.

Ljunggren, C., and Amadei, B., 1989, Estimation of virgin rock stresses from horizontal hydrofractures: *International Journal of Rock Mechanics and Mining Sciences,* 26:69–78.

Ljunggren, C., Amadei, B., and Stephansson, O., 1989, Use of the Hoek and Brown failure criterion to determine in situ stresses from hydraulic fracturing measurements: Proceedings of the Conference on Applied Rock Engineering (CARE '88) Newcastle-upon-Tyne, pp. 133–42.

Ljunggren, C., and Stephansson, O., 1986, Sleeve fracturing—A borehole technique for in situ determination of rock deformability and rock stresses: in Stephansson, O., ed., *Rock Stress and Rock Stress Measurements:* CENTEK Pub., Luleå, Sweden. pp. 323–30.

Lorenz, J.C., Teufel, L.W., and Warpinski, N.R., 1991, Regional Fractures I: A mechanism for the formation of regional fractures at depth in flat-lying reservoirs: *American Association of Petroleum Geologists Bulletin,* 75:1714–37.

Love, A.E.H., 1934, *A treatise on the mathematical theory of elasticity:* Cambridge, Cambridge University Press.

Luton, M.J., and Sellars, C.M., 1969, Dynamic recrystallization in nickel and nickel-iron alloys during high temperature deformation: *Acta Metallurgica,* 17:1033–43.

Lutton, R.J., 1971, Tensile fracture mechanics from fracture surface morphology: in G.B., Clark, ed., Dynamic Rock Mechanics: Proceedings of the 12th U.S. Symposium on Rock, Mechanics, Baltimore, Maryland, Port City Press Inc., pp. 561–71.

Mackwell, S.J. and Kohlstedt, D.L., 1990, Diffusion of hydrogen in olivine: Implications for water in the mantle: *Journal of Geophysical Research,* 95:5079–88.

Mackwell, S.J., Bai, Q, and Kohlstedt, D.L., 1990, Rheology of olivine and the strength of the lithosphere: *Geophysical Research Letters,* 17:9–12.

Madariaga, R., 1977, implications of stress-drop models of earthquakes for the inversion of stress drop from seismic observations: *Pure and Applied Geophysics,* 115:301–16.

Magara, K., 1978, *Compaction and fluid migration:* New York, Elsevier.

Mandl, G., 1988, *Mechanics of Tectonic Faulting: Models and Basic Concepts:* New York, Elsevier.

Mao, N., Sweeney, J., Hanson, J., and Costantino, M., 1985, Using a sonic technique to estimate in situ stresses: in Dowding, C.H., and Singh, M.M., Rock Mechanics in Productivity and Protection: Proceedings of the 25th U.S. Symposium on Rock Mechanics, Evanston, Illinois, Northwestern University Press, pp. 167–75.

Mardia, L., 1972, *The statistics of orientation data:* London, Academic Press.

Marshak, S., Geiser, P.A., Alvarez, W., and Engelder, T., 1982, Mesoscopic fault array of the northern Umbrian Apennine fold belt, Italy: Geometry of conjugate shear by pressure-solution slip: *Geological Society of America Bulletin,* 93:1013–22.

Martel, S.J., Pollard, D.D. and Segall, P., 1988, Development of simple strike-slip fault zones, Mount Abbot Quadrangle, Sierra Nevada, California: *Geological Society of America Bulletin,* 100:1451–65.

Martin, C.D., Read, R.S., and Chandler, N.A., 1990, Does scale influence is situ stress measurements?—Some findings at the underground research laboratory: in de Cunha, P., ed., *International Symposium on Scale effects in Rock Masses,* A. A. Balkema, Rotterdam., pp. 307–16.

Martin, R.J., and Durham, W.B., 1975, Mechanisms of crack growth in quartz: *Journal of Geophysical Research,* 80:4837–44.

Martna, J., Hiltscher, R., and Ingevald, K., 1983, Geology and rock stresses in deep boreholes at Forsmark in Sweden: Proceedings of the 5th International Society Rock Mechanics Congress, Melbourne, Australia, 2:F111–16.

Maury, V.M., Santarelli, F.J., and Henry, J.P., 1988, Core discing: a review: Proceedings of the 1st African Conference on Rock Mechanics., Swaziland.

May, A.N., 1958, Determination of stresses in the solid: Dept. of Mines and Tech. Surveys, Ottawa, Technical Memoir 106/58-MIN.

Mayer, A., Habib, P., and Marchand, R., 1951, Conférence internationale sur les pressions de terrains et al soutenement dans les chantiers d'exploration, Liège, du 24 au 28 avril:217–21.

McAdoo, D.C., and Sandwell, D.T., 1985, Folding of oceanic lithosphere: *Journal of Geophysical Research,* 90:8563–68.

McAdoo, D.C., Caldwell, J.G., and Turcotte, D.L., 1978, On the elastic-perfectly plastic bending of the lithosphere under generalized loading with application to the Kurile trench: *Geophysical Journal of the Royal Astronomical Society,* 54:11–26.

McClintock, F.A., and Walsh, J.B., 1962, Friction of Griffith cracks under pressure: Proceedings, 4th U.S. National Congress of Applied Mechanics, pp. 1015–21.

McCutchen, W.R., 1982, Some elements of a theory for in situ stress: *International Journal of Rock Mechanics and Mining Sciences,* 19:201–3.

McGarr, A., 1980, Some constraints on levels of shear stress in the crust from observations and theory: *Journal of Geophysical Research,* 85:6231–38.

———, 1988, On the state of lithospheric stress in the absence of applied tectonic forces: *Journal of Geophysical Research,* 93:13,609–18.

McGarr, A., and Gay, N.C., 1978, State of stress in the earth's crust: *Annual Reviews of Earth and Planetary Science,* 6:405–36.

McGarr, A., Pollard, D.D., Gay, N.C., and Ortlepp, W.D., 1979, Observations and analysis of structures in exhumed mine-induced faults: Proceedings of Conference VIII—Analysis of actual fault-zones in bedrock: U.S. Geological Survey Open-file Report 79–1239, pp. 101–20.

McGarr, A., Spottiswoode, S.M., and Gay, N.C., 1975, Relationship of mine tremors to induced stresses and to rock properties in the focal region: *Bulletin of the Seismological Society of America,* 65:981–93.

McGarr, A., Spottiswoode, S.M., Gay, N.C., and Ortlepp, W.D., 1979, Observations relevant to seismic driving stress, stress drop, and efficiency: *Journal of Geophysical Research,* 84:2251–61.

McGarr, A., Zoback, M.D., and Hanks, T.C., 1982, Implications of an elastic analysis of in situ stress measurements near the San Andreas fault: *Journal of Geophysical Research,* 87:7797–7806.

McHone, J.G., 1978, Distribution, orientations, and ages of mafic dikes in central New England: *Geological Society of America Bulletin,* 89:1645–55.

McKenzie, D.P. 1969a, The relation between fault plane solutions for earthquakes and the directions of the principal stresses: *Bulletin of the Seismological Society of America,* 59:591–601.

McKenzie, D.P., 1969b, Speculations on the consequences and causes of plate motions: *Geophysical Journal of the Royal Astronomical Society,* 18:1–32.

———, 1972, Plate tectonics: in Robertson, E.C., ed., *The Nature of the Solid Earth,* McGraw-Hill, New York, pp. 323–60.

McKenzie, D.P., and Bowin, C., 1976, The relationship between bathymetry and gravity in the Atlantic ocean: *Journal of Geophysical Research,* 81:1903–15.

McNutt, M., 1979, Compensation of oceanic topography: An application of the response function technique to the Surveyor area: *Journal of Geophysical Research,* 84:7589–98.

———, 1980, Implications of regional gravity for state of stress in the earth's crust and upper mantle: *Journal of Geophysical Research,* 85:6377–96.

McNutt, M., and Menard, H.W., 1978, Lithospheric flexure and uplifted atolls: *Journal of Geophysical Research,* 83:1206–12.

McNutt, M., Diament, M., and Kogan, M.G., 1988,. Variations of elastic plate thickness at continental thrust belts: *Journal of Geophysical Research,* 93:8825–38.

Means, W.D., 1976, *Stress and Strain—Basic Concepts of Continuum Mechanics for Geologists:* New York, Springer-Verlag.

Meglis, I.L., Engelder, T., and Graham, E.K., 1991, The effect of stress-relief on ambient porosity in core samples from the Kent Cliffs (New York) and Moodus (Connecticut) scientific boreholes: *Tectonophysics,* 186:163–74.

Meglis, I.L., Engelder, T., and Greenfield, R.J., 1990, Ambient microcrack porosity, ultrasonic wave velocity and attenuation in cores from the Appalachian Plateau, N.Y.: *EOS,* 71:635.

Meier, D., 1984, Zur Tektonik des schweizerischen Tafel- und Faltenjura (regionale und lokale Strukturen, Kluftgenese, Bruch- und Faltentektonik, Drucklösung) *Clausthaler Geowiss. Diss.,* 14.

Meissner, R. and Strehlau, J., 1982, Limits of stresses in continental crust and their relation to the depth-frequency distribution of shallow earthquakes: *Tectonics,* 1: 73–89.

Melin, W., 1982, Why do cracks avoid each other? *International Journal of Fracture,* 23:37–45.

Menard, H.W., 1986, *The Ocean of Truth:* Princeton, New Jersey, Princeton University Press, 353 pp.

Mendiguren, J.A., 1971, Focal mechanism of a shock in the middle for the Nazca plate: *Journal of Geophysical Research,* 76:3861–79.

Mercier, J-C., 1980a, Single-pyroxene thermobarometry: *Tectonophysics,* 70:1–37.

———, 1980b, Magnitude of the continental lithospheric stresses inferred from rheomorphic petrology: *Journal of Geophysical Research,* 85:6293–6303.

Mercier, J-C., Anderson, D.A., and Carter, N.L., 1977, Stress in the lithosphere: Inferences from the steady state flow of rocks: *Pure and Applied Geophysics,* 115: 199–226.

Merrill, R.H., Willianson, J.V., Ropchan, D.M., and Kruse, G.H., 1964, Stress determinations by flatjack and borehole-deformation methods: Report of Investigations 6400, Washington, D.C., U.S. Bureau of Mines.

Michael, A. J., 1984, Determination of stress from slip data: Faults and folds: *Journal of Geophysical Research,* 89:11517–26.

———, 1987, Stress rotation during the Coalinga aftershock sequence: *Journal of Geophysical Research,* 92:7963–79.

Michelson, A.A., 1920, The laws of elastico-viscous flow: *Journal of Geology,* 28: 18–24.

Miles, A.J., and Topping, A.D., 1949, Stresses around a deep well: *Transactions of the AIME,* 179:186.

Minster, J.B., and Jordan, T.H., 1978, Present-day plate motions: *Journal of Geophysical Research,* 83:5331–54.

Minster, J.B., Jordan, T.H., Molnar, P., and Haines, E., 1974, Numerical modeling of instantaneous plate tectonics: *Geophysical Journal of the Royal Astronomical Society,* 36:541–76.

Mogi, K., 1972, Fracture and flow of rocks: *Tectonophysics,* 13:541–68.

Mohr, H.F., 1956 Measurement of rock pressure: *Mine and Quarry Engineering,* 22:178–89.

Mohr, O., 1900, Welche Umstände bedingen die Elastizitätsgrenze und den Bruch eines Materiales?: *Zeitschrift des Vereines Deutscher Ingenieure,* 44:1524–30; 1572–77.

Molnar P. and Wyss, M., 1972, Moments, source dimensions and stress drops of shallow-focus earthquakes in the Tonga-Kermadec Arc: *Physics Earth and Planetary Interiors,* 6:263–78.

Moos, D., 1986, Drilling for Earthquake Potential in the Northeast: 1985 Lamont-Doherty Geological Observatory Yearbook, Palisades, New York.

Morgan, W.J., 1972, Deep mantle convection plumes and plate motions: *American Association of Petroleum Geologists Bulletin,* 56:203–13.

Morlier, P., 1971, Description de l'etat de Fissuration d'une Rocke a Partir d'Essais Non-Destructifs Simples: *Rock Mechanics,* 3:125–38.

Mount, V.A., and Suppe, J., 1987, State of stress near the San Andreas fault: Implications for wrench tectonics: *Geology,* 15:1143–46.

Muller, O.H., and Pollard. D.D., 1977, The stress state near Spanish Peaks, Colorado determined from a dike pattern: *Pure and Applied Geophysics,* 115:69–86.

Mullis, J., 1979, The system methane-water as a geologic thermometer and barometer from the external parts of the Central Alps: *Bulletin of Mineralogy,* 102:526–36.

———, 1988, Rapid subsidence and upthrusting in the Northern Apennines, deduced by fluid inclusion studies in quartz crystals from the Porretta Terme: *Schweizerische Mineralogische und Petrographische Mitteilungen,* 68:157–70.

Murray, W.M., and Stein, P.K., 1956, *Strain gauge techniques:* Cambridge, Massachusetts, MIT Press.

Murrell, S.A.F., 1971, Micromechanical basis of the deformation and fracture of rocks: in Te'eni, M., ed., *Structure, Solid Mechanics and Engineering Design,* London, Wiley-Interscience, pp. 239–48.

Nakamura, K., Jacob, K.H., and Davies, J.N., 1977, Volcanoes as possible indicators of tectonic stress orientation—Aleutians and Alaska: *Pure and Applied Geophysics,* 115:87–112.

Narr, W., and Currie, J.B., 1982, Origin of fracture porosity—Example from Altamont field, Utah: *Bulletin of American Association of Petroleum Geologists,* 66: 1231–47.

Nelson, W.J., and Bauer, R.A., 1987, Thrust faults in sourthern Illinois basin—Result of contemporary stress? *Geological Society of America Bulletin,* 98:302–07.

Nemat-Nasser S., and Horii, H., 1982, Compression-induced nonplanar crack extension with application to splitting, exfoliation, and rockburst: *Journal of Geophysical Research,* 87:6805–21.

Newmark, R.L., Zoback, M.D., and Anderson, R.N., 1984, Orientation of in situ stresses in the oceanic crust: *Nature,* 311:424–28.

Nichols, T.C., 1975, Deformations associated with relaxation of residual stresses in a sample of Barre Granite from Vermont: U.S. Geological Survey Professional Paper 875.

Nichols, T.C., Abel, J.F., and Lee, F.T., 1968, A solid-inclusion borehole probe to determine three-dimensional stress changes at a point in a rock mass: *U.S. Geological Survey Bulletin 1258-C,* pp. C1-C12.

Nichols, T.C., and Farrow, R.A., 1973, Discussion of R.de la Cruz and C.B. Raleigh's paper Absolute Stress Measurements at the Rangely Anticline, Northwestern Colorado: *International Journal of Rock Mechanics and Mining Sciences,* 10:751–55.

Nicholson, R., and Pollard, D.D., 1985, Dilation and linkage of echelon cracks: *Journal of Structural Geology,* 7:583–90.

Nickelsen, R.P., 1966, Fossil distortion and penetrative rock deformation in the Appalachian plateau, Pennsylvania: *Journal of Geology,* 74:924–31.

———, 1979, Sequence of structural stages of the Allegheny orogeny at the Bear Valley Strip Mine, Shamokin, Pennsylvania: *American Journal of Science,* 279:225–71.

Nickelsen, R.P., and Gross, G.W., 1959, Petrofabric study of Conestoga limestone from Hanover, Pennsylvania: *American Journal of Science,* 257:276–86.

Nickelsen, R.P., and Hough, V.D., 1967, Jointing in the Appalachian Plateau of Pennsylvania: *Geological Society of America Bulletin,* 78:609–30.

Nissen, H.U., 1964, Dynamic and kinematic analysis of crinoid stems in a quartz graywacke: *Journal of Geology,* 72:346–60.

Nolte, K.G., 1982, Fracture design considerations based on pressure analysis: SPE paper 10911, May 1982.

Nolting, R.M., and Goodman, R.E., 1979, In situ stress measurement by overcoring cast-in-place epoxy inclusions: Proceedings of the 4th International Congress of Rock Mechanics, Montreux, France.

Nur, A, 1972, Dilatancy, pore fliuids, and premonitory variations in t_s/t_p travel times: *Bulletin of the Seismological Society of America,* 62:1217–22.

Nur, A., and Byerlee, J. D., 1971, An exact effective stress law for elastic deformation of rocks with fluids: *Journal of Geophysical Research,* 76:6414–19.

Nur, A., and Simmons, G., 1969, Stress-induced anisotropy in rock: An experimental study: *Journal of Geophysical Research,* 74:6667–74.

Nye, J.F., 1957, *Physical Properties of Crystals:* Oxford, Clarendon Press.

Obert, L., 1962, In situ determination of stress in rock: *Mining Engineering,* 14:51–58.

Obert, L., and Stephenson, D.E., 1965, Stress conditions under which core discing occurs: *American Institute of Mining, Metals, and Petroleum Engineering Transactions,* AIME, 232:227–34.

Obert, L., Merrill, R.H., and Morgan, T.A., 1962, Borehole deformation gauge for determining the stress in mine rock: Reports of Investigations 5978, Washington, D.C., U.S. Bureau of Mines.

Ode, H., 1957, Mechanical analysis of the dyke pattern of the Spanish Peaks area, Colorado: *Geological Society of America Bulletin,* 68:567–76.

Ohnaka, M., 1973, The quantitative effect of hydrostatic confining pressure on the compressive strength of crystalline rocks: *Journal of Physics of the Earth,* 21:285–303.

Olsen, O.J., 1957, Measurement of residual stress by the strain relief method: *Quarterly Colorado School of Mines,* 52:183–204.

Olson, J., and Pollard, D.D., 1989, Inferring paleostress from natural fracture patterns: A new method: *Geology,* 17:345–48.

Olsson, W. and Peng, S., 1976, Microcrack nucleation in marble: *International Journal of Rock Mechanics and Mining Sciences,* 18:53–59.

Oppenheimer, D.H., Reasonberg, P.A., and Simpson, R.W., 1988, Fault plane solutions for the 1984 Morgan Hill, California, earthquake sequence: Evidence for the state of stress on the Calaveras fault: *Journal of Geophysical Research,* 93:9007–26.

Ord, A., and Hobbs, B.E., 1989, The strain of the continental crust, detachment zones, and the development of plastic instabilities: *Tectonophysics,* 158:269–89.

Orkan, N., and Voight, B., 1985, Regional joint evolution in the Valley and Ridge Province of Pennsylvania in relation to the Alleghany Orogeny: *Guidebook for the 50th Annual Field Conference of Pennsylvania Geologists,* Bureau of Topographic and Geological Survey, Harrisburg, Pa., pp. 144–64.

Overby, W.K., and Rough, R.L., 1968, Surface studies predict orientation of induced formation fractures: *Producers Monthly,* 32:16–19.

Paillet, F.L., and Kim, K., 1987, Character and distribution of borehole breakouts and their relationship to in situ stresses in deep Columbia River basalts: *Journal of Geophysical Research,* 92:6223–34.

Pallister, G.F., 1969, The measurement of virgin rock stress [M.S. thesis]: Johannesburg, South Africa, University of Witwatersrand.

Panek, L.A., 1961, Measurement of rock pressure with a hydraulic cell: *American institute of Mining Metals, and Petroleum Engineering Transactions,* AIME, 220: 287–90.

Panek, L.A., and Stock, J.A., 1964, Development of a rock stress monitoring station based on the flat slot method of measuring existing rock stress: Report of Investigations 6537, Washington, D.C., U.S. Bureau of Mines.

Parsons, B., and Sclater, J.G., 1977, An analysis of the variation of ocean floor bathymetry and heat flow with age: *Journal of Geophysical Research,* 82:803–27.

Paterson, M.S., 1958, Experimental deformation and faulting in Wombeyan marble: *Geological Society of America Bulletin,* 69:465–76.

———, 1978, *Experimental Rock Deformation—The Brittle Field:* New York, Springer-Verlag.

Pavlis, T.L., and Bruhn, R.L., 1988, Stress history during propagation of a lateral fold-tip and implications for the mechanics of fold-thrust belts: *Tectonophysics,* 145:113–27.

Pecher, A., Lespinasse, M., and Leroy, J., 1985, Relations between fluid inclusion trails and regional stress field: *a tool for fluid chronology—an example of an in-*

tragranitic uranium ore deposit (northwest Massif Central, France): *Lithos,* 18:229–37.

Petit, J.-P., 1987, Criteria for sense of movement on fault surfaces in brittle rocks: *Journal of Structural Geology:* 9:597–608.

Pfiffner, O.A., 1982, Deformation mechanisms and flow regimes in limestones from the Helvetic zone of the Swiss Alps: *Journal of Structural Geology,* 4:429–42.

Philpotts, A.R., 1990, *Principles of Igneous and Metamorphic Petrology:* Englewood Cliffs, New Jersey, Prentice Hall.

Pincus, H.J., 1966, Capabilities of photoelastic coating for study of strain in rock: *Testing Techniques for Rock Mechanics,* A.S.T.M., STP 402, American Society for Testing Materials, pp. 87–105.

Pine, R.J., Ledingham, P., Merrifield, C.M., 1983, In situ stress measurement in Carnmenellis granite —II, Hydrofracture tests at Rosemanowes quarry to depths of 2000m: *International Journal of Rock Mechanics and Mining Sciences,* 20:63–72.

Plumb, R.A., 1983, The correlation between the orientation of induced fractures with in situ stress or rock anisotropy: in Zoback, M.D., and Haimson, B.C., eds., *Hydraulic Fracturing Stress Measurements,* Washington, D.C., National Academy Press, pp. 221–34.

———, 1989, Fracture patterns associated with incipient wellbore breakouts: in Maury V., and Fourmaintraux, D., eds., *Rock at Great Depth,* A.A. Balkema, Rotterdam, pp. 761–68.

Plumb, R.A., and Cox, J., 1987, Stress directions in eastern North America determined to 4.5 km from borehole elongation measurements: *Journal of Geophysical Research,* 92:4805–16.

Plumb, R.A., and Hickman, S.J., 1985, Stress-induced borehole elongation: A comparison between the four-arm dipmeter and the borehole televiewer in the Auburn Geothermal well: *Journal of Geophysical Research,* 90:5513–22.

Plumb, R.A., and Luthi, S.M., 1986, *Application of borehole images to geological modeling of an eolian reservoir:* SPE 15487, 61st Annual Fall Tech. Conf., New Orleans, LA., October 5–8, 1986.

Plumb, R.A., Engelder, T., and Sbar, M., 1984b, Near-surface in situ stress, 2. A comparison with stress directions inferred from earthquakes, joints, and topography near Blue Mountain Lake, New York: *Journal of Geophysical Research,* 89:9333–49.

Plumb, R.A., Engelder, T., and Yale, D., 1984a, Near surface in situ stress, 3. Correlation with microcrack fabric within the New Hampshire granites: *Journal of Geophysical Research,* 89:9350–64.

Plumb, R.A., Evans, K. F., and Engelder, T., 1991, Geophysical log analysis of lithology dependent stress contrasts in Paleozoic rocks of the Appalachian Plateau, New York: *Journal of Geophysical Research,* 96:14,509–28.

Plumb, R.A., Fang, W., and deBoer, J., 1988, Preliminary wellbore breakout and fracture orientations in the Moodus #1 well: EOS, 69:492.

Poirier, J-P., 1985, *Creep of crystals:* Cambridge, Cambridge University Press.

Pollard, D.D., 1989, Quantitative interpretation of joints and faults: Geological Society of America Short Course Notes, Annual Meeting 1989, Geological Society of America, Boulder, Colorado

Pollard, D.D., and Aydin, A., 1988, Progress in understanding jointing over the past century: *Geological Society of America Bulletin,* 100:1181–1204.

Pollard, D.D., and Segall, P., 1987, Theoretical displacements and stresses near fractures in rock: with applications to faults, joints, veins, dikes, and solution surfaces: in Atkinson, B., ed., *Fracture Mechanics of Rock:* London, Academic Press, pp. 227–350.

Pollard, D.D., Segall, P., and Delaney, P.T., 1982, Formation and interpretation of dilatant echelon cracks: *Geological Society of America Bulletin,* 93:1291–1303

Pomeroy, P.W., Simpson, D.W., and Sbar, M.L., 1976, Earthquakes triggered by surface quarrying—Wappingers Falls, New York sequence of June, 1974: *Bulletin of the Seismological Society of America,* 66:685–700.

Post, R.L., 1977, High-temperature creep of Mt. Burnet dunite: *Tectonophysics,* 42:75–110.

Potts, E.L., 1954, Stress distribution, rock pressures and support loads: *Colliery Engineering* (August 1954), pp. 333–39.

Powell, R. L. 1976. Some geomorphic and hydrologic implications of jointing in carbonate strata of Mississippian age in south-central-Indiana: [Ph.D. thesis]: Lafayette, Indiana, Purdue University.

Powers, M.C., 1967, Fluid-release mechanisms in compacting marine mudrocks and their importance in oil exploration: *American Association of Petroleum Geologists Bulletin,* 51:1240–54.

Pratt, H.R., Black, A.D., Brown, W.S., and Brace, W.F., 1972, The effect of specimen size on the strength of unjointed diorite: *International Journal of Rock Mechanics and Mining Sciences,* 9:513–29.

Preston, D.A., 1968, Photoelastic measurement of elastic strain recovery in outcropping rocks: EOS, 49:302.

Price, N.J., 1966, *Fault and Joint Development in Brittle and Semi-brittle Rock:* London, Pergamon Press.

Pulli, J.J., and Toksoz, M.N., 1981, Fault plane solutions for northeastern United States earthquakes, *Bulletin of the Seismological Society of America,* 71:1875–82.

Quittmeyer, R.C., Statton, C.T., Mrotek, K.A., and Houlday, M., 1985, Possible Implications of Recent microearthquakes in southern New York: *Earthquake Notes,* 56:35–44.

Rabinowicz, E., 1965, *Friction and wear of materials:* New York, John Wiley.

Raleigh, C.B., 1968, Mechanisms of plastic deformation of olivine: *Journal of Geophysical Research,* 73:5391–5406.

Raleigh, C.B., and Evernden, J., 1981, Case for low deviatoric stress in the lithosphere: in Carter, N.L., Friedman, M., Logan, J.M., and Stearns, D.W., eds. *Mechanical Behavior of Crustal Rocks:* Geophysical Monograph 24, Washington, D.C., American Geophysical Union, pp. 173–86.

Raleigh, C.B., and Talbot, J.L., 1967, Mechanical twinning in naturally and experimentally deformed diopside: *American Journal of Science,* 265:151–65.

Raleigh, C.B., Healy, J.H., and Bredehoeft, J.D., 1972, Faulting and crustal stress at Rangely, Colorado: in Heard, H.C., Borg, I.Y., Carter, N.C., and Raleigh, C.B., eds., *Flow and Fracture of Rocks:* Geophysical Monograph 16, Washington, D.C., American Geophysical Union, pp. 275–84.

———, 1976, An experiment in earthquake control at Rangely, Colorado: *Science,* 191:1230–37.

Ramsay, J.G., 1980, The crack-seal mechanism of rock deformation: *Nature,* 284:135–39.

Ramsay, J.G., and Huber, M.I., 1983, *The techniques of modern structural geology: Volume 1: Strain analysis:* Orlando, Academic Press, Inc.

Ranalli, G., and Chandler, T.E., 1975, The stress field in the upper crust as determined from in situ measurements: *Geologische Rundschau,* 64:653–74.

Ranalli, G., and Murphy, D.C., 1987, Rheological stratification of the lithosphere: *Tectonophysics,* 132:281–96.

Ratcliffe, N.M., 1980, Brittle faults (Ramapo Fault) and phyllonitic ductile shear zones in basement rocks of the Ramapo seismic zones, New York and New Jersey, and their relationship to current seismicity: in Manspeizer, W., ed., *Field studies of New Jersey geology and guide to field trips,* 52nd Annual Meeting of the New York Geological Association: Newark, New Jersey, Rutgers University, pp. 278–312.

Reasenburg, P., and Aki, K., 1974, A precise, continuous measurement of seismic velocity for monitoring in situ stress: *Journal of Geophysical Research,* 79:399–406.

Reches, Z., 1987, Determination of the tectonic stress tensor from slip along faults that obey the Coulomb yield condition: *Tectonics,* 6:849–61.

Reches, Z., and Lockner, D., 1990, Self-organized cracking—A mechanism for brittle faulting: *EOS,* 71:1586.

Reid, H.F., 1910, The mechanics of the earthquake: Report of the State Earthquake Investigation Commission, vol. 2: The California earthquake of April 18, 1906, Washington, D.C., Carnegie Institution.

Ren, N.-K., and Roegiers, J.-C.,1983, Differential strain curve analysis—A new method for determining the pre-existing in situ stress state from rock core measurements: Proceedings of the 5th International Congress of Rock Mechanics, 5:F117–27.

Ren, N.K., and Hudson, P., 1985, Predicting the in situ state of stress using differential wave velocity analysis: in Ashworth, E., ed., Proceedings of the 26th U.S. Symposium on Rock Mechanics, Rotterdam, A.A. Balkema, pp. 1235–44.

Rice, J.R., and Cleary, M.P., 1976, Some basic diffusion solutions for fluid-saturated elastic porous media with compressible constituents: *Reviews of Geophysics and Space Physics,* 14:227–41

Richardson, R.M., 1987, Modeling the tectonics of the Indo-Australian plate: *EOS,* 68:1466.

———, 1992, Ridge forces, absolute plate motions, and the intraplate stress field: *Journal of Geophysical Research,* vol. 97 (forthcoming).

Richardson, R.M., and Cox, B.L., 1984, Evolution of oceanic lithosphere: A driving force study of the Nazca plate: *Journal of Geophysical Research,* 89:10043–52.

Richardson, R.M., Solomon, S.C., and Sleep, N.H., 1976, Intraplate stress as an indicator of plate tectonics driving forces: *Journal of Geophysical Research,* 81: 1847–56.

———, 1979, Tectonic stress in the plates: *Reviews of Geophysics and Space Physics,* 17:981–1019.

Richter, F., and McKenzie, D., 1978, Simple plate models of mantle convection: *Journal of Geophysics,* 44:441–71.

Rickard, D., 1988, Account of a deep hole: *Geological Magazine,* 125:659–61.

Riley, P.B., Goodman, R.E., and Nolting, R.E., 1977, Stress measurement by overcoring cast photoelastic inclusions: in R.Wang and Clark, A.B., eds., Proceedings of the 18th Symposium on Rock Mechanics, Colorado School of Mines, Golden, paper 4C4.

Ritsema, A.R., 1957, Earthquake-generating stress systems in southeast Asia: *Bulletin of the Seismological Society of America,* 57:267–79.

———, 1959, On the focal mechanism of southeast Asian earthquakes: in Hodgson, J.H., ed., *The Mechanics of Faulting with Special Reference to the Fault-plane Work,* Publ. Dominion Obs., Ottawa, 20:253–418.

Roberts, A., 1973, Discussion of R.De La Cruz and C.B. Raleigh's Paper, Absolute Stress Measurements at the Rangely Anticline, Northwestern Colorado: *International Journal of Rock Mechanics and Mining Sciences,* 10:375–76.

———, 1977, Geotechnology: *An introductory text for students and engineers:* Oxford, Pergamon Press.

Robertson, E.C., 1964, Viscoelasticity of rocks: in Judd, W.R., *State of stress in the Earth's Crust:* New York, Elsevier, pp. 181–224.

Robin, P.-Y., 1978, Pressure solution at grain-to-grain contacts: *Geochimica Cosmochimica Acta,* 42:1383–89.

Rocha, M., Silverio, A., Pedro, J., Delgado, J., 1974, A new development of the LNEC stress tensor gauge: Proceedings of the 3rd International Society of Rock Mechanics Congress, Denver, 2A, pp. 464–67.

Roedder, E., 1984, *Fluid inclusions: Reviews in mineralogy,* vol.12, Mineralogical society of America.

Ross, J.V., and Nielsen, K.C., 1978, High-temperature flow of wet polycrystalline enstatite: *Tectonophysics,* 44:233–61.

Ross, J.V., Avé Lallement, H.G., and Carter, N.L., 1980, Stress dependence of recrystallized grain and subgrain size in olivine: *Tectonophysics,* 70:39–61.

Ross, J.V., Bauer, S.J., and Hansen, F.D., 1987, Textural evolution of synthetic anhydrite-halite mylonites: *Tectonophysics,* 140:307–26.

Rummel, F., 1987, Fracture mechanics approach to hydraulic fracturing stress measurements: in Atkinson, B.K., ed., *Fracture Mechanics of Rocks:* London, Academic Press, pp. 217–39.

Rummel, F., and Hansen, J., 1989, Interpretation of hydrofrac pressure recordings using a simple fracture mechanics simulation model: *International Journal of Rock Mechanics and Mining Sciences,* 26:483–88.

Rummel, F., Baumgartner, J., and Alheid, H.J., 1983, Hydraulic fracturing stress measurements along the eastern boundary of the SW-German Block: in Zoback, M.D., and Haimson, B.C., eds., *Hydraulic Fracturing Stress Measurements,* Washington, D.C., National Academy Press, pp. 3–17.

Rundle, T.A., Singh, M.H., and Baker, C.H., 1985, In situ stress measurements in the Earth's crust in the eastern United States: Nuclear Regulatory Commission Report, NUREG/E1-1126.

Russell, J.E., and Hoskins, E.R., 1973, Residual stresses in rock: in *New Horizons in Rock Mechanics,* Proc. 14th U.S. Symposium on Rock Mechanics, American Society of Civil Engineers, New York, pp. 1–24.

Russell, W. L. 1972. Pressure-depth relations in the Appalachian region: *American Association of Petroleum Geologists Bulletin,* 56:528–36.

Rutter, E.H., 1976, The kinetics of rock deformation by pressure solution: *Philosophical Transactions of the Royal Society of London, A,* 283:203–19.

———, 1983, Pressure solution in nature, theory, and experiment: *Journal of the Geological Society of London,* 140:725–40.

Sandwell, D., 1986, Thermal stress and the spacing of transform faults: *Journal of Geophysical Research,* 91:6405–18.

Sanford, A.R., 1959, Analytical and experimental study of simple geologic structures: *Geological Society of America Bulletin,* 70:19–52.

Saull, V.A., and Williams, D.A., 1974, Evidence for recent deformation in the Montreal area: *Canadian Journal of Earth Sciences,* 11:1621–24.

Savage, W.Z., 1978, The development of residual stress in cooling rock bodies: *Geophysical Research Letters,* 5:633–36.

Savage, W.Z., Swolfs, H.S., and Powers, P.S., 1985, Gravitational stresses in long symmetric ridges and valleys: *International Journal of Rock Mechanics and Mining Sciences,* 22:291–302.

Sbar, M.L., 1982, Delineation and interpretation of seismotectonic domains in western North America: *Journal of Geophysical Research,* 87:3919–28.

Sbar, M.L., and Sykes, L.R., 1973, Contemporary Compressive stress and seismicity in eastern North America: An example of intra-plate tectonics: *Geological Society of America Bulletin,* 80:1231–64.

Sbar, M.L., Armbruster, J., and Aggarwal, Y.P., 1972, The Adirondack, New York, earthquake swarm of 1971 and tectonic implications: *Seismological Society of America Bulletin,* 62:1303–18.

Sbar, M.L., Engelder, T., Plumb, R.A., and Marshak, S., 1979, Stress pattern near the San Andreas fault, Palmdale, California, from near-surface in situ measurements: *Journal of Geophysical Research,* 84:156–64.

Sbar, M.L., Richardson, R.M., Flaccus, C., and Engelder, T., 1984, Near-surface in situ stress: 1. Strain relaxation measurements along the San Andreas Fault in Southern California: *Journal of Geophysical Research,* 89:9323–32.

Schapery, R.A., 1972, Viscoelastic behavior and analysis of composite materials: Report MM 72-3, Mechanics and Materials Research Center, Texas A&M University, College Station, Texas.

Schedl, A., Kronenberg, A.K., and Tullis, J., 1986, Deformation microstructures of Barre Granite: An Optical SEM and TEM Study: *Tectonophysics,* 1986:149–64.

Scheidegger, A.E., 1959, Statistical analysis of recent fault plane solutions of earthquakes: *Bulletin of the Seismological Society of America,* 49:337–47.

Schlanger, S.O., 1964, Petrology of the limestones of Guam: U.S. Geological Survey Professional Paper, 403-D.

Schmid, S.M., 1983, Microfabric studies as indicators of deformation mechanisms and flow laws operative in mountain building: in Hsu, K.J., ed., *Mountain Building Processes,* New York, Academic Press, pp. 95–110.

Schmitt, D., and Zoback, M.D., 1989, Poroelastic effects in the determination of the maximum horizontal principal stress in hydraulic fracturing tests—a proposed breakdown equation employing a modified effective stress relation for tensile failure: *International Journal of Rock Mechanics and Mining Sciences,* 26:499–506.

Schmitt, T., 1979, In situ stress profile through the Alps and foreland: *Allgemeine Vermessungs-Nachrichten,* 86:367–70.

Scholz, C.H., 1968, Microfracturing and the inelastic deformation of rock in compression: *Journal of Geophysical Research,* 73:14–17, 1432.

———, 1982, Scaling laws for large earthquakes: consequences for physical models: *Bulletin of the Seismological Society of America,* 72:1–14.

———, 1990, *The mechanics of earthquakes and faulting:* Cambridge, Cambridge University Press.

Scholz, C.H., and Engelder, T., 1976, The role of asperity indentation and ploughing in rock friction: I Asperity creep and stick-slip: *International Journal of Rock Mechanics and Mining Sciences,* 13:149–54.

Scholz, C.H., Aviles, C.A., and Wesnousky, S.G., 1986, Scaling differences between large interplate and intraplate earthquakes: *Bulletin of the Seismological Society of America,* 76:65–70.

Scholz, C.H., Molnar, P., and Johnson, T., 1972, Detailed studies of frictional sliding in granite and implications for earthquake mechanism: *Journal of Geophysical Research,* 77:6392–6406.

Scholz, C.H., Sykes, L.R., and Aggarwal, Y.P., 1973, Earthquake prediction: A physical basis: *Science,* 181:803–10.

Schubert, G., Yuen, D.A., Froidevaux, C., Fleitout, L., and Souriau, M., 1978, Mantle circulation with partial shallow return flow: Effects on stresses in oceanic plates and topography of the sea floor: *Journal of Geophysical Research,* 83:745–58.

Scott, P.A., Engelder, T., and Mecholsky, J.J., 1992, The correlation between fracture toughness anisotropy and surface morphology of the siltstones in the Ithaca Formation, Appalachian Basin: in Evans, B., ed. *The W.F. Brace Volume on Geological Rock Mechanics,* Academic Press (forthcoming).

Seagar, J.S., 1964, Pre-mining lateral stresses: *International Journal of Rock Mechanics and Mining Sciences,* 1:413–19.

Seborowski, K.D., Williams, G., Kelleher, J.A., and Statton, C.T., 1982, Tectonic implications of recent earthquakes near Annsville, New York: *Bulletin of the Seismological Society of America,* 7:1601–9.

Secor, D.T., 1965, Role of fluid pressure in jointing: *American Journal of Science,* 263:633–46.

———, 1969, Mechanics of natural extension fracturing at depth in the earth's crust: Geological Survey of Canada Paper 68–52.

Sedgwick, A., 1835, Remarks on the structure of large mineral masses and especially on the chemical changes produced in the aggregation of stratifical rocks during different periods after their deposition: *Transactions, Geological Society of London, 2nd Ser.,* 3:461–86.

Seeburger, D.A., and Zoback, M.D., 1982, The distribution of natural fractures and joints at depth in crystalline rock: *Journal of Geophysical Research,* 87:5517–34.

Segall, P., 1989, Earthquakes trippered by fluid extraction: *Geology,* 17:942–46.

Segall, P., and Pollard, D.D., 1983, Joint formation in granitic rock of the Sierra Nevada: *Geological Society of America Bulletin,* 94:563–75.

Shamir, G., and Zoback, M.D., 1988, Stress orientation profile in the Cajon Pass, California, scientific drillhole, based on detailed analysis of stress induced breakouts: in Maury, V., and Fourmaintraux, D., eds, *Rock at Great Depth,* Rotterdam, A.A. Balkema, pp. 1041–48.

Shamir, G., Zoback, M.D., and Barton, C.A., 1988, In situ stress orientation near the San Andreas fault: Preliminary results to 2.1 km depth from the Cajon Pass scientific drillhole: *Geophysical Research Letters,* 15:989–92.

Shelton, G.L., 1981, Experimental deformation of single phase and polyphase crustal rocks at high pressures and temperatures [Ph.D. thesis]: Providence, Rhode Island, Brown University.

Sibek, V., 1960, Contribution to Paper D.5, International Strata Control Conference: Paris Paper, D.5, May, 1960, pp. 295–311.

Sibson, R.H., 1974, Frictional constraints on thrust, wrench, and normal faults: *Nature,* 249:542–44.

———, 1977a, Fault rocks and fault mechanisms: *Journal of Geological Society of London,* 133:191–213.

———, 1977b, Kinetic shear resistance, fluid pressures, and radiation efficiency during seismic faulting: *Pure and Applied Geophysics,* 115:387–400.

———, 1986, Earthquakes and rock deformation in crustal fault zones: *Annual Reviews of Earth and Planetary Sciences,* 14:149–75.

Siegfried, R., and Simmons, G., 1978, Characterization of oriented cracks with differential strain analysis: *Journal of Geophysical Research,* 83:1269–77.

Sieh, K., 1977, Study of Holocene displacement history along the southcentral reach of the San Andreas fault [Ph.D. thesis]: Stanford, California, Stanford Univ.

Sih, G.C., 1973, *Handbook of stress intensity factors:* Institute of Fracture and Solid Mechanics: Bethlehem, Pennsylvania, Lehigh University Press.

Simmons, G., and Nur, A., 1968, Granites: Relation of properties in situ to laboratory measurements: *Science,* 162:789–91.

Simmons, G., and Richter, D., 1976, Microcracks in rock: in Strens, R.G.J., ed., *The Physics and Chemistry of Minerals and Rocks:* New York, John Wiley, pp. 105–37.

Simmons, G., Siegfried, R.W., and Feves, M., 1974, Differential strain analysis: A new method of examining cracks in rocks: *Journal of Geophysical Research,* 79:4383–85.

Simmons, G., Siegfried, R.W., and Richter, D., 1975, Characteristics of microcracks in lunar samples: *Proceedings of the 6th Lunar Science Conference,* 3:3227–54.

Simpson, D.W., 1976, Seismicity changes associated with reservoir impounding: *Engineering Geology,* 10:371–85.

———, 1986, Triggered Earthquakes: *Annual Review of Earth and Planetary Sciences,* 14:21–42.

Simpson, D.W., and Negmatullaev, S. Kh., 1981, Induced seismicity at Nurek reservoir, Tadjikistan, USSR: *Bulletin of the Seismological Society of American,* 71:1561–86.

Simpson, D.W., Leith, W.S., and Scholz, C.H., 1988, Two types of reservoir induced seismicity: *Bulletin of the Seismological Society of American,* 78:2025–40.

Sleep, N.H., and Biehler, S., 1970, Topography and tectonics at the intersections of fracture zones and central rifts: *Journal of Geophysical Research,* 75:2748–52.

Slobodov, M.A., 1958, Test application of the load-relief method for investigating stresses in deep rock: *Ugal,* 7:30–35. (D.S.I.R. Russian Translation RTS 1068).

Smith, A.T., and Toksöz, M.N., 1972, Stress distribution beneath island arcs: *Geophysical Journal of the Royal Astronomical Society,* 29:289–318.

Smith, D.L., and Evans, B., 1984, Diffusional crack healing in quartz: *Journal of Geophysical Research,* 89:4125–35.

Sokolnikoff, I.S., 1956, *Mathematical theory of elasticity:* 2d ed., New York, McGraw-Hill.

Solomon, S.C., 1972, Seismic wave attenuation and partial melting in the upper mantle of North America: *Journal of Geophysical Research,* 77:1483–1502.

Solomon, S.C., Sleep, N.H., and Jurdy, D.M., 1977, Mechanical models for absolute plate motions in the early Tertiary: *Journal of Geophysical Research,* 82:203–12.

Solomon, S.C., Sleep, N.H., and Richardson, R.M., 1975, On the forces driving plate tectonics: Inferences from absolute plate velocities and intraplate stress: *Geophysical Journal of the Royal Astronomical Society,* 42:769–801.

Spang, J.H., 1972, Numerical method for dynamic analysis of calcite twin lamellae: *Geological Society of America Bulletin,* 83:467–72.

Spang, J.H., and Groshong, R.H., 1981, Deformation mechanisms and strain history of a minor fold from the Appalachian Valley and Ridge Province: *Tectonophysics,* 72:323–42.

Spang, J.H., and Van Der Lee, J., 1975, Numerical dynamic analysis of quartz deformation lamellae and calcite and dolomite twin lamellae: *Geological Society of America Bulletin,* 86:1266–72.

Spiers, C.J., and Rutter, E.H., 1984, A calcite twinning paleopiezometer: in Henderson, C.M.B., ed., *Progress in Experimental Petrology:* Natural Environment Research Council Publication Series D., no. 25, pp. 241–45.

Spottiswoode, S.M., and McGarr, A., 1975, Source parameters of tremors in a deep-level gold mine: *Bulletin of the Seismological Society of America,* 65:93–112.

Srivastava, D., Engelder, T., and Lacazette, A., 1992, Fluid inclusions as a stressmeter: II. Overburden thickness during emplacement of the Yellow Springs Thrust Sheet, Appalachian fold-thrust belt, Pennsylvania, (forthcoming).

Stacey, T.R., 1982, Contribution to the mechanism of core discing: *Journal of South African Institute of Mining and Metallurgy,* 83:269–74.

Stauder, W., 1962, The focal mechanism of earthquakes: *Advances in Geophysics,* 9:1–76.

Stearns, D.W., 1972, Structural interpretation of the fractures associated with Bonita fault: *New Mexico Geological Society Guidebook,* 23:161–64.

Stein, C.A., Cloetingh, C., and Wortel, R., 1989, SEASAT-derived gravity constraints on stress and deformation in the northeastern Indian Ocean: *Geophysical Research Letters,* 16:823–26.

Stein, R.S., and King, G.C.P., 1984, Seismic potential revealed by surface folding: 1983 Coalinga, California, earthquake: *Science,* 224:869–72.

Stein, R.S., and Yeats, R.S., 1989, Hidden earthquakes: *Scientific American,* 260:48–57.

Stein, S., and Okal, E.A., 1978, Seismicity and tectonics of the Ninetyeast Ridge area: Evidence for internal deformation on the Indian plate: *Journal of Geophysical Research,* 83:2233–46.

Stein, S., Sleep, N.H., Geller, R.J., Wang, S.C., and Kroeger, G.C., 1979, Earthquakes along the passive margin in eastern Canada: *Geophysical Research Letters,* 6:537–40.

Steketee, J.A., 1958, Some geophysical applications of the elasticity theory of dislocations: *Canadian Journal of Physics,* 36:1168–98.

Stephansson, O., and Ångman, P., 1986, Hydraulic fracturing stress measurements at Forsmark and Stidsvi in Sweden: *Bulletin of the Geological Society of Finland,* 58:307–33.

Stephansson, O., Särkkä, P., and Myrvang, A., 1986, State of stress in Finnoscandia: in Stephansson, O., ed., *Rock Stress and Rock Stress Measurements:* CENTEK Pub., Luleå, Sweden, pp. 21–32.

Stesky, R.M., Brace, W.F., Riley, D.K., and Robin, P.-Y., 1974, Friction in faulted rock at high temperature and pressure: *Tectonophysics,* 23:177–203.

Stock, J.M., and Healy, J.H., 1988, Continuation of a Deep Borehole Stress measurement profile near the San Andreas Fault 2. Hydraulic fracturing stress measurements at Black Butte, Mojave Desert, Ca.: *Journal of Geophysical Research,* 93:15,196–206.

Stock, J.M., Healy, J.H., Hickman, S.H., and Zoback, M.D., 1985, Hydraulic fracturing stress measurements at Yucca Mountain, Nevada, and relationship to the regional stress field: *Journal of Geophysical Research,* 90:8691–706.

Stocker, R.L., and Ashby, M.F., 1973, On the rheology of the upper mantle: *Reviews of Geophysics and Space Physics,* 11:391–426.

Strickland, F.G., and Ren, N.K., 1980, Use of differential strain curve analysis in predicting in situ stress state of deep wells: in Summers, D.A., ed., Rock Mechanics, A State of the Art: Proceedings of the 21st U.S. Symposium on Rock Mechanics, pp. 523–32.

Suppe, J., 1985, *Principles of Structural Geology:* Englewood Cliffs, New Jersey, Prentice-Hall, Inc.

Swenson, A., 1989, Joint patterns in Mississippian and Pennsylvanian rocks in western Kentucky and southern Illinois: *Southeastern Geology* (forthcoming).

Swolfs, H.S., 1975, Determination of in situ stress orientation in a deep gas well by strain relief techniques: Technical Report 75-43, Terra Tek, Salt Lake City, Utah.

———, 1976, Field investigation of strain relaxation and sonic velocities in Barre Granite, Barre, Vermont: Terra Tek Final Report TR 76-13.

Swolfs, H.S., and Brechtel, C.E., 1977, in Wang, R. and Clark, A.B., eds., Proceedings of the 18th U.S. Symposium on Rock Mechanics, Paper 4C4, Golden, Colorado, Colorado School of Mines.

Swolfs, H.S., and Handin, J., 1976, Dependence of sonic velocity on site and in situ stress in a rock mass. In investigation of stress in rock: ISRM, Symp., Sydney, Australia.

Swolfs, H.S., Brechtel, C.E., Brace, W.F., and Pratt, H.R., 1981, Field mechanical properties of a jointed sandstone: in Carter, N.L., Friedman, M., Logan, J.M., and Stearns, D.W., eds. Mechanical Behavior of Crustal Rocks: Geophysical Monograph 24, Washington, D.C., American Geophysical Union, pp. 161–72.

Swolfs, H.S., Handin, J., and Pratt, H.R., 1974, Field measurements of residual strain in granitic rock masses: Proceedings of the 3rd Congress of the International Society of Rock Mechanics, 2:563–68.

Sykes, L.R., and Sbar, M.L., 1974, Focal mechanism solution of intraplate earthquakes and stresses in the lithosphere: in Kristjansson, L., ed., *Geodynamics of Iceland and the North Atlantic Area,* Hingham, Mass., D. Reidel. pp. 207–24.

Talwani, P., and Acree, S., 1985, Pore-pressure diffusion and the mechanism of reservoir-induced seismicity: *Pure and Applied Geophysics,* 122:947–65.

Talwani, P., and Rastogi, B.K., 1979, Mechanism for reservoir induced seismicity: *Earthquake Notes,* 49:59.

Tapponnier, P., and Brace, W.F., 1976, Development of stress induced microcracks in Westerly granite: *International Journal of Rock Mechanics and Mining Sciences,* 13:103–12.

Terzaghi, K, 1943, *Theoretical Soil Mechanics,* New York, John Wiley.

Terzaghi, K., and Richard, R.E., 1952, Stresses in rock about cavities: *Geotechnique,* 3:57–90

Teufel, L.W., 1976, The measurement of contact areas and temperature during frictional sliding of Tennessee Sandstone [M.S. thesis]: College Station, Texas, Texas A&M University.

———, 1982, Prediction of hydraulic fracture azimuth from anelastic strain recovery measurements of oriented core: in Goodman, R.E., and Heuze, F.E., eds., Issues in Rock Mechanics: Proceedings of the 23rd U.S. Symposium on Rock Mechanics, New York, Society of Mining Engineers, pp. 238–46.

———, 1983, Determination of in situ stress from anelastic strain recovery measurements of oriented cores: SPE 11649, presented at 1983 SPE/DOE Symposium on Low Permeability Reservoirs, Denver, Colorado. March 14–16, 1983.

———, 1985, Insights into the relationship between wellbore breakouts, natural fractures and in situ stress: in Ashworth, E., ed., Proceedings of the 26th U.S. Symposium on Rock Mechanics, Rotterdam, A.A. Balkema, pp. 1199–1206.

———, 1989, Acoustic emissions during anelastic strain recovery of cores from deep boreholes: in Khair, A.W., ed., Rock Mechanics as a guide for efficient utilization of natural resources: Proceedings of the 30th U.S. Symposium on Rock Mechanics, Rotterdam, A.A. Balkema, pp. 269–76.

Teufel, L.W., and Farrell, H.E., 1990, In-situ stress and natural fracture distribution in the Ekofisk Field, North Sea: in Proceedings of the Third North Sea Chalk Symposium, June 11–12, 1990, Copenhagen, Denmark.

Teufel, L.W., and Warpinski, N.R., 1984, Determination of in situ stress from anelastic strain recovery measurements of oriented core: Comparison to hydraulic fractures stress measurements in the Rollins sandstone, Piceance Basin, Colorado: in Dowding, C.H., and Singh M.M., eds., Rock Mechanics in Productivity and Protection: Proceedings of the 25th Symposium of Rock Mechanics, New York, Society of Mining Engineers of AIME. pp. 176–85.

Thiercelin, M.J., Hudson, P.J., Ren, H.-K., and Roegiers, J.-C., 1986, Laboratory determination of the in situ stress tensor. International Symposium on Engineering in Complex Rock Formations (ECRF), Beijing, China, November 3–7, 1986.

Tincelin, M.E., 1951, Conférence internationale sur les pressions de terrains et al soutenement dans les chantiers d'exploration: Liège, du 24 au 28 avril, p. 158–175.

Tissot, B.P. and Welte, D.H., 1978, *Petroleum formation and occurrence:* Springer, Berlin.

Toksöz, M.N., Sleep, N.H., and Smith, A.T., 1973, Evolution of the downgoing lithosphere and mechanisms of deep focus earthquakes: *Geophysical Journal of the Royal Astronomical Society,* 25:285–310.

Tourneret, C., and Laurent, P., 1990, Paleostress orientations from calcite twins in the North Pyrenean foreland, determined by the Etchecopar inverse method: *Tectonophysics,* 180:287–302.

Tôth, J., 1980, Cross-formational gravity-flow of groundwater: A mechanism of the transport and accumulation of petroleum (The generalized hydraulic theory of petroleum migration): *American Association of Petroleum Geologists Bulletin,* 64:121–67.

Trurnit, P., 1968, Pressure solution phenomena in detrital rocks: *Sedimentary Geology,* 2:89–114.

Tsai, Y.B., and Aki, K., 1970, Precise focal depth determination from amplitude spectra of surface waves: *Journal of Geophysical Research,* 75:5729–43.

Tse, S., and Rice, J., 1986, Crustal earthquake instability in relation to the depth variation of frictional slip properties: *Journal of Geophysical Research,* 91:9452–72.

Tsenn, M.C., and Carter, N.L., 1987, Upper limits of power law creep in rocks: *Tectonophysics,* 136:1–26.

Tsukahara, H., 1983, Stress measurements utilizing the hydraulic fracturing technique in the Kanto-Tokai Area, Japan: in Zoback, M.D., and Haimson, B.C., eds., *Hydraulic Fracturing Stress Measurements,* Washington, D.C., National Academy Press, pp. 18–27.

Tullis, J., and Yund, R.A., 1977, Experimental deformation of dry Westerly Granite: *Journal of Geophysical Research,* 82:5705–18.

Tullis, T.E., 1977, Reflections on measurement of residual stress in rock: *Pure and Applied Geophysics,* 115:57–68.

———, 1980., The use of mechanical twinning in minerals as a measure of shear stress magnitudes: *Journal of Geophysical Research,* 85:6263–68.

———, 1981, Stress measurements via shallow overcoring near the San Andreas fault: in Carter, N.L., Friedman, M., Logan, J.M., and Stearns, D.W., eds., *Mechanical Behavior of Crustal Rocks:* Geophysical Monograph 24, Washington, D.C., American Geophysical Union, pp. 199–214.

Tullis, T.E., and Tullis, J., 1986, Experimental rock deformation: in Hobbs, B.E., and Heard, H.C., eds., *Mineral and Rock Deformation: Laboratory Studies:* Geophysical Monograph 36, Washington, D.C., American Geophysical Union, pp. 297–324.

Tullis, T.E., and Weeks, J.D., 1986, Constitutive behavior and stability of frictional sliding of granite: *Pure and Applied Geophysics,* 124:383–414.

Tunbridge, L., 1988, Interpretation of the shut-in pressure from the rate of pressure decay: *International Journal of Rock Mechanics and Mining Sciences,* 26:457–60.

Turcotte, D.L., 1974a, Are transform faults thermal contraction cracks? *Journal of Geophysical Research,* 79:2573–77.

———, 1974b, Membrane tectonics: *Geophysical Journal of the Royal Astronomical Society,* 36:33–42.

Turcotte, D.L., and Schubert G., 1982, *Geodynamics—Applications of Continuum Physics to Geological Problems:* New York, John Wiley.

Turcotte, D.L., McAdoo, D.C., and Caldwell, J.G., 1978, An elastic-perfectly plastic analysis of the bending of the lithosphere at a trench: *Tectonophysics,* 47:193–206.

Turner, F.J., 1948, Note on the tectonic significance of deformation lamellae in quartz and calcite: *Transactions of AGU,* 29:556–69.

———, 1953, Nature and dynamic interpretation of deformation in calcite of three marble: *American Journal of Science,* 251:276–98.

Turner, F.J., and Ch'ih, C.S., 1951, Deformation of Yule marble: Part III—observed fabric changes due to deformation at 10,000 atmospheres confining pressure, room temperature, dry: *Geological Society of America Bulletin,* 62:887, 906.

Turner, F.J., Griggs, D.T., and Heard, H.C., 1954, Experimental deformation of calcite crystals: *Geological Society of America Bulletin,* 65:883–934,

Tuttle, O.F., 1949, Structural petrology of planes of liquid inclusions: *Journal of Geology,* 57:331–56.

Twiss, R.J., 1977, Theory and applicability of a recrystallized grain size paleopiezometer: *Pure and Applied Geophysics,* 115:227–44.

———, 1986, Variable sensitivity piezometric equations for dislocation density and subgrain diameter and their relevance to olivine and quartz: in Hobbs, B.E., and Heard, H.C., eds., *Mineral and Rock Deformation: Laboratory Studies:* Geophysical Monograph 36, Washington, D.C., American Geophysical Union, pp. 247–62.

Twiss, R.J., and Sellars, C.M., 1978, Limits of applicability of the recrystallized grain-size geopiezometer: *Geophysical Research Letters,* 5:337–40.

Van Hise, C.R., 1896, Principals of North American Precambrian Geology, 16th Annual Report, U.S. Geological Survey (1894–1895), U.S. Government Printing Office, pt. 1, pp. 571–843.

Varnes, D.J., and Lee, F.T., 1972, Hypothesis of mobilization of residual stress in rock: *Geological Society of America Bulletin,* 83:2863–66.

Vassiliou, M.S., and Hager, B.H., 1988, Subduction zone earthquakes and stress in slabs: *Pure and Applied Geophysics,* 128:547–624.

Ver Steeg, K. 1942. Jointing in the coal beds of Ohio: *Economic Geology,* 37:503–9.

Vernik, L., and Zoback, M.D., 1992, Estimation of maximum horizontal principal stress magnitude from stress-induced wellbore breakouts in the Cajon Pass scientific research borehole: *Journal of Geophysical Research* (forthcoming).

Vigny, C., Ricard, Y., and Froidevaux, C., 1991, The driving mechanism of plate tectonics: *Tectonophysics,* 187:345–60.

Voight, B., 1966, Interpretation of in situ stress measurements: Proceedings of the 1st Congress of International Society of Rock Mechanics, 3:332–48.

———, 1968, Determination of the virgin state of stress in the vicinity of a borehole from measurements of a partial anelastic strain tensor in drill cores: *Felsmechanik und Ingenieurgeologie,* 6:201–15.

———, 1974, A mechanism for "locking-in" orogenic stress: *American Journal of Science,* 274:662–65.

Voight, B., and St.Pierre, B.H.P., 1974, Stress history and rock stress: in Advances in Rock Mechanics: Proceedings of the 3rd Congress International Society of Rock Mechanics, 2:580–82.

von Karmon, T., 1911, Festigkeitsversuche unter allseitige Druck, *Zeitschrift des Vereines Deutscher Ingenieure,* 55:1749–57.

von Schoenfeldt, H., 1970, An experimental study of open hole hydraulic fracturing as a stress measurement method with particular emphasis on field tests [Ph.D. thesis]: Minneapolis, Minnesota, University of Minnesota.

Vrolijk, P., 1987, Tectonically driven fluid flow in the Kodiak accretionary complex, Alaska: *Geology,* 15:466–69.

Vrolijk, P., Georgianna, M., and Casey, M., 1988, Warm fluid migration along tectonic melanges in the Kodiak accretionary complex, Alaska: *Journal of the Geophysical Research* 93:10313–324.

Walcott, R.I., 1970a, Flexure of the lithosphere at Hawaii: *Tectonophysics,* 9:435–466.

———, 1970b, Isostatic response to leading of the crust in Canada: *Canadian Journal of Earth Science,* 7:716–27.

———, 1970c, Flexural rigidity, thickness, and viscosity of the lithosphere: *Journal of Geophysical Research,* 75:3941–54.

Walcott, R.I., 1972, Gravity, flexure and the growth of sedimentary basins at a continental edge: *Geological Society of America Bulletin,* 83:1845.

Walsh, J.B., 1965, The effect of cracks on the compressibility of rock: *Journal of Geophysical Research,* 70:381–89.

Wang, C., Mao, N., and Wu, F.T., 1980, Mechanical properties of clays at high pressure: *Journal of Geophysical Research,* 85:1462–68.

Wang, H.F., and Heard, H.C., 1985, Prediction of elastic moduli via crack density in pressurized and thermally stressed rock: *Journal of Geophysical Research,* 90: 10342–50.

Wang, H.F., and Simmons, G., 1978, Microcracks in crystalline rock from 5.3-km depth in the Michigan Basin: *Journal of Geophysical Research,* 83:5849–56.

Warpinski, N.R., 1984, Investigation on the accuracy and reliability of in situ stress measurements using hydraulic fracturing in perforated, cased holes: Proceedings of 24th U.S. Symposium of Rock Mechanics, Texas A&M University, College Station, Texas. pp. 773–86.

———, 1989, Determining the minimum in situ stress from hydraulic fracturing through perforations: *International Journal of Rock Mechanics and Mining Sciences,* 26:523–32.

Warpinski, N.R., and Teufel, L.W., 1989a, In-situ stresses in low-permeability, nonmarine rocks: *Journal of Petroleum Technology,* 41:405–14.

———, 1989b, A viscoelastic constitutive model for determining in situ stress magnitudes from anelastic strain recovery of core: *SPE Production Engineering,* 5:272–80.

Warpinski, N.R., Branagan, P., and Wilmer, R., 1983, In-situ stress measurements at DOE's Multiwell Experiment site, Mesa Verde Group, Rifle, CO.: SPE #12142, presented at 58th Annual Technical Meeting of SPE, October 5–8th, San Francisco, CA.

———, 1985, In-situ stress measurements at U.S. DOE's Multiwell experiment site, Mesaverde Group, Rifle Colorado: *Journal of Petroleum Technology,* 37:527–36.

Warpinski, N.R., Finley, S.J. Vollendorf, W.C., O'Brien, M., and Eshom, E., 1982, The interface test series: an in situ study of factors affecting the containment of hydraulic fractures: Sandia National Laboratories Report SAND81-2408.

Warpinski, N.R., Schmidt, R.A., and Nothrup. D.A., 1981, In situ stresses: The predominant influence on hydraulic fracture containment: *Journal of Petroleum Technology,* 34:653–64..

Warren, W. E., 1981, Packer induced stresses during hydraulic well fracturing: *Journal of Energy Resources Technology,* 105:336–43.

Warren, W. E., and Smith, C. W., 1985, In situ stress estimates from hydraulic fracturing and direct observation of crack orientation: *Journal of Geophysical Research,* 90:6829–39.

Watts, A.B., 1978, An analysis of isostasy in the world's oceans, I, Hawaiian-Emperor Seamount Chain: *Journal of Geophysical Research,* 83:5989–6004.

Watts, A.B., and Ribe, N.M., 1984, On geoid heights and flexure of the lithosphere at seamounts: *Journal of Geophysical Research,* 89:11152–70.

Watts, A.B., and Talwani, M., 1974, Gravity anomalies seaward of deep-sea trenches and their tectonic implications: *Geophysical Journal of the Royal Astronomical Society,* 36:57–90.

Watts, A.B., Bodine, J.H., and Steckler, M.S., 1980, Observations of flexure and the state of stress in the oceanic lithosphere: *Journal of Geophysical Research,* 85:6369–76.

Watts, A.B., Cochran, J.R., and Selzer, G., 1975, Gravity anomalies and flexure of the lithosphere: A three-dimensional study of the Great Meteor Seamount, northeast Atlantic: *Journal of Geophysical Research,* 80:1391–98.

Watts, A.B., Karner, G.D., and Steckler, M.S., 1982, Lithospheric flexure and the evolution of sedimentary basins: in Kent, P., et al., eds., *The Evolution of Sedimentary Basins: Philosophical Transactions,* Royal Society of London, 305A:249–81.

Watts, N., 1983, Microfractures in chalks of Albuskjell Field, Norwegian Sector, North Sea: Possible origin and distribution: *American Association of Petroleum Geologists Bulletin,* 67:201–34.

Watts, N., Larpe, J., van Schijndel-Goester, F., and Ford, A., 1980, Upper Cretaceous and Lower Tertiary chalks of the Albuskjell area, North Sea: Disposition in a slope and a base slope environment: *Geology,* 8:217–21.

Weathers, M.S., Cooper, R.F., Bird, J.M., and Kohlstedt, D.L., 1979, Deformation-induced microstructures in the Ralston Buttes-Idaho Springs shear zone: *EOS,* 60:948.

Weertman, J, and Weertman, J.R., 1970, Mechanical properties, strongly temperature dependent: in Cahn, R.W., ed., *Physical Metallurgy.*

Weissel, J.K., Anderson, R.N., and Geller, C.A., 1980, Deformation of the Indo-Australian plate: *Nature,* 278:284–91.

Wernicke, B.P., 1981, Low-angle normal faults in the Basin and Range province—Nappe tectonics in an extending orogen: *Nature,* 291:645–48.

Westbrook, J.H., and Jorgensen, P.J., 1968, Effects of water desorption on indentation microhardness anisotropy in minerals: *American Mineralogist,* 53:1899.

White, O.L., and Russell, D.J., 1982, High horizontal stresses in southern Ontario: in *Proc. 4th Cong. Int. Ass. Eng. Geology,* 5:39–51.

White, O.L., Karrow, P.K., and MacDonald, J.R., 1973, Residual stress relief phenomena in southern Ontario: in Proceedings of the 9th Canadian Rock Mechanics Symposium, Ottawa, Canada, Department of Energy, Mines, and resources, pp. 323–48.

Whitehead, W., Gatens, J., and Holditch, S., 1989, Determination of in situ stress profiles through hydraulic fracturing measurements in two distinct geologic areas: *International Journal of Rock Mechanics and Mining Sciences,* 26:637–46.

Wiens, D.A., and Stein, S., 1984, Intraplate seismicity and stresses in young oceanic lithosphere: *Journal of Geophysical Research,* 89:11442–64.

———, 1985, Implications of oceanic intraplate seismicity for plate stresses, driving forces, and rheology: *Tectonophysics,* 116:143–62.

Williams, H. R., Corkery, D. and Lorek, E. G. 1985. A study of joints and stress-release buckles in Paleozoic rocks of the Niagara Peninsula, southern Ontario: *Canadian Journal of Earth Science,* 22:296–300.

Willis, B., 1923, *Geologic Structures:* New York, McGraw-Hill.

Wise, D.U., 1964, Microjointing in basement, Middle Rocky Mountains of Montana and Wyoming: *Geological Society of America Bulletin,* 75:287–306.

Wise, D.U., Dunn, D.E., Engelder, T., Geiser, P.A., Hatcher, R.D., Kish, S.A., Odom, A.L., and Schamel, S., 1984, Fault-related rocks: Suggestions for terminology: *Geology,* 12:391–94.

Withjack, M., and Scheiner, C., 1982, Fault patterns associated with domes—An experimental and analytical study: *American Association of Petroleum Geologists Bulletin,* 66:302–16.

Wong, T.F., and Walsh, J.B., 1985, A theoretical analysis of tectonic stress relief during overcoring: *International Journal of Rock Mechanics and Mining Sciences,* 22:163–71.

Woodworth, J.E., 1896, On the fracture system of joints, with remarks on certain great fractures: *Boston Society of Natural History, Proceedings,* 27:63–183.

Wooten, C., and Elbel, J., 1988, State of the art determination of fracture closure pressure: in Haimson, B.C., Rogiers, J.C., and Zoback, M.D., eds., Proceedings of the 2nd International Workshop on Hydraulic Fracture Stress Measurements, June 15–18, 1988, Minneapolis, Minn., Geological Engineering Program, Madison, Wisconsin, University of Wisconsin Press, pp. 916–52.

Worotnicki, G., and Walton, R., 1976, Triaxial hollow inclusion gauges for measurements of triaxial rock stresses: in Proceedings I.S.R.M. Symposium on Investigation of stress in rock, Institution of Engineers, Australia National Conference Publication No. 76/4, pp. 1–8.

Wortel, M.J.R., and Vlaar, N.J., 1988, Subduction zone seismicity and the thermomechanical evolution of downgoing lithosphere: *Pure and Applied Geophysics,* 128:625–59.

Wyss, M., and Brune, J.N., 1968, Seismic moment, stress, and source dimensions for earthquakes in the California-Nevada region: *Journal of Geophysical Research,* 73:4681–94.

Wyss, M., and Molnar, P., 1972, Efficiency, stress drop, apparent stress, effective stress, and frictional stress of Denver, Colorado, earthquakes: *Journal of Geophysical Research,* 77:1433–38.

Yang, J.P., and Aggarwal, Y.P., 1981, Seismotectonics of northeastern United States and Canada: *Journal of Geophysical Research,* 86:4981–98.

Yuster, S.T., and Calhoun, J.C., 1945, Pressure parting of formations in water flood operations: *The Oil Weekly* (March 12), pp. 38–42.

Zemanck, J., and Caldwell, R.L., 1969, The borehole televiewer—A new logging concept for fracture location and other types of borehole inspection: *Journal of Petroleum Technology,* 25:762–74.

Zhang, D.L., 1982, Use of the Kaiser effect for the estimation of the previous state of stress in rock: Internal report RML-IR/82-4, Dept. Mineral Engineering, University Park, Pennsylvania, Pennsylvania State University.

Zheng, Z., Cook, N.G.W., and Myer, L.R., 1988, Borehole breakout and stress measurements: in Cundall, P. A., Sterling, R.L., and Starfield, A.M., eds., *Key Questions in Rock Mechanics,* Rotterdam, A.A. Balkema, pp. 471–78.

Zheng, Z., Klemeny, J., and Cook, N.G.W., 1989, Analysis of borehole breakouts: *Journal of Geophysical Research,* 94:7171–82.

Zoback, D., and Zoback, M.L., 1991, Tectonic stress field of North America and relative plate motions, in Slemmons, D.B., Engdahl, E.R., Zoback, M.D., and Blackwell, D.D., eds., *Neotectonics of North America: Boulder, Colorado,* Geological Society of America, Decade Map Volume 1.

Zoback, M.D., 1986, In-situ stress measurements in the Kent Cliffs research well: Report to Empire State Electric Energy Corporation, New York, New York.

Zoback, M.D., and Healy, J.H., 1984, Friction, faulting, and in situ stress: *Annales Geophysicae,* 2:689–98.

———, 1988, In situ stress measurements in the Cajon Pass scientific research well: Preliminary results, equipment development and evolution of methodology: in Haimson, B.C., Rogiers, J.C., and Zoback, M.D., eds., Proceedings of the 2nd International Workshop on Hydraulic Fracture Stress Measurements, June 15–18, 1988, Minneapolis, Minn., Geological Engineering Program, Madison, Wisconsin, University of Wisconsin Press.

Zoback, M.D., and Hickman, S., 1982, In situ study of the physical mechanisms controlling induced seismicity at Monticello Reservoir, South Carolina: *Journal of Geophysical Research,* 87:6959–74.

Zoback, M.D., and Moos, D., 1988, In situ stress, natural fracture and sonic velocity measurements in the Moodus, Connecticut scientific research well; in Final Report EP86–42 to Empire State Electric Energy Research Corporation, Appendix C.

Zoback, M.D., Moos, D., Mastin, L.G., and Anderson, R.N., 1985, Wellbore breakouts and in situ stress: *Journal of Geophysical Research,* 90:5523–30.

Zoback, M.D., Tsukahara, H., and Hickman, S., 1980, Stress measurements at depth in the vicinity of the San Andreas fault: *Journal of Geophysical Research,* 85:6157–73.

Zoback, M.D., Zoback, M.L., Mount, V.S., and others, 1987, New evidence on the state of stress of the San Andreas fault system: *Science,* 238:1105–11.

Zoback, M.L., 1983, Structure and Cenezoic tectonism along the Wasatch fault zone, Utah: in Miller, D., ed., *Tectonic and Stratigraphic Studies in the Eastern Great Basin: Geological Society of America Memoir* 157:3–27.

———, 1992, First and second order patterns of stress in the lithosphere: The World Stress Map Project: *Journal of Geophysical Research* (forthcoming).

Zoback, M.L., Nishenko, S.P., Richardson, R.M., Hasegawa, H.S., and Zoback, M.D., 1986, Mid-plate stress, deformation, and seismicity: in Vogt, P.R., and Tucholke, B.E., eds., *The Geology of North America,* M., The Western Atlantic Region, Geological Society of America, Boulder Colorado, pp. 297–312.

Zoback, M.L., and Zoback, M.D., 1980, State of stress in the conterminous United States: *Journal of Geophysical Research,* 85:6113–56.

———, 1989, Tectonic stress field of the Continental United States: in Pakiser, L., and Mooney, W., eds., *Geophysical Framework of the Continental United States: Geological Society of America Memoir* 172, pp. 523–39.

Zoback, M.L., et al., 1989, Global patterns of tectonic stress: *Nature,* 341:291–98.

Zuber, M.T., Bechtel, T.D., and Forsyth, D.W., 1989, Effective elastic thickness of the lithosphere and mechanisms of isostatic compensation in Australia: *Journal of Geophysical Research,* 94:9353–67.

Index